THE HISTORY OF TUNNELING
IN THE UNITED STATES

THE HISTORY OF TUNNELING
IN THE UNITED STATES

EDITORS
Michael F. Roach | Colin A. Lawrence | David R. Klug

GRAPHICS EDITOR
W. Brian Fulcher

FOREWORD
Doug Most

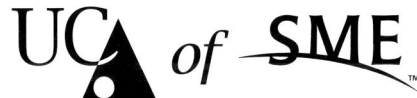

PUBLISHED BY SOCIETY FOR MINING, METALLURGY & EXPLORATION | ENGLEWOOD, COLORADO

Society for Mining, Metallurgy & Exploration (SME)
12999 E. Adam Aircraft Circle
Englewood, Colorado, USA 80112
(303) 948-4200 / (800) 763-3132
www.smenet.org

The Society for Mining, Metallurgy & Exploration (SME) is a professional society whose more than 15,000 members represents all professionals serving the minerals industry in more than 100 countries. SME members include engineers, geologists, metallurgists, educators, students and researchers. SME advances the worldwide mining and underground construction community through information exchange and professional development.

Copyright © 2017 Society for Mining, Metallurgy & Exploration
All Rights Reserved. Printed in the United States of America.

Information contained in this work has been obtained by SME from sources believed to be reliable. However, neither SME nor its authors and editors guarantee the accuracy or completeness of any information published herein, and neither SME nor its authors and editors shall be responsible for any errors, omissions, or damages arising out of use of this information. This work is published with the understanding that SME and its authors and editors are supplying information but are not attempting to render engineering or other professional services. Any statement or views presented herein are those of individual authors and editors and are not necessarily those of SME. The mention of trade names for commercial products does not imply the approval or endorsement of SME.

No part of this publication may be reproduced, stored in a retrieval system, or transmitted in any form or by any means, electronic, mechanical, photocopying, recording, or otherwise, without the prior written permission of the publisher.

ISBN 978-0-87335-430-1

Library of Congress Cataloging-in-Publication Data has been applied for.

CONTENTS

Foreword vii

Preface xi

Tunneling Milestones in the U.S. .... Foldout

Chapter 1 | The Building of a Nation 1

Chapter 2 | Societal Benefits 19

Chapter 3 | Railroad Tunnels 51

Chapter 4 | Transit Tunnels 125

Chapter 5 | Highway Tunnels 233

Chapter 6 | Water Tunnels 279

Chapter 7 | Wastewater Tunnels 355

Chapter 8 | Innovations in Tunneling 383

Chapter 9 | The Future of Tunneling 457

Appendix | U.S. Constructed Tunnel Archive 489

Essential Underground Publications 527

Contributors and Acknowledgments 529

Illustration Credits 533

Index 539

FOREWORD

BY DOUG MOST

On a freezing cold day, deep in the lush green Berkshire Hills of Western Massachusetts, a gentle rumbling began. It was late in the afternoon on February 9, 1875—a day, and a moment, that had been anticipated for a quarter century. As the ground began to shake, a puff of smoke emerged from the freshly cut hole in the side of a mountain, and then at precisely 4:34 p.m., a locomotive appeared from the darkness, its red, white, and blue bunting covered in black soot. Behind it came a train of open flatcars carrying more than a hundred choking, cheering, and coughing dignitaries.

Coated in the same soot from head to toe, they had made history. By emerging safely from a 36-minute journey through a 4¾-mile-long tunnel that cut straight through Hoosac Mountain, the group had traveled through the longest tunnel in the United States. And it did more than merely connect Boston with the West. It put a merciful end to a project that had been interrupted by so many explosions, fires, and drownings that it came to be known tragically as "The Bloody Pit." The workers had traveled from around the world to complete it, lured by promises of fortune, only to be squeezed into filthy shack villages and sent out to dig from dawn to dusk for $1.40 a day.

But this was how tunnels got built, and so build them they did.

From the early days of civilization when tunnels were dug by slaves to avoid torture—using sticks, rocks, and their bare fingers—right up to the modern-day 5,000-ton tunnel boring machine, known today simply by its acronym, one common bond has tied together generations of tunneling in the United States: the workers.

Anonymous. Invisible. Forgotten. They wade deep into the earth to put their muscles and skills to use, and their lives in danger, until one hole links up with a second hole from the opposite direction. And a tunnel is born. It might let passengers drive their cars through it one day, or it might move trains, boats, cargo, water, or sewage. But no matter what size, shape, depth, or length a tunnel takes, they are all, indisputably, marvels of civil engineering because of this simple achievement: People feel safe inside them. And that was no easy feat.

For centuries, the underground was viewed as the underworld, where the devil lived and where evil spirits lurked. The only reason people saw fit to go beneath their sidewalks and streets was if they were dead. The great explorers in the fifteenth, sixteenth, and seventeenth centuries helped shake those fears, but it was not until 1818 when true tunneling came to America. The men who owned the Schuylkill and Susquehanna Canal in Pennsylvania wanted to create a right angle through a hill in the path of their canal, just to prove it could be done. The Auburn Tunnel, used to move coal down the river, was the first transportation tunnel built in the United States.

Canal tunnels quickly proved to be an efficient way for commercial transportation to pass through rolling hills and steep mountains. No longer did workers have to trudge up, over, or around dangerous terrain, relying on horses and donkeys. And no tunnel proved its worth more than the opening of the Erie Canal on October 25, 1825.

Only the super rich could venture north of New York City in the early 1800s, and when they did, it was with horses pulling them up narrow Post Road, past brooks and ponds, deer and rabbits. When the canal opened, the Hudson Valley was connected to the Great Lakes region, and the time it took for milk or coal or wood or eggs to travel from Gotham to the upper Mississippi Valley dropped from 26 days to 6. Overnight, New York became the nation's hub of imports and exports, and the economic value of tunneling was suddenly clear.

But if anyone thought that all tunnel projects would suddenly be welcomed and embraced for the rest of eternity, he or she was mistaken. The upheaval and disruption caused by tunneling, along with the economic barriers, unearthed opponents from all corners of society—from the corner store businessman who feared construction would reduce foot traffic to his shop, to the taxpayer who considered the spending to be frivolous, to the politician, dirty or clean, who had his own private agenda to worry about.

Half a century after the Erie Canal opened, a skinny, mustachioed, opera-loving inventor in New York City named Alfred Ely Beach proposed to build America's first subway right under Broadway, and power it pneumatically with giant fans. Skeptical at first, New Yorkers grew excited at the thought of seeing the congestion on their streets relieved. But standing in Beach's way was a 300-pound crime boss. The ensuing David-and-Goliath battle between Beach and "Boss" Tweed pushed Beach to dig a secret tunnel beneath Manhattan. What he unveiled to the public in 1870 was an underground station as opulent as any Vanderbilt living room, complete with settees and goldfish and glimmering chandeliers. Alas, his tunnel stretched only 312 feet and his fan technology was comically flawed. New Yorkers would have to wait another three decades for relief, and watch with envy as a little town to the north opened America's first subway in 1897.

Boston achieved its feat using the new technique of the day, appropriately called cut-and-cover. A massive trench was dug into the street, the tunnel box then framed off inside, and the street was covered over, allowing traffic to flow above while the finishing touches to the tunnel were applied below. Boston's two-year, 2-mile project was impressive, but it would quickly be overshadowed by the 21 miles of shallow trenches and deep tunnels through rock and water that New York opened in 1904, America's second subway, designed by the man whose name became synonymous with canals and tunnels, William Barclay Parsons. The firm that he founded, and that thrives to this day, Parsons Brinckerhoff, refined and improved the cut-and-cover method throughout the twentieth century to lessen the impact on the streets, the buildings, and the entire community affected.

If the rudimentary cut-and-cover approach defined the first half of twentieth-century tunneling, the improvements that followed took the industry to new heights—or rather depths. In the early 1900s, rather than slather concrete onto rock and brick in a tedious and time-consuming fashion, a nozzle gun to spray wet mixed concrete, or shotcrete as it came to be known, gained in popularity. And in 1931, while the Hoover Dam was being built, a rock-drilling machine called a jumbo was used to help divert the Colorado River. And as recently as 1970, a novel approach to solidifying and protecting a tunnel route was perfected by injecting a liquid bonding agent, or a grout-like material, into the dirt or fractured rock along the tunnel route.

Each new advancement proved how the industry wanted to learn from its past mistakes and find ways to tunnel faster, deeper, farther. Ultimately, it was the evolution of the TBM that took tunneling into a more modern era, just as the jet engine propelled aviation forward and the elevator propelled skyscrapers higher.

No longer viewed as just another piece of technology, today's tunnel boring machines are so much a part of the fabric of tunneling that they are given nicknames and adopted by their work crews like children. Very big children. There is "Bertha" in Seattle, there was "Molina" in New York, and there is "Lady Bird" in Washington, D.C. And while the earliest TBM was invented by a British engineer in the early 1800s, and Alfred Beach even invented his own boring creature for his secret tunnel in the 1860s, it was not until James Robbins built a rock tunnel boring machine to complete the Oahe Dam diversion project outside Pierre, South Dakota, in 1952 when the future of tunneling was realized.

Half a century later, a TBM nicknamed "Adi," named after the granddaughter of a high-ranking construction official in the Metropolitan Transit Authority, reached East 63rd Street in New York, completing its boring work on the Second Avenue Subway project. It was a monumental feat for a project half a century in the planning. But there was no time to rest for Adi. She was shipped off to Indianapolis to dig a water tunnel.

Spinning and whirling and cutting and chopping and grinding through the earth, the TBM is as much a symbol of new-world technological wizardry as it is good old-fashioned shoveling. It's a machine. There is no mistaking that. It's automated. It's controlled from a central power station. And it eats through 30, 40, or 50 feet of earth a day. But even the biggest TBM of them all, Seattle's Bertha, needed 25 and sometimes 50 people a day to keep her humming.

In the end, technology is useless without the brains of its designers, the hands of its operators, and the know-how of its engineers to point it in the right direction. As the workers in the Hoosac Tunnel learned after 25 years of bloody, painful, and exhaustive digging, it is only when the last inch is breached, the final wall comes down, and a beam of light can shine through from end to end that a tunnel officially becomes a tunnel.

Doug Most is a journalist in Boston and the author of *The Race Underground: Boston, New York, and the Incredible Rivalry That Built America's First Subway* (New York: St. Martin's Press, 2014).

PREFACE

On many occasions throughout a typical year, individuals from the tunnel industry congregate, whether it is at conferences, forums, tours, or more likely, working together on tunneling projects. Over the years, one topic that has consistently emerged from such discussions is that the tunnel industry is obscure and therefore often taken for granted by the general public.

Typically, when a person takes water from a faucet, uses the bathroom, takes a trip on a train, drives through a mountain, or sees the streets drained of water following a heavy storm, little or no thought is given to what makes this all possible. And although there seems to be both a fascination for and an appreciation of the challenges of these significant underground accomplishments, most people have no concept of the skill, sacrifice, and technology involved or even how and when such facilities were constructed.

After much discussion, industry experts concluded that maybe it was the industry itself that had not effectively informed society about the heritage and essential nature of underground infrastructure. So the tunneling industry has now challenged itself to begin to make a societal change from a general lack of understanding to a better informed and therefore more supportive public.

The main organization that represents all tunneling within the United States—engaging owners, engineers, contractors, and suppliers—is the Underground Construction Association of the Society for Mining, Metallurgy & Exploration, Inc. (UCA of SME). The UCA represents the United States within the International Tunnelling and Underground Space Association (ITA), which is the leading international organization that promotes the use of tunnels and underground space through knowledge sharing and application of technology. The UCA recognized that the importance and achievements of the underground industry needed to be revealed so they could be more fully appreciated by society. Further discussions were held within the UCA executive committee, and as a first step in what was anticipated to be a very long process, the editors of this book envisioned creating a concise history of tunneling in the United States. As with all good ideas, those who had foreseen the outcome became the volunteers. And so this book was born.

Although other books have covered the historic underground achievements of specific tunnel projects or those that formed part of a much larger system, no recent attempts have been made to capture the entire underground heritage of this country in such depth. As such, the story of our proud history has been largely untold. The information that supports this chronicle is extensive and exhibits the collaboration of many authors, without which this book would not have been possible. Our sincere thanks extends to all the individuals referenced within the book, those gladiators of our underground industry who have worked much of their spare time to help produce this rich history. What has surprised most of us within the underground world is actually the success story that has unfolded as we unearthed surprises and unexpected achievements that were long forgotten by the industry.

The history of tunneling in the United States has a heritage spanning more than two centuries and is comparable to, and in many cases exceeds, underground achievements elsewhere in the world. One theme that has recurred throughout every decade of the 200 years is clear: It is evident that tunneling and underground infrastructure plays a vital role in sustaining and developing society. Without it, our world would be a very different place, given the positive effect of tunnels on health, quality of life, connectivity, and supporting the modern expectations and requirements of our many cities and towns across the country.

Past generations have periodically overlooked or neglected infrastructure development for many reasons, whether it is a lack of available funding or the diversion of funds or investments to other areas of society. This has inevitably caused generation gaps to occur within tunneling, given that this industry relies on specialist skills and experience to deliver projects successfully. Over the last decade, these generation gaps are being addressed in several ways, but vital to these efforts is the establishment of an ongoing process to encourage the younger generation to engage in our industry as an exciting career. In what was once a male-dominated industry, many—both male and female—have taken up the calling of this challenging but rewarding profession. Our future tunneling generations appear to be driven and as passionate about the importance of the underground as the old-timers.

People are beginning to understand that the infrastructure around them needs to be maintained and further developed to improve basic life conditions. As an industry, we recognize that our critical infrastructure needs to be sustained through rehabilitation and maintenance, as well as expanded to accommodate the population it serves. From the early canal tunnels that supplied food from agriculture, to improving health through dedicated water supply and better sanitation, to supporting the industrial revolution by building railroad tunnels, to extending the road network within and beyond cities with highway tunnels, each type of tunnel has provided a beneficial impact in improving everyone's quality of life and will continue to become ever more applicable in the future.

We would like to dedicate this book to all of those individuals—past, present, and future—who often go unnamed or unrecognized in their feats but have dedicated their lives to this vital underground industry. Many in the past have paid the ultimate price of this dedication with unfortunate consequences while performing their work. We also recognize that enacting our collective passion through the excessive hours worked away from our homes would not be possible without the dedication, support, and tolerance that our families have maintained and for which we are eternally grateful.

The industry as a whole does not engage in tunneling to be famous in the media, as this is not our focus. As a result, we may be victims of our own making in terms of selling the importance of our industry to society. The editors and authors of this book hope that the details contained herein are thought provoking for readers of all ages and in a small or more significant way provide you with a new and refreshing opinion of our long-forgotten story that is an important cornerstone in the building of this great nation.

Michael F. Roach
Colin A. Lawrence
David R. Klug
W. Brian Fulcher

TUNNELING MILESTONES IN THE U.S.

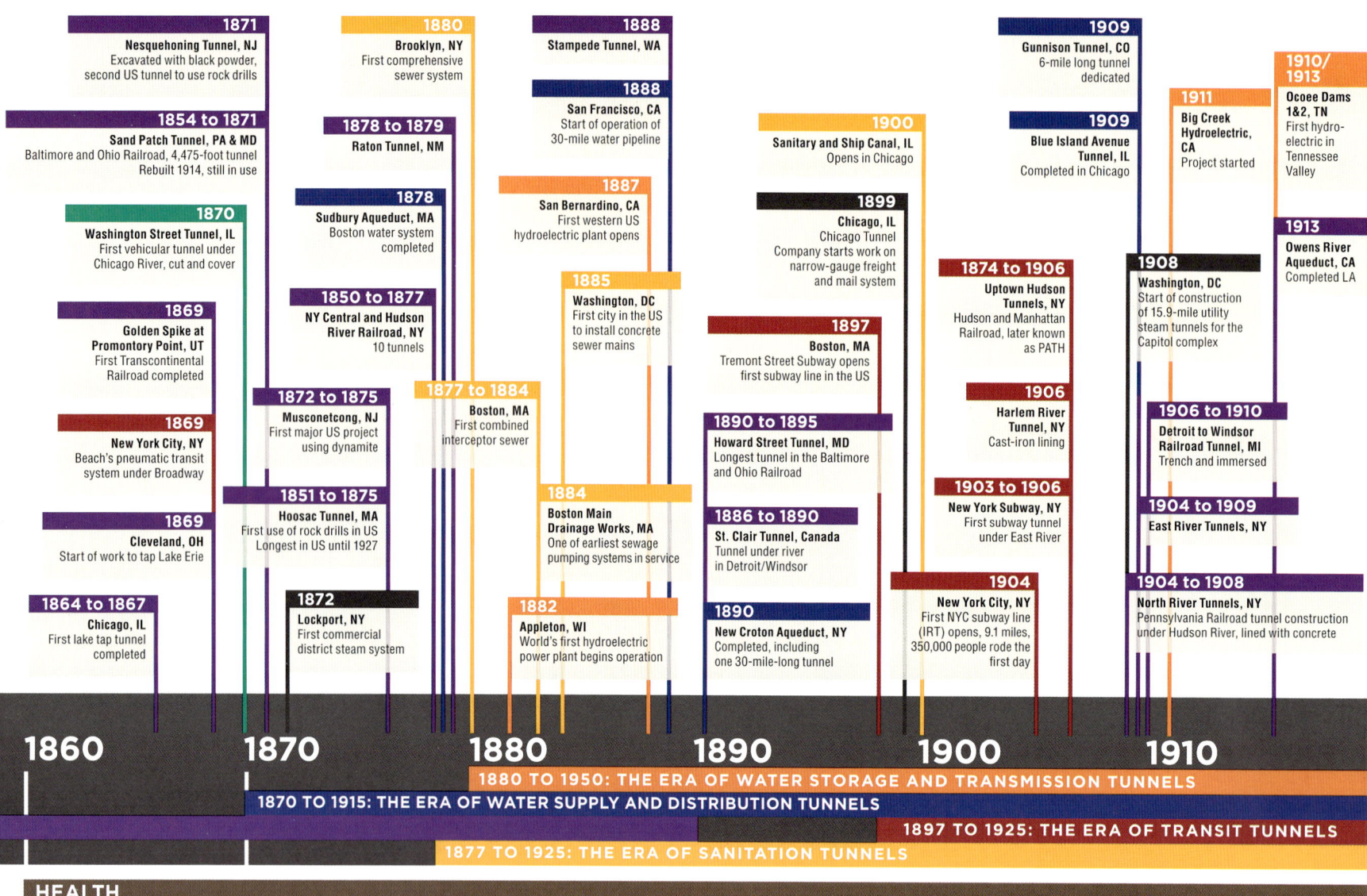

TUNNELING MILESTONES IN THE U.S.

Legend:
- CANALS
- RAILROADS
- WATER SUPPLY/DISTRIBUTION
- SANITATION
- TRANSIT
- WATER STORAGE/TRANSMISSION
- HIGHWAYS
- CSO ENVIRONMENTAL
- OTHER

Tunneling Milestones (1800–1858)

1801 — Salem, MA — Smuggling tunnels

1818 to 1821 — Canal tunnel near Auburn, PA — First completed in US (1863-1888 Converted to rail tunnel)

1819 — Montgomery Bell (or Patterson Forge) Tunnel, TN — Water power for local iron industry. Built by slave labor

1824 to 1827 — "Hacklebernie" Tunnel near Mauch Chunk, PA — First mining tunnel in US

1825 to 1827 — Union Canal, PA — Also known as Schuylkill and Susquehanna Canal, work started 1791. Earliest canal tunnel in the US still in existence

1828 — Baltimore and Ohio Railroad, PA — First use of steam engines 1830, 44 tunnels by 1866

1831 to 1833 — Staple Bend (or Portage) Tunnel, PA — First railroad tunnel in US

1833 — Wadesville Tunnel, PA — Danville and Pottsville Railroad opens as second railroad tunnel in US

1835 to 1837 — Black Rock Tunnel, PA — Philadelphia and Reading Railroad. First US tunnel project with shafts sunk

1837 — Taft Tunnel, CT — Oldest railroad tunnel still in use

1837 to 1842 — Croton Aqueduct / Dam, NY — Completed, including 16 tunnels

1836 to 1850 — Paw Paw Tunnel, MD — Chesapeake and Ohio Canal

1843 — Cincinnati and Whitewater Canal — Opens including 1,782-foot tunnel

1844 — Cobble Hill Tunnel, Brooklyn, NY — Oldest railroad tunnel under a city street

1849 to 1850 — Chetoogeta Tunnel, GA — First major railroad tunnel in the South

1850 — Henryton Tunnel, MD — Opens on the Baltimore and Ohio Railroad

1853 — US Naval Academy, MD — First steam network still in use

1858 — Blue Ridge Tunnel, VA — Longest in the US at completion

1858 — Chicago, IL — First comprehensive waste/water management systems designed

---

1800 TO 1850: THE ERA OF CANAL TUNNELS

1824 TO 1890: THE ERA OF RAILROAD TUNNELS

WORLD EVENTS

HEALTH

Pre-1800: Poor sanitation, variable water quality.

1816-1823: European cholera pandemic.

1832: Cholera outbreak in US.

1848-1849: Cholera and typhoid outbreaks in US (President Polk dies of cholera).

1850s: Cholera, dysentery, typhoid outbreak in US. First comprehensive sewer systems in the US built in the late 1850s (Chicago, Brooklyn).

POLITICAL/GOVERNMENTAL

1775: US Army Corps of Engineers founded.
1802: First federal appropriation to fund improvement on inland waterways.
1806: Law passed to start building the "National Road," known as the Cumberland Road, which opens in 1818.
1808: "Report of the Secretary of the Treasury on the Subject of Public Roads and Canals" released.

1824: *Gibbons v. Ogden* – Congressional control of interstate commerce.
1825: Erie Canal opens (authorized by New York State Legislature in 1817).

1830: President Jackson vetoes bill to build a Kentucky turnpike.
1837: Major panic causes pause in railway construction.

1850: Illinois Central Land Grant Act.

TECHNOLOGICAL

Pre-1800: Industrial Revolution.
1807: Robert Fulton invents the steamboat.
1809-1810: First railway (the experimental Leiper Railway) in the US at Crum Creek, PA.

1810s: Black powder the only viable method for excavation (advance rates approximately 40 ft/month).

1820s: Portland cement developments. First large-scale cement use in US (Erie Canal, completed 1825). Brunel shield patents.
1825: Indoor plumbing in the White House.
1829: Peter Cooper's Tom Thumb was the first American-built locomotive to run in the US.

1831: Safety fuse patented (Bickford). The first flanged T rail (also called T-section) arrived in America from Britain and was laid into the Pennsylvania Railroad. Cochrane files patents for use of compressed air in underground and underwater construction.
1838: Singer drop drill patented for drill/blast operations (Illinois).

1844: Brunton suggests using compressed air for drill machines.
1845: Heavy iron T rails first manufactured in US at Mt. Savage, MD (Baltimore and Ohio Railroad).
1847: Nitroglycerin discovered (Sobrero).
1848-1851: First percussion (steam) rock drill invented (Couch and Fowle, Philadelphia).
1849: Gold in California.

1853: First use of a tunnel boring machine in the US (Hoosac Tunnel). Miners use hydraulic jets for placer gold.
1856: Bessemer process brings in modern era in steelmaking. Centrifugal pumps developed.
1859: Comstock Lode discovered. First oil well in Titusville, PA.

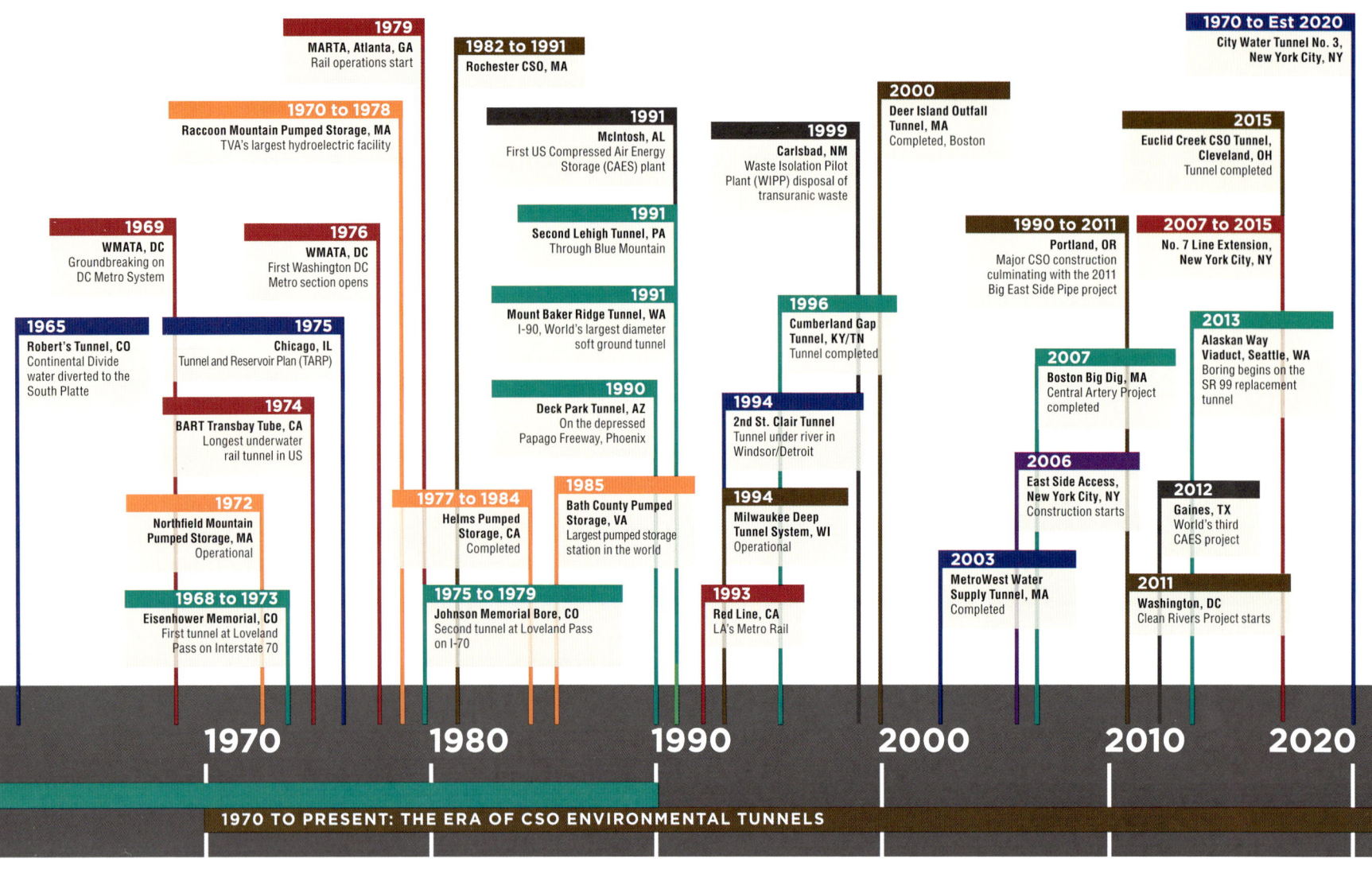

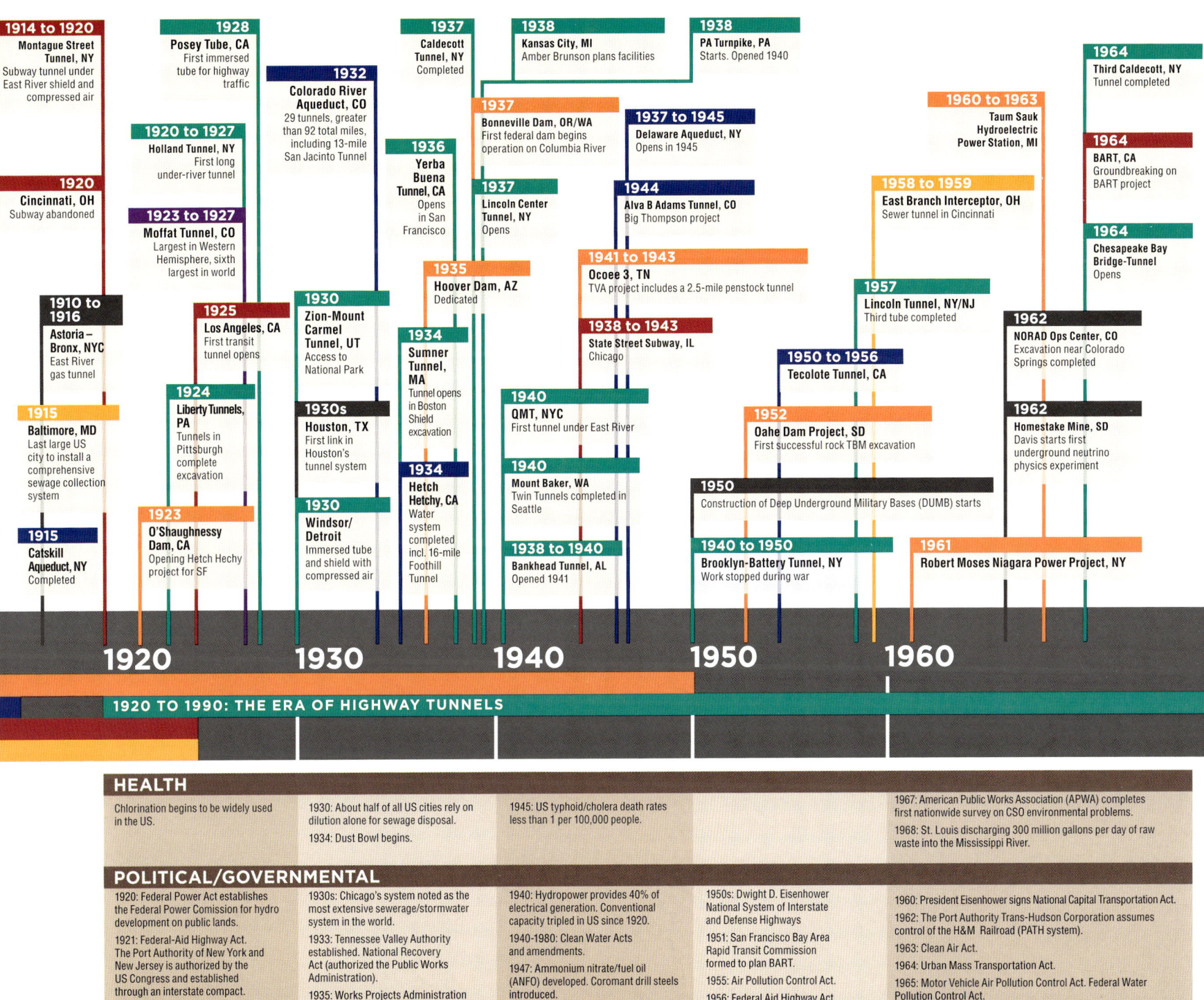

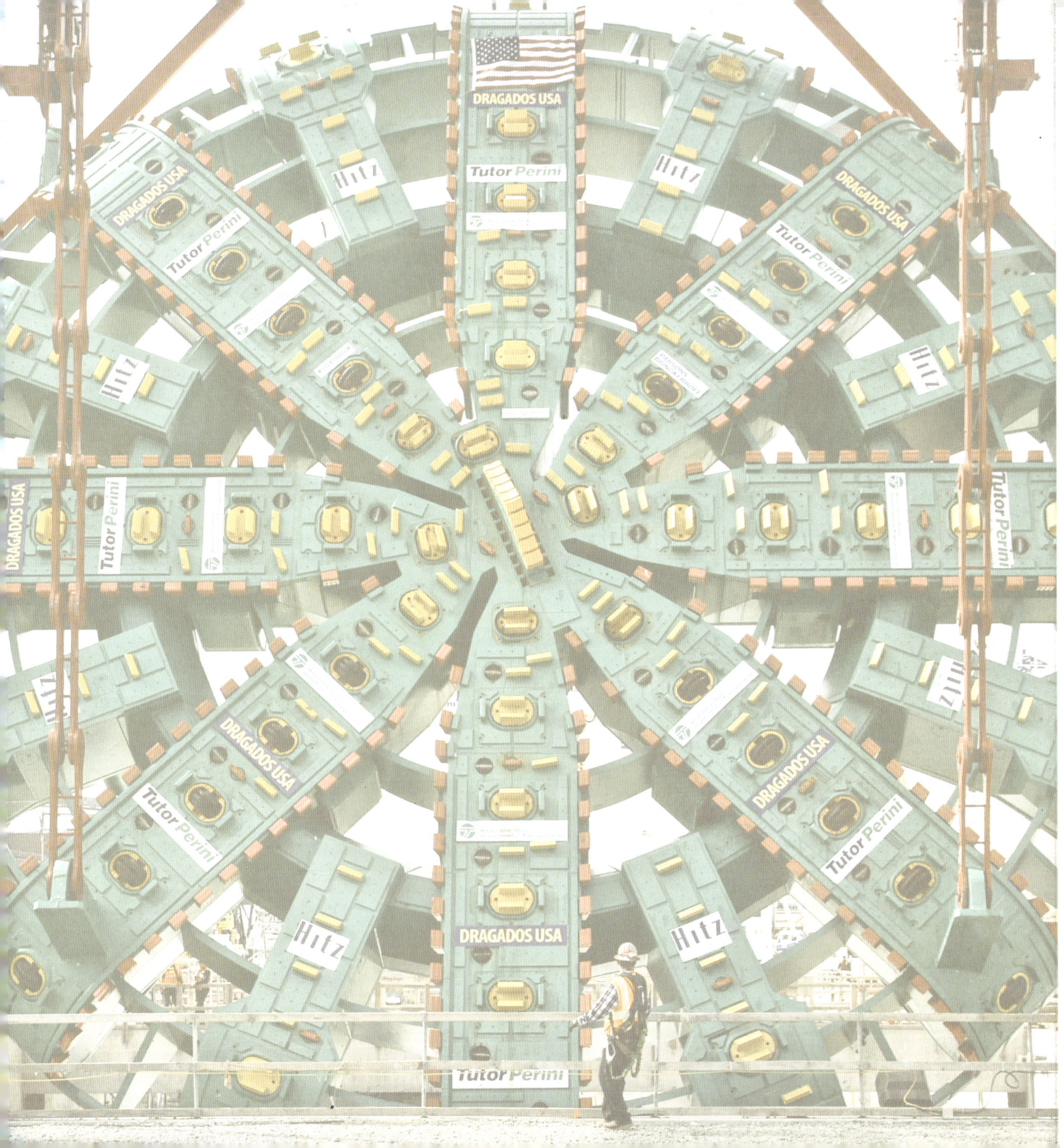

1
THE BUILDING OF A NATION

2 THE HISTORY OF TUNNELING IN THE UNITED STATES

The United States has a proud 200-year heritage of tunneling—an industry that grew within the same era as much of the tunneling performed in Europe during the early 1800s.

This is no surprise, given that America collaborated closely with European countries and shared lessons learned that served to push what was achievable underground. America continues to be at the forefront of tunneling technology, constructing some of the world's most challenging projects.

Over the past two centuries, more than 1,600 tunnels have been constructed in the United States, representing over 1,900 miles, with several world-class achievements having been accomplished during this time.

Compared with other facets of society, most people have little knowledge about tunneling or the extent and importance of its fascinating story. When considering infrastructure, most people typically think of prominent architecture or elegant bridges rather than underground infrastructure such as transportation portals (Figure 1.1), sewer systems, or water tunnels. Much of our work, particularly with water and wastewater tunnels, tends to be out of sight, totally overlooked, or not understood by the general public (Figures 1.2 and 1.3). Those tunnel shafts and caverns that are accessible, such as subway stations, do not always capture public admiration as an underground marvel, since the full scale and extent of the structure containing the station are not fully visible to those using the subway.

Previous Page | **New Irvington Water Tunnel near San Francisco**

Figure 1.1 | **Washington, D.C., Metro System Station**

CHAPTER ONE | THE BUILDING OF A NATION | 3

As such, the benefits of underground construction for society are typically not fully evident or appreciated. The underground industry tends to focus on overcoming the challenges of tunneling through the ground while protecting the public rather than broadcasting its many achievements, and even those are often understated. Being famous is not really a goal for the majority of those engaged in underground work. The reality is that society would not have evolved and grown to what it is today without extensive underground facilities (Figure 1.4). Whether it is improving health through better sanitation and reliable water supplies, enhancing the quality of life through subways and highway tunnels, or building up the economy through reliable rail transport of goods, tunnels connect and affect everyone every single day.

With this impact, it is difficult to imagine a world without tunnels. Any developed society relies on this construction because its population grows and everyday life support depends on it. With that comes a challenge to deliver larger, longer, quicker, and more affordable tunnels for a wide variety of uses. Tunneling is a specialized, high-risk industry with many traditions and practices. The underground industry has grown through lessons learned from its past projects and, out of necessity, has continually adapted tradition with the introduction of innovation to address new and increased challenges that have occurred over the last 200 years in the United States. This approach will continue well into the future as the industry continues to "blaze the trail" of what is possible.

WHAT DRIVES USE OF UNDERGROUND SPACE?

As society moved from an agricultural-based economy into the Industrial Revolution, population centers started to grow significantly as people moved to urban centers for work. This in turn stimulated the needs and demands of a quickly growing population.

The development of underground space in the United States has closely followed the development of society and the growth of populations living in the major cities across the country (Figure 1.5). Concurrent with the evolution of tunneling in Europe, the modern era of tunneling started more than 200 years ago.

Although early tunnels tended to be for the illegal purpose of smuggling, the Industrial Revolution drove the most significant demand for the construction of canals that were used to transport large volumes of raw materials and manufactured goods reliably and more quickly than horse-drawn transport. Inevitably, canal tunnels were needed as part of this new transportation network to pass through hilly and mountainous topography. The era of canal tunnels occurred from 1800 to 1850, and these early underground construction skills became the genesis for the tunneling industry that we know today.

As society advanced, so did the forms of transportation. Canals were overtaken by the railroads as a more efficient means of transporting goods and people across the large continent between the Atlantic and Pacific coasts. The era of railroad tunnels from 1824 to 1890 saw significant undertakings in terms of scale, achievements,

Figure 1.2 | **Construction Progress on the Croton Aqueduct Tunnel, Circa 1840**

Figure 1.3 | **Chicago Sewer Tunnel Construction in the 1920s**

Figure 1.4 | **Hand-Dug Rock Tunnel with Masonry Lining**

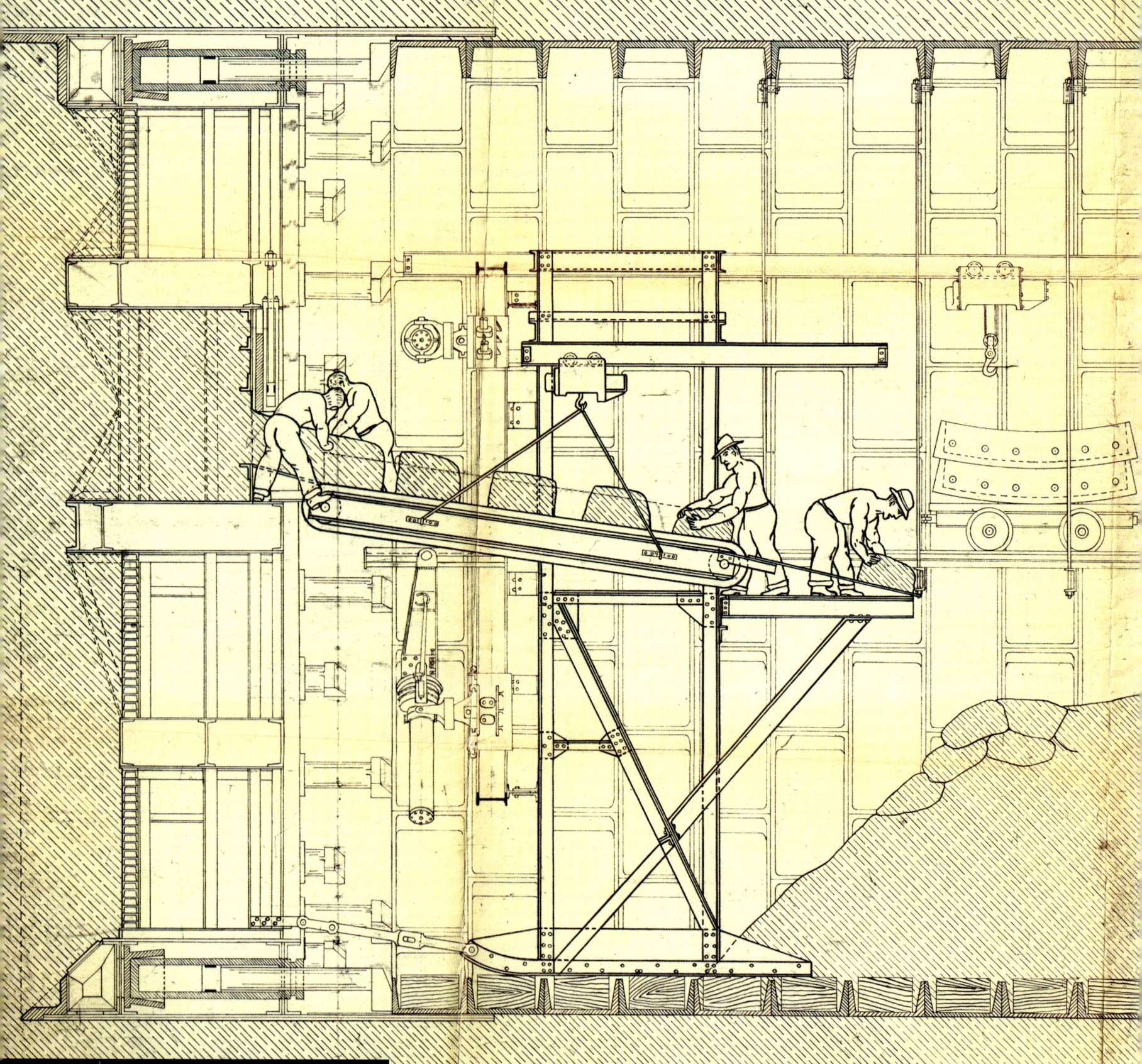

Figure 1.5 | **Lincoln Tunnel Shield Under the Hudson River**

LONGITUDINAL SECTION

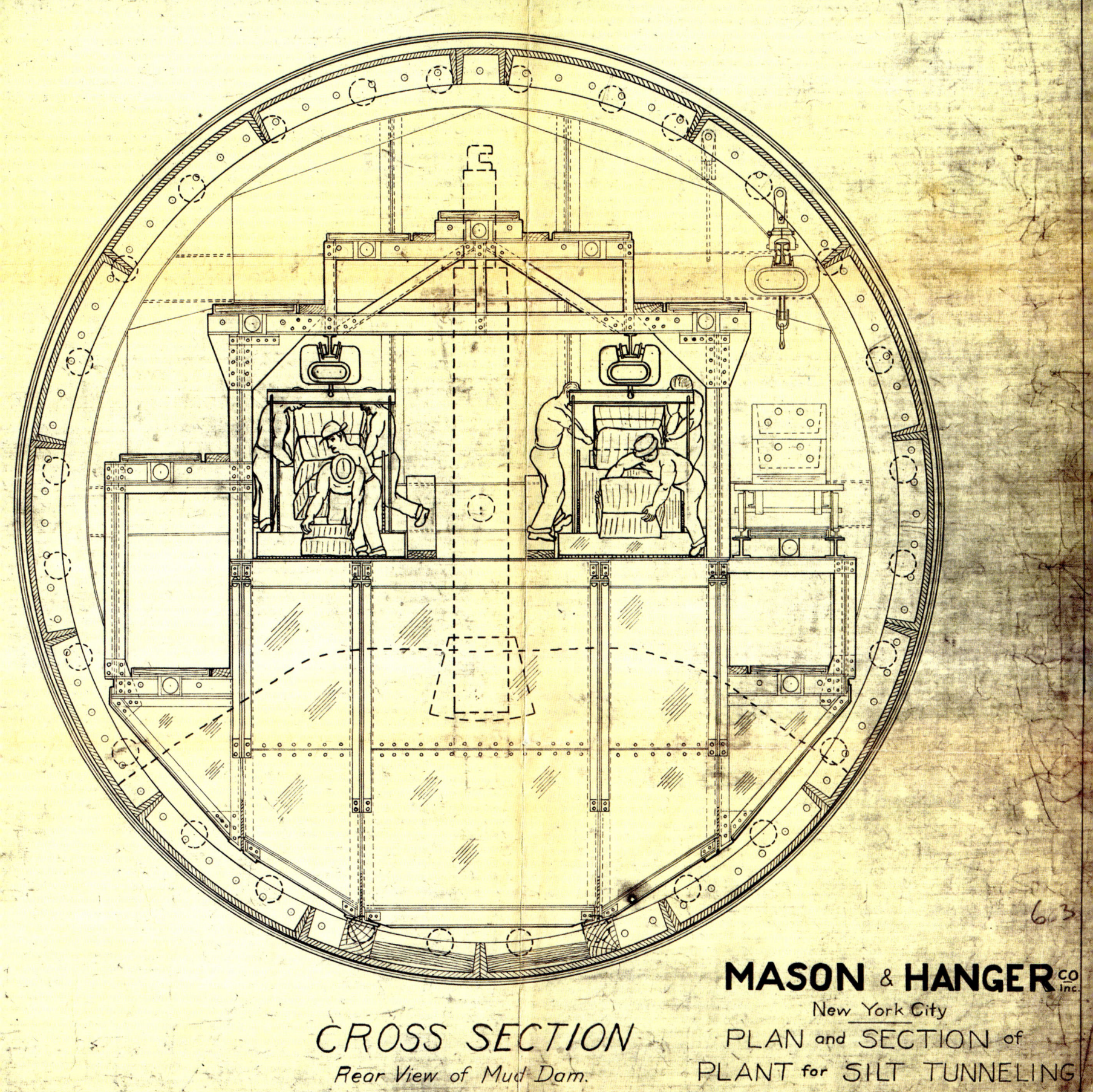

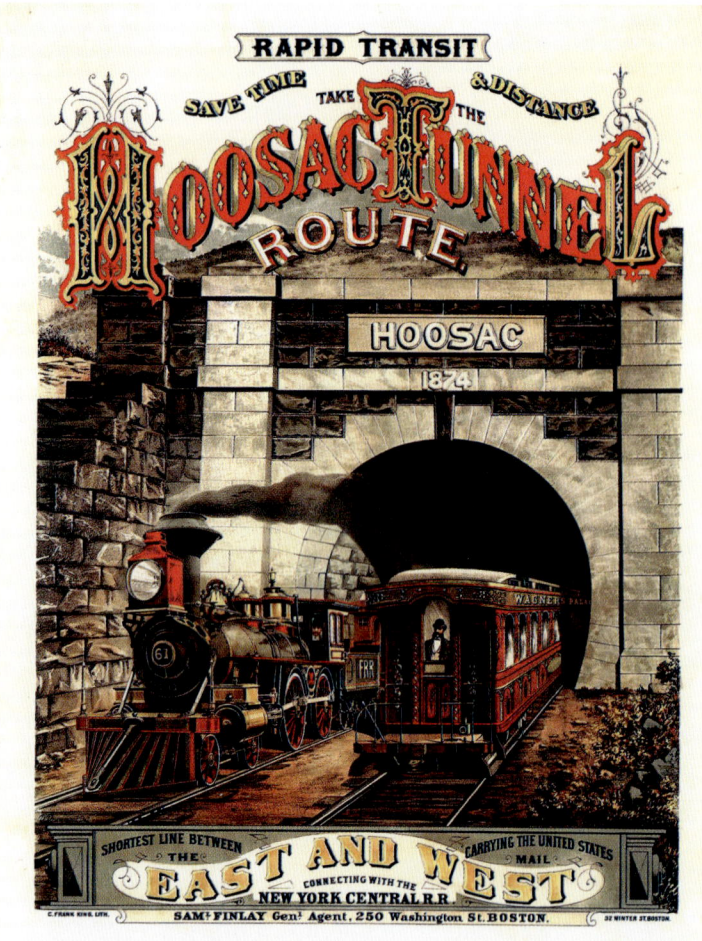

and a significant cost to human life while building a vast network of rail infrastructure through some of the most challenging terrain of the day. The Hoosac Tunnel, for example, was intended to improve trading and transportation from the industrialized Boston area westward to northern New York state (Figure 1.6).

With the population growth that rail and road systems encouraged, cities across America were imposing greater demands on natural resources, the most vital of which for life support was a fresh and reliable water supply. The era of water supply and distribution tunnels from 1870 to 1915 seamlessly followed the era of railroad tunnels. By the end of the nineteenth century, most major cities in America were enjoying reliable access to clean water, travel by railroad, and efficient access to a wide variety of manufactured goods. In many cases, most of these benefits were achievable in part through underground engineering and construction. But as urban populations continued to grow, additional means to address health concerns and improve the quality of life were needed (Figure 1.7).

It was only in the latter half of the 1800s, with the implementation of water and railroad tunnels, that extensive urban tunneling was contemplated. The successful implementation of water and railroad tunneling gave the industry confidence that extensive urban tunneling would be achievable. At the turn of the twentieth century, cholera outbreaks and poor sanitation required a comprehensive plan to develop large sanitary sewage collection systems and no longer allow sewage to collect and congregate at the street surface. Between 1877 and 1925, the era of sanitation tunnels saw many systems being implemented in major cities and the discharge of sewage away from the urban centers and into nearby streams, rivers, lakes, and oceans (Figure 1.8). Many of these gravity-operated systems were constructed by excavating large trenches, installing sewers, and backfilling above; however, where this was not achievable, tunnels were excavated beneath city streets.

Figure 1.6 | **Hoosac Tunnel Promotional Brochure**

Figure 1.7 | **New York City Street Cleanup** Figure 1.8 | **Early Chicago Sewer Tunnel Construction**

Concurrent with the implementation of the sanitation tunnels was another demand from society to address the horrendous congestion that had developed from horse-drawn traffic within the urban centers. This was particularly prevalent in Boston and New York, where the combination of horse-drawn traffic and pedestrians made commuting unbearable for everyone (Figures 1.9 and 1.10). With this background, the era of subway tunnels was implemented, starting in Boston in 1897, with much of the early subway systems in operation by 1920. By this time, life in the major cities had never been better. Society experienced improved health through better sanitation, reliable and clean water supplies, and efficient means by which the urban workforce could travel within the city, and by railroad beyond it, with a well-supplied industry and commercial market to cater to the demands and needs of the population—all made possible by the use of tunnels and underground facilities.

As cities continued to grow, the original water supplies proved to be inadequate. Expanded and more reliable systems were needed. The planning for such systems was visionary and required a new look at water storage by means of reservoirs, together with aqueducts, to transmit and distribute water into and throughout the growing cities. Many aqueducts were constructed by means of tunneling within the era of water storage and transmission tunnels between 1880 and 1950.

Figure 1.9 | **New York City Street in Mixed Use**

During the first half of the twentieth century, development of the automobile replaced horse-drawn vehicles. Roads had been constructed that crossed rivers and valleys, but tunnels through mountainous terrain simply had not been contemplated. By the 1950s, vehicle traffic demands together with urban development constraints within cities made highway tunnels inevitable. The era of highway tunnels occurred from 1920 to 1980 with construction of some of the largest and longest tunnels ever envisioned.

By 1970, population growth within cities had progressed to the stage where pollution levels of streams, rivers, lakes, and oceans had become unacceptable to the public, affected the quality of life that was expected, and once again led to concerns regarding public health. With the help of the Environmental Protection Agency (EPA), the era of CSO (combined sewer overflow) environmental tunnels started in 1970 and continues today and will persist well into the future. By intercepting and storing combined sewage flows during storm events, watercourses are improved for the benefit of everyone. Today, fish and water life can now be seen where once-contaminated sludge prevailed.

And even now, a new era is dawning within the tunneling industry. Integrated transportation hubs enabling railroads, subways, trams, and buses to connect to airports are being contemplated for the future. Most will likely require underground solutions. High-speed rail is being developed for travel between cities, thereby providing a viable,

Figure 1.10 | **New York City Traffic Before Subways**

efficient alternative to air and long-distance road travel. Water supply infrastructure is being made more reliable and resilient for our cities. Combined-use tunnels containing multiple utilities will save space and construction costs. These are but a few of the many ideas currently in planning and development that we can expect to see in the near future.

As an industry, we believe that a civilized society relies on the robust underground industry to continue to meet the many challenges that will inevitably lie ahead to fullfill our desire to improve the quality of life for everyone. Our industry is confident that we will continue to meet those challenges far into the future.

KEY PEOPLE AND POLITICS IN THE MAKING OF A TUNNEL

Projects would not be possible without a project sponsor. A project sponsor is an organization or a person who is promoting or leading the project, someone who directs all organizations engaged with the project and follows it from inception through to completion. Abraham Lincoln, for example, signed into law in 1862 the transcontinental railway legislation that initiated massive railroad construction to connect the country. In effect, he became the overall project sponsor that stimulated the era of the railroad tunnels (Figure 1.11). For large projects, the sponsor may hand over this role to a successor, and the new sponsor will continue in this role. Project sponsors take many forms and functions; however, the general tenet remains the same: to successfully implement the project. With most large projects, a project sponsor cannot guarantee delivery of the project unless he or she has the support of the community, authority to act, and essential funding sources. The elected representatives of the affected community also need to provide support for the project so that together with the project sponsor they can collectively guarantee a successful outcome (Figure 1.12).

Figure 1.11 | **Abraham Lincoln, 16th President of the United States**

In the past, several underground projects had not successfully progressed beyond the drawing board, to use an old term, because of a breakdown in alignment between the project sponsors and elected representatives. This breakdown may have been the result of any number of valid reasons, the most common of which is lack of project funding. A lack of financial support may be identified at the planning stage of a project or at any point thereafter. During construction, financial backing can be withdrawn or funding of subsequent phases of a large program may be temporarily withdrawn or completely canceled.

New tunnel projects and programs require funding levels that can extend beyond the range of a public budget but are nevertheless essential and critical to the future development of a city or region (Figures 1.13 and 1.14). The economy of a region cannot continue to grow indefinitely if that society does not have the infrastructure to support it. To address this shortfall, some owners venture into private investment under public–private partnerships (Figure 1.15). With this arrangement, the public enables a new facility to be financed, temporarily owned, and operated by a private entity over a fixed period, yet over the long term, the public tends to pay more for this approach. This arrangement is not new; for example, many of the subways in New York and Boston were financed and constructed using the same mechanism at the turn of the nineteenth century. Society moved on from this form of investment for capital projects to a preference for public financing and control of the facilities. Our tunneling world seems to have come full circle.

Figure 1.12 | **Sunken Tube Conceptual Plan for BART in San Francisco**

Given the societal challenges that our nation will face in the future, decisions from an underground perspective need to be made on a much longer wavelength than may have been enacted within the previous 60 years, albeit borne out of necessity for many reasons. As history has already shown, the best legacy we can create for the benefit of future generations is that of continuing to improve the development of infrastructure that ultimately improves the quality of life. For large projects, funding can be managed through strategic program phasing and

Figure 1.13 | **Tightening Segment Bolts in Lincoln Tunnel, 1935**

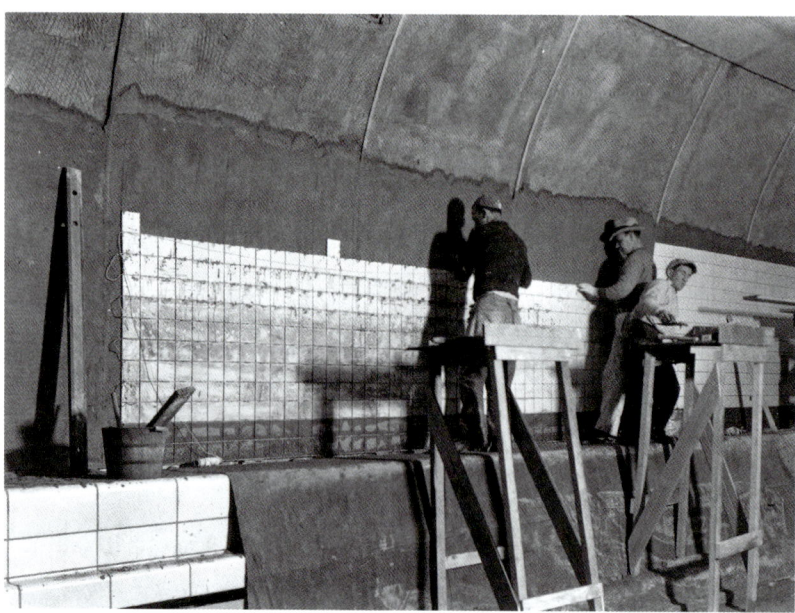

Figure 1.14 | **Setting Wall Tiles in Lincoln Tunnel, 1937**

Figure 1.15 | **Hudson & Manhattan Railroad Stock Certificate for the First Rail Tunnel Under the Hudson River in New York**

with coordinated prioritization. The underground industry has an important part to play, but so do the public and political groups. By working together with communities, elected representatives, the many project sponsors, and the other key decision makers for our critical underground projects, we can achieve this goal.

THE BACKBONE OF DEVELOPMENT IN SOCIETY

Throughout the past 200 years, tunneling has continually been used to address some of the most significant problems facing society (Figure 1.16). Society relies tremendously on underground infrastructure for daily living, often without much thought about how it got there, what it does, or even how long it will last. To contemplate an age before tunneling would be unthinkable in the modern era in terms of how far society has developed in just two centuries.

The underground construction industry has remained active during this entire time. The continuous use of tunneling during this period demonstrates its value to developing and shaping the modern world. As populations continue to grow and the "built environment" on the surface provides less and less available space for expansion, inevitably the future will demand more and more underground space solutions to provide life-supporting infrastructure (Figure 1.17).

Although the end use of the tunnel may have changed from one decade or era to another, the techniques and equipment employed to complete these challenging projects have been applied in the same way. A common saying within the industry is that "the ground does not know what you are putting inside the tunnel," and to some degree this is valid. The greatest challenge for most tunnels and underground facilities is with the actual ground conditions encountered and controlling the ground during excavation to protect the surrounding infrastructure and environment.

The public may see and experience the benefit of tunnels, but there is less of an appreciation for the scale or complexity of tunnel construction. In addition, tunnels are often negatively portrayed in the media with the typical topics being about technical issues leading to cost and schedule overruns, but this is a small fraction compared to the many successes that are built almost anonymously and without great public awareness. Often the basis by which a tunnel is judged within the media does not recognize the scale or difficulty of what is being undertaken. If astronauts landed on Mars a week or two late, would the media and the public still consider this to be a success?

Having a greater public appreciation of tunneling solutions is important for society, as all of these underground projects require funding. Lessons from the past show that although tunneling would appear relatively continuous throughout the past 200 years in the United States, the number of tunnels under construction has not been uniform. Since the mid-1970s, for example, many essential tunnel projects have been delayed for a variety of reasons, such as a lack of funding or a change in political direction or opinion. The net result is that society suffers from the lack of growth in its supporting infrastructure. The consequences of such inaction are that many important projects have been delayed, shelved, or completely canceled, placing an unhealthy reliance on aging infrastructure developed at a time when the population was much smaller. This places a greater burden and operational risk on the existing infrastructure, which in turn may reduce the service life of those tunnels. History has shown that many tunneling projects that are delayed or canceled will eventually be built. Whether the wait is a few years, a decade, or longer, the projects or some modern variant of it will ultimately be constructed.

As a tunnel comes to the end of its useful life, closer attention needs to be paid to rehabilitating, upgrading, or replacing this essential infrastructure. It is often more economically feasible to replace aging facilities rather than repair them, in which case tunneling will continue to thrive and help shape our society long into the future.

Figure 1.16 | **City Street Excavation for New Subway Construction**

Figure 1.17 | **New York City East Side Access Caverns to Grand Central Terminal**

PROTECTING AND EXTENDING THE LIFE OF A TUNNEL

Many early tunnels were not designed with a specific design life in mind; rather they were "built to last" in the best opinion of the day. This approach in turn created rules of thumb in terms of best practices and applied rudimentary design for excavation and ground support. Through ongoing research and analysis, engineering became more scientific, which served to make the designs more efficient and cost-effective. A better understanding of the theory, materials, and limitations of construction techniques helps engineers determine a design life during which the tunnel structure can be relied on to support the ground conditions throughout its anticipated service life.

Figure 1.18 | **Early Rock Drill Jumbo Design**

To extend the life of a tunnel beyond original design expectations, a process of maintenance and rehabilitation has developed over the years. Inspections of tunnels performed at regular intervals help to identify any deterioration that needs to be addressed and confirms that the tunnel remains sound and intact. Tunnel inspections typically involve looking for groundwater ingress through the lining and any structural deterioration, such as cracking, that may have developed. Such defects can be quickly and effectively addressed to restore the lining to a fully functioning, low-maintenance condition.

This process of inspection and tunnel rehabilitation would usually be performed after a complete shutdown of normal operations within the tunnel until work had been completed. In modern society, the expectation is that tunnel operations remain unaffected as much as possible and all work needs to fit in and around a continuously functioning tunnel. This can be done by working at off-peak times, but it is not always possible or cost-effective to perform the work without some interruption to tunnel service. This drives a need to be efficient in all inspection and rehabilitation operations and to minimize any shutdown periods. Once again, a societal need drives innovation within the underground industry.

An example of the progress that has been made in this area is the use of digital photography and laser profiling that have been implemented as a means of capturing existing tunnel conditions as accurately and efficiently as possible. There are many innovations in tunnel rehabilitation that will continue to develop with the sole aim of protecting our underground assets. Technology will continue to evolve in this area and will serve to minimize the impacts to the public while maximizing the use of existing tunnels into the future.

LEARNING FROM HISTORY

The many tunneling accomplishments in the United States over the last few centuries rival those of any country in the world. Indeed, the first air-powered rock drills were invented by Burleigh in the United States and successfully used in 1866 in the Hoosac Tunnel (Figure 1.18). The skills and experiences achieved over a considerable period of time have provided a wealth of information and lessons learned upon which the tunneling industry is founded. This substantial record of underground achievements has

Figure 1.19 | **Multi-Level Rock Drill Jumbo Excavating a Railway Tunnel, 1958**

provided the skills and confidence to advance the industry to ever greater achievements, including larger tunnels excavated with drilling and blasting methods (Figure 1.19).

Looking back through history, it is clear that the tunneling industry exists today because of our ability to assess what has been previously attempted, identify what has and has not worked, and determine how best to solve the problems encountered by others. Out of necessity, we have become an industry that focuses on identifying and managing the risks, which comes from comprehending the multitude of lessons learned from the past. This has led to continuous innovative thinking to overcome the hurdles that stand between project failure and success.

It is continually satisfying to those working in the industry to look back at what was attempted using state-of-the-art technology just 10 to 15 years ago that seems to be no longer a challenge because it has since been successfully overcome many times over. This indicates the true progression of the industry in helping to serve our society with increased use of underground space (Figure 1.20).

Given that many cities currently have a fully developed infrastructure, available space for access and construction of a new tunnel becomes limited. Also, existing underground infrastructure, such as deep basements, utilities, subways, and building foundations, moves tunnels ever deeper or requires the new tunnel to "thread the needle" within a very limited gap underground.

Figure 1.20 | **Eisenhower Tunnel Construction, Colorado**

Many of the near-surface tunnels of the past, such as subway tunnels, were constructed by closing off and tearing up entire streets with large excavations, building subway tunnels and stations, and then backfilling and reopening the street to the public. In the modern era, such disruption is considered unacceptable in most downtown areas of large cities; consequently, alternative methods for tunnel excavation have been encouraged.

Burrowing under rivers, lakes, and oceans is always one of the most difficult forms of tunneling that can be undertaken. Dealing with groundwater during tunneling is usually the greatest challenge of any tunnel project. In the past, a common method was to use compressed air in the tunnel to withstand the groundwater pressure and to excavate the tunnel by hand using tools at the face. Sadly, in the early days, there were many recorded deaths using this method because too much pressure risked a blowout at the face where the air would be lost and a direct connection with the river or sea would ensue. The workers also risked suffering from "the bends," similar to that experienced by deep-sea divers.

Once again, another method of tunnel excavation was developed using a tunnel boring machine (TBM) that could effectively be sealed off from the tunnel face during excavation, enabling crews to work safely within the tunnel. Compressed air and other pressurized gases are still used in tunnels to this day; however, there have been many advances in their use, including better decompression tables and hyperbaric chambers (Figure 1.21). As a result, the safety of this method of tunneling is much improved.

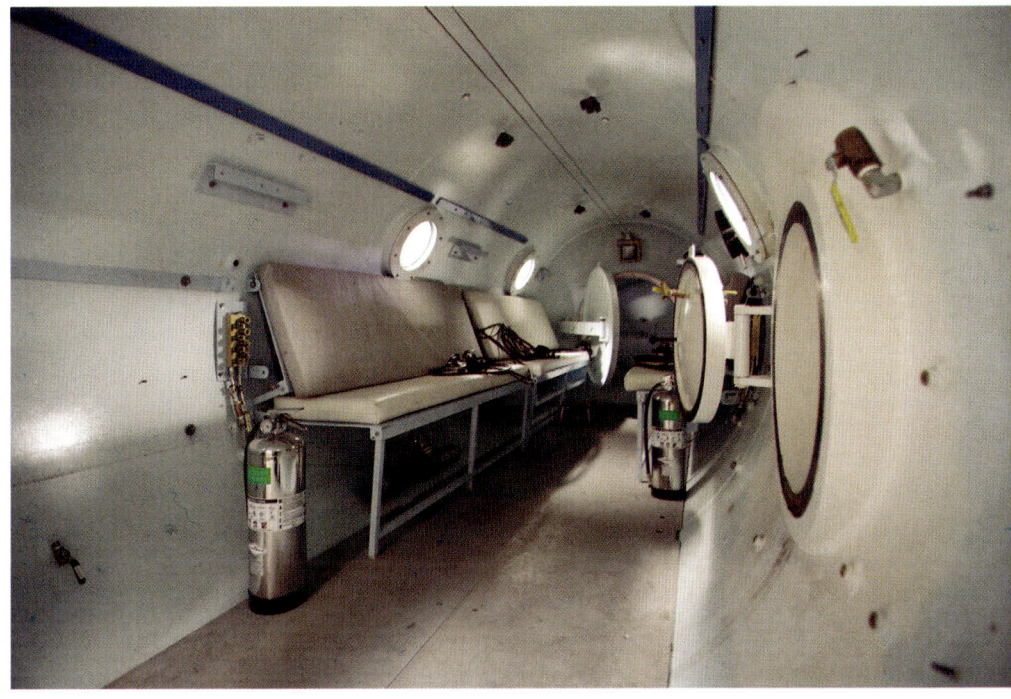

Figure 1.21 | **Hyperbaric Chamber for Tunnel Crews**

14 THE HISTORY OF TUNNELING IN THE UNITED STATES

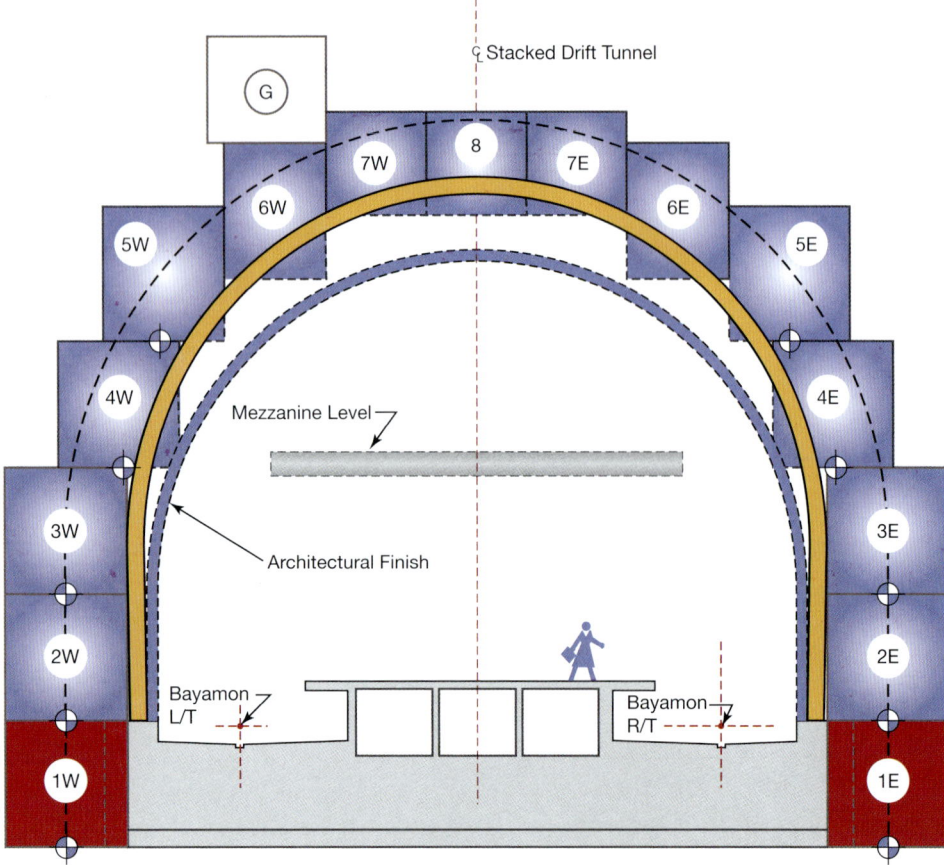

Figure 1.22 | **Tren Urbano, Rio Piedras Station Cavern Design in Puerto Rico**

Over the decades, the industry has amassed a tremendous knowledge base of lessons learned for all types of underground projects. This experience is continually shared through the various tunnel conferences that are held on an annual basis in the United States. This has helped to develop an industry that is knowledgeable, has strong networking for information sharing, and can learn and advise very quickly when things do or do not go as planned on a tunnel project.

NECESSITY DRIVES INNOVATION

There are many drivers of innovation. The most serious priority is safety, for both the public and the workforce during construction and operation of the tunnel. Although history shows that tunnels were an innovation that improved public health through better sanitation and a freshwater supply, there is a continual need to keep building on those earlier achievements (Figures 1.22 and 1.23). In the modern era, a greater emphasis on the quality of life and sustainability linked to environmental and planetary concerns is becoming more prominent than ever before. In striving to improve society and make further advances, the tunneling industry confronts ever greater adverse ground conditions. Finally, with the current generation there is always a continual, unending pressure to implement projects better, faster, and cheaper and with fewer impacts to the environment. All of these elements combine to stimulate innovation within our industry, and history demonstrates that we have made significant progress in this area. To meet the challenges of the tunneling industry, we find technological advances in all areas, such as equipment, materials, design, and safety.

Figure 1.23 | **Tren Urbano, Rio Piedras Station Cavern Construction**

Over the last 200 years, many underground workers sacrificed their lives during tunnel construction. Fatalities occurred from a combination of excavating through highly challenging ground conditions that had not previously been attempted using little geotechnical information and rudimentary equipment, hiring untrained workers, and following the accepted but unsafe practices of the day.

Since the 1960s, we have moved from an industry that once carried a high risk to human life—where some projects recorded significant numbers of worker injuries and each project had an acceptable accident rate—to one that has no tolerance for any accident during construction. That is not to say that accidents do not occur today, as tunneling continues to be a risky undertaking. However, the goal is always to achieve *zero accidents* on each and every project. Loss of life is now extremely rare within the industry.

The advances in underground equipment have gone hand in hand with the development of new and improved materials (Figures 1.24 and 1.25). This is not surprising, as tunnel excavation always needs an appropriate tunnel lining to support the ground as the tunnel advances. Many methods are employed to excavate a tunnel, but all of the tunneling equipment and materials in use are continually being developed with innovative advances to improve performance.

Manufacturing a TBM or a large rock-drilling machine is not similar to car manufacturing. Cars are mass produced so that by the time one year or so has elapsed, most defects have been fixed and the subsequent cars continuously improve in reliability. With underground equipment, although they may look the same as previous versions, most are uniquely made due to continual development and improvements. When considered in conjunction with frequent demands for increased diameters, new lining configurations, and performing in ever more challenging ground conditions, it is admirable that tunneling equipment can meet those requirements.

Figure 1.24 | **Early TBM Development**

Tunneling was once highly labor intensive, with several hundred workers engaged in daily activities. As the industry has developed, there are now, in general, considerably fewer workers involved during construction. Mechanization has improved tunnel safety and productivity, and continues to be more cost-effective with each improvement.

Tunnel design has come a long way since the days of the drawing board, the blueprint, and the slide rule. Computerization has enabled underground designs to be developed using three-dimensional (3-D) computer models of the ground loads and the structural tunnel lining that supports it. With this technology, ground behavior can be predicted and linings have become more efficient, leading to a reduction in cost and an extension of the design-life performance. Using building information modeling technology, complex underground facilities can be modeled in 3-D (Figures 1.26 and 1.27). This modeling can provide a fly-through of the completed facility prior to construction as well as generate all drawings needed to build the project. This technology can also generate construction schedules and costs for the project.

Figure 1.25 | **Modern Day Hard-Rock Tunnel Boring Machine**

16 THE HISTORY OF TUNNELING IN THE UNITED STATES

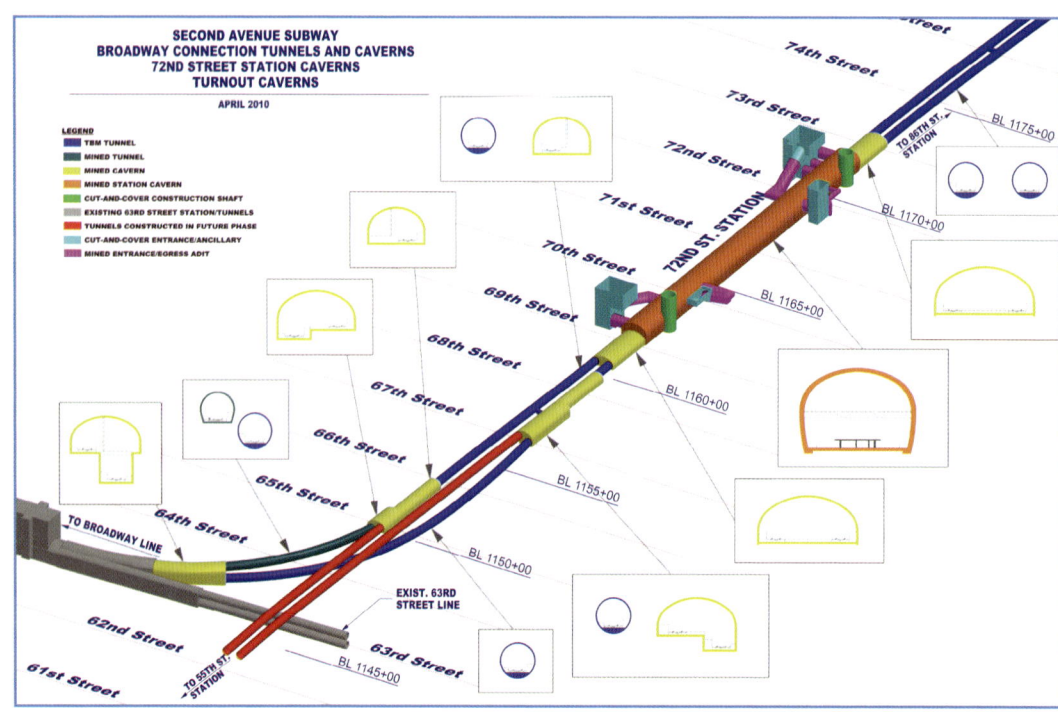

Figure 1.26 | **New York City Second Avenue Subway**

Figure 1.27 | **Seattle's Mount Baker Ridge Highway Tunnel**

The early tunnelers within the industry were effectively good risk managers of their day; they just did not realize that they were applying an informal form of risk management in successfully completing their projects. Today, risk is taken very seriously and has been formalized and quantified to capture project risks, determine what measures need to be taken to address those risks, and assess and develop the need for contingencies in both cost and time in case those risk events do occur. The industry has come a long way in identification and catering for risks on tunnel projects, both in design and construction. Through respecting the lessons learned from the past, this process will continue to evolve successfully into the future.

THE FUTURE IS UNDERGROUND

Applying lessons learned from our past to the future of underground construction is always difficult to do with any degree of certainty. Our industry has learned that securing funding for the entire project helps to increase project certainty of delivery. Many of the so-called megaprojects are so large that project financing needs to be phased with respect to scope, time, and budget to enable the project to commence as soon as possible and often before the entire funding package is in place. Such financial phasing needs careful management to avoid undue delays to the project between phases while awaiting the next funding appropriation. In the future, funding requirements will need to be more reliable to continue to build the critical underground infrastructure required without adversely impacting quality of life.

Often we hear about politicians canceling infrastructure investments while others restart them at a later date. It may be difficult to garner political support if the duration of project construction outlasts the typical term that a politician is in office. But the reality is that the big, long-term decisions for building underground infrastructure need to be made for everyone's benefit (Figure 1.28). These decisions were most certainly made in our long history to develop the impressive tunnel inventory that exists today. Yet the trend since 1975 has been to delay much-needed development in favor of short-term solutions while diverting funding to other areas of society. Going forward, we need to rekindle the old habits of having the vision to make a decision and to do something that will benefit not just this generation, but future generations, so that society can continue to prosper (Figure 1.29). Working together and having continual investment and support for a well-thought-out and ratified plan will help us achieve this goal.

Although investment is critical, those investments must be timely and make economic sense, and they need to be sustainable for generations to come. Also key will be the built-in capacity for future growth, extending the operating life of the tunnel, and looking closely at the environmental impacts of both the tunnel construction and the final operation of the facility with respect to the environment (Figure 1.30). Energy measures such as having a low carbon footprint from the manufacture of the concrete through to the generation of the electrical power may prove to be important future requirements that will need to be met when contemplating a tunnel.

Making the best use of the existing underground infrastructure has always been a consideration in the recent past, and this will likely continue. We see that regular investment in this area is a cost-effective way to extend the operational life. Having a complete understanding of the existing tunnel conditions and proactively implementing appropriate rehabilitation measures will continue to develop and improve.

Figure 1.28 | **Chicago's Extensive, Multiple-Phase TARP Tunnel Project**

Figure 1.29 | **Los Angeles Metro Gold Line Tunnel near Completion**

THE HISTORY OF TUNNELING IN THE UNITED STATES

Figure 1.30 | **Washington, D.C., Metro Station, Judiciary Square**

Figure 1.31 | **Underground Subway Cavern Final Lining Formwork in New York City**

Tunneling is a thriving and challenging career. For many who have been engaged within the underground industry, it is a lifelong commitment. Although the final product can often look simple, the engineering, analysis, and construction that are behind the successful completion of a tunnel project can be intricate and substantial. Much of the industry requires state-of-the-art technology that is continually under development through lessons learned and a never-ending requirement to "push the envelope" beyond what has previously been attempted (Figure 1.31). As a result, underground construction is a highly specialized, high-risk undertaking that has many facets and specialties within the realm that is referred to as "tunneling."

Considering the last four decades, it is surprising that far fewer new tunnels were constructed in comparison to the first 70 years of the 1900s. The sporadic nature of the tunneling industry throughout this period has created generational gaps of people possessing the specialist skills of underground engineering. Given that underground engineering and construction is an inevitable solution for the continued development and future growth of society, we as an industry are preparing to meet this challenge.

There has been significant investment across the many facets of this industry to engage the younger generation in having a career in tunneling. Going forward, this needs to be a priority and to occur continuously despite the peaks and troughs of activity in the underground business. Specialist training continues to be developed to keep pace with an ever-changing industry. Young tunneling professionals possessing the energy and passion to get involved with tunneling have engaged in peer groups within companies and also within various national industry organizations. By working together with the senior members of the industry throughout the various projects that are under planning, design, and construction, future leaders can gain considerable wisdom and technical knowledge from lessons learned. Consequently, these young tunnel professionals will be engaged in helping to shape the future of our industry.

As tunneling continues to be a highly specialized industry, it is essential that our knowledge and resulting individual skills continue to be passed on for future generations to learn and advance further. It is this heritage that we continually look to preserve and develop to meet our society's demands for the future of our country's infrastructure.

2 SOCIETAL BENEFITS

Tunnels and underground infrastructure helped build twenty-first-century American civilization. The benefits to society have been enormous.

To demonstrate this, we begin with some truisms:

- No major American city could exist in its current form without extensive underground tunnels and systems for water, sewage, and transportation.

- When economic necessity drives the need for a tunneled solution, the underground industry has come through successfully time and again throughout the history of the United States.

- Now and in the future when citizens demand a higher quality of life, safer and quicker transportation, and expanded green spaces, the urban planners, entrepreneurs, and political figures will once again turn to the underground construction industry to innovate and provide cost-effective solutions to solve the issues presented.

It is also true that the tunnel business has a mixed history of budget and schedule controls on major projects. Although this is undeniable, it is equally irrefutable that the long-term societal value of underground infrastructure and facilities goes far beyond the initial construction costs. Even cities like Boston, Massachusetts, that have struggled with these short-term budget problems are now reaping the benefits of the grand vision with fully functioning underground facilities, such as the Central Artery/Tunnel project.

When the history of our country is examined, it is true that every great advance in shaping the development of the modern United States of America as we know it has been driven by the vision, engineering, and construction of underground infrastructure. For example, there would be no Transcontinental Railroad without the tunnels through the Appalachians and Rocky Mountains; no functioning commercial activity in New York City in the absence of the extensive subway system; and no San Francisco without the Hetch Hetchy water system shown in Figure 2.1 that brings life to its inhabitants from more than 150 miles away. Figure 2.2 shows workers with the San Francisco Public Utilities Commission inspecting the New Irvington Tunnel.

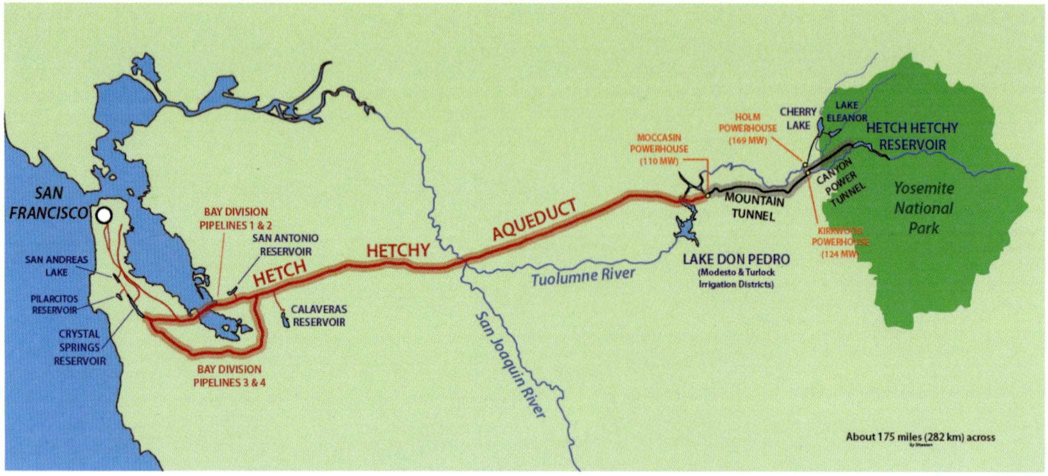

As we explore the historic importance of underground infrastructure to the development of nineteenth- and early twentieth-century civilization in the United States, and explore its further development into the commercial superpower that exists today, we see a clear path forward. Further investment in better sanitation, cleaner waterways, and movement of goods and people is a key factor in maintaining a vibrant and civilized society.

Figure 2.1 | **Extent of the Hetch Hetchy Aqueduct**

Previous Page | **Tunnel Construction at Dulles International Airport, Washington, D.C.**

Looking into the future, we see further population growth, urbanization, and a growing demand for materials, energy, and space on the earth's surface as serious challenges facing society. This is recognized around the world as well as in the United States. For this reason, infrastructure has been and will continue to be a focus of the U.S. political vision up to and including the presidential level, from Abraham Lincoln's Transcontinental Railroad, to Dwight D. Eisenhower's Interstate Highway System, to the creation of the Environmental Protection Agency, and on into the future.

CHAPTER TWO | SOCIETAL BENEFITS

Figure 2.2 | Inspecting the New Irvington Water Tunnel near San Francisco

A testament to the importance of tunneling and underground construction to society is encapsulated by Henry Sturgis Drinker (1850–1937). Drinker was a mechanical engineer who helped engineer the Musconetcong Tunnel in the 1870s, which enabled railroad travel between Easton (Pennsylvania) and New York City. He also studied law and became Solicitor General for the railroad for 20 years before accepting the presidency of Lehigh University in 1905. Drinker wrote a comprehensive treatise on tunneling in 1883 and in it, he stated the following:

It is worthy of note that this art of tunneling has gone in past ages hand in hand with the higher civilization of each era. As a people become more civilized, its civilization can be gauged by its progress in tunnel construction, and whatever is the particular motive, the result is always the same.

Finally, with modern civilization, we have tunneling in its last and greatest development, and we see that in proportion to the civilization of a people will be found their development in this art. This is most natural, for, of all branches of construction, it is one of the most difficult. A barbarous people may, perhaps, develop a high degree of perfection in the mere art of open-air building, where stone can be piled on stone, and rafter fitted to rafter, in the light of day; but it takes the energy, knowledge, experience, and skill of an educated and trained class of person to cope with the unknown dangers of the dark depths that are to be invaded by the tunneler. (Drinker 1883)

As we plan to rebuild America's infrastructure, we must first understand how and why it came to be. A dominant theme of this book is that America's history is in large part a story of its infrastructure, innovation, and continuous improvements.

Abraham Lincoln (1809–1865)

Lincoln was the 16th president of the United States and a true advocate of the development of America's railroads. After his election to the presidency in 1860, Lincoln made his mark on railroad history and signed the Pacific Railway Act, authorizing land grants and government bonds, which amounted to $32,000 per mile of track laid, to two companies, the Central Pacific Railroad and the Union Pacific Railroad. No meeting point had been set for the two rail lines when President Lincoln signed the act in 1862. Following Lincoln's assassination in 1865, work on the Transcontinental Railroad continued with bitter disputes between the two rival railroads as to the meeting point. On April 9, 1869, Congress established the meeting point in an area known as Promontory Summit in Utah. One month later, Central Pacific's engine Jupiter and the Union Pacific's engine No. 119 met on May 10, 1869.

Figure 2.3 | Diversion Tunnel Construction for the Hoover Dam

CHAPTER TWO | SOCIETAL BENEFITS | 23

This chapter presents the story of America's underground infrastructure and provides insight into how the underground industry has developed to serve the advancement of the nation for the past two centuries (Figure 2.3).

BUILDING OUR CITIES

Development of the nation's urban infrastructure required construction of public works systems that delivered sufficient quality and quantity of water and controlled and treated wastewater. As cities grew in the nineteenth century, increasing concerns were raised about public health, as water and sewers changed from a private luxury to a widespread public necessity. The use of storm drains to convey sewage had been common practice. This led to untreated sewage being flushed into rivers and lakes adjacent to the rapidly growing cities that withdrew drinking water from the same sources.

It is somewhat counterintuitive that in order to have an efficient sewage system, there must first be an adequate supply of fresh water. The two go hand in hand. Soon after water supply systems were established that enabled more effective firefighting and to bring fresh water to every home (beginning in the 1840s), the use of water closets and flushing toilets grew dramatically and the need to convey wastewater became an imperative. For example, as water systems were developed, per capita water usage increased dramatically: from 5 to 15 gallons per day before the presence of municipal water systems, to volumes ranging from 75 gallons per capita per day to more than 150 gallons per capita per day. This water, once supplied, also has to be removed in the form of wastewater before treatment and final disposal.

The mid-nineteenth century in the United States also saw rapid urbanization. Urban area populations across the country more than doubled between 1840 and 1880, from 11% to 28%. It was eventually recognized that better public sanitation designs were needed. It was evident that cities and towns rather than individuals and private entities had to play an ever-increasing role in the design and construction of sewers.

Figure 2.4 | **Tunnel Works for the New York City Water Supply System**

WATER IS LIFE

Many major U.S. cities simply would not exist without the major water tunnels that bring safe and clean drinking water into every home. Countless drinking water systems come from remote sources, such as the 319 miles of the New York City's water tunnel system shown in Figure 2.4, or the 167 miles from the Hetch Hetchy Reservoir system

Figure 2.5 | **Drill-and-Blast Excavation for New York City's Water Tunnel No. 1, 1914**

Figure 2.6 | **TBM Mobilization for New York City's Water Tunnel No. 3, 1994**

into San Francisco. These epic underground engineering feats are taken for granted now and are largely unknown achievements compared to signature surface structures such as the Transamerica Pyramid in San Francisco or the Empire State Building in New York that have a much less positive and perpetuating impact on society.

The New York State Legislature passed the Water Supply Act of 1905 (the bill originally written and promoted by New York City Mayor George B. McClellan Jr.) that set up the State Commission and Bureau of Water Supply that then set aside the necessary investment that brought New York City the water it needed to continue to grow. The resulting 163-mile Catskill Aqueduct (Figure 2.5) joined the existing 41-mile Croton Aqueduct. This was supplemented by the 85-mile Delaware Aqueduct (begun in 1937) to supply some of the highest-quality drinking water anywhere in the world to New York City. This marvel of underground engineering fulfills all of the city's water supply needs with more than 95% of the system working by gravity alone (Figure 2.6).

The achievements of New York City are mirrored on the West Coast where both Los Angeles and San Francisco transport their water supply over long distances. The Mulholland water supply system in Los Angeles includes the 238-mile-long Los Angeles Aqueduct (constructed between 1908 and 1913 and completed 20 months ahead of schedule). This city- and state-defining achievement consists of more than 55 miles of tunneling (in 142 tunnels, including the 5-mile-long Elizabeth Tunnel) with distribution systems continued, as shown in Figure 2.7, under the auspices of the Metropolitan Water District of Southern California. The Colorado River Aqueduct and distribution system extends more than 200 miles, from the Nevada/Arizona border to the city of Los Angeles. The growth of present-day Los Angeles would be inconceivable without the vision executed by the superintendent of the Los Angeles Department of Water and Power, William Mulholland (Figure 2.8), and the politicians who supported this vision.

Figure 2.9 shows workers celebrating the completion of a Pasadena water tunnel in 1937. This 3.5-mile-long tunnel was constructed to supply Los Angeles with Colorado River water from the Colorado River Aqueduct.

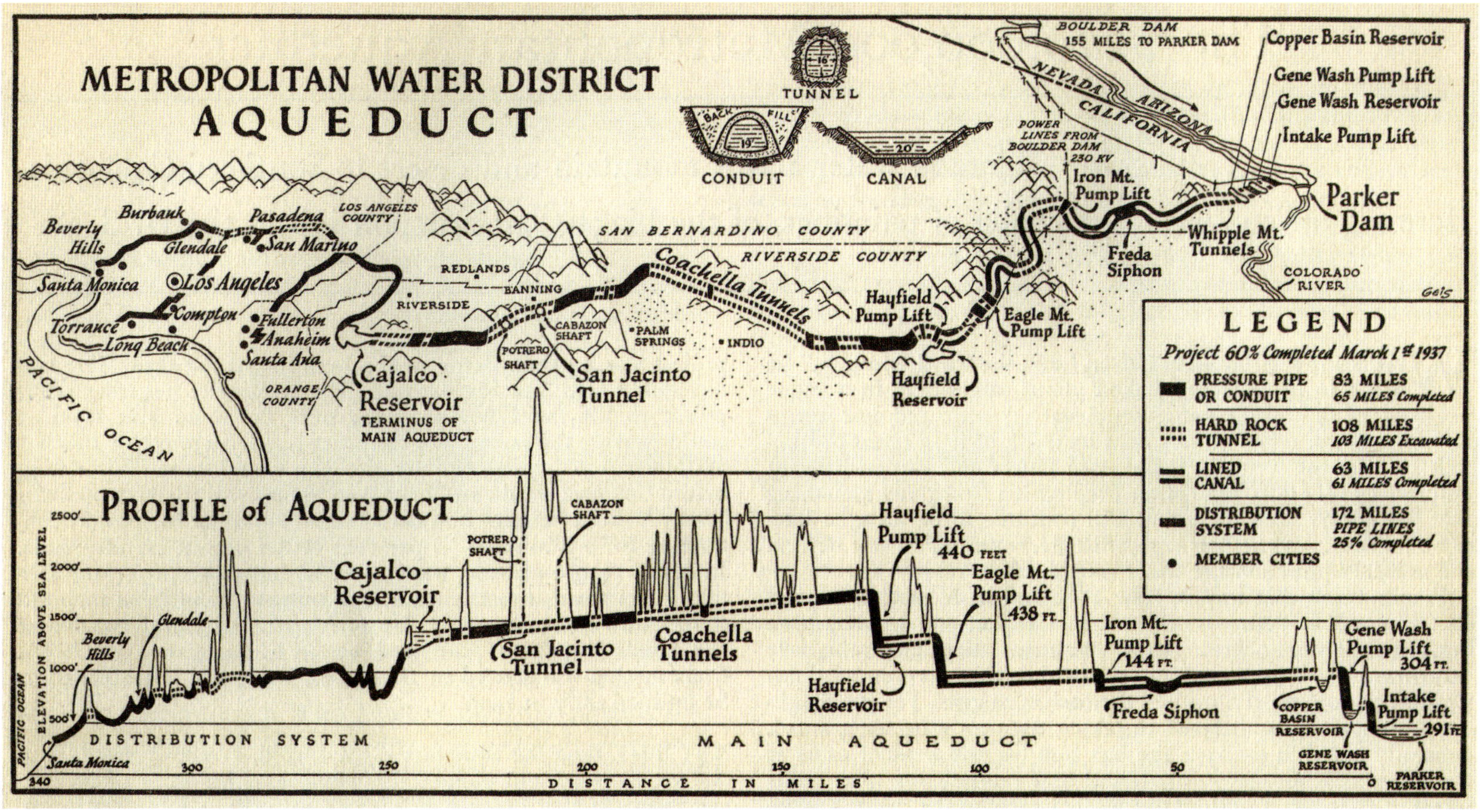

Figure 2.7 | **Map and Profile of the Colorado River Aqueduct and Distribution System**

The Hetch Hetchy water supply system for San Francisco consists of a 167-mile aqueduct to transfer water, again by gravity, from federal land near Yosemite Valley and transports it to the San Francisco Bay Area. Despite opposition to the project, the Raker Act of 1913 was passed by the U.S. Congress and signed by President Woodrow Wilson. This act authorized and funded the Hetch Hetchy project. In a forward-thinking testament to maximizing the energy efficiency of infrastructure assets, the Hetch Hetchy system also produces 2 billion kilowatt-hours of electricity a year from in-line power generation facilities.

Other water transfer tunnels have been built across the Continental Divide of the Rocky Mountains and include the Gunnison Tunnel, the Big Thompson water diversion tunnel, and the Harold D. Roberts Tunnel. The Roberts Tunnel (originally known as the Blue River Tunnel) is a 23.3-mile-long aqueduct that was planned as a key element for the development of Front Range communities in Colorado (Figure 2.10). Water diverted from the Blue River Basin in Summit County provides nearly 40% of Denver Water's supply. At the time the tunnel was built, it was the third-longest water supply tunnel in the world (Figure 2.11).

Figure 2.8 | **William Mulholland, Visionary Engineer for the City of Los Angeles**

Figure 2.9 | **Workers Celebrate Completion of the Pasadena Water Tunnel, 1937**

26 | THE HISTORY OF TUNNELING IN THE UNITED STATES

Figure 2.10 | **Roberts Tunnel (Formerly Called the Blue River Tunnel), 1954**

Figure 2.11 | **Roberts Tunnel Hole-Through When the East and West Tunnel Crews Met in 1960**

Denver, New York City, Los Angeles, and San Francisco required significant underground infrastructure to provide reliable and clean water to their inhabitants. The benefits of tunneling to these communities include the reason for their very existence.

PUBLIC HEALTH IN AMERICAN CITIES

Development of underground infrastructure to provide fresh water and to remove wastewater via underground conveyances was critical for public health in the nineteenth century. In 1854, the New York Common Council ruled that homes had to be connected to municipal sewer lines. Progress was slow, but the common interest in pure drinking water and sanitation was spurred by frequent epidemics of typhoid and cholera. As a result of this investment in underground infrastructure, a cholera epidemic that swept from Europe to North America in the 1860s killed fewer inhabitants of New York City than it did in other cities. In New York, with its relatively advanced sewer systems, 500 people died; whereas 1,200 persons died in Cincinnati, Ohio, and 3,500 in St. Louis, Missouri, in much smaller urban populations where there were no comparable sewer systems at that time.

A trend developed in municipal sanitation program construction in the mid-nineteenth and early-twentieth centuries. Many cities constructed extensive sewer systems to help successfully control outbreaks of typhoid and cholera.

Chicago, for example, was struck by cholera for six consecutive years in the 1850s, leading to establishment of its own Board of Sewerage Commissioners in February 1855. Progressive development of its water and wastewater system ensued, including a new intake in Lake Michigan and the 29,000-foot-long, 8-foot-diameter Blue Island Avenue water tunnel. By 1930, Chicago's water and sewer systems were the most extensive in the world.

In the 1870s, Americans began to study European methods to determine whether to combine or separate the stormwater and sewer systems. For urban areas where densities were high, typical designs were to convey the waste through existing stormwater drainage sewers to receiving water bodies, where it was thought that there was enough dilution to render it harmless. Thus arose the practice of combining sanitary wastewater with stormwater in one pipe, or "combined sewer" system. Combined trunk or main sewers were large-diameter conveyances installed underground (some by tunneling) and often designed to convey sizable quantities of stormwater whose return frequency was as rare as once in 10 years. By 1905, nearly all U.S. towns and cities with a population of more than 4,000 had municipal sewers. The Baltimore city sewer system, begun in 1915, was the last to be built in a major U.S. city.

The benefits to society of underground water and wastewater systems are obvious when the number of typhoid and cholera deaths per 100,000 people is mapped. Figure 2.12 shows the decline of these diseases during the same time period as sanitation systems improved by use of underground conveyances for clean water and sewage. This achievement is one of the key benchmarks of modern civilization, and significant investment in underground infrastructure over many decades was the principal way that the United States achieved this milestone.

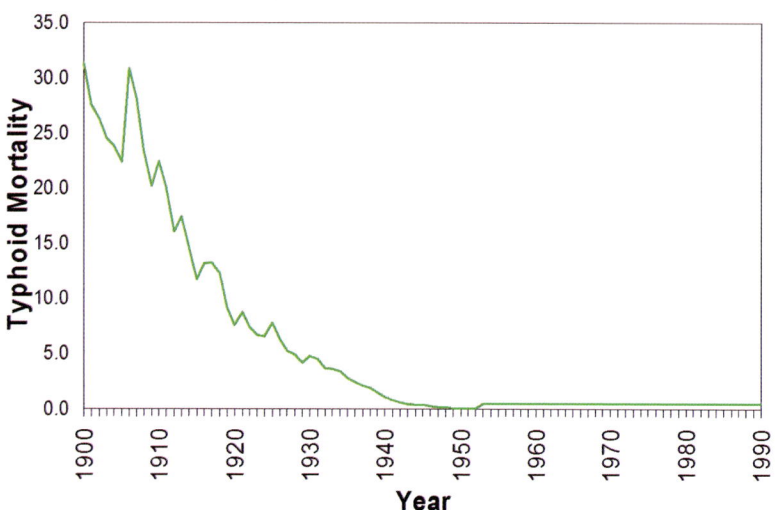

Figure 2.12 | **Typhoid and Cholera Deaths per Year per 100,000 People in the United States**

MOVING WEST AND UNIFYING THE NATION BY RAIL

Development of canals, such as the Erie Canal or the Chesapeake and Ohio (C&O) Canal with its Paw-Paw Tunnel, opened up the possibilities of swift transportation of heavy goods between the Midwest and major ports of the East Coast. Although using canals was an important step, it was the development of railroads—and specifically, the vision of a single, uninterrupted rail journey across the North American continent—that fueled the westward expansion and created the United States as we know it today. While canals used the tedious process of locks and gates to gain and lower elevation, railroads were able to develop and use new technology for underground construction more effectively to allow faster speeds on tracks laid through and under mountains. Because of this technology, most canal systems built in the late eighteenth and early nineteenth centuries, were not used commercially after 1860, with many canal companies declaring bankruptcy.

One of the early ventures for railroads to directly compete with canals resulted in a rail alignment for Boston to grow its harbor trade in the same way the Erie Canal had boosted trade in New York. Western Railroad struggled with steep grades over the mountains in western Massachusetts, and to mitigate this, in 1848, a significant tunnel was proposed by railroad president Alvah Crocker (Figure 2.13). Called the "father of modern tunneling" by some, Crocker initiated the use of geologists, pneumatic tools, and explosives during excavation of the Hoosac Tunnel, which started in 1851 but was not completed until 1875. At 4.8 miles in length, it was the nation's longest tunnel and the second-longest tunnel in the world at the time it opened.

When construction began on the Hoosac Tunnel in 1851, excavation was by hand-drilling and blasting with gunpowder. But the Hoosac Tunnel is a landmark in hard-rock tunneling. During its construction, virtually every kind of tunnel innovation was used. In March 1853, one of the earliest tunnel boring machines bored 10 feet into the Hoosac Mountain and broke down, never to run again. In 1866, two tunnel blasting tools, nitroglycerin and the compressed-air-powered drill, were used in North America for the first time. Workers blasted faster than ever before, but not without risk and appalling results. Many workers lost their lives in accidental explosions, and it took 20 years and numerous refinements and further innovations to finally complete excavation of the Hoosac Tunnel in 1875.

In 1862, President Lincoln signed the Pacific Railway Act into law. The act said that there were to be two main railroad lines: the Central Pacific Railroad would come from California, and the Union Pacific Railroad would come from the Midwest. The two railroads would meet at an unspecified point in the middle. The act gave the railroad companies land where they could build the railroad. It also paid them for each mile that they built at different rates for different terrain.

The Central Pacific Railroad began laying track eastward from Sacramento, California, in 1863, and the Union Pacific started laying track westward from Omaha, Nebraska, two years later in July 1865. Construction crews had the formidable task of laying the track crossing California's rugged Sierra Nevada and the Rocky Mountains and had to blast 15 tunnels. On May 10, 1869, after completing 1,776 miles of new track, the two rail lines met in a famous ceremony on flat and open country at Promontory Summit, Utah (CPRR 1999), shown in Figure 2.14.

In 1868, the Central Pacific Railroad's 1,659-foot Summit Tunnel (Tunnel 6) at Donner Pass in California's Sierra Nevada was opened, permitting the establishment of the commercial mass transportation of passengers and freight over the Sierras for the first time. It remained in daily use until 1993 when the Southern Pacific Railroad closed it and transferred all rail traffic through the 10,322-foot-long and rather generically named Tunnel 41, built a mile to the south in 1925 (Wikipedia 2015).

Figure 2.13 | **Alvah Crocker, Builder of the Hoosac Tunnel**

In spite of the famous photograph of the "Golden Spike ceremony" in Utah, it was the sacrifice of untold numbers of lives of Chinese and Irish immigrants in conjunction with the hard-rock tunneling technology first used at the Hoosac Tunnel and developed during the latter part of the nineteenth century that enabled the Transcontinental Railroad to meet its objective of uniting the country and allowing development of California and the western United States.

Predictably perhaps, once California began to see such huge commercial benefits from an east–west fixed rail link, a concerted effort was made to provide similar connections from Chicago to the Pacific Northwest through the Cascade Range.

The first Cascade Tunnel was excavated at Stevens Pass through the Cascade Mountains of Washington state (Figure 2.15). This was a 2.63-mile-long single-track railroad tunnel built by the Great Northern Railway and opened in 1900 to avoid problems caused by heavy winter snowfalls on the original line that had eight exposed

Figure 2.14 | **Golden Spike Ceremony in Utah for Completion of the Transcontinental Railroad**

Figure 2.15 | **Old Cascade Tunnel Portal in Stevens Pass, Washington**

Figure 2.16 | **New Cascade Tunnel in Stevens Pass Still in Service**

switchbacks. A second Cascade Tunnel was constructed at a lower elevation and was opened in 1929. The first tunnel was abandoned soon thereafter. The second tunnel (Figure 2.16) is a 7.8-mile single-track tunnel, which is still in operation and is the longest railroad tunnel in the United States (McClary 2014).

John Frank Stevens was the principal engineer on this section of track and was chief engineer of the Great Northern Railroad during construction of more than 1,000 miles of railroad, including the original Cascade Tunnel (Figure 2.17). Stevens Pass, located above the tunnels, was named after him. The commercial benefit of the second tunnel was significant, cutting travel times in half over that route (Figure 2.18). It is ironic and a sign of the lack of recent investment in modern rail infrastructure that today the second Cascade Tunnel is the choke point and controlling factor for freight and passenger movements on this line. Stevens later became chief engineer for the successful completion of the Panama Canal in 1914.

Once the North American continent had been conquered, attention shifted to focus on connections between commercial centers and the expanding markets of cities on the east and west coasts. The solution to urban connectivity and crossing under water was the subject of much discussion and investigation during the nineteenth century. Early attempts at subaqueous tunneling rarely achieved success. Not until the completion of the St. Clair River Tunnel (Figure 2.19), which connected the U.S. city of Port Huron, Michigan, with Sarnia in Canada, were subaqueous tunneling problems overcome. With the three most important advances in mid-nineteenth-century tunneling technology (tunnel shields, the cast-iron tunnel lining, and compressed air), the St. Clair River Tunnel was the first full-size tunnel in North America to conclusively demonstrate the feasibility of subaqueous tunneling. Many similar tunnels soon followed, such as the twin Hudson River Tunnels in New York City in 1906, illustrated on a stock certificate shown in Figure 2.20.

Figure 2.17 | **John Frank Stevens, Builder of the Cascade Tunnel**

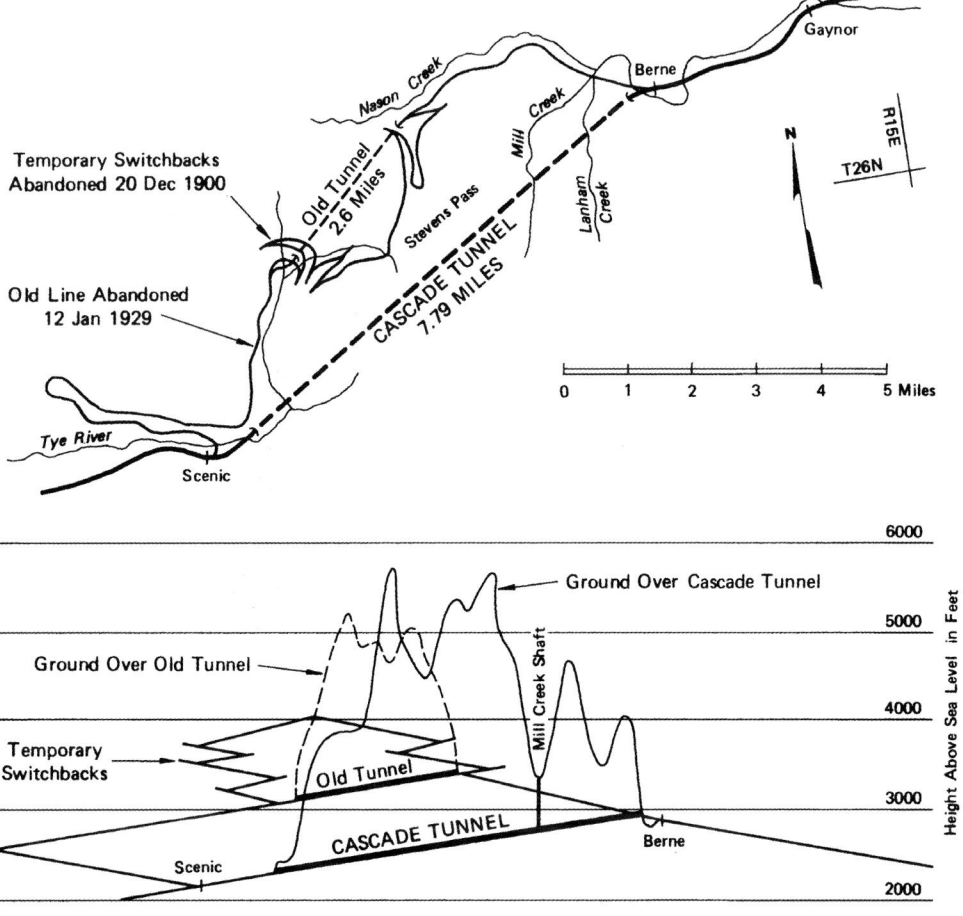

Figure 2.18 | **Map and Profile of the Old and New Cascade Tunnels in Washington**

Figure 2.19 | Port Huron Entrance to St. Clair River Tunnel, Circa 1915

34 | THE HISTORY OF TUNNELING IN THE UNITED STATES

Figure 2.20 | **Common Stock Certificate for the H&M Railroad Tunnel Under the Hudson River**

Rail connections through dense urban areas include the Howard Street Tunnel in Baltimore, Maryland (Figure 2.21), and the Virginia Avenue Tunnel in Washington, D.C. These tunnels are typical for providing heavy-rail access directly to downtown stations. Rail tunnels provide the means for functional freight and passenger transportation between cities and across the country. The expanding and reliable rail system was the industrial boost that catapulted the United States to industrial superpower status.

URBANIZATION AND MAKING CITIES WORK

As a secondary function to the public health needs for water supply and wastewater removal, effective transportation is nonetheless an important part of urban life. The ability to allow large numbers of people to easily commute to and from work and generally move around the urban area is essential if the city is to function and grow in any meaningful way. One of the best ways to move large numbers of people is by rail. Transit systems, be they heavy- or light-rail systems, have been built for most major cities in the world. As an example of the people-moving power of transit, the New York City subway system currently transports more than 4.5 million people per day. This translates to over 2,000 miles of bumper-to-bumper traffic on a four-lane highway—clearly illustrating how the city inhabitants could not get to work without its extensive transit system.

Early transit development was aboveground, but the number of buses and trains in the late nineteenth and early twentieth centuries made further development a significant safety hazard with numerous accidents and even more delays. Elevated train lines (L trains) in New York and Chicago improved safety, but the quality of life significantly deteriorated for residents and workers in the city with noise and vibration affecting adjacent properties. It was eventually recognized that significant benefits existed when rapid transit infrastructure was placed underground. Underground subway transit systems have proven to be remarkably safe, reliable, and durable (being shielded from the elements).

Since the mid-1860s, a vast increase in urban population has resulted in rapid development of public transportation, from the horse-drawn omnibus through Alfred Beach's privately funded pneumatic subway tunnel in New York City in 1873, to subway trains powered by an electric traction system invented by Frank Sprague. This electric traction first ran trains aboveground in Richmond, Virginia, but was clean and sufficiently powerful to allow the underground transit systems of Boston and New York to develop and flourish. City planners eagerly integrated the underground transit systems with city development and building plans (Figure 2.22).

The first underground subway system in the United States was opened in Boston in 1897 (Figures 2.23 and 2.24). The Boston subway carried more than 100,000 riders on its opening day and more than 50 million passengers in its first year. Boston and later the New York subway, which opened in 1904, were

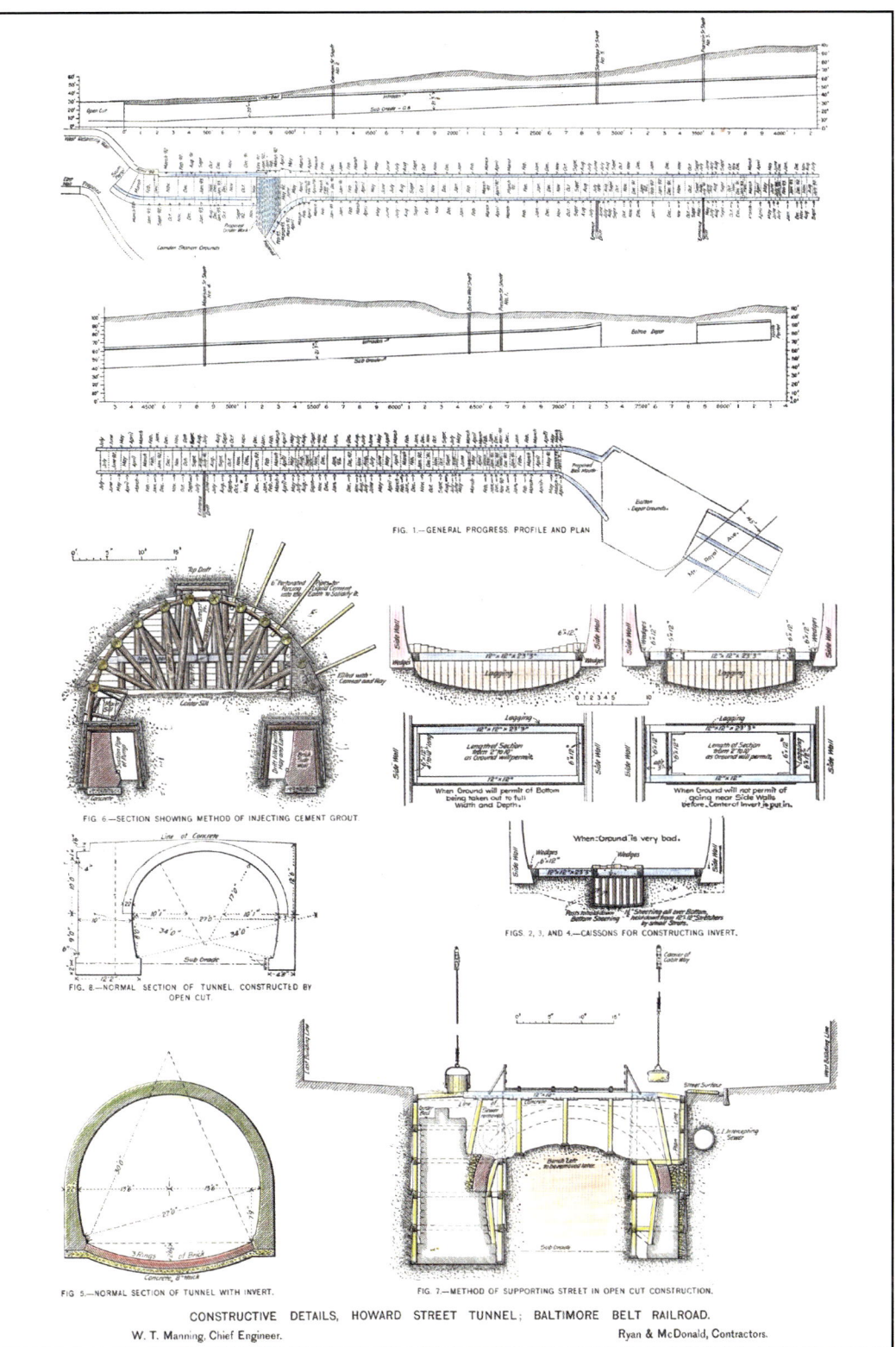

Figure 2.21 | **Howard Street Tunnel Construction Sequence, Baltimore, 1893**

Figure 2.22 | **Futuristic Urban Planning with Integrated Underground Transit in 1913**

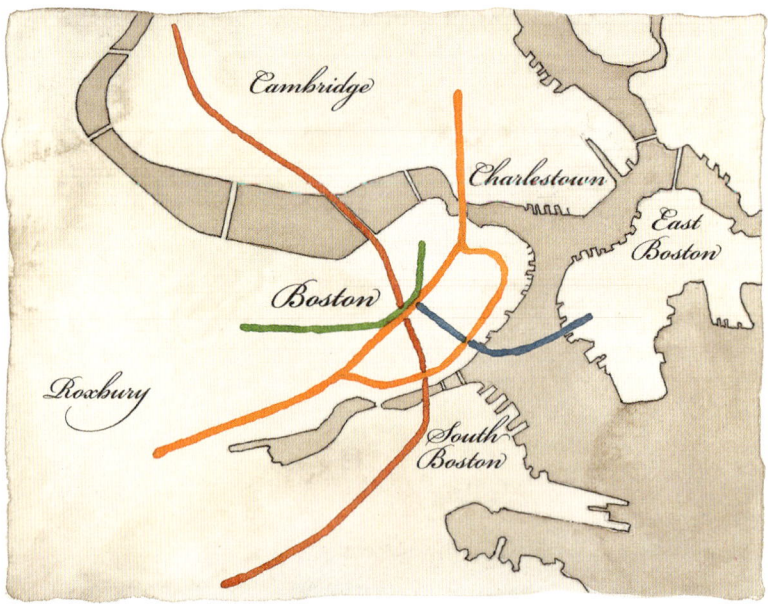

Figure 2.23 | **Boston Subway Routes, Circa 1897**

Figure 2.24 | **Jacked Arch Construction of Tremont Street Tunnel, Boston**

heralded as clean and reliable forms of public transportation (Figure 2.25). The New York Subway featured the Interborough Rapid Transit system that drew criticism in satirical cartoons (Figure 2.26) when first opened but was a monumentally successful project that changed the character of the city. As time moved on into the 1920s, many cities recognized underground rapid transit as the solution to surface congestion. Figure 2.27 shows the grandeur of the original City Hall Station on the New York subway (but is now closed to the public).

Alongside the success stories of New York and Boston, cities that developed their systems further, there were other cities whose efforts were frustrated by a variety of external factors. Cincinnati built 7 miles of underground tunnels before its money ran out and the Stock Market Crash of 1929 prevented further investment and the subway

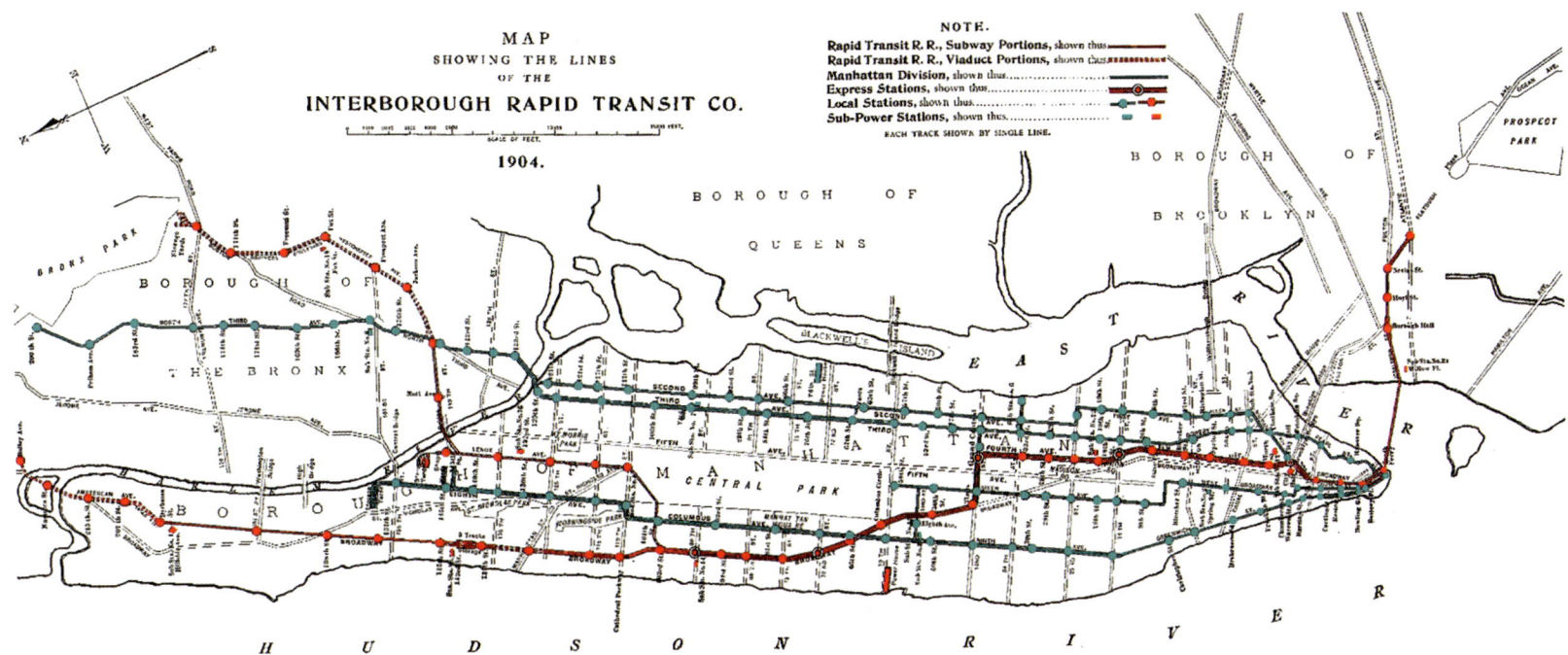

Figure 2.25 | **New York City Subway Map in 1904**

was never completed (Figure 2.28). In a similar tale, construction of the Los Angeles subway system first began in the 1920s and accepted the first passengers through the "Hollywood Subway" in December 1925. The line was popular, and ridership increased to more than 65,000 people per day in 1944. The investment in the Los Angeles freeway system in the 1950s led to the demise of the subway system until, of course, the surface highway system was itself overwhelmed. With no remaining surface area to expand into, underground infrastructure, again, showed its worth from the 1980s onward with the modern incarnation of the rapidly expanding Los Angeles Metro Rail system (Figure 2.29).

What could be characterized as a second wave of major investment in U.S. transit began in the second half of the twentieth century with the Washington Metropolitan Area Transit Authority (WMATA) in Washington, D.C. (Figure 2.30); the Metropolitan Atlanta Rapid Transit Authority (MARTA) in Atlanta, Georgia; and the Bay Area Rapid Transit (BART) in San Francisco. These systems were the fruits of President Lyndon Johnson's "Great Society" of the 1960s. Underground urban transportation infrastructure was seen as a good way to meet the objectives of this domestic program to eliminate poverty and racial injustice. Although achieving these lofty goals remained elusive, there was no doubt that the transportation systems benefited these cities and their citizens substantially. A global trend continues in which city planners deliberately seek out underground space (and reserve additional underground space for future growth) to complement and benefit their commercial and residential infrastructure on the surface while not taking up critical space or impacting the surface with noise, dust, or traffic.

Figure 2.26 | **Satirical Cartoon of New York City Subway, Early 1900s**

Figure 2.27 | Original City Hall Station on the New York City Subway

CHAPTER TWO | SOCIETAL BENEFITS | 39

Figure 2.28 | Unfinished and Abandoned Cincinnati Subway Station

Figure 2.29 | Modern Los Angeles Metro Train at Union Station

Figure 2.30 | Washington, D.C., Metro Station

By the time the initial phase of the Washington, D.C., system was completed in the mid-1980s, the revitalized interest in transit spread to New York City with significant projects such as East Side Access, the No. 7 Line extension, and, finally, the long-awaited Second Avenue Subway. From there, the investment spread across the country from Boston to Seattle and San Francisco; and from Los Angeles to Dallas and Atlanta—each city with its own specific needs and geologic issues to overcome but a common thread of needing and wanting a better experience for commuters, tourists, and residents so that the city's downtown area could be revitalized and redeveloped in those areas where people want to live, work, and play.

The implementation of extensive, reliable, and widespread underground transit has positively impacted life in American cities. One of the foremost delineating factors of a world-class city in the twenty-first century is whether the city has an extensive underground transit system to increase mobility and improve quality of life.

THE HISTORY OF TUNNELING IN THE UNITED STATES

Figure 2.31 | **Scenic Keystone Tunnel near Mount Rushmore in South Dakota**

Figure 2.32 | **Holland Tunnel in New York City**

Figure 2.33 | **President Dwight D. Eisenhower, 34th President**

RECONNECTING THE NATION BY CAR

When automobiles were first introduced in the United States in about 1892, the nation was just beginning to move from an agrarian to an industrial economy. Distances between cities and other points of commerce were great, and the car was not an economic mode of transport for inter-city or long-distance commercial travel. In 1900, the United States had only a few hundred miles of paved roadway. By 1976, that number had risen to more than 4 million miles of paved road.

Road builders have punched a hole through promontories and bluffs where necessary to create any number of short highway tunnels, such as the Keystone Tunnel near Mount Rushmore in South Dakota (Figure 2.31). Some of the most famous highway tunnels in the United States were built in the 1920s after a government-sponsored research project was completed to determine effective ways to ventilate the tunnels, thereby removing automobile exhaust. The knowledge gained in this study helped in the design of the ventilation system for the Holland Tunnel in New York City and all vehicular tunnel projects undertaken thereafter (Figure 2.32). The majority of tunnels in the 1930s were constructed in western states, providing access to the growing system of national parks. There were also more highway tunnels under the rivers in New York, providing access to the commercial center of Manhattan, such as the Lincoln, Queens–Midtown, and Brooklyn-Battery Tunnels.

In the 1950s, President Eisenhower put forward his vision of an interstate highway system to reconnect the nation by road (Figure 2.33), the highest point of which is the Rocky Mountain tunnel that bears his name. He did this almost 100 years after President Lincoln stated the same necessity to connect the commercial centers of the East and the West across the continent by rail. The commercial necessity remained the same. The nation could only build and ship goods efficiently if there was an extensive transportation network available to businesses. On this highway system, even individuals could move freely across the country to change their fortunes or find greater opportunity.

Although the commercial and societal benefits of interstate travel were the same, and possibly even greater, by the 1950s over those advantages realized by the Transcontinental Railroad, the geographical challenges of crossing the three major mountain ranges of the nation still existed.

Once again, without the tunneled sections of the Interstate Highway System, the grand vision of a fully interconnected and truly united country would not be realized. Construction technology had moved forward since the Transcontinental Railroad tunnel days. These advances were impacted, however, because cross-country highway tunnels require much larger excavations than railroad tunnels.

Figure 2.34 | **Eisenhower–Johnson Memorial Tunnel in Colorado**

One of the final sections of the originally planned Eisenhower Interstate Highway System to be completed in 1979, due to its complex geology and geometry, was the Dwight D. Eisenhower–Edwin C. Johnson Memorial Tunnel. Originally known as the Straight Creek Tunnel during construction (Figure 2.34), it is a twin bore, four-lane vehicular tunnel located about 50 miles west of Denver, Colorado. The tunnel carries Interstate 70 under the Continental Divide in the Rocky Mountains. With a maximum elevation of 11,158 ft, it was the highest vehicular tunnel in the world at the time of completion and is the highest point on the Interstate Highway System.

Connecting commercial centers did not stop with the Interstate system though. Several toll roads also provide important connections for people and businesses. The use of these roads raises funds to help pay for their own maintenance costs through tolling—from the first limited-access superhighway, the Pennsylvania Turnpike (seven tunnels totaling 4.5 miles) that opened in 1940 (Figures 2.35 and 2.36), to the Chesapeake Bay Bridge-Tunnel that is 17.6 miles from shore to shore and contains two tunnels, each approximately 1 mile long. The concept of toll roads is nothing new but remains important into the future with many projects using toll revenues for operation and maintenance or as part of capital construction funds.

The PortMiami Tunnel in Florida is an example of a different way of paying for public infrastructure, a public–private partnership project where public funding is deferred by using private financing and debt that is structured over a fixed concession period. This type of financing is anticipated to be a significant source of funds for a new generation of highway tunnel work in the twenty-first century. Figure 2.37 shows the hole-through in 2012 of the 42-foot-diameter TBM after excavation of the PortMiami Tunnel.

Figure 2.35 | **Andrew Carnegie Visiting the Original Rays Hill Railroad Tunnel During Construction**

No discussion of U.S. highway tunneling is complete without mentioning the Big Dig in Boston. Late twentieth-century developments involving increasingly sophisticated tunneling technology led to the feasibility of a scheme to beautify and reconnect a city that had previously been divided by a major arterial highway. Indeed, the Central Artery/Tunnel has probably provided more benefits to society in the Boston area than any other single transportation infrastructure program anywhere in the country (Figures 2.38 and 2.39).

The experience of visiting the city of Boston now is markedly superior to what it was in the early 1980s before tunnel construction began. A 15-year construction period was both longer and more expensive than originally anticipated, but the benefits of the facility are dramatic. Waterfront and city reunited just as the local businesses wanted. A chronic eyesore, the elevated I-93 structure was removed. Numerous technological and tunnel construction "firsts" were recorded during this project—too many to list here—but this facility achieved its originally planned objectives. When other cities follow this example, as they undoubtedly shall, they will receive similar benefits for considering their underground space as an important part of the city that can be used to benefit their citizens.

Figure 2.36 | **Old and New Lehigh Tunnels on the Pennsylvania Turnpike**

CLEANING UP THE ENVIRONMENT

Although early sewer systems used the same pipelines for stormwater and sewage, it was only late in the twentieth century as cities grew larger that this was recognized as a big problem for pollution of our waterways. Overflows of untreated sewage and pollution into rivers have caused some legendary outcomes, such as rivers catching fire. Unfortunately, a river catching fire was common in the early- to mid-twentieth century as industrial pollution increased. Figure 2.40 shows the Cuyahoga River on fire in Cleveland, Ohio, in 1951. The last such fire in the United States took place in 1969, again on the Cuyahoga River. Extensive media coverage, including a *Time Magazine* article, eventually inspired the passing of the Clean Water Act of 1972 (introduced by Senator Edmund Muskie of Maine) to improve the cleanliness of lakes and rivers along with their urban waterfronts (Figure 2.41). Major combined-sewer-overflow (CSO) tunnels were constructed across the United States in the latter part of the twentieth century.

A total of 772 communities in the United States have combined sewer systems, serving about 40 million people (EPA 2011). Most of these combined sewer systems are in the Northeast and Great Lakes region, and the Pacific Northwest (Figure 2.42).

By 1994, the U.S. Environmental Protection Agency (EPA) reported that individual combined sewer overflow (CSO) discharges occurred at an average rate of 50 to 80 times per year. Nationwide, this resulted in the annual spillage of about 1.2 trillion gallons of raw sanitary wastewater, untreated industrial wastes, and stormwater runoff into receiving waters.

Figure 2.37 | **TBM Hole-Through of the PortMiami Tunnel, 2012**

Figure 2.38 | **Overall Plan of the Central Artery Project in Boston**

Figure 2.39 | **Dramatically Improved Downtown Surface Walking Experience of the Central Artery Project**

In 1994, a CSO policy was developed by EPA from the Clean Water Act to address the specific concerns related to CSOs and their effects on waters of the United States. The national goal is an 85% capture and clean rate for the more than 700 cities with combined sewer systems. This policy requires each community with CSOs to develop a long-term control plan designed to reduce or eliminate the effects of the community's CSOs.

As a result of this legislation, many cities began to develop plans to capture and treat their CSOs. The only feasible solution for capturing several billion gallons (in some cases) of polluted stormwater was to construct a substantial subsurface storage and conveyance system of tunnels. These tunnels would be periodically inundated, preventing untreated sewage from entering waterways, and later would be pumped to a sewage plant at a more controlled rate for treatment and discharge. Again, we can see how the effective use of underground space beneath a major city can dramatically improve the quality of life for the residents of that city.

One of the first cities to complete their tunnel program was Milwaukee, Wisconsin. The positive results from this system, as well as the lessons learned, are benefitting many communities currently undertaking completion of their own long-term control plans. Milwaukee achieved its objective of more than 95% elimination of raw sewage CSOs and saw a dramatic fall in CSO volume from 2.8 billion gallons in 2010 to only 170 million gallons in 2011 after commissioning Phase 2 of its overall overflow reduction program. Figure 2.43 shows the final tunnel lining construction.

The volumes of polluting sewage for CSO systems are huge. Chicago, Atlanta, Cleveland, St. Louis, and Washington, D.C., are at various stages of their own major tunnel programs. Chicago's extensive Tunnel and Reservoir Plan (TARP) system can store and convey approximately 17.5 billion gallons of combined sewage and prevent it from overflowing into Lake Michigan.

Figure 2.40 | **The Cuyahoga River Fire of 1951 in Cleveland, Ohio**

Figure 2.41 | Rehabilitated Cuyahoga Riverfront in Cleveland, Ohio

Figure 2.44 shows the unique intersection of two TBM-bored tunnels constructed for the Torrence Avenue segment of the Chicago TARP system. Figure 2.45 illustrates the connection between a large drop shaft and the main conveyance tunnel on the Chattahoochee CSO project in Atlanta. Some of these underground programs are running in parallel with surface inflow minimization programs, commonly known as green infrastructure.

When these massive volumes of pollution are considered, prudent use of deep-tunnel linear storage and conveyance systems is the most cost-effective and maintenance-free solution. Green infrastructure can help shave the peak flows from CSOs, but it takes longer to build and is more expensive on a per-gallon-captured basis. It is also significantly more costly to maintain and operate than a deep-tunnel CSO solution. It is ironic to many in the tunnel industry that the underground remedy that prevents environmental pollution at lower cost and with a significantly reduced overall carbon footprint is labeled by some environmental advocates as the "gray" solution!

SUMMARY

The underground construction industry has consistently risen to the occasion of providing the nation with the necessary infrastructure to meet society's needs. The critical step is when political figures look at a developing or existing need and have the long-range vision to provide action and funding to put in place the essential infrastructure solution.

Project financing is an area where the underground industry has come full circle, from the mid-nineteenth century with public–private partnerships, through the twentieth century of publicly funded schemes, to the twenty-first century looking at alternative forms of delivery and seeking private investment again in major infrastructure programs.

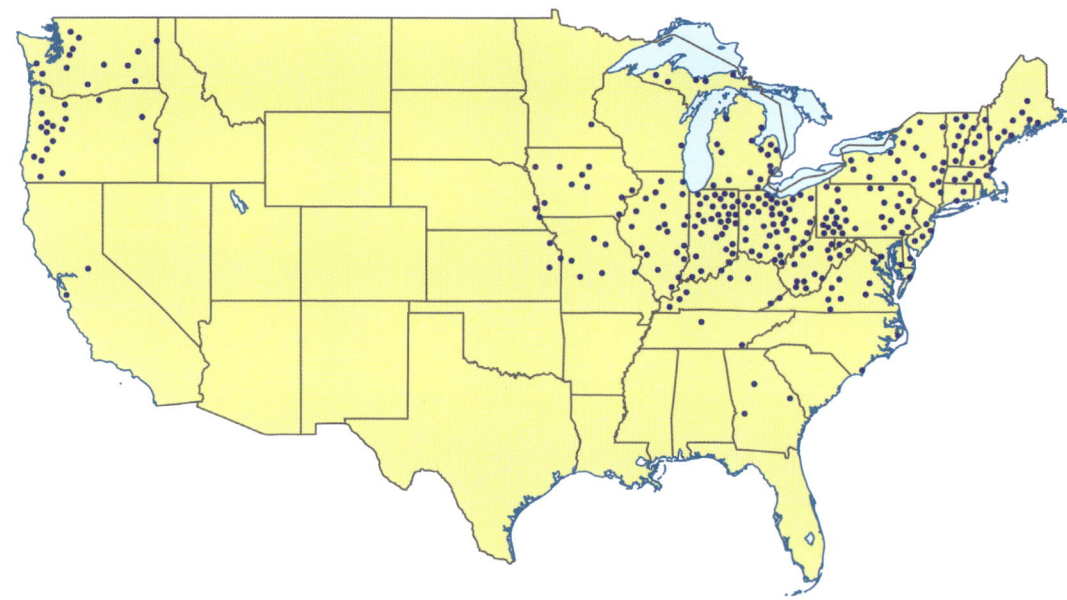

Figure 2.42 | **U.S. Communities with Combined Sewage and Stormwater Systems**

Figure 2.43 | **Final Concrete Lining Construction in a Milwaukee CSO Tunnel**

46 THE HISTORY OF TUNNELING IN THE UNITED STATES

Figure 2.44 | **Holing Through a Connecting Tunnel for the Torrence Avenue TARP Segment**

Figure 2.45 | **Connection Between a Drop Shaft and Tunnel on the Chattahoochee CSO Project in Atlanta, Georgia**

Superior urban planning involves the use of underground space and specifically how this space can be used to benefit society. As a result of this planning, there are better uses of surface areas, with recreational green space replacing surface or elevated infrastructure. Underground solutions reduce urban sprawl, create green space, and improve the lives of every citizen.

Public Health and Life Support Systems

When the need arrived to clean up our waterways, the tunnel industry stepped up and developed technologies and the means to provide a solution, under ever more challenging geologic conditions, to reduce and eliminate polluting CSO events in the Great Lakes of Chicago and Milwaukee. CSOs have been built throughout the country, from Atlanta and Washington, D.C. (Figure 2.46), to the cities of Ohio, across the Mississippi Valley in St. Louis and Kansas City, as well as in Seattle, Los Angeles (Figure 2.47), and Portland, Oregon. It is tunnels, not the trendy buzz word of so-called "green infrastructure," that have achieved the vision and cleaned up our rivers and lakes, providing massive environmental benefits in areas where people live and work and now may also play.

Water supply tunnel systems bring life to our major cities—from New York to San Francisco and Los Angeles. These cities would not exist without the hundreds of miles of water tunnels that bring clean water to every household and business. The challenges get bigger as extended droughts around the country further stretch the need for reliable water supplies. These supply networks are likely to expand in the future.

Relying on Our History to Provide Examples for the Future

The history of the tunneling industry is one of successfully overcoming all obstacles thrown in its path. The industry has always been and remains responsive and responsible and stands ready to continue fulfilling our important responsibility to the public and the nation's economy, bringing huge benefits to society.

Whatever the future holds for our nation, there will be a need to build and rebuild our underground infrastructure. Additional transit systems across the Midwest, generation and transmission of power, and ever-improved access to ports and recreational facilities by road and rail will be required. Tunnels will be built to accommodate these needs. The technology required for tunneling in difficult geological conditions, expanded and new parameters for both size and complexity of underground space, and pressures to achieve these objectives with less and less disturbance to the public and adjacent property owners either already exist or are in development awaiting the opportunity to be implemented.

Figure 2.46 | The DC Water Clean Rivers "Lady Bird" TBM

THE HISTORY OF TUNNELING IN THE UNITED STATES

Figure 2.47 | **Los Angeles East Central Interceptor Sewer (ECIS) Pipe Lining**

Figure 2.48 | **Hard-Rock TBM for Chicago's TARP System**

Figure 2.49 | **Lake Mead TBM for the World's Highest Water Pressure Bored Tunnel**

The Continuing Evolution of Underground Construction Technology and Innovation

If the vision to improve our country's infrastructure remains, our industry will extend the boundaries of technology to find a way to achieve that vision as we have always done. Development of technology from the past and into the future has seen the industry achieve larger and longer tunnels (Figure 2.48), in harder rock or softer soil (equally challenging), in ever more congested urban environments, and by constantly developing better equipment and technologies. Whether by commissioning the biggest tunnel boring machine in the world to revitalize the waterfront in Seattle, tunneling under the highest water pressure in the world to secure the water supply for Las Vegas from Lake Mead (Figure 2.49), or conveying CSOs deep beneath the streets of Atlanta (Figure 2.50), the underground industry seeks to make good on the visionary promises of political figures who are serving the nation's interests.

Protecting and Extending the Life of Underground Infrastructure

Once the importance of underground infrastructure has been established, it becomes obvious that we must maintain it and keep it in serviceable condition. It is a challenging task and will not get easier, because rehabilitation is simply not as glamorous as cutting a ribbon upon completion of new construction. Regardless of the glamour, however, rehabilitation of existing underground infrastructure is expected to be a large and expanding part of the underground industry into the future.

With changing climate, both meteorological and political, come new challenges, but whether the country needs a new oil pipeline or a new network of water distribution tunnels from the Great Lakes to the increasingly dry South and West of the nation, the underground industry remains as ready as it has always been to turn these grand visions into a reality that provides economic and social benefits to the nation.

Figure 2.50 | **Final Concrete Lining Construction in Atlanta's Chattahoochee CSO Tunnel**

And Finally...

Visionary political leaders have proposed programs and schemes that have, in many cases, saved their communities from the ruin of drought, disease, and economic stagnation. In other cases they have dramatically improved the standard of living in their city, state, or indeed across the nation. The tunneling industry has been instrumental in allowing these visions to become reality, whether in the water, wastewater power, or transportation field.

Tunnels are the arteries that have built major American cities, from New York and Boston to Los Angeles and San Francisco. It is the tunnels that have allowed this great nation to be connected for the free movement of labor and goods, first by rail and later by road. Tunnels have unified the nation and allowed the export of crops and manufactured goods in the heartland of Iowa, Texas, Colorado, and elsewhere, and have provided a safe and reliable transcontinental route for imports from overseas.

Without extensive and ever-expanding underground infrastructure, the United States would not have developed into the nation we know. Without the visionary civic and political leadership, there would be no infrastructure sufficient to support growth as well as improvements to society and quality of life. The economic value of this underground network is infinite, both in terms of commercial development and public health. Tunnelers have literally done the groundwork, and America has prospered in so many ways from their ingenuity and achievements.

Richard M. Nixon (1913–1994)

As 37th President of the United States, Nixon orchestrated significant policy change that affected the construction industry. In 1970, Nixon announced the formation of the Environmental Protection Agency, which began operation on December 2, 1970. By signing the National Environmental Policy Act, federal projects were required to have environmental impact statements prior to approval. Nixon broke new ground by discussing environment policy in his State of the Union speech. He also signed the Occupational Safety and Health Act in 1970, which ushered in the formation of the Occupational Safety and Health Administration, more commonly known as OSHA.

REFERENCES

CPRR (Central Pacific Railroad). 1999. Transcontinental Railroad: Central Pacific Railroad Photographic History Museum. http://cprr.org/Museum/faster.html (accessed Dec. 2, 2015).

Drinker, H.S. 1883. *A Treatise on Explosive Compounds, Machine Rock Drills and Blasting.* New York: John Wiley & Sons.

EPA (United States Environmental Protection Agency). 2011. News Releases from Region 2: EPA releases new report on sewage pollution in New York and New Jersey. July 8. http://yosemite.epa.gov/opa/admpress.nsf/d10ed0d99d826b068525735900400c2a/34fa6322de1ddeaf852578c70053c1f2!opendocument (accessed Dec. 2, 2015).

McClary, D.C. 2014. The Great Northern Railway Eight-Mile Tunnel is dedicated on January 12, 1929 (Essay 10705). In *HistoryLink, The Free Online Encyclopedia of Washington State History.* www.historylink.org/index.cfm?DisplayPage=output.cfm&file_id=10705 (accessed Dec. 2, 2015).

Wikipedia contributors. 2015. Examples of tunnels—In history. In *Wikipedia, The Free Encyclopedia.* 01 Dec. 2015. Web. 02 Dec. 2015. https://en.wikipedia.org/wiki/Tunnel (accessed Dec. 2, 2015).

3 RAILROAD TUNNELS

THE HISTORY OF TUNNELING IN THE UNITED STATES

The early 1830s was a time when railroads and canal systems were competing for the transport of freight and passengers from industrialized eastern cities into the midwestern and western wilderness areas.

These early tunnels have been an integral part of the railroad system in the United States since that time and enabled the canal or railroad alignments to remain relatively level when crossing through hilly to mountainous terrain by tunneling. As it became apparent that railroads were quicker to build, less expensive, and easier to maintain than canal systems, and provided a faster mode of travel, particularly once steam locomotives were introduced, then the era of canal tunnels ended and the era of railroad tunnels began.

Railroads, and their associated tunnels, enabled the industrial revolution to expand rapidly between 1830 and 1850, from the industrial centers on the East Coast to the north and south, and as far west as Chicago and St. Louis, along the east side of the Mississippi River. The expansion of the railroad system spurred westward expansion of American civilization with the rapid transfer of freight and people to territories west of the Mississippi River and to the West Coast between 1850 and 1890.

Although railroads are able to navigate over hilly terrain at grades of up to about 4%, this generally requires additional motive power or helper engines, and shorter trains; therefore, every attempt is made to limit grades to less than 1%. Consequently, the construction of tunnels and bridges enabled railroads to pass through mountains and thus reduce the grades of the alignment. Tunnels also made it possible for railroads to pass under rivers and beneath cities with a reduced impact on whatever lay above. Railroads and their tunnels also helped the Union Army in the Civil War to rapidly deploy soldiers and supplies to the southward advancing front, and thereby more swiftly defeat the Confederate Army. In addition, the United States could quickly move troops and supplies to major coastal ports for shipment overseas. Because of their critical importance to the transport of raw materials, manufactured goods, and people, the railroads, and particularly their tunnels and bridges, were prime targets during the Civil War. They have been targeted and protected during subsequent wars.

Railroad tunnels have also proven to be sustainable resources that can be periodically rehabilitated to extend their usefulness and in many cases enlarged to accommodate larger railcars, particularly double-stack container cars and tri-level automobile carriers. The introduction of new construction materials has further improved the longevity and usefulness of tunnels that were initially supported with timber ribs and lagging more than 100 years ago. These new materials include shotcrete and resin-grouted steel and fiberglass rock bolts for rock reinforcement, chemical and permeation cement grouting to stabilize and reinforce soil and rock, PVC (polyvinyl chloride) and HDPE (high-density polyethylene) membranes to create waterproof and even watertight tunnels, and steel and plastic fiber-reinforced concrete for final cast-in-place liners.

New railroad alignments, most including tunnels, continue to be evaluated to shorten routes and reduce grades by passing beneath rivers, through mountains, and under major cities. The evolution of tunnel boring machines for soil and rock excavation, resin-grouted rock bolts, steel fiber-reinforced shotcrete, and gasketed precast concrete and polymer segmental linings have made feasible the ever longer and larger-diameter tunnels at increasingly greater depths. The Second St. Clair River crossing and the East Side Access Tunnels in New York are examples of applications of recent tunneling technology for a new generation of railroad tunnels. A number of other innovations have made increasingly more challenging tunnel projects possible with advances in metallurgy, chemistry, hydraulics, mechanical engineering, and computer simulations, to name a few applicable and relevant scientific fields. Riding this wave of innovations in tunneling, a new generation of high-speed rail alignments designed to carry

Previous Page | **Drano Tunnel in Hood, Washington**

CHAPTER THREE | RAILROAD TUNNELS

passengers at speeds of 100 to 200 mph, and compete with cars and airlines on trips of a few hundred miles, is in the process of being evaluated across the United States. The State of California has already implemented a high-speed rail design and has awarded design–build contracts for the final design and construction of initial segments of railroad alignment between San Francisco and Los Angeles.

THE HISTORY OF RAILROADS

Track-mounted transport began in the United States in the early 1800s with the construction of wooden rails in underground mines for the removal of excavated spoils by mule, horse, or human-powered muck carts. Through the 1820s the transport of raw materials, manufactured goods, and people was either by land or by water utilizing horse- or mule-drawn wagons and barges or wind- or paddle-driven boats. However, canals cannot readily traverse hilly terrain without the use of locks and mountainous terrain without either tunneling through the mountains or portaging the barges over mountain passes. Consequently, canals were primarily limited to the East Coast by the presence of the Appalachian Mountains running through the states bordering the Atlantic Ocean. Between 1816 and 1840, more than 3,300 miles of canals were constructed in the eastern states (University of Richmond 2015) and only 2,818 miles of railroad tracks (Drinker 1893). However, in a head-to-head battle for new alignments in the 1830s and 1840s, it soon became apparent that it was much quicker and less expensive to construct new railroad alignments, to maintain and repair them, to negotiate uneven terrain, and to tunnel beneath mountains and cities. In addition, with the introduction of wood- and then coal-fired steam engines, the costs to operate were less and the speed of travel was considerably greater for railroads than for canals (Gallamore and Meyer 2014).

Figure 3.1 | **Auburn Canal Tunnel, the First Transportation Tunnel in the United States, Circa 1821**

First Transportation Tunnels in the United States

The first documented transportation tunnel in the United States was the Auburn Tunnel (Drinker 1893), excavated through red shale bedrock (Figure 3.1) between 1818 and 1821 and used for transport on the Schuylkill Canal through Auburn, Pennsylvania, as located on the U.S. railroad map shown in Figure 3.2. The tunnel was 18 feet high, 20 feet wide, and 450 feet long and was used for canal transport. Poor ground conditions resulted in eventual open-cut excavation or "daylighting" of the total length of tunnel.

The first U.S. tunnel used for rail transport was the Staple Bend Tunnel constructed from 1831 to 1833 near Johnstown, Pennsylvania (NPS 2015). This tunnel was excavated through slate as part of the Allegheny Portage

Railroad and was 21 feet high, 25 feet wide, and 901 feet long—sufficient for double-track operations. Construction of this "rail" tunnel began November 21, 1831, with excavation by hand drilling of nine or ten 3-foot-long, 1-inch-diameter holes, taking up to 3 hours per hole. Black powder cartridges were inserted into 18 inches of each hole and ignited with a fuse. The tunnel was advanced about 18 inches per day, with headings advanced from both ends. The Staple Bend Tunnel was used from 1834 through 1854 to transport barges loaded onto flatbed railcars through the Allegheny Mountains to the Pennsylvania Mainline Canal system between Philadelphia and Pittsburgh. It was the third transportation tunnel built in the United States, with the other two being for canals. However, this tunnel was abandoned in 1857 in favor of the rapidly expanding and more cost-effective railroad system. It is now part of a U.S. National Historic Site commemorating construction of the first rail transportation tunnel. This was not a true railroad in that no railroad engines were employed; rather there were 10 stationary steam-powered engines that pulled the wheel-mounted barges up an incline and through the tunnel to be lowered down inclines on the other side of the mountain.

Tunnel Construction Explodes in the United States

By 1850 at least 17 true railroad tunnels had been constructed in the eastern United States on more than 9,000 miles of track (Table 3.1). Of these 17 early tunnels, only 5 are still in use after 165 years, having been enlarged over several phases to provide clearance for larger cars and re-supported with modern steel and shotcrete or concrete linings to replace the original timber, brick, and ashlar stone masonry linings. The Black Rock Tunnel, constructed between 1835 and 1837, was the first true operating railroad tunnel in the United States and is located on the Philadelphia & Reading Railroad line in Phoenixville, Pennsylvania. The Flat Rock Tunnel was constructed by 1840 and is also on the Philadelphia & Reading Railroad line, just west of the Schuylkill River across from Philadelphia. Both tunnels are still in operation as single-track tunnels, having been enlarged in 1994 for double-stack freight cars and supported with grouted rock dowels and shotcrete.

It is notable that the last canal tunnel was built in 1838. By 1850, there were 9,021 miles of railroad track linking much of the eastern United States from south to north and to the Great Lakes region, including Chicago, and south to St. Louis. Table 3.2 shows the railroad expansion over time by decade. A major leap in the construction of railroad alignment occurred in the 1850s with more than 21,000 miles of new track between 1850 and 1860, and at least 122 new tunnels. The total railroad alignments in the United States through 1860 are shown in Figure 3.3.

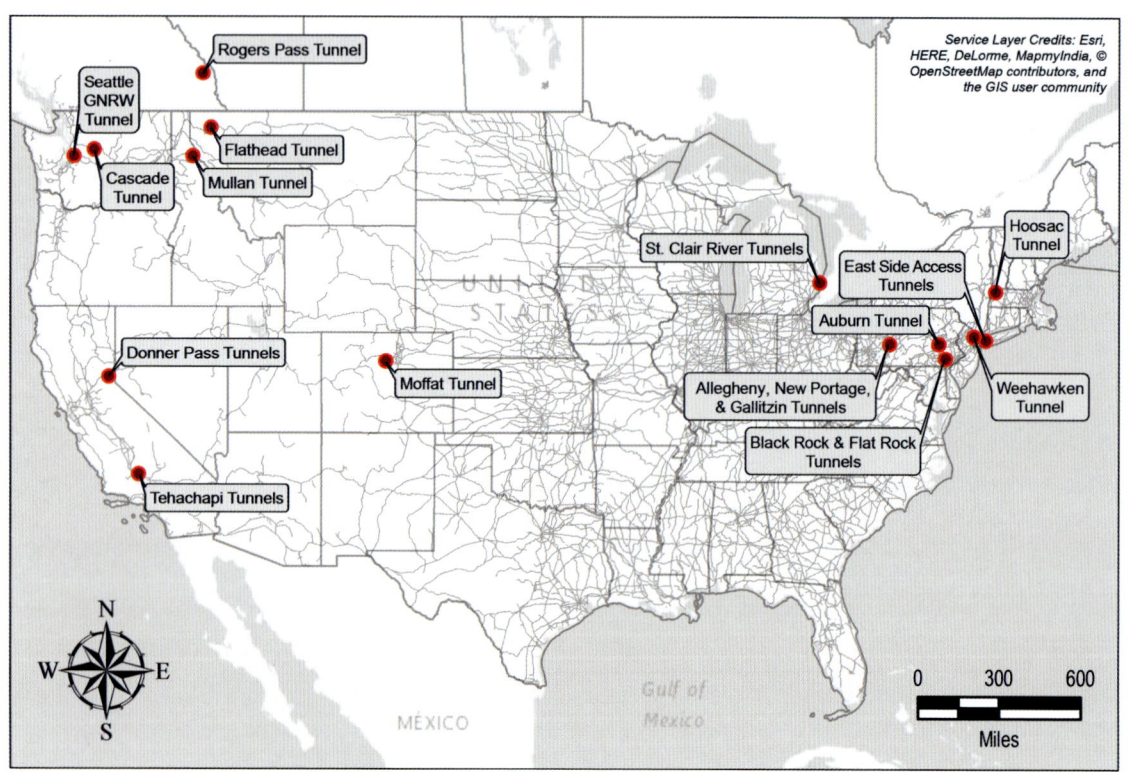

Figure 3.2 | **Selected Tunnels on Railroad Map of the United States**

Table 3.1 Selected Railroad Tunnels Constructed by 1850

Name of Tunnel	Location	Railroad	When Built	Still in Service in 2015	Length, ft*	Single or Double Track*	Width, ft*	Height, ft*	Ground	Current Lining
Staple Bend	Near Johnstown, PA	Allegheny Portage	1831–1833	No	901	D	25	21	Sandstone	Cut stone & concrete portals
Black Rock	Phoenixville, PA	Philadelphia & Reading	1835–1837 (Enlarged & relined in 1859 & 1994)	Yes	1,932	D	19	17.3	Limestone	Rock bolts & shotcrete
Elizabethtown	Elizabethtown, PA	Pennsylvania	1835–1838	No	900	S	14.9	15.5	Sandstone	Masonry
"Old" Harlem	4th Avenue in New York City, NY	New York & Harlem	1836–1837	No	844	D	24	21	Schist	Unknown
Summit	Summit Station, PA	Philadelphia & Reading	1838	No	1,050	S	18.5	Unknown	Conglomerate	None
Harper's Ferry	Harper's Ferry, WV	Baltimore & Ohio	1839–1840 (Enlarged in 1931 & 2011)	Yes	86	D	24	22	Horneblend slate	None. Concreted portals 1931
Pulpit Rock	Port Clinton, PA	Philadelphia & Reading	1839–1841	No	1,637	D	19.5	17.3	Quartzite	Unknown
Doe Gully	Doe Gully, WV	Baltimore & Ohio	1839–1841 (Open cut in 1914)	No	1,207	D	22	20	Clay slate	Masonry
Flat Rock	West Manayunk, PA	Philadelphia & Reading	1840 (Enlarged & relined in 1859, 1889, 1936, & 1994)	Yes	937	D	19	Unknown	Gneiss	Rock bolts & shotcrete
Paw Paw	Paw Paw, WV	Baltimore & Ohio	1840–1841 (Open cut in 1914)	No	250	D	22	20	Clay slate	Masonry
One Tunnel	White Haven, PA	Lehigh & Susquehanna	1841	Yes	1,800	S	Unknown	Unknown	Unknown	Masonry
One Tunnel	35 miles from Albany, NY	Albany & W. Stockbridge	1841–1842	No	530	D	26	22	Slate & limestone	None
Canaan	Canaan, MA	Boston & Albany	1842	No	547	D	26	18	Clay slate	None
Phipps Hill	1 mile south of Holliston, MA	Boston & Albany	1846	No	92	S	12.3	16	Clay & sand	Masonry
Walpole	Norfolk County, MA	New York & New England	1848	No	201	D	24.6	19.3	Cut & cover	Cut granite block
Henryton	Marriotsville, MD	Baltimore & Ohio	1848–1849 (Enlarged & relined in 1865 & 1903)	Yes	419	D	24	22	Mica slate	Brick
Everett's	3 miles west of Piedmont, WV	Baltimore & Ohio	1849–1850	No	350	D	22	20	Clay slate, sandstone	Masonry

Source: Adapted from Drinker 1893 and data from Federal Transit Administration.
*Dimensions and whether single or double track are given for the time of initial construction. Several tunnels have been enlarged and relined to provide clearance for larger railcars.

Table 3.2 Expansion of Railroad System over Time

Year	Miles of Track In Use	Number of Tunnels
1830	23	0
1840	2,818	16
1850	9,021	34
1860	30,635	156
1870	52,898	228
1880	93,267	374
1890	163,597	456
1900	190,000	512
1910	240,000	766
1920	254,000	917
1930	250,000	981
1940	240,000	1,002
1950	224,511	1,023
1960	220,000	1,032
1970	205,000	1,045
1980	165,000	1,048
1990	155,000	1,050
2000	144,500	1,051
2010	139,679	~500 still in use

Source: Data from Drinker 1893, AAR 1951, Hallberg 2009, Federal Railroad Administration 2015, and Rodrigue 2015.

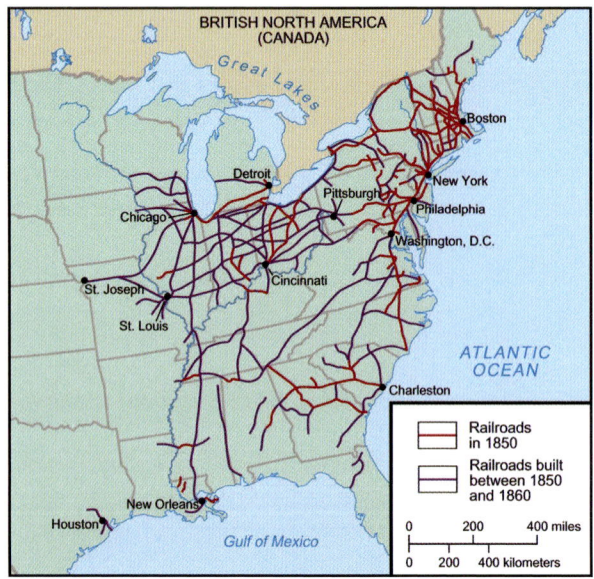

Figure 3.3 | **Railroad Alignments in the United States Through 1860**

Railroads in the Civil War

The American Civil War (April 12, 1861, to April 9, 1865) greatly slowed the construction of new railroad alignment and the associated tunnels. However, the existing railroads were a vital element in President Lincoln's Civil War strategy. By 1860, about 21,300 miles of track (American-Rails.com 2015b), including about 120 tunnels, were available to the Union (North) Army to rapidly deploy and replenish troops and supplies to the front lines, thereby bringing swift and heavy damage to the Confederacy (South). During the war, the North added 4,000 miles of new alignment. In 1862 when the Civil War was at its peak, the Union Army was able to deploy 2,400 fresh troops to the front line by rail in 11 days, where more than a month would have been required by road, resulting in the Civil War also being called the "first railroad war" (AAR n.d.). President Lincoln understood the value of rapid troop deployment, and therefore struck a deal to put railroads under government control for more efficient operation of rail traffic and to better preserve the security of the system against sabotage by the Confederate Army. Some rail lines carried more than 800 tons per day of supplies, equipment, and men to the front.

The Confederate South had a much smaller network of railroads under private control, with just 9,000 miles of rail alignment along with five tunnels. An additional 400 miles of track were added in the south during the Civil War. The Confederacy also made good use of its railroads and tunnels to move troops rapidly to either side of the Blue Ridge Mountains through the Blue Ridge Tunnel (also known as the Crozet Tunnel), which when completed in 1858 was the longest tunnel in the United States (Drinker 1893) at a length of 4,273 feet (Figure 3.4). Because of size limitations on increasingly larger rail cars and engines, the old Blue Ridge Tunnel was replaced in 1944 with a new concrete-lined, larger tunnel, which is also called the Blue Ridge Tunnel.

During the five years following the Civil War, additional tunnels and another 22,263 miles of new alignment were constructed for totals of 52,898 miles of rail alignment and at least 228 tunnels (Table 3.2).

Continued Expansion of the Railroads

A peak in railroad tunnel construction took place from the mid-1860s through 1916 when major westward expansion of the railroads occurred. Prior to this period, nearly all of the railroad development and the associated tunneling was confined primarily to east of the Mississippi River. However, based in part on the discoveries and mapping by the Lewis and Clark exploratory expedition between 1804 and 1806, development expanded westward from St. Louis, Missouri, through the newly acquired Louisiana Purchase to the Oregon Territory and Pacific Ocean. Following the reunification of the United States in 1865 after the Civil War ended, there was rapid westward expansion of the railroads to the Pacific Ocean, as well as north–south growth along the Pacific Coast. By 1880, more than 93,000 miles of track were in operation with almost 400 tunnels. At the turn of the twentieth century, there were about 190,000 miles of track in service, and by 1916, at the peak of railroad expansion, 254,000 miles of track had been constructed, including an estimated 1,000 railroad tunnels.

Unfortunately, a number of factors resulted in a major drop in the miles of railroad alignment and a corresponding reduction in the number of tunnels that were in use, including World War I (1914–1918), the Great Depression (1929–1939), the introduction and rapid growth in the number of automobiles (early 1900s), and the rapid development of a road and highway system from the 1920s onward (Gallamore and Meyer 2014). This major slump in railroad traffic and usage was further exacerbated from the 1950s through the 1970s by development of the modern U.S. Interstate Highway System that enabled efficient and highly competitive truck freight. The railroads also lost much of their passenger revenue with the introduction of inexpensive commercial airline transportation following World War II. These competing forces for freight and passenger transport resulted in a much leaner and more efficient freight rail system in the United States. The current railroad system involves about 140,000 miles of actively used track with an estimated 500 tunnels. Concurrent with this nearly 45% reduction in rail alignment, the tons of freight carried per mile of track is at the highest levels in railroad history, having ballooned from 414 billion ton-miles of transport in 1920 (at the peak in miles of track) to 1,741 billion ton-miles of freight transport in 2013 (AAR 2014). Some of the unused tunnels and associated rail alignments have now been converted in the Rails-to-Trails program for public use. However, much of the abandoned 114,000 miles of alignment and the associated tunnels are now overgrown and deteriorated after nearly a century of disuse.

TRANSCONTINENTAL RAILROADS

Perhaps the most rapid expansion of the railroads, with the greatest impact on growth of the United States, occurred with the building of the first of more than five transcontinental railroads. Preliminary planning for a transcontinental railroad had begun in the 1830s, and in 1845 a central route was explored by Asa Whitney. Northern and southern routes had also been proposed and initially evaluated, but were considered to have less benefit. Theodore D. Judah, a businessman and owner of 26 miles of railroad track in Sacramento, California, drafted the original Pacific Railroad charter, and lobbied Congress and the President to fund the First Transcontinental Railroad. President Abraham Lincoln signed a bill on May 6, 1862, to provide funding, allocate federal land grants, and use the two selected companies (Union Pacific and Central Pacific) to build the railroad. Lincoln recognized that a transcontinental railroad provided many opportunities. It would unify the very remote and poorly connected west, rich in natural resources and poor in available labor, with the east, rich in labor and manufacturing facilities. And it would allow finished goods to be moved more rapidly westward from the manufacturing hubs in the east, such as New York, Chicago, and St. Louis. The railroad would also enable much more rapid colonization and development of the west and accelerate incorporation of the territories west of the Mississippi River into the Union (Ambrose 2001; Wikipedia 2015b).

Figure 3.4 | **Original Blue Ridge Tunnel**

Much of the leadership and organizational skills for construction of the Transcontinental Railroad became available after the Civil War ended. Military leaders from the Civil War had learned and honed their skills in planning and rapidly deploying men, equipment, and materials over long distances. They had also learned how to organize and coordinate a large labor force, and rapidly construct earthworks such as embankments, cuts, and tunnels (Ambrose 2001). Construction began in late 1863 with laying of the first sections of rail outside of Sacramento. The railroad was completed on May 10, 1869, when the Union Pacific Railroad met the Central Pacific Railroad at Promontory Summit, north of Salt Lake City, Utah. The Union Pacific had constructed 1,087 miles of track westward over prairies and rolling terrain from Omaha, Nebraska; the Central Pacific constructed 690 miles of track eastward from Sacramento through the Rocky Mountains. After six years of construction, the Transcontinental Railroad was completed, with track extending from New York, though Chicago, and westward to Sacramento. A ceremony was held where the two great

Figure 3.5 | **The Last Spike at Promontory Summit, Utah, in 1869**

railroads met at Promontory Summit, Utah. The hammering in of the last spike, a ceremonial golden spike to anchor the last lengths of rail, may have been the world's first live, mass media event using the telegraph to transmit the hammer contacts with the rail to the eastern United States (Figure 3.5).

In passing through the Sierra Nevada, the Central Pacific had built 15 tunnels, collectively called the Donner Pass Tunnels. The longest of these was Summit Tunnel with a length of 1,659 feet, excavated through relatively solid granite (Figure 3.6). The construction of these tunnels was made possible by using a large labor force of Chinese and Irish immigrants (Figure 3.7). The tunnels were completed in six years, using chisel bits and sledgehammers, and initially blasting with black powder, and subsequently with a new and innovative blasting agent, nitroglycerin. Nitroglycerin was introduced in the United States by James Howden from England and used for the first time in the United States while excavating the Donner Pass Tunnels. On-site manufacturing of the unstable nitroglycerine was required following several mishaps that occurred while transporting it, including the demolition of a post office building and the death of a dozen bystanders. Subsequently, Howden manufactured up to 100 pounds of nitroglycerin per day in an old kettle, located in a basement space beneath a shed near Donner Pass. The use of nitroglycerin nearly doubled the advance rates, from 1 foot per day per tunnel heading using black powder to nearly 2 feet per day with nitroglycerin (Drinker 1893).

Completion of the First Transcontinental Railroad in the late 1860s had many major impacts on the inhabitants of the United States. It allowed travel from coast to coast in as little as one week and at a fraction of the cost—on the order of $100—versus a year of time and more than $4,000 that an overland trip by covered wagon or a sailing voyage through the Strait of Magellan around South America's Cape Horn would have taken. It enabled much more rapid expansion of towns and cities into the western states and provided much easier access to minerals, timber, farm goods, furs, and other natural resources and provided a means for shipping manufactured goods rapidly to the west. East-to-west travel became so fast and efficient that Charles Dowd, a schoolmaster in Saratoga Springs, New York, recognized the challenges of arranging train schedules and undertaking telegraph communication between the east and west coasts. Dowd subsequently proposed the first U.S. standardized transcontinental time

Figure 3.6 | **Sierra Grade at Donner Summit in 1869 and 2003**

zone map, with four time zones initially referenced to Washington, D.C., and the U.S. Naval Observatory. Dowd's time zone concept was adapted by the U.S railroads in 1870, but referenced to the Greenwich Meridian in England, as an essential means for developing train schedules (White 2004).

The second transcontinental line was a southern alignment. It was originally conceived to be the first transcontinental railroad because of the Gadsden Purchase of about 30,000 square miles in 1853 from Mexico. The Gadsden Purchase established the border between Mexico and the United States and the southern borders of New Mexico and Arizona. But construction of this second rail alignment across the southern United States was not authorized and incorporated by the U.S. Congress until 1866 after the conclusion of the Civil War. Initial work consisted of a series of locally owned lines in Southern California, constructed without the benefit of federal land grants or funding, from San Francisco to Modesto (1870), and south to Delano (1873). Other segments of the line from San Francisco to Los Angeles, including construction of the 6,975-foot-long San Fernando Tunnel (the longest transportation tunnel in the United States at the time) that took 18 months to complete because of numerous cave-ins of the water- and oil-saturated sandstone, were similarly progressing (Chinese Historical Society of America 1969). On September 5, 1876, a golden spike was driven at Lang Station near Palmdale, California, by the president of the Southern Pacific Railroad to complete the Los Angeles to San Francisco segment (Best 1976). Completion of this 400-mile stretch of railroad launched a land boom in Southern California and enabled the shipping of produce and manufactured goods that in a single day could travel from Los Angeles to San Francisco over a distance that used to take two to three weeks by rough road or ocean travel. By March 8, 1881, the Southern Pacific Railroad from the west met with the Atchison, Topeka and Santa Fe Railway from the east at Deming, New Mexico, thus completing the Second Transcontinental Railroad that linked the East Coast to the West Coast via a southern route through Louisiana, Texas, New Mexico, Arizona, and Southern California. The First and Second Transcontinental Railroads are shown in Figure 3.8. Construction of the second alignment required the construction

Figure 3.7 | **Chinese Railroad Workers at Donner Pass**

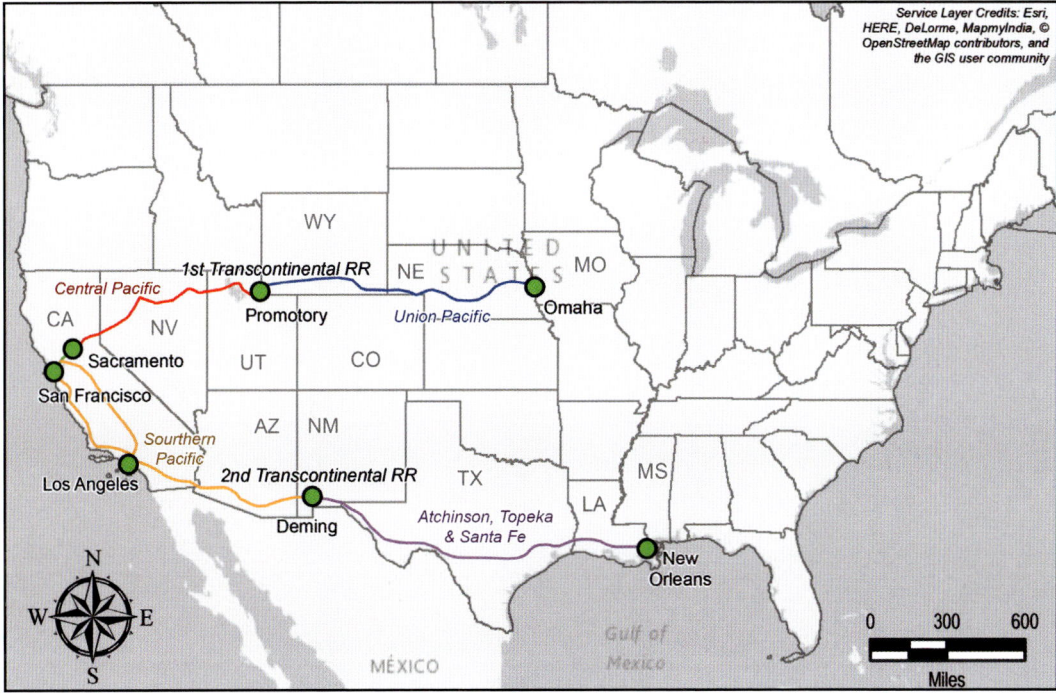

Figure 3.8 | **Map of First and Second Transcontinental Railroads**

Figure 3.9 | Tehachapi Loop Where the Track Passes Beneath Itself Through a Tunnel

CHAPTER THREE | RAILROAD TUNNELS | 61

of at least 20 tunnels. The portion of the alignment through the Tehachapi Mountains in Southern California includes a unique 0.73-mile-long spiral loop in which the track passes beneath itself in a short tunnel, as shown in Figure 3.9, to ascend through the Tehachapi Mountains at altitudes of up to 7,000 feet through fractured to massive granite.

A third transcontinental route was proposed by the Northern Pacific Railroad Company. Chartered by the U.S. Congress in 1864 and given nearly 40 million acres of land grants, the Northern Pacific raised funds in Europe and America by selling inexpensive land and transportation packages to German and Scandinavian farmer immigrants who settled along the alignment and planted crops of wheat and other grains between 1881 and 1890 in Minnesota and North Dakota. Eventually the Northern Pacific built more than 6,400 miles of track to connect the Great Lakes industrial area near Chicago, Illinois, and Duluth, Minnesota, through North Dakota, Montana, and Idaho with the Pacific Northwest, as shown in Figure 3.10, thus linking the Oregon Territory with the rest of the country.

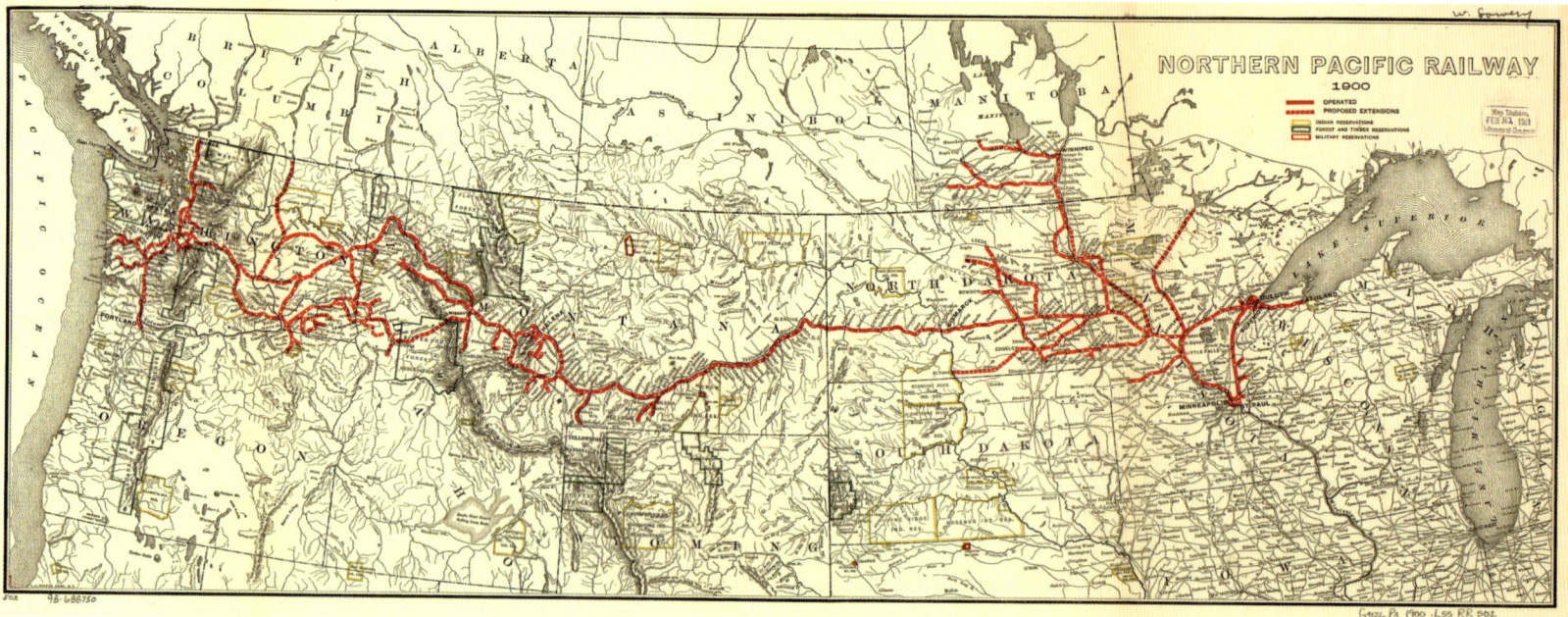

Figure 3.10 | **Northern Pacific Railway Routes, 1900**

Construction began in 1870 near Duluth, Minnesota, and in 1871 on the Pacific Coast from Seattle, Washington. By the mid-1870s, rail had been constructed through North Dakota Territory and into Montana Territory. The Seventh U.S. Cavalry under the command of Lieutenant Colonel Armstrong Custer was brought in to protect the railroad survey and construction crews, and allow settlers to enter the territories. Meanwhile along the West Coast, the railroad had extended northward from Tacoma nearly to Seattle, where it could provide access to and haul coal from rich coal fields south to San Francisco. The railroad initially bypassed the Cascade Mountains by taking a longer circuitous route along the Columbia River and then to Kalama, Washington, which enabled completion in 1883 when former President Ulysses S. Grant drove the ceremonial "golden spike" in Western Montana. A shorter route to Seattle was initiated in 1887 via a series of very steep switchbacks with grades of up 5.6% across Stampede Pass. By May 1888, a 9,850-foot-long tunnel was completed with a 2.2% grade, shown in Figure 3.11, beneath Stampede Pass in the Cascade Mountains to the east of Seattle. In the late 1990s, after being unused for much of the last 40 years of the twentieth century, Stampede Pass Tunnel and two nearby tunnels were rehabilitated with new drainage and the excavation of minor notches in the tunnel quarter arches to accommodate freight container trailers loaded onto flat cars. The tunnels are currently being used for shipping much of the agricultural crops from eastern Washington to West Coast ports, since these tunnels are not large enough for double-stack container cars.

The fourth—and certainly the most northerly—transcontinental railroad was constructed by the Great Northern Railway between 1880 and 1893 (Figure 3.12). The Great Northern Railway began as an agglomeration of several small railroads, including the St. Paul & Pacific Railroad, which held extensive land grants in the Midwest and Pacific Northwest, but had not yet constructed much track. James Jerome Hill pulled together a number of wealthy investors to buy rights-of-way, purchase inexpensive land, and develop spur lines to haul iron and copper ore from mining districts in Minnesota and Montana to the steel mills in the Midwest. The railroad was largely constructed with private funds and without the benefit of federal land grants. However, the railroad was able to buy ample inexpensive land and resell it at a substantial profit to European immigrants who settled and farmed along the alignment. The railroad's best known engineer was John Frank Stevens, who explored and selected the alignment

Figure 3.11 | **Stampede Pass Tunnel, 1890**

across Montana and through Marias Pass (shared with the Northern Pacific Railroad), which is the lowest crossing of the Rocky Mountains in the United States. Stevens also picked out a pass in the Cascade Mountains of Washington, which was eventually named Stevens Pass, and was crossed via a series of eight switchbacks at grades of up to 4%. The railroad was completed in 1893, including more than a dozen tunnels, with the final spike driven at Scenic, Washington (Hidy et al. 1988).

In part, the Great Northern Railway was built to provide access to the mineral- and timber-rich mountains of the northwest and to provide access for immigrants to the rich farm land and forests along the alignment. It was also built to enable tourists to travel across the scenic northern United States and was instrumental in lobbying congress to establish Glacier National Park in 1910 (Hidy et al. 1988). In 1900, the switchbacks were eliminated with the construction of a 2.63-mile-long tunnel beneath Stevens Pass at a grade of only 1.7%. The railroad line was electrified beginning in 1915 through the Rocky Mountains and Cascade Mountains to reduce the toxic smoke and fumes that railroad staff and passengers had to endure in passing through some of the longer tunnels (Figure 3.13). The railroad remained partially electrified until 1956, when the line was fully converted to diesel-electric locomotives, and a ventilation system was installed along with closable portal doors to enable rapid flushing of diesel fumes from the tunnel. On March 1, 1910, a major landslide near the west portal of the Stevens Pass Tunnel near Wellington, Washington, swept over a train, killing at least 96 people. Subsequently, the town was moved and renamed Tye, and a new, lower-elevation and longer tunnel was constructed beneath Stevens Pass in 1925. Cascade Tunnel, at 7.8 miles long, became the longest U.S. railroad tunnel (Bauhof 1989).

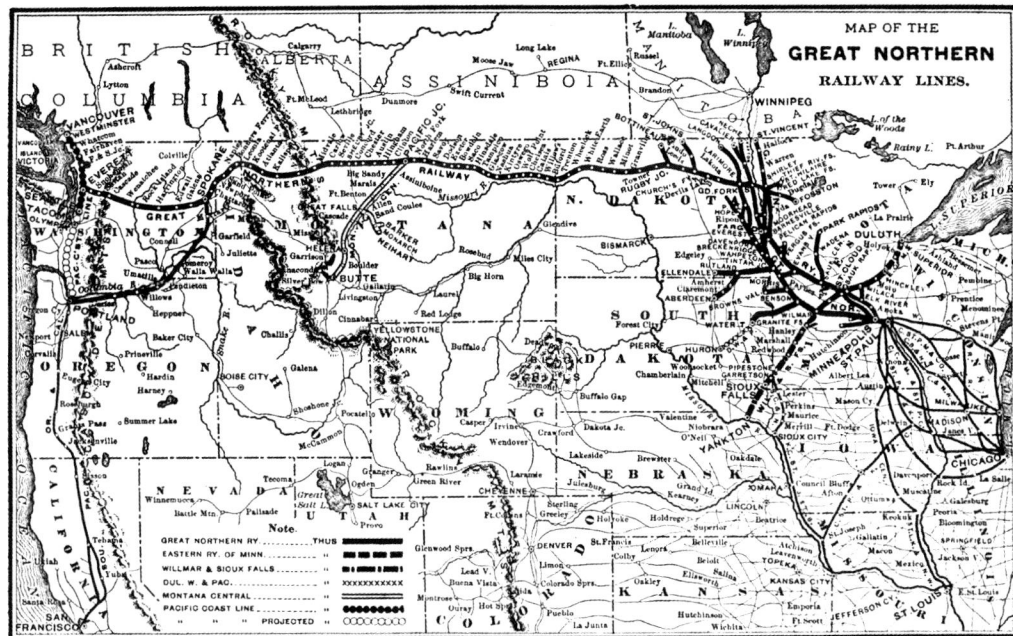

Figure 3.12 | **Great Northern Railway, 1897**

Figure 3.13 | **Postcard of Great Northern Railway Electric Locomotive, Circa 1924**

LOCOMOTIVES

The evolution of motive power systems for railroad trains was a critical and essential link in the development of an efficient, fast freight and passenger railroad system in the United States. The Baltimore & Ohio (B&O) Railroad was the first commercial railroad, incorporated in 1827 and officially opened in 1830, using horses, mules, and oxen for pulling the track-mounted cars along. Wind power had also been tried, unsuccessfully, on various lines. The first test of a steam locomotive was conducted in 1829, using the prototype Stourbridge Lion steam engine (American-Rails.com 2015a) built by Peter Cooper in England and imported to be used briefly on the Delaware & Hudson Canal Company rail line (Figure 3.14). The engine could move at speeds of up to 18 mph, but it was too heavy for the roadway and was eventually warehoused and then destroyed (Wikipedia 2015c). Subsequently, Phineas Davis in 1831 improved Cooper's design, which spawned a rapid increase in railroad line construction and ever greater improvements in the efficiency, pulling power, and speed of coal-fired, steam-powered engines. The first railroad engine built in the United States for an American railroad was the "Tom Thumb," a small steam locomotive manufactured for demonstration purposes on the B&O (Wikipedia 2015a).

Figure 3.14 | **Stourbridge Lion Steam Engine**

Figure 3.15 | **Electric Locomotive for St. Clair Tunnel**

In the 1830s electric locomotives, powered by batteries, were experimented with in Scotland and Europe for a 7-ton car, capable of hauling 6-ton loads at 4 mph. However, it wasn't until the mid-1880s that the world's first electric tram line was opened in Berlin, Germany, and Vienna, Austria, to be followed in 1888 with electric trolleys in the eastern United States. The toxic fumes produced by the wood- or coal-fired steam locomotives, particularly in tunnels, prompted the electrification of 4 miles of the B&O Railroad through Baltimore in 1895. Legislation in New York forced the electrification of several railroads entering the city in the early 1900s (Wikipedia 2015d). The Grand Trunk Railroad converted the 6,032-foot long St. Clair Tunnel to electric-powered locomotives by 1904, as shown in Figure 3.15, following several deaths caused by asphyxiation due to the coal-burning locomotives (Middleton et al. 2007). By the 1920s, many of the railroads had introduced electric locomotives as substitutes for coal-burning steam locomotives at major tunnels such as the 4.8-mile-long Hoosac Tunnel and the 7.8-mile-long Cascade Tunnel. And by the 1930s the Pennsylvania Railroad had electrified its entire system east of Harrisburg, Pennsylvania, including the Black Rock Tunnel in Phoenixville and the Flat Rock Tunnel in Manayunk, west of Philadelphia, as located on Figure 3.2. These two tunnels, completed in 1837 and 1840, respectively, were enlarged in 1859 to accommodate larger freight cars. Both tunnels were enlarged with the first application of electric detonation of multiple explosive charges for tunnel excavation. These two tunnels are still in service today, after having been enlarged for double-stack freight cars and re-supported in 1994.

In 1918 a trio of companies—the American Locomotive Company, Ingersoll-Rand, and General Electric—joined forces to produce an initial diesel-electric motor car for a small connecting railroad in New York City. In 1925, General Electric and Ingersoll-Rand built a prototype diesel-electric engine and then produced several diesel-electric motor railcars that evolved into a powered 60-ton boxcar. By 1837, the B&O bought its first multiple unit diesel-electric locomotive from General Motors' Electro-Motive Division and made the first coast-to-coast trip of a diesel-driven passenger train (Wikipedia 2015a). Diesel-electric locomotives produced less toxic smoke, required fewer stops for fueling and water, and required less maintenance. By the late 1950s, diesel-electric locomotives had replaced steam locomotives on most major railroads. Various models of these highly efficient, high speed diesel-electric locomotives are used throughout the United States (Figure 3.16). The much less toxic gases exhausted by these locomotives has eliminated the need for active ventilation systems in all but a few of the longest railroad tunnels in the United States, such as the Cascade Tunnel (Washington), Flathead Tunnel (Montana), and Mullan Tunnel (Montana), located on Figure 3.2.

Figure 3.16 | **Diesel Locomotive at Cascade Tunnel**

ADVANCES IN ROCK TUNNELING

Following the Civil War, which had absorbed much of the manpower and natural resources of the United States, technology improved and new innovative tools were developed for surveying, drilling, blasting, excavation, and ground support. Consequently, it was in the latter half of the 1800s that something resembling today's rock tunneling drill-and-blast excavation methods had evolved. The Hoosac Tunnel was constructed intermittently from 1851 to 1875 in the northwest corner of Massachusetts, as located on Figure 3.2, and, more than any other project of the time, benefited from the rapid transition from hand drilling to vastly faster steam-driven percussion rock drills or power drills developed by J.J. Couch of Philadelphia in 1849. These drills ultimately led to the development of Charles Burleigh's compressed-air or pneumatic-powered percussion drills in 1866 that were predecessors to the pneumatic percussion rock drills in use today (Drinker 1893).

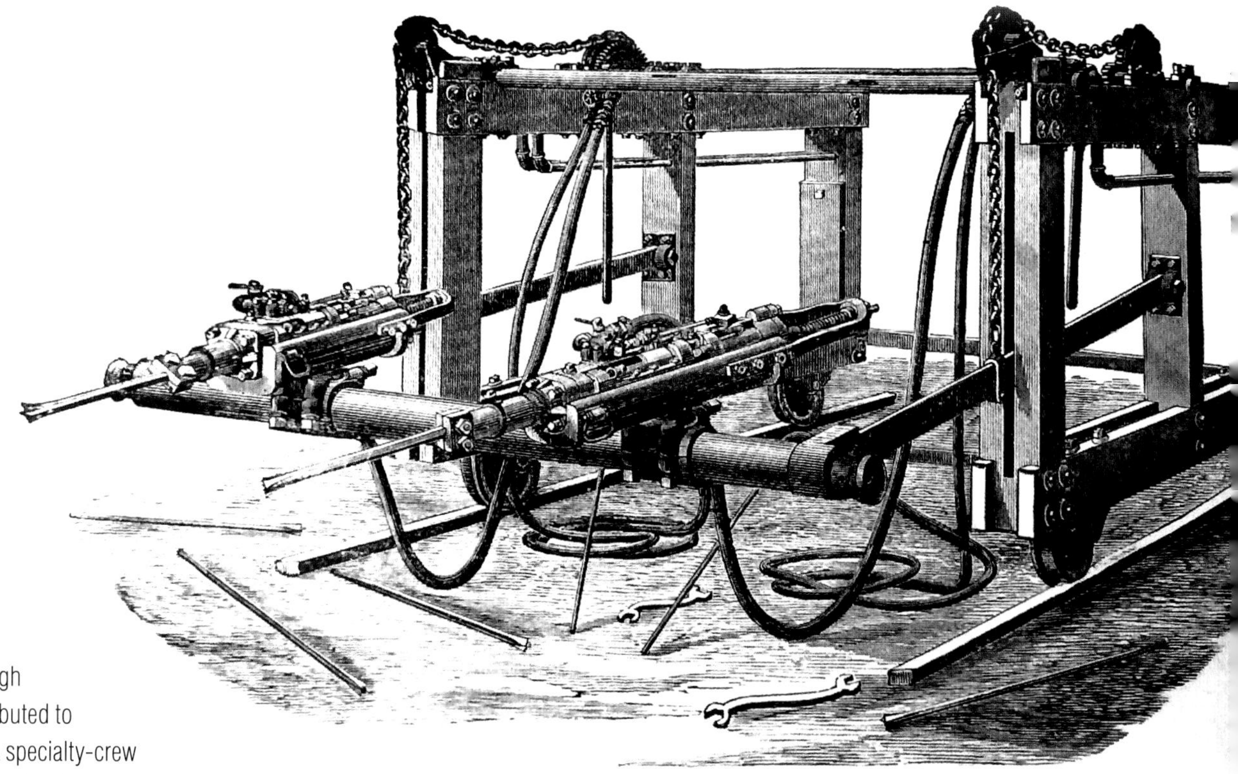

CHAPTER THREE | RAILROAD TUNNELS

Although Couch's steam-driven drill was a remarkable advance at a time when drilling holes of up to 4 feet deep was accomplished with a sharpened chisel and a sledgehammer, the steam introduced and vented into the tunneling environment could not be efficiently exhausted by the primitive ventilation systems of the time and actually worsened overall working conditions. By 1866 Burleigh had introduced a new and vastly improved pneumatically powered rock drill that was capable of drilling holes over 48 inches deep. The drill was often mounted in groups of two to eight drills on a wheeled drill carriage (Figure 3.17) to more rapidly move to the construction heading for drilling and then retreat down the tunnel prior to blasting. The development of the Burleigh drill along with several other innovations contributed to the eventual completion of the Hoosac Tunnel. A specialty-crew approach was developed for the Hoosac Tunnel, which has been utilized on most tunnel projects since then. With this method, drilling crews used the new Burleigh compressed-air-driven percussion rock drills, and a specialty blasting crew loaded the newly developed center-hole blasting pattern of holes with the much safer Nobel dynamite cartridges that were patented in 1868 and initiated with safer detonators. A muck handling crew excavated the soft rock and removed blasted rock in the Great Northern Cascade Tunnel with the newly introduced and faster compressed-air-powered shovels, as shown in Figure 3.18. Electric-powered muck trains (used in 1905 for the Great Northern Tunnel in Seattle) and other excavators also enabled more rapid construction of railroad grades of up to about 2% via cuts and fills and tunnels through soil and rock. Although grades of more than 5% were used on some mountainous slopes, grades of less than 1% were preferred for lower power consumption (Vogel 2012).

Figure 3.17 | **Burleigh Pneumatic Percussion Drills Mounted on a Wheeled Drill Carriage**

Although a mechanical tunneling machine was tried and had failed in 1853 at the Hoosac Tunnel in very hard igneous and metamorphic rock, it was not until the early 1950s that tunnel boring machines (TBMs) became a viable option for constructing railroad tunnels through rock. In 1952, the Robbins Company built a TBM that was used to carve out the 3,650-foot-long water intake tunnel through volcanic rocks at the Oahe Dam diversion project near Pierre, South Dakota (Roby et al. 2008). The first railroad tunnel in North America excavated through rock with a TBM was the 9.1-mile-long Mount Macdonald railroad tunnel constructed in the vicinity of Rogers Pass in the Selkirk Mountains (northern extension of the Rocky Mountains) of central British Columbia, Canada (see Figure 3.2). For this tunnel, a 22.3-foot-diameter Robbins main beam TBM was used to construct the top heading of 27,400 feet of the tunnel, with the bench, or the lower 7 to 10 feet, excavated by drill-and-blast to form a keyhole-shaped tunnel opening. The finished tunnel had a centerline height above top of rail of

Figure 3.18 | **Compressed-Air-Powered Shovel, 1928**

24.9 feet and a clear width of 17 feet and was designed for double-stack container cars and tri-level auto-carrier cars (Bickel et al. 1996). Other than tunneling in soil, TBMs have not been used for railroad tunneling in the United States in part because most of our tunnels were constructed prior to the 1950s and because a TBM-driven railroad tunnel would have to be at least 30 feet in diameter to provide the necessary clearance for double-stack container cars.

Through the 1930s, where unstable rock conditions required support, tunnels were lined with timber ribs, cribbing, and lagging, or in some instances with large cut stone blocks (ashlar stone), cobble-sized rock and mortar, or multiple courses of brick. A good example of all of these lining types are the Allegheny, New Portage, and Gallitzin Tunnels, as located on Figure 3.2. All were initially supported with timber, and later supported with a variety of materials. The Allegheny Tunnel and neighboring New Portage Tunnel are still in service in western Pennsylvania. Both have been enlarged and re-supported for double-stack container cars. The Allegheny Tunnel has been widened for twin tracks, and both tunnels have been relined with grouted rock dowels and steel fiber-reinforced shotcrete or concreted steel ribs to accommodate the modern double-track clearances.

Up until World War II, tunnels were normally supported with timber ribs and lagging. A few tunnels, such as the Moffat Tunnel in north-central Colorado, finished in 1927, was primarily supported with timber ribs, including about 2,000 feet or about 6% of the 6.2-mile-long tunnel to be supported with steel ribs, similar to the tunnel shown in Figure 3.19. The steel ribs were only used in very poor ground, due to the very high ground loads and the need for rapid installation of ground support (McMechen 1927). During World War II, steel was primarily committed to military applications. However, following the war, the publication of the Proctor and White handbook in 1946 titled *Rock Tunneling with Steel Supports* prompted much wider usage of steel ribs for tunnels in the United States. Consequently, in the 1950s and 1960s, many of the tunnels that had been supported with timber sets and lagging 50 to 100 years prior were re-supported with steel sets and either timber lagging or gunite/shotcrete. The 7-mile-long Flathead Railroad Tunnel in Montana, completed in 1969, was supported with steel ribs, shotcrete, rock bolts, and a final cast-in-place concrete lining (Skinner 1974). Beginning in the 1970s and 1980s, shotcrete was used as both initial and final support in most rock tunnels, generally combined variously with grouted rock bolts, steel ribs, and lattice girder ribs to replace deteriorated timber lagging and ribs in older railroad tunnels (Parker et al. 2001).

Ungrouted, tensioned steel rock bolts were introduced in the 1950s to reinforce the fractured rock arch and walls of old and new rock tunnels. In the 1960s, fully cement-grouted rock bolts, and subsequently resin-grouted rock dowels, were used to reinforce the fractured rock around rock tunnel perimeters.

Single-track tunnels constructed prior to about 1900 were generally constructed with timber supports to clear rail loads that were 14 to 16 feet wide by 16 to 19 feet high at tunnel centerline, with a 3- to 6-inch clearance gap between the load and timber ribs. However, these tunnels are generally too small to accommodate today's larger railcars as shown by the required clearance envelopes on Figure 3.20 for double-stack container cars, tri-level automobile carriers, and other assorted transport cars.

ADVANCES IN SOIL TUNNELING

Although railroad tunnels for the most part have been constructed in rock, in many cases the portals, and in a few instances entire tunnel lengths, have been constructed in soil. Much like tunnels through rock, tunnels constructed in soil have used a wide range of evolving excavation means and methods. Early tunnels were excavated in soil with pick and shovel, mucked out with wheelbarrows and mule-drawn carts, and supported with an initial lining of timber ribs and lagging, as shown in Figure 3.21 for the 35-foot-wide Great Northern Tunnel in Seattle that was

---

David Halliday Moffat (1839–1911)

Moffat was one of Denver's most important financiers and industrialists in late-nineteenth and early-twentieth-century Colorado. He served as president, treasurer, and a board member of railroads, banks, and city government posts. Over the years, he had claims to more than one hundred Colorado mines and nine railroads. His brainchild was a tunnel through the Continental Divide west of Denver. Construction of the 6.2-mile-long Moffat Tunnel took place from 1923 to 1927. It was officially opened on February 28, 1928, with much fanfare and several trainloads of special guests in attendance at the east portal.

Figure 3.19 | Primarily Timber-Rib-Supported Tunnel with Steel Ribs, 1990s

excavated through a wide range of glacial and interglacial sand, silt, clay, and till. Tunnels in soil almost always had a final lining of stone masonry or multiple courses of brick and mortar and, after about 1900, were lined with concrete, generally cast directly against the initial timber lining.

Bolted, segmental, ¼-inch-thick rolled iron liner plate lining was first introduced by Dewitt Clinton Haskin for use on the Hudson River Railroad Tunnel, between New York and New Jersey, in 1879 (Vogel 2012). Haskin elected to construct the twin-track tunnel through wet silt and sand using only iron liner plate and compressed-air pressure of 35 psi to provide temporary support of the ground, with a final 2.5-foot-thick brick-and-mortar liner constructed within 20 to 50 feet of the advancing face. Two serious blowouts and loss of life halted construction for nearly 20 years until an iron/steel shield, augmented with compressed air, was implemented in 1902 to complete the 5,000-foot-long tunnel by 1905 (Cudahy 2002). It was not until the late 1930s that the pressed, corrugated, boltable, flanged steel liner plates were developed that are still used today for initial support in some tunnels in soil, as shown in Figure 3.22.

In the mid-1800s, steam shovels and then pneumatically powered, tracked shovel/loader/excavators and pneumatic handheld spaders, along with gasoline, diesel, and electric (Seattle's Great Northern Tunnel in 1905) locomotive-powered muck trains were used to excavate tunnels.

The development of a number of ground improvement methods since the 1940s has enhanced the ability to build tunnels through soil in challenging ground conditions. These techniques include a wide range of dewatering options (deep wells, vacuum augmented wells, well points, eductor wells, and horizontal vacuum lances), grouting options (cement and chemical permeation grouting, jet grouting, compensation grouting, compaction grouting, and fracture grouting), and soil freezing.

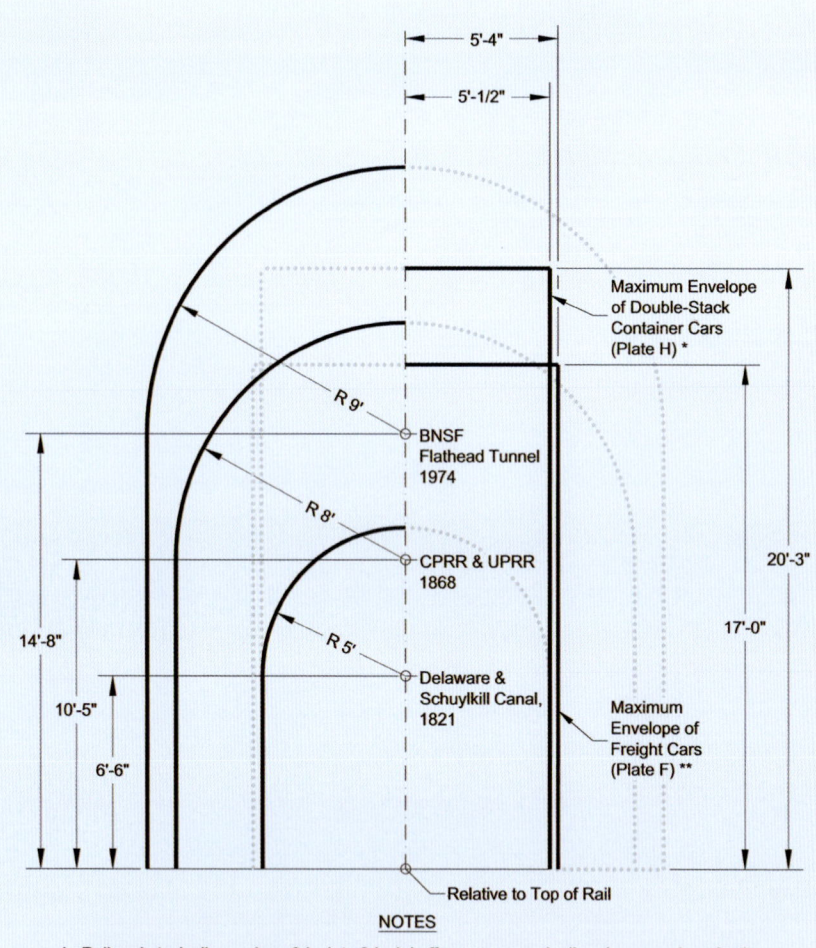

Figure 3.20 | **Railroad Tunnel Clearance Requirements**

Figure 3.21 | **Excavating the Core Soils and Upper Drifts of Seattle's Great Northern Tunnel, Circa 1903**

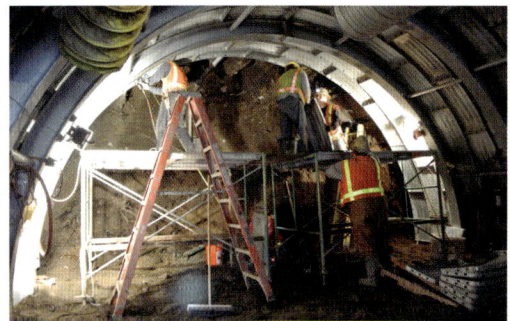

Figure 3.22 | **Excavation Through Embankment, Supported with Bolted Liner Plates**

Introduction of Shield Tunneling

Shield tunneling in the United States closely followed successful shield tunneling in England. The first documented and patented iron shield for excavating through wet sand, silt, and clay was a 38-foot-wide by 22.5-foot-high rectangular tunneling shield consisting of 12 independently advanced rectangular working chambers and a masonry lining developed in 1818 by Sir Marc Brunel in England for the first undercrossing of the River Thames in London. The Thames Tunnel took more than 20 years to complete, in part due to several rapid inflows of water and mud from shallow soil cover below the river bottom. In 1864 Peter W. Barlow patented the first cylindrical tunnel shield and in 1869 constructed a 1,300-foot-long, 7.3-foot-diameter tunnel. His student, James Henry Greathead, further refined and patented the circular iron shield concept in 1864. Greathead utilized ship's screw jacks to advance the cylindrical shield and used compressed-air pressure, which was maintained in the tunnel with an air lock, to stabilize the face in wet silt and sand. He enhanced his design with cast-iron segments to support the ground and hydraulic jacks to advance the shield, patented in 1868 (Vogel 2012).

Almost concurrently with Greathead's success on the Tower Subway Tunnel in England, Alfred Ely Beach was constructing a trial 8-foot-diameter tunnel that exited from the basement of a clothing store on Warren Street in New York City, made a 90-degree turn beneath Broadway, and ran for 312 feet. The tunnel

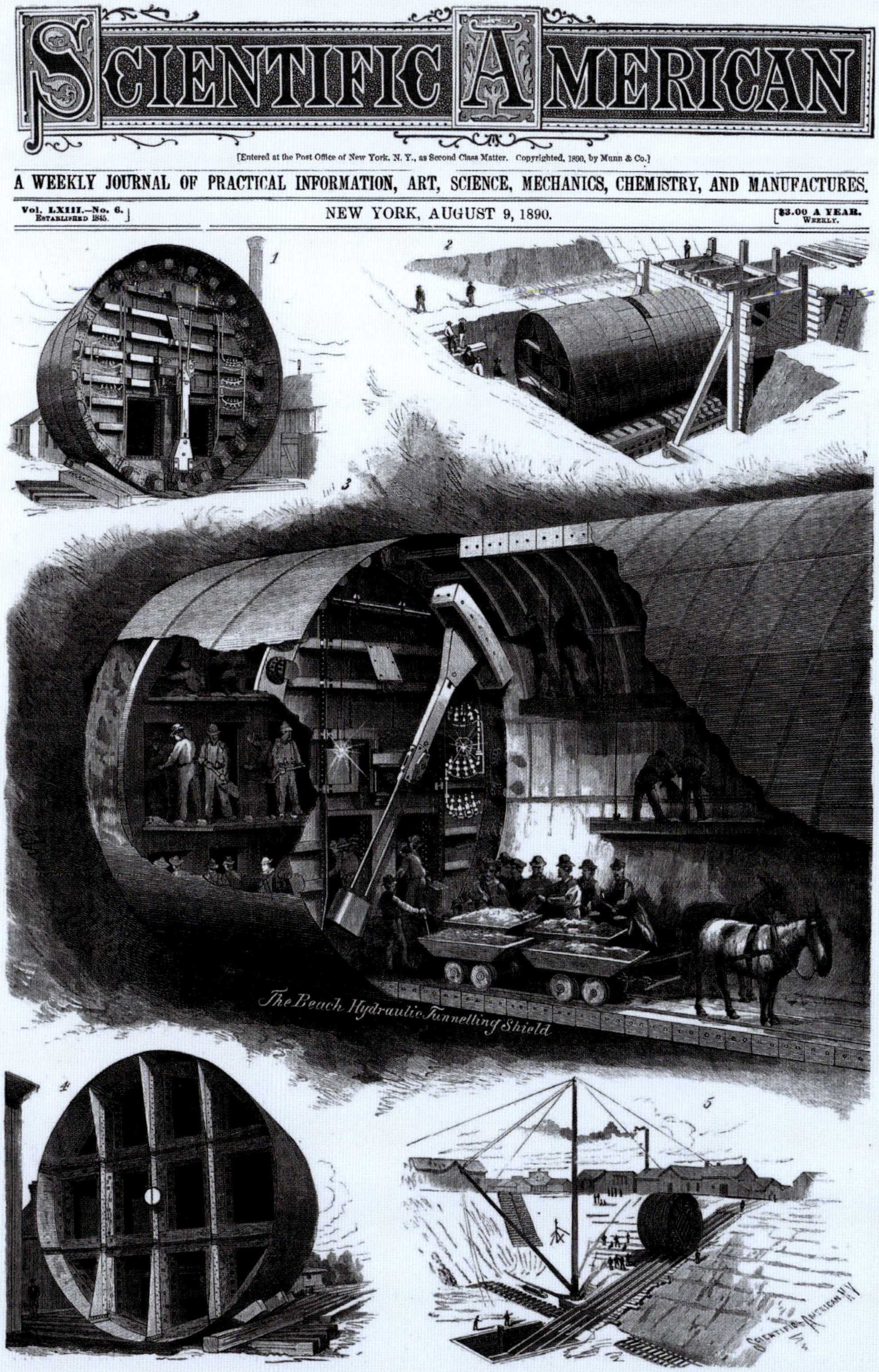
SCIENTIFIC AMERICAN

[Entered at the Post Office of New York, N. Y., as Second Class Matter. Copyrighted, 1890, by Munn & Co.]

A WEEKLY JOURNAL OF PRACTICAL INFORMATION, ART, SCIENCE, MECHANICS, CHEMISTRY, AND MANUFACTURES.

Vol. LXIII.—No. 6. Established 1845. | NEW YORK, AUGUST 9, 1890. | $3.00 A YEAR. Weekly.

Figure 3.23 | **Beach Hydraulic Tunneling Shield**

was constructed over a period of 58 nights at an average rate of about 6 feet per night. Beach, a patent attorney and publisher of *Scientific American*, patented his invention of the Beach hydraulic tunneling shield in 1869 (Figure 3.23). The Beach shield was an iron cylinder like Barlow's, but included a number of unique features, such as 18 hydraulic thrust rams with a 16-inch stroke powered by a single hand pump, and horizontal metal sand shelves that cut into the loose, sandy soil. The shield could be steered by controlling pressure to individual rams. Permanent support was provided by a cast-iron liner plate in the curve and brick masonry in the straightaway. A single passenger car was thrust through the circular tunnel by 0.25-psi air pressure (a total thrust of nearly one ton) exerted by a blower fan.

Subsequently, Chief Engineer Joseph Hobson of the Grand Trunk Railway designed a shield with an outside diameter of 21.5 feet, based on Greathead's London shield and Beach's Broadway Subway shield for constructing the 6,025-foot-long St. Clair River Tunnel between 1888 and 1890, making it North America's first subaqueous transportation tunnel. The shield incorporated 28 hydraulic jacks used to advance it forward in 17-inch cycles to enable the installation of bolted cast-iron segments with a mechanical segment erector. Compressed air up to 28 psi was used to help maintain face stability in wet soft clay, flowing silt, and sand. Two shields were used to excavate the tunnel from both ends, and meeting near the middle of the alignment (Vogel 2012). Today's use of Hobson's adaption of the Beach hydraulic shield system of tunneling, although improved by a number of innovations over the last 125 years, is still basically the same in concept and application as applied in 1890.

An earth pressure balance machine (EPBM) is a shield with a rotating cutterhead and a pressurized chamber within the cutterhead that maintains sufficient backpressure soil in the cuttings to counteract natural soil and groundwater pressure. An EPBM was used in 1995 for the construction of a second larger-diameter tunnel parallel to the original St. Clair River Tunnel. This second 27.5-foot inside diameter tunnel was needed to provide adequate clearance for the double-stack freight cars that are used to greatly increase freight volumes on major railroads in the United States. In addition to an EPBM, which allowed excavation of the tunnel through wet soils beneath a river while allowing the construction staff to work at normal air pressure, the tunnel was also lined, concurrent with excavation, with a gasketed, bolted concrete segmental lining (Harrison 1995). The new St. Clair River Tunnel is North America's largest diameter tunnel completed by an EPBM in soil to date.

REHABILITATION AND CLEARANCE IMPROVEMENT

Most of the tunnels built between the 1830s and the 1940s have been rehabilitated or recycled by undergoing several phases of re-support to replace rotting timbers, deteriorating brick and mortar, failing portals, and deficient drainage. In some cases, these same tunnels may have also been enlarged or cleared to accommodate larger freight cars. Many of the railroad tunnels constructed in the United States prior to about 1940 are still partially supported with timber sets, at least where dry tunnel conditions exist. Although cedar and redwood timber sets are generally expected to have a life span of more than 50 years, inadequate drainage in the tunnel can shorten this natural life, while good drainage or dry tunnel conditions can substantially increase the functional life span to 100 years or more. However, where timber sets, lagging, and backpacking have been subjected to dripping or flowing water, the wood typically rots over time, often before the normal 50-year life span, and additional "jump" sets of either timber or steel have to be placed between the rotting sets.

Beginning in the late 1940s, after World War II, there were rapid advances in steel manufacturing technology and the more widespread use of steel rib tunnel support. In addition, pneumatically sprayed and chemically accelerated

---

Joseph Hobson (1834–1917)

Hobson was a Canadian land surveyor, civil engineer, and railway design engineer. He served as the resident engineer for construction of the first St. Clair railway tunnel connecting Sarnia, Ontario, and Port Huron, Michigan. Constructed in less than a year, the first underwater railway tunnel between Canada and the United States opened for train service in 1891. A little over 6,000 feet in length, portal to portal, it was the longest subaqueous tunnel and the first undersea tunnel constructed between two countries. The first St. Clair Tunnel remained in service until the opening of the Second St. Clair tunnel in 1994. Following completion of the St. Clair Tunnel, Hobson became chief engineer of the Grand Trunk Railway in 1896.

concrete was developed to seal and reinforce the excavated rock surface. *Gunite* is used when sand aggregate is added, and *shotcrete* when a sand and gravel aggregate is added. Rock bolting was introduced in which a ¾- to 1-inch-diameter threaded steel dowel is anchored into a drilled 1- to 1.5-inch-diameter by 5- to 15-foot-deep borehole, initially using mechanically expanded steel-shell anchors that were subsequently replaced with more permanent cement and resin grouts to encase and anchor the bolts and reinforce the fractured rock.

A major challenge in the rehabilitation of timber-supported tunnels has been the rapid increase in freight volumes from the 1930s to present that has made it increasingly difficult to shut down a tunnel or series of tunnels for several weeks or months. Consequently, most railroad tunnel rehabilitation projects have been accomplished under "live track" conditions in work windows of 1 to 8 hours between freight trains. Tunnels in mountainous terrain are often located several miles from the nearest yard or siding. Consequently, rehabilitation work has often had to be accomplished incrementally through a tunnel or series of tunnels by the incremental removal of one or more old timber ribs and wood lagging support (Figure 3.24) and any backpacking or rock rubble, and the rapid installation of replacement support before the next train arrives. Re-support has often included the application of shotcrete and grouted rock dowels, as shown in Figures 3.25 and 3.26, respectively. In poor ground quality, the timber ribs may be replaced with steel ribs and shotcrete or concrete, as shown in Figure 3.27. To minimize delays to rail traffic, the demolition and re-support operations must be safely and quickly accomplished by using self-contained, fully equipped and supplied work trains, as shown in Figure 3.28. The work trains include a locomotive or car mover; multiple flat cars with excavation equipment, rock drills, shotcrete mixers and pumps, materials storage and supply, and a maintenance shop; and a gondola car for muck storage and transport in and out of the tunnel (Parker et al. 2001).

Figure 3.24 | **Timber Removal to Place New Lining**

Figure 3.25 | **Shotcrete Application**

Figure 3.26 | **Installing Rock Bolts and Applying Shotcrete**

Figure 3.27 | **Steel Arch Supports in Donner Pass Tunnel**

Figure 3.28 | **Work Train**

Figure 3.29 | **Grinding Down the Gallitzin Tunnel Invert**

Tunnel rehabilitation has often gone hand in hand with tunnel enlargement to provide improved clearances for larger locomotives and cars. In the last 30 years, increased clearances have been needed in many mainline tunnels for double-stack container cars and tri-level automobile carriers. Flat car-mounted truck trailer containers were introduced in the 1950s, and double-stack container railcars with intermodal container boxes stacked two high were introduced in the United States in 1984. These new low-slung flat cars and double-stack containers allowed freight trains to carry twice the tonnage for a given train length; however, the required vertical clearance for a double-stack railcar is as much as 20 feet at the tunnel quarter arches versus a normal freight car clearance of about 16 feet. The clearance envelopes shown in Figure 3.20 for the various types of railcars are typically increased by each railroad to include a 3- to 6-inch cushion to compensate for sway of the cars, misalignment of containers, misalignment of track relative to tunnel centerline, and other geometric variations between tunnels and train cars. Additional clearance on the order of 1 inch per degree of tunnel track curvature is added to the clearance envelope to accommodate the inside chord of rigid cars between wheel assemblies (called "trucks") and the overhang of cars beyond the ends of the trucks, as well as the super-elevation and tilt of railcars on curves.

Where it is feasible to shut down a tunnel for several months for rehabilitation and clearance improvement, it is generally safer and less expensive to excavate the invert and lower the sidewall support. For the Gallitzin Tunnel near Altoona, Pennsylvania, which required an additional 4 to 5 feet of vertical clearance, the invert was ground down in the mid-1990s, as shown in Figure 3.29, and the sidewalls were extended down.

CHAPTER THREE | RAILROAD TUNNELS | 75

Where it is not feasible to take a tunnel out of service, as is the case in most instances, then the old timber is incrementally removed and the tunnel is locally enlarged using blasting, roadheaders, and hoe rams (a type of hydraulic hammer), depending on the strength of the rock and amount of clearance required. As the clearance operation advances, the old timber supports are replaced with steel ribs, grouted rock dowels, and shotcrete and concrete lining as needed.

Many tunnels constructed after 1900 were large enough that sufficient clearance could be obtained for the larger railcars by cutting a notch that is a few inches to 1 foot deep near the tunnel quarter arches. A variety of methods have been tried and found to be useful for notching. In hard rock, blasting may be required, but in medium-strength rock, roadheaders, rock/concrete saws, and hoe rams have variously proven effective (Figure 3.30).

Between 1950 and 2015, more than 400 tunnels have undergone major rehabilitation, and most of these tunnels have also been enlarged to provide sufficient clearance for current double-stack container cars and tri-level automobile carriers.

Figure 3.30 | **Roadheader Notching for Additional Railcar Clearance**

RAILROADS TODAY

Since the beginning of the twenty-first century, a more efficient rail system has evolved based on high-capacity, long-distance corridors connecting major maritime gateways and inland terminals. These corridors are almost all double-tracked. Additionally, rail freight has faced a surge in demand linked with globalization and a level of de-industrialization of the North American economy, as well as rising energy prices making rail more competitive. The three most important factors behind the recent growth of rail traffic involve increased (1) intermodal containerized trade, (2) quantities of coal being shipped to power plants (namely, from the Powder River Basin in Montana and Wyoming), and (3) trade with Canada and Mexico (Rodrigue 2015). A new wave of investments for double or triple tracking of long-distance corridors and intermodal rail terminals has improved the efficiency and capacity of the system. Prospects for the future of rail transportation appear positive.

As of 2000, the American network mileage was standing at 160,000 miles, a major reduction from the much less efficient peak of more than 254,000 miles of track around 1916. The decline in the amount of rail likely relates to the introduction of the automobile and trucks, and the rapid expansion of the nation's highway system. However, the consolidation of railroad companies in the United States has enabled more efficient railroad operation. In the early 1900s, there were more than 100 railroad companies. By 1980 there were nine major or Class I railroads, and now there are just seven Class I railroads (CSX Transportation, Norfolk Southern, Union Pacific/Southern Pacific, Burlington Northern Santa Fe, Kansas City Southern, Canadian National, and Canadian Pacific), augmented with the 16 regional Class II railroads (such as the Alaskan Railroad and Long Island Railroad), and the more than 500 short lines or Class III railroads that operate relatively short segments of track (Central Oregon & Pacific Railroad and the Illinois Central) or control local switching yards and terminals. The deregulation of the railroads following

Figure 3.31 | **First and Second St. Clair Tunnels, Sarnia, Ontario**

passage of the Staggers Rail Act in 1980 has allowed the railroads to set their own freight rates, abandon unprofitable lines, and rebound from near collapse to an efficient and profitable railroad system that expanded from about 900 billion freight ton-miles in 1980 to more than 1,850 billion freight ton-miles in 2010 (Gallamore and Meyer 2014).

Although many railroad tunnels have been rehabilitated and enlarged or cleared for larger railcars, including double-stack container cars and tri-level automobile carriers, very few new railroad tunnels have been constructed since 1950. Most notable of the relatively new railroad tunnels are the Flathead, Second St. Clair, and East Side Access Tunnels (as located on Figure 3.2). The 7-mile-long Flathead Tunnel was constructed as part of the railroad relocations around the new Libby Dam and reservoir in 1968. The new 27.5-foot inside diameter St. Clair River Tunnel was constructed beneath the St. Clair River and the U.S./Canada border in 1993 from Sarnia, Ontario, to Port Huron, Michigan, to replace the original 19-foot-10-inch-diameter St. Clair River Tunnel opened in 1891 (Figure 3.31). Currently under construction are the East Side Access Tunnels in New York.

These three tunnel projects epitomize the state of the art of tunneling in the late twentieth century for railroads as well as for tunnel applications on other large civil projects. The Flathead Tunnel was constructed between 1966 and 1968 on the Burlington Northern Railway line though the Salish Mountains (northern Rocky Mountains). Construction involved drill-and-blast excavation using a unique four-level drilling gantry jumbo designed for full-face excavation of the nearly 30-foot-high by 21-foot-wide tunnel and also to allow mucking through the bottom of the drill jumbo. Rock support consisted of a combination of steel ribs, rock bolts, and a cast-in-place concrete lining.

At the time of its construction, the Flathead Tunnel was the second longest railroad tunnel in the United States and the thirteenth longest railroad tunnel in the world (Skinner 1974). It was also a proving ground for the Rock Structure Rating, or RSR, classification system developed by Wickham et al. (1972).

The new St. Clair River Tunnel was constructed between 1993 and 1994 using the largest-diameter TBM in the United States at the time. The new St. Clair Tunnel exemplifies the state of the art of soil tunneling with the application of a 30-foot-diameter earth pressure balance TBM to construct the 6,129-foot-long single-track freight tunnel. Because of very high track usage and the need for a larger tunnel to carry double-stack freight cars, a third St. Clair River Tunnel is currently in the planning stages.

The East Side Access Tunnel project in New York City is using modern drill-and-blast methods, roadheaders, two 22-foot-diameter TBMs to excavate through Manhattan Schist, two slurry pressure balance TBMs to excavate through glacial till, and the sequential excavation method to construct a 60-foot-wide cavern beneath Northern Boulevard in Queens.

Although as many as 1,000 tunnels may have been constructed over the last 180 years, about half of these have been abandoned, along with railroad lines that are no longer in use, because of consolidation of railroad companies, the addition of new ports, and modifications of rail routes to make them more efficient. More than 100 of the tunnels on abandoned railroad lines have been converted in the Rails-to-Trails program for hikers and bicyclists, creating a unique outdoors experience. It is estimated that about 500 railroad tunnels remain in use today on the various railroad lines.

FUTURE RAILROAD EXPANSION

The United States appears poised to launch into a new phase of railroad tunnel construction with the planned implementation of the High-Speed Rail system. President Barack Obama and the U.S. Congress introduced the American Recovery and Reinvestment Act of 2009 to provide more than $10 billion in initial funding for high-speed intercity passenger rail. The system is currently focused on setting alignments in the northeast coast from Charlotte, North Carolina, to Boston, Massachusetts; in the Midwest from Chicago to St. Louis; and along the West Coast from Vancouver, B.C., to Eugene, Oregon, and from San Francisco to San Diego, as shown in Figure 3.32. Several thousand miles of new high-speed railroad alignment are currently proposed across the United States. Some of these proposed routes will require substantial quantities and lengths of tunnels to minimize grades by constructing tunnels through mountains, and are likely to require tunnels beneath rivers and into and under major cities to reduce the environmental impacts. The California High-Speed Rail system consists of two phases, with Phase 1 extending from San Francisco to Los Angeles with a projected finish date of 2022. The project will be constructed in multiple segments, primarily under design–build contracts, at a current estimated cost of $68 billion. A second phase of the project will extend the alignment north to Sacramento and south to San Diego, resulting in more than 800 miles of new track, dozens of bridges, and at least 25 stations.

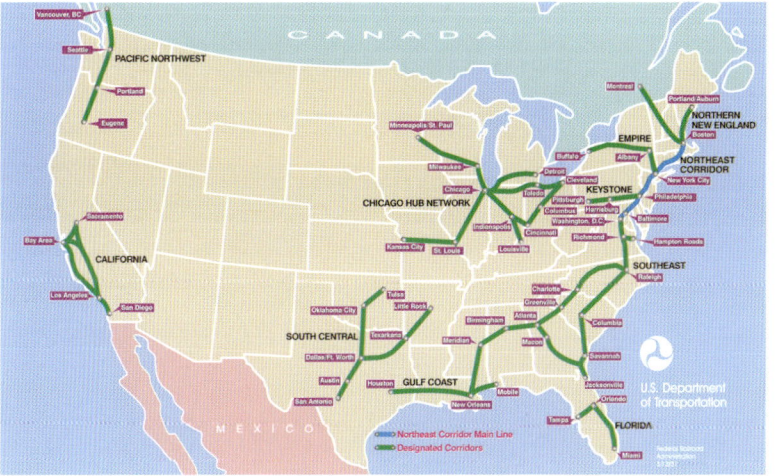

Figure 3.32 | **High-Speed Rail Corridor Designations**

In addition, most of the freight railroads are in the process of assessing revised alignments, often involving tunneling to reduce grades and shorten existing routes. Consequently, the future appears to be fairly optimistic with regard to the design and construction of ever longer and larger railroad tunnels.

GROUND-BREAKING TUNNELS

Several notable tunnels epitomized the evolution of ground-breaking technology that made the construction of more than 1,000 railroad tunnels possible in a 100-year period. It is estimated that about 500 of these tunnels have been abandoned or converted to hiking and biking trails. During construction, unique issues were encountered that provided innovative solutions for completing these tunnels, without which modern railroading would not have been possible.

BLACK ROCK AND FLAT ROCK TUNNELS—Phoenixville and West Manayunk, Pennsylvania

The Black Rock Tunnel, constructed from 1835 to 1837, was the first true railroad tunnel (using railroad locomotives and freight or passenger cars on rail), and the Flat Rock Tunnel, constructed in 1840, was the ninth railroad tunnel constructed in the United States. As noted previously in this chapter, the Staple Bend Tunnel, constructed from 1831 to 1833, was the first tunnel to include tracks on which barges on flatcars were transported through the Appalachian Mountains from one canal to another, but the motive power needed to move the wheel-mounted barges up and down 10 stretches of steeply inclined track was provided by winches rather than with self-propelled locomotives. The Black Rock and Flat Rock Tunnels were primarily intended for transporting coal from the mines to major cities. For this purpose, the Philadelphia & Reading Railroad (P&R) developed the first robust iron coal cars and also developed its own port and coal transport shipping fleet (HAER 1993c, 1993d). Both tunnels were widened from 19 feet to about 23 feet in 1859 for double-track operation and to handle wider gauge track and wider cars. It is likely that these tunnels have been enlarged at least a few other times as larger engines and freight and passenger cars were developed. The most recent enlargement for clearance improvement for double-stack container cars was in 1994.

The railroad was in direct competition with the barge and canal system. This competition prompted numerous acts of vandalism on both sides, including the burning of two railroad bridges near Mill Creek in 1842. Railroads eventually proved to be faster to build and more cost effective and efficient to operate than the canals.

The P&R developed and implemented a number of major innovations in the railroad and transportation industry. It was the first to design and build an iron railway bridge (1845), the first to use stone ballast for rail track beds (1836), the first with a double-track rail line (1843), the first to use iron coal cars (1843), the first to build an armored car for use in the Civil War (1862), the first railroad to operate its own fleet of coal transport ships (1870), and the first to install a fixed signal system (1886). By 1871 the P&R had grown to be the largest corporation in the world and, despite several downturns in the railroad industry, remained one of the top 10 largest freight carriers in the United States through the 1950s (HAER 1993c, 1993d).

Black Rock Tunnel is currently 1,925 feet long, 23 feet high at centerline above the top of rail, and 25 to 27 feet wide after several phases of enlargement to meet railcar clearance requirements. The tunnel is roughly horseshoe shaped, but with a flat crown due to the near horizontally bedded sedimentary rock. The tunnel is located on the northeast edge of the town of Phoenixville, approximately 30 miles west of Philadelphia, and was constructed in relatively massive, flat-lying limestone and argillite. It is the first tunnel in which shafts were excavated down to tunnel grade to enable multiple headings to be advanced and also to provide positive survey control from the ground surface. Most of the tunnel is unlined, except for the portal areas. The north portal is lined for 101 feet and south portal for 7.5 feet. Six short sections of 17 to 35 feet long are lined with stone masonry and concrete for a total lined length of 252 feet. Most of the lining was likely installed in 1894; however, rehabilitation and clearance improvements in 1994 resulted in the current lengths of lined tunnel and new portals (Shannon & Wilson 1993).

Flat Rock Tunnel is currently 901 feet long, 23 feet high above top of rail, and 26 to 28 feet wide. It is located in West Manayunk on the west side of the Schuylkill River, just west of Philadelphia. This tunnel is also horseshoe shaped and was constructed in relatively massive gneiss and schist. Most of the tunnel is unlined except for a total of 35 feet at the portals that are lined with stone masonry and concrete.

Tunnel Construction

Both tunnels were constructed by hand drilling 18-inch to 2-foot-deep holes and filling them with black powder in paper cylinders. Both tunnels were also originally constructed for single-track operation. The Black Rock Tunnelced at the rate of 40 feet per month from the two portals. It was the first tunnel in the United States to use intermediate shafts to provide the ability to excavate the tunnel from more headings than just the two portals. Five shafts were constructed to allow for 10 more excavation headings that were advanced at the average rate of 33 feet per month. The shafts were also used for survey control and ventilation once the tunnel had holed through. Total cost for construction of the Black Rock Tunnel was $178,992 (Drinker 1893) and was started by James Appleton Construction and completed by Mr. O'Moriarty. The cost for construction of the Flat Rock Tunnel was estimated by W.J. Nicolls (1897) at $130 per foot, or about $117,130.

In 1859 both tunnels were enlarged from the original width of 19 feet to a width of 23 to 24 feet. Widening involved essentially the same construction methods, except that electric detonation with a galvanic battery had been introduced and was used, with the black powder delivered and inserted into the drill holes in tin cartridges to keep the powder dry, enhance ignition, and reduce the number of misfires (Drinker 1893).

Figure 3.33 | **Old Clearance Notches Cut into Lined Section of Black Rock Tunnel**

Long-Term Performance, Upgrades, and Clearance Improvements

The tunnel widths and heights were increased several times to accommodate the increasing width and height of passenger and freight cars and locomotives with clearance enlargements in 1859, 1889, and 1936. During one of these enlargements, the inverts were lowered in both tunnels by a foot or more, clearance notches were cut into the upper sidewalls and quarter arches of the lined portions of Black Rock Tunnel (Figure 3.33) for single-track operations, and new portals were constructed on both tunnels. The flat arch in the Black Rock Tunnel had been supported with mechanically anchored bolts and steel straps.

In 1994 both tunnels were enlarged to accommodate double-stack container cars. Black Rock Tunnel was shortened by 6.5 feet, and the exposed rock portal shown on Figure 3.34 was covered with a formed concrete portal designed in the style of the original rock portal to stabilize the rock surface, as shown in Figure 3.35. The Flat Rock Tunnel was shortened by 37 feet during the most recent portal reconstruction (Figure 3.36). To gain the necessary clearance for double-stack container cars, the arches of both tunnels were smooth-wall blasted and re-supported with resin-grouted rock bolts and shotcrete. The five original construction shafts at the Black Rock Tunnel were also sealed with 10-foot-thick plugs of reinforced concrete and rock rubble fill. Insulation shields were installed in the Flat Rock Tunnel at six locations, totaling about 250 feet of the tunnel length, to reduce the potential for the formation of ice during the colder winter periods, as can be seen at the edge of the portal structure in Figure 3.37.

Summary

The Black Rock and Flat Rock Tunnels comprised the first and ninth true railroad tunnels constructed in the United States. They were constructed at a time when the sizes and configurations of railroad engines and the cars they pulled were undergoing rapid changes as the technology advanced. Consequently, it is quite an accomplishment that these and many other pre-1900s tunnels are still in use. Nearly all of these early tunnels have undergone several phases of enlargement and re-support to accommodate the changing nature and configuration of the rail loads. The introduction of new construction materials over the last 150 years, such as steel, shotcrete, grouted rock bolts, and chemical and cement grouts as well as new techniques and technologies for the enlargement and re-support for these old freight tunnels, has greatly enhanced the tunneling industry's ability to recycle and reuse tunnels that would otherwise have likely been abandoned.

Figure 3.34 | **Unsupported Rock Portal in Horizontally Bedded Sedimentary Rock of Black Rock Tunnel**

Figure 3.36 | **Formwork at North Portal of Flat Rock Tunnel in 1994**

Figure 3.37 | **Reconstructed North Portal of Flat Rock Tunnel with Icing Shield Visible Inside Tunnel in 1994**

Figure 3.35 | Reconstructed Concrete Portal of Black Rock Tunnel in 1994

ALLEGHENY, GALLITZIN, AND NEW PORTAGE TUNNELS—Gallitzin, Pennsylvania

The New Portage Tunnel, along with the neighboring Allegheny Tunnel (initially called the Summit Tunnel), were both completed in the 1850s (HAER 1993a). The New Portage Tunnel was constructed by the Commonwealth of Pennsylvania to provide a second tunnel to augment the Staple Bend Tunnel on the Portage Allegheny Railroad that connected a network of canals and rivers to the east and west of the Allegheny Mountains, enabling travel and shipping by barge from Philadelphia to Pittsburgh, a distance of more than 400 miles. The Summit Tunnel (now called the Allegheny Tunnel) was constructed by the Pennsylvania Railway in direct competition with and to replace the canal transportation system and to provide a 2,167-foot elevation crossing of the Allegheny Mountains (Drinker 1893). Although constructed as a single-track tunnel in the early 1850s, the Allegheny Tunnel was enlarged in the mid-1990s to accommodate twin tracks, and both the Allegheny and New Portage tunnels were enlarged to accommodate double-stack freight cars (Figure 3.38). These two tunnels provided a critical link through the mountains to the west and, therefore, were carefully guarded by the railroad police during the Civil War against the Confederates, since the Union Army was effectively using the rail lines to move troops and supplies, and during World War II when German agents were captured on the Atlantic Coast on the way to sabotage these tunnels. A third tunnel, the single-tracked Gallitzin Tunnel, was completed in 1904 (HAER 1993b).

Tunnel Construction

The 3,605-foot-long single-track Allegheny Tunnel was constructed from two central shafts and both portals. The 1,625-foot long New Portage Tunnel was constructed from both portals. The Allegheny Tunnel was excavated as an 8-foot-high by 20-foot-wide top heading and bench with timber support. The arch excavation was ultimately widened to about 30 feet to provide a clear width of 24.5 feet and height of 22 feet. The narrower New Portage Tunnel was constructed as a top heading and bench excavation. The Gallitzin Tunnel was constructed utilizing pneumatic drilling and dynamite with electric blasting initiation.

Figure 3.38 | **Installation of Icing Shields in Enlarged Allegheny Tunnel, 1995**

Geological-Geotechnical Conditions

The three tunnels were constructed through the crest of the Appalachian Mountains in near horizontally bedded sandstone, underclay, coal, slate, and limestone. The tunnels passed through 18 different strata and several sections of fractured, unstable rock that required substantially increased support. The coal was about 4.5 feet thick and was located about 20 feet above tunnel crown at the east portals of the Allegheny and Gallitzin Tunnels, and dropped to below invert about 1,800 feet east of the west portals. Tunnel spoils may have been excavated and removed by steam-driven, coal-fired equipment that necessitated the partial excavation of the coal seam between the tunnels, leaving a 10- to 15-foot-wide pillar of coal in the sidewalls of all three tunnels.

Construction Method

Excavation and support for the Allegheny Tunnel was accomplished by 400 Irish immigrants from County Cork by hammering in 4-foot-long drill steel and loading holes with ordinary black blasting powder (Drinker 1893). The average tunnel excavation and support rate was about 30 feet per week with a maximum rate of 127 feet per month with a flat-arched top heading and lower bench, and a final arched crown. All but 900 feet of the tunnel was supported with temporary timbers. The underclay layer tended to swell and slake. High train-traffic demands required that the tunnel arch be excavated above the temporary timber support and the final brick arch lining and sandstone ashlar masonry stone walls be constructed under live traffic conditions.

Very little information is available on construction of the New Portage Tunnel. However, it was constructed through essentially the same geological conditions, within a couple of years of the Allegheny Tunnel, and therefore was excavated with the same drill-and-blast techniques and used timber sets and lagging for initial support.

The Gallitzin Tunnel, constructed 80 feet center-to-center to the south of the Allegheny Tunnel between 1902 and 1904, involved pneumatic drills, modern blasting, and a locomotive-powered muck train. One blast near the portal showered the central part of town with large rocks, killing one person and injuring another. Blasting also caused damage to the school building directly above the tunnel, which prompted the construction of a new school in 1906.

Construction Progress and Improvements

The Allegheny Tunnel was initially constructed over a period of 30 months at an average rate of advance of about 120 feet per month, utilizing a tunneling staff of more than 400 laborers and as many as 10 headings from the two portals and four shafts. By comparison, the Allegheny Tunnel was later nearly doubled in size with 90 workers over a period of about 1 year (HAER 1993a).

The New Portage Tunnel was enlarged in 1993 to accommodate double-stack container cars by grinding down the tunnel invert using a large tracked rotary grinder. The Allegheny Tunnel was enlarged from 1993 to 1995 to accommodate the twin tracks of double-stack freight cars (Figures 3.39 and 3.40). Enlargement of the Allegheny Tunnel was accomplished with drill jumbos, hydraulic hoe rams, and blasting cartridges with nonelectric blasting caps, with support provided by fully resin-grouted steel rock dowels, steel ribs, and steel fiber-reinforced shotcrete.

The Gallitzin Tunnel was constructed about 50 years after the neighboring Allegheny Tunnel at an average rate of 50 feet per week, with no intermediate shafts (HAER 1993b). Consequently, construction in this tunnel is reported to have advanced at more than three times the rate of advance in the Allegheny Tunnel. However, the Gallitzin Tunnel was not enlarged and was taken out of service in 1995 after the improvements to the other two tunnels.

Figure 3.39 | **Allegheny Tunnel Enlargement Showing Steel Ribs**

Figure 3.40 | **West Portal of Allegheny Tunnel After Enlargement in 1995**

HOOSAC TUNNEL— Hoosac Mountain, Massachusetts

The Hoosac Tunnel in Western Massachusetts is one of the most important tunneling projects ever accomplished. Largely as a result of its size and its timing at the beginning of the industrial revolution, a significant number of technological and construction management practices came into being for this project that set the stage for rock tunnel construction throughout the world for more than 100 years. To quote Gosta E. Sandstrom from his famous book about tunnel history published in 1963: "Out of the Hoosac mess would ascend the American compressed-air industry which took an unchallenged leadership in developing and providing the mining and construction industries with the only types of tools and machines capable of mechanized work underground." It was not until the late 1960s that tunnel boring machines became a viable option for rock tunnel construction as compared to what was accomplished at the Hoosac Tunnel.

During the 1850s, the State of Massachusetts wanted access to the enormous economic developments taking place in Upstate New York largely as a result of the highly successful Erie Canal. Upon completion of the Erie Canal in 1825, all economic development in Upstate New York was being directed to the Hudson River and, as a result, fueling the economy of New York City. In an effort to capture a piece of that commerce, a plan was developed by the merchants and politicians in Massachusetts to build a tunnel through Hoosac Mountain that would provide access to the Hudson River near Troy, New York.

Early in this planning effort, it became obvious that the required tunnel would be very large by contemporary standards, as shown in Figure 3.41. As proposed, the Hoosac Tunnel would be 26 feet in diameter, almost 5 miles long, and would have anywhere from 1,000 to more than 1,200 feet of rock cover above the tunnel crown. In addition, controversies developed about how the tunnel should be contracted and who would pay for it. Early attempts to

Figure 3.41 | **Hoosac Tunnel Plans**

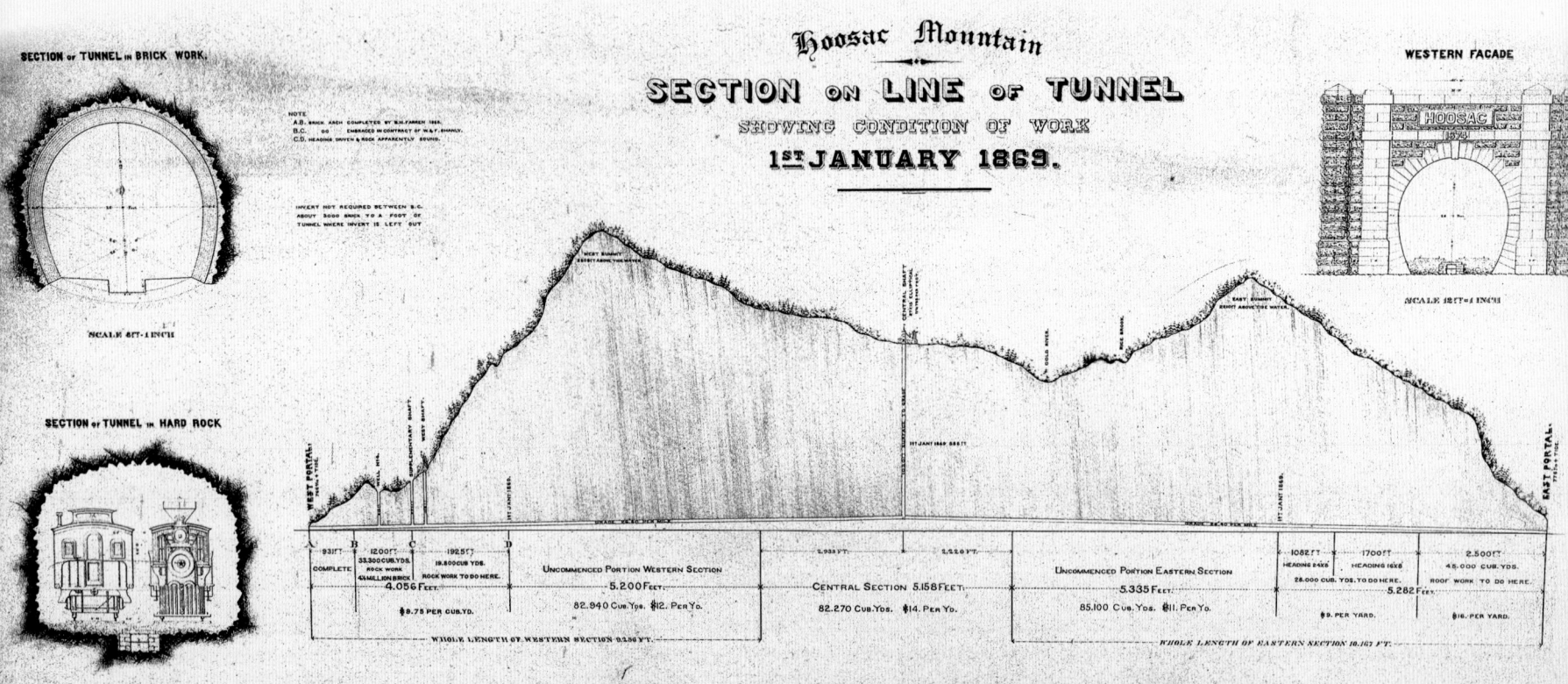

build the tunnel resulted in bankruptcies and extremely contentious political battles in the Massachusetts state legislature. In addition, the Civil War greatly complicated construction activities at the tunnel that further delayed tunnel progress.

Construction Improvements

Finally in 1868, the State of Massachusetts awarded a contract for $4.5 million to the Shanley Brothers from Montreal, Quebec, at which time construction activities began in earnest. Rock tunnels constructed before 1868 involved a beehive of activity with workers throughout the tunnel drilling holes, loading and igniting black powder, and removing broken rock. If loose rock was encountered, it was supported by heavy wood bracing followed by a thick brick lining. In addition, all work in the tunnel would have been performed solely by human beings and mules because there was no other source of power available to work deep underground.

Figure 3.42 | **Drill Carriage**

The Shanleys immediately realized that the only way to increase the rate of advance of the tunnel was to excavate and support the top heading as quickly as possible. Once the top heading was completed, it was relatively easy for other portions of the tunnel to be finished. The two most important keys to success relative to advancing the top heading were drilling holes and blasting rock.

The Shanleys became convinced early on that compressed air could be used as a power source for drilling rock. Many scientists at the time believed that compressed air would lose its "elastic force" if transmitted for long distances through small-diameter pipes. However, this was shown not to be the case, and research began immediately for developing air compressors and compressed-air rock drills, of which both developments had to overcome numerous technological challenges. Eventually Charles Burleigh did invent a drill that could be used underground in hard rock. Since each of these drills weighed several hundred pounds, it became necessary to develop a "drill carriage," as shown in Figure 3.42, to support the drills that could be moved to and from the face on rails. Figure 3.43 is a photograph of one of the tunnel workers actually operating a drill at the tunnel heading.

With this equipment in place, the Shanleys also introduced the drill/load/blast/muck cycle of rock tunneling. Instead of a busy-beehive atmosphere, all workers at the heading were employed in stages drilling holes, loading explosives, detonating the blast, and then removing the broken rock. This cycle was then repeated as quickly as possible for each round and resulted in an increase in the rate of heading advance of more than five times as compared to the ancient methods. Interestingly, the compressed air from the rock drills was also used to help ventilate the face and provide fresh air for the workers.

Figure 3.43 | **Tunnel Worker Operating a Drill at Tunnel Heading**

Another huge improvement at the Hoosac Tunnel was the introduction of nitroglycerine as the blasting agent. Since black powder was barely able to fracture the hard granitic rock along the Hoosac Tunnel, the Shanleys hired George Mowbray to construct a nitroglycerine manufacturing facility at the west portal of the tunnel, as shown in Figure 3.44. Initially, numerous accidents occurred using this highly volatile explosive compound, but the workers learned to use the nitroglycerine safely and appreciated that it could fracture a great deal more rock with fewer fumes. Since the Hoosac Tunnel was being built before the invention of dynamite, the workers actually placed tin cans filled with liquid nitroglycerine into the drill holes, which were then detonated using electric blasting caps.

To increase production, the Shanleys also constructed a 1,000-foot-deep shaft near the center of the tunnel to provide two additional working faces. The head house for this shaft is shown in Figure 3.45. To remove rock from this shaft, the Otis Elevator Company developed a 1,000-foot-long hoisting apparatus that was unprecedented at that time. During one annual report, it was estimated that this hoisting apparatus actually removed ten times the weight of water from the shaft as rock since large quantities of water were also loaded into the hoisting buckets.

Construction Challenges

Numerous other problems also had to be overcome at the Hoosac project, including tunneling through a massive thrust fault near the west portal and being able to accurately survey a 5-mile-long tunnel. At that time it was expected that a tunnel heading for that length of tunnel would only be accurate to within several feet.

Figure 3.44 | **West View of Nitroglycerin Factory**

Figure 3.45 | **Head House at Hoosac Tunnel**

However, the Hoosac surveyors brought the headings to completion within a few inches. Even more interesting is that two tunnel boring machines were invented and used at the Hoosac project.

Although other projects around the world were experimenting with various aspects of rock tunneling during this period, it was the Hoosac Tunnel project that brought all the characteristics of drill-and-blast rock technology together on one project. Hence, as pointed out by Sandstrom (1963), it is correct to credit the Hoosac Tunnel as having established the demarcation from the ancient to the "modern" drill-and-blast tunneling practices that are still in use today.

DONNER PASS TUNNELS—
Nevada County, California

Original Construction from 1866 to 1868

After the passing of the Pacific Railway Act of 1862, which provided for the construction of a (transcontinental) railroad line from Missouri to the Pacific Ocean, the Central Pacific Railroad was authorized to build the western portion of the line from the Pacific Coast to the California state line and, eventually, farther eastward until meeting up with Union Pacific's construction that was advancing westward from Nebraska. Central Pacific's portion included 150 miles of rugged mountain terrain in the Sierra Nevada and, ultimately, through the longest tunnel on the alignment at the time, the 1,659-foot-long Tunnel 6 ("Summit Tunnel") across Donner Pass at an elevation of 7,056 feet.

The original railroad line, which was essentially planned by Theodore D. Judah and built under the supervision of his former assistant, Lewis M. Clemente (Judah passed away in 1863) included Tunnels 1 through 15 and numerous snow sheds and galleries, which were constructed essentially within a 24-month period between 1866 and 1868 (Figure 3.46).

Tunnel 0 was added in 1873 to bypass a 90-foot-high trestle in the foothills. This first line across the mountains in the West would later become the "No. 1 Track" after the construction of a second track that in part paralleled the original track or was built on an entirely new alignment.

It is estimated that during the full swing of construction, it took a labor force of perhaps as many as 9,000 men to build the railroad line in the mountains. Almost all were from China since the white laborers, who were mostly miners, persistently left the workforce and used their employment with the railroad to get transportation over the mountains to pursue a stake in prospecting for gold (Signor 1985).

The Chinese were willing to perform the dangerous and often deadly blasting and excavation work for the required rock cuts and tunnels. Most of the cuts and tunnels were constructed in various granitic rocks of the Sierra Nevada batholith. These consisted of more than 100 separate granite plutons of Mesozoic Age that intruded the earth's crust between about 225 and 80 million years ago. Blasting of the extremely hard granite with black powder proved to be very slow and difficult, with many instances where "explosives shot out of the holes without disturbing the solidity of the surrounding rock" (Signor 1985). The use of newly invented nitroglycerin for the construction of Tunnels 6 and 8 improved the blasting results significantly and almost doubled daily progress rates in the hard rock, from 2.5 feet per day with black powder to 4.4 feet with nitroglycerin, according to Gilliss (1872). But its use was abandoned after several fatal accidents due to the high risks of handling the highly explosive material. The tunnels were typically excavated from two headings starting at the portals (Figure 3.47). Only at Tunnel 6 was a shaft driven at the center of the tunnel to add two additional headings and thereby increase production rates in the hard granitic bedrock.

Because of the high strength and the generally favorable properties of the granite for tunnel stability, most of the tunnels only required temporary timber support at the portals during excavation and otherwise were unsupported, with the exception of a few tunnels that required permanent timber support where the rock was of poorer quality.

Figure 3.46 | **Erecting Snow Sheds at the East Approach of Tunnel 10**

Construction of the No. 2 Track— 1909 to 1925

Soon after the two rail lines built by the Central Pacific and the Union Pacific met at Promontory Summit, Utah, on May 10, 1869, with the ceremonial driving of the "Last Spike" for the first transcontinental railway, plans were underway to construct a second track at an "easy" grade across the mountains, and thereby eliminate a major bottleneck on the alignment. Beginning in 1884, the track across Donner Pass was leased to the Southern Pacific Company (SPRR), which itself eventually came under the control of the Union Pacific (UPRR) in 1901. (Note that this merger predates the more recent merger between these two railroad giants in 1996. The 1901 merger was subsequently dissolved by an anti-trust suit brought by the federal government in 1912 against both companies, also leading eventually to the Supreme Court–ordered separation of the Central Pacific from the Southern Pacific in 1914). Edward H. Harriman was a skillful financier and railway executive who came from the Union Pacific. It was during the first merger attempt in 1901, under Harriman's administration, when the Southern Pacific applied a visionary and ambitious program to coordinate and fund the modernization of its operations, including equipment and the construction and reconstruction of rail lines. Consequently, work on a second track, the so-called "No. 2 Track," eventually started in the spring of 1909, double-tracking the section between the cities of Rocklin and Colfax. This new line initially required the construction of 20 new tunnels, including Tunnels 15 through 34, which were completed by 1913. The final push for the completion of the double-track line across the rugged terrain of the summit of the Sierra Nevada occurred in the early to mid-1920s after the Southern Pacific prevailed in 1923 against the government's required breakup of SPRR and UPRR in 1914 and again regained control over the Central Pacific. Double-tracking the line continued by enlarging the original Tunnel 1 to provide for a second track, and constructing seven additional tunnels across the Sierra summit from 1924 to 1925. This included a new summit tunnel—Tunnel 41 or "The Big Hole"—that was almost 2 miles in length and the longest of the new tunnels (and the third longest in the continental United States at the time). The new "No. 2 Track" was completed and ready for operation on October 15, 1925.

Figure 3.47 | **Granitic Bedrock of Tunnel 8, East Portal**

During construction of the No. 2 Track, concrete was the preferred construction material. It was used for supporting the new tunnels where needed and to replace existing timber-lined sections and wooden portal structures in tunnels along the No. 1 Track. Concrete was also used for the construction of new snow sheds and to slowly replace many miles of original wooden snow sheds that were established to protect the line from the heavy snowpacks and day-long blizzards during the harsh winters. About 23 miles of snow sheds were in place when the second track across the Sierra Nevada was completed.

Clearance Improvements in 1967/1968

Since their construction, the tunnels had remained largely unchanged from the standards and dimensions required at the time they were built. The first clearance improvements were undertaken in 1967/1968. The clearance

Figure 3.48 | **Lowering the Track in a Snow Shed**

improvement program involved tunnels on the original No. 1 Track, and the enlargements were typically accomplished by excavations in the quarter arches and crowns of the tunnels. Only in Tunnel 6 at the summit of the pass was the additional clearance gained by lowering the invert of the tunnel approximately 3 feet (Figure 3.48).

The Long Way to Clearance Improvements in 2009

The rail line was under the control of the SPRR when considerations to enlarge the tunnels across Donner Pass for the largest double-stack container cars led to the first geotechnical studies and exploration programs in the late 1980s to accomplish this. Then in 1993, due to difficult and costly maintenance primarily during the harsh winter months, the Southern Pacific closed the section of the original No. 1 Track that leads over the summit, thereby abandoning the famous Summit Tunnel (Figure 3.49) and the six other tunnels on this section of the line. Consequently, all eastbound and westbound train traffic was now routed through Tunnel 41, The Big Hole, with the railroad essentially giving up on plans for a continuous second track on the alignment.

The initial clearance improvement plans and studies were revisited by the new owner in 1998, two years after the merger between the Union Pacific and the Southern Pacific, but it took until 2009 when construction of the clearance improvements was finally accomplished. The clearance project involved a group of 14 tunnels, located on both the No. 1 and No. 2 Tracks, to provide passage for double-stack container cars and tri-level auto-carrier cars over the Sierra Nevada by switching back and forth between the two tracks, resulting in one cleared single line. The clearance improvements primarily included track realignment and notching of the quarter arches of the concrete-lined tunnels, and in a few places required the notching of existing steel sets and the blasting of rock tights in unlined tunnel sections (Figures 3.50–3.53). Because of the fire hazard presented by remaining timber sets spaced between concrete ribs, the timber sets were removed and replaced with shotcrete. Notching of the existing concrete liner required grout injection through the liner to fill gaps between the liner and the rock, and rock bolt installations for additional support in places where the liner was either very thin or deep notches were required for the clearance improvements. The project also included the rehabilitation of deteriorated concrete liner in the tunnels typically using steel fiber-reinforced shotcrete for the repairs.

Figure 3.49 | **Tunnel 6 After Closure**

It would require additional clearance improvements in a second group of 11 tunnels—located on the No. 2 Track—to reestablish a true double-tracked alignment across the mountains, except for the remaining single-tracked section through Tunnel 41 at the summit of Donner Pass. The eventual elimination of this single-track bottleneck would require reconstruction and clearance improvements along the approximately 3-mile-long section of the No. 1 Track—including Tunnels 6 through 12—which was abandoned in 1993. Table 3.3 lists the tunnels across Donner Pass.

Figure 3.50 | Removal of Remaining Timber Sets Between Concrete Ribs in Tunnel 17

Figure 3.51 | Notching of Steel Sets in Tunnel 20

Figure 3.52 | Using a Roadheader for Notching the Concrete Liner

Figure 3.53 | Completed Notches in Double-Tracked Tunnel 18

Table 3.3 Summary of Tunnels on the Rail Line Across Donner Pass

Tunnel No.	Track No.	Year Built	Original Length (ft)	Lining	1967/1968 Clearance Improvement	2009 Clearance Improvement	Comments
0	1	1873	711	Masonry			Abandoned in 1942
1	1	1866	514	Concrete		Yes	Widened for a second track during the double tracking of the line between 1909 and 1925
2	1	1866	271	n/a			Daylighted during the double tracking of the line between 1909 and 1925
3	1	1866–1868	280	Unlined			
4	1	1866–1868	92	Unlined		Yes	
5	1	1866–1868	128	Unlined			Daylighted in 1895
6	1	1866–1868	1,659	Unlined	Yes		"Summit Tunnel"; invert was lowered by approximately 3 feet during clearance improvements in 1967/1968; out of service after track removal in 1993
7	1	1866–1868	100	Unlined	Yes		Daylighted; out of service after track removal in 1993
8	1	1866–1868	375	n/a	Yes		Out of service after track removal in 1993
9	1	1866–1868	216	n/a	Yes		Out of service after track removal in 1993
10	1	1866–1868	509	n/a	Yes		Out of service after track removal in 1993
11	1	1866–1868	577	n/a	Yes		Out of service after track removal in 1993
12	1	1866–1868	342	n/a	Yes		Out of service after track removal in 1993
13	1	1866–1868	870	Concrete & brick		Yes	
14	1	1866–1868	200	n/a			Abandoned in 1913 during the double tracking of the line between 1909 and 1925
15	1	1866–1868	96	n/a			Daylighted in 1895; abandoned in 1913 during the double tracking of the line between 1909 and 1925
15	2	1912	1,904	Concrete		No work required	After abandoning original 15 tunnels on Track 1, numbering of new tunnels on Track 2 continued from last existing at the time on Track 1 (14)
16	2	1912	777	Concrete		No work required	
17	2	1912	1,648	Concrete, gunite, timber sets with concrete ribs		Yes	
18	2	1912	1,000	Concrete		Yes	Double-tracked
19	2	1912	377	n/a			Daylighted between 1974 and 1976
20	2	1912	1,248	Gunite, gunite with steel sets		Yes	

Tunnel No.	Track No.	Year Built	Original Length (ft)	Lining	1967/1968 Clearance Improvement	2009 Clearance Improvement	Comments
21	2	1912	1,210	Concrete, gunite, gunite with steel sets, timber sets with concrete ribs		Yes	
22	2	1912	984	Gunite, gunite with steel sets			
23	2	1912	843	Concrete			Next to abandoned Tunnel 0 on Track 1
24	2	1912	300	Concrete			
25	2	1912	771	Concrete			
26	2	1912	149	Concrete			
27	2	1912	855	Concrete, timber sets with concrete ribs			Partially daylighted and shortened to 686 feet between 1959 and 1968
28	2	1912	3,208	Concrete			
29	2	1912	1,009	Concrete, timber sets with concrete ribs			
30	2	1912	780	Concrete, gunite			
31	2	1912	443	Concrete, gunite with steel sets			
32	2	1912	769	Concrete			
33	2	1912	1,331	Concrete		No work required	Double-tracked
34	2	1913	410	Concrete		Yes	Double-tracked
35	2	1924	737	Concrete, unlined		No work required	
36	2	1924–1925	325	Concrete		Yes	
37	2	1924–1925	410	Concrete, unlined		Yes	
38	2	1924–1925	920	Concrete, unlined		Yes	
39	2	1924–1925	279	Concrete		Yes	
40	2	1924–1925	315	Concrete, unlined			Daylighted between 1976–78
41	2	1924–1925	10,325	Concrete, unlined		Yes	"The Big Hole"; longest tunnel of the alignment; third longest tunnel of the continental United States at the time
42	2	1924–1925	892	Concrete		Yes	

Courtesy of Klaus Winkler, Shannon & Wilson, Inc.

TEHACHAPI TUNNELS— California

The rail route across the Tehachapi Mountains was built by Southern Pacific between 1875 and 1876 to link existing transcontinental rail lines to the north and to link the San Joaquin Valley eventually with Los Angeles in the south, as located on Figure 3.2. The 28-mile stretch climbed nearly 3,000 feet, leading from the valley floor of the San Joaquin Valley near Caliente to the top of Tehachapi Pass at an elevation of 4,025 feet. It curved along steep and narrow canyon walls, passing through 17 tunnels, and including, arguably, the most famous portion of the route, the Tehachapi Loop (see Figure 3.9). The Tehachapi Loop is where the railroad crosses over itself above the 428-foot-long Tunnel 9, which was designed by Southern Pacific's civil engineer William Hood and is considered one of the greatest engineering accomplishments of the time.

Seventeen Tunnels

The segment between Caliente in the north and Tehachapi was built at a ruling grade of 2.2% and, to accomplish this, it required the excavation of 17 single-track tunnels ranging in length from 110 feet (Tunnel 11) to 1,175 feet (Tunnel 5), as shown in Figure 3.54, in addition to the construction of the famous Tehachapi Loop. Construction on the route through the north slope of the Tehachapi Mountains began in the spring of 1875. The segment including Tunnels 1 through 5 was completed on April 6, 1875. Two month later, construction reached the approach of the Tehachapi Loop, thereby completing three more tunnels, Tunnels 6 through 8. After completing Tunnel 9 at the loop in June of that same year, eight more tunnels (Tunnels 10 through 17) were required to complete the remaining 10-mile-long section to reach the station of Tehachapi Summit near the top of the pass. On July 10, 1876, the first trains began operating the new line to this point.

Initial Construction and General Ground Conditions

Between Caliente and Tehachapi, bedrock underlying the alignment generally consisted of Paleozoic and Mesozoic plutonic and metasedimentary rocks. All the tunnels except Tunnels 7, 8, 10, and 11 were excavated through "soft rock" (Signor 1983), apparently including decomposed and "brittle" granitic rock and metasedimentary rock comprised of interbedded schist, limestone, dolomite, and quartzite, as opposed to the "hard granitic" rock encountered in the aforementioned four tunnels (Figure 3.55).

The tunnels were excavated by the drill-and-blast method and mucked by means of picks and shovels, and horse-drawn carts. The tunnels were 22 feet high and trapezoidal in cross section with a 16.5-foot-wide bottom and an 18-foot-wide top (Hansen 2010). Ten of the 17 tunnels required temporary timber support during construction; later, a permanent timber lining was installed in all of the tunnels.

Soon after the 1915 devastating tunnel fire in Tunnel 15, a program was undertaken to replace the original timber lining with a reinforced cast-in-place concrete lining in all of the tunnels on Tehachapi Pass (Figure 3.56).

When Disaster Struck—The 1952 Earthquake

The magnitude 7.3 Kern County earthquake (also called the "Tehachapi Quake" in various newspaper accounts) struck the line in 1952, shifting tracks and damaging portals and concrete linings in all of the tunnels. The earthquake occurred along the

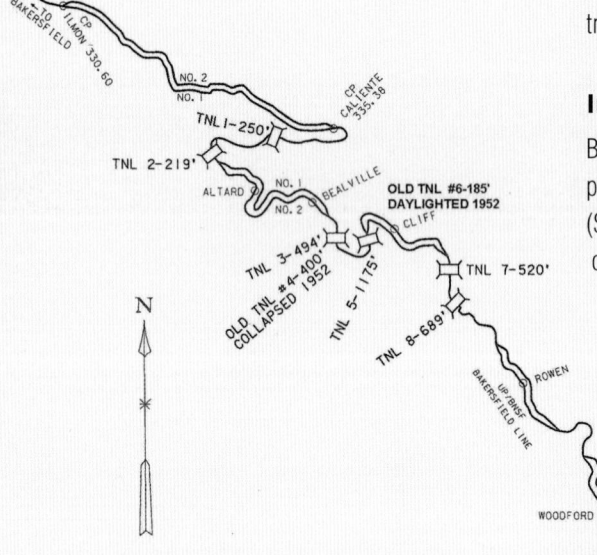

Figure 3.54 | **Tunnels on the Tehachapi Route**

Figure 3.55 | Southern Pacific Train Approaching Completed Tunnel 10

Figure 3.56 | Replacing the Original Timber Lining with a Rebar-Reinforced Cast-in-Place Concrete Liner

White Wolf Fault, which crosses the railroad alignment in the vicinity of Clear Creek Canyon, where the railroad follows the perimeter of the canyon. The heaviest damage occurred to this section of the rail line, including Tunnels 3, 4, 5, and 6. Of these four tunnels, Tunnel 5 experienced the most severe damage and required re-mining of large portions of the tunnel that had collapsed as a result of the earthquake (Figure 3.57). Earthquake-related landslides affected and severely damaged both portals of the tunnel. The entire tunnel was relined and new portal structures were constructed. Approximately 200 feet of tunnel, including the portal structure, were daylighted at the east end of Tunnel 3 following the earthquake. Tunnel 4 collapsed and was abandoned, and Tunnel 6, which in large portions had also collapsed, was daylighted as a result of the earthquake. Most of the other tunnels required substantial repairs because of shifted and damaged segments of the concrete liner.

Figure 3.57 | **Inspecting Severely Damaged Concrete Lining**

Clearance Improvements

Because of increased demand for higher freight volumes and the implementation of double-stack container cars and tri-level auto-carrier cars, tunnel clearance improvements were authorized by Southern Pacific at the beginning of the 1990s. Through a trial of several alternative methods for cutting the notches (concrete saw, hoe ram, roadheader, etc.), it was determined that a roadheader riding on a temporary work deck mounted to the top of a gondola car was the most efficient means for cutting the generally 3- to 12-inch-deep notches in the quarter arches. Clearance improvements to the tunnels, primarily consisting of track realignment and notching of the concrete liner quarter arches, were accomplished in 1991. Notching of the existing 1- to 2-foot-thick concrete liner required grout injection through the liner to fill gaps between the liner and the rock, and rock-bolt installations for additional support in places where the liner was either very thin or deep notches were required for the clearance improvements. In some of the tunnels, the invert was undercut and lowered to avoid cutting all the way through the existing liner during the notching process (Figure 3.58).

Summary

A joint track usage agreement for the use of the rail line across the Tehachapi Mountains was first signed between the Southern Pacific and the Santa Fe railroad companies in 1899. Today the successor companies, the Union Pacific Railroad and the Burlington Northern Santa Fe Railway, respectively, are still operating this railroad section jointly. From the 17 originally constructed tunnels, only 12 remain in service. Today's total aggregate length of the 12 tunnels is about 5,425 feet. Besides the aforementioned clearance improvements in the early 1990s, no significant rehabilitation work has been required or has been performed on the tunnels. In addition, the same concrete liners that were installed 100 years ago are still performing well today, except for Tunnel 5 which had to be relined after the 1952 earthquake.

Figure 3.58 | **Deep Notch in Concrete Liner of Tunnel 8 to Clear Double-Stack Container Cars**

WEEHAWKEN TUNNEL—Weehawken, New Jersey

Constructed in the 1880s, the Weehawken Tunnel was initially used for operation as a two-track passenger and freight rail line, providing access to the Hudson River waterfront where ferry service was provided to West 42nd Street in Manhattan. When passenger service was suspended in 1959, the tunnel then served freight trains, with Conrail running trains until 2002 on a single track along its 4,156-foot passageway. After 2002, the tunnel was repurposed by New Jersey (NJ) Transit with a new mid-tunnel passenger station to service several densely populated Hudson County cities to points south along the Hudson River on the Hudson-Bergen Light Rail Transit line.

Construction for the retrofitting started in June 2002 and was completed in December 2006. The design components included

- Demolition and reconstruction of portals;
- Enlarging and retrofitting the existing tunnel with dual trackways from the portals to the underground station;
- Construction of an underground station with a 280-foot-long center platform and shaft for rider access, emergency egress, ventilation, and utilities; and
- Construction of a passenger plaza, shaft head-house, ventilation stacks, and utility building at street level.

Figure 3.59 | **Three Headings in the Cavern Construction**

Tunnel Background and Geology

The Weehawken Tunnel traverses three rock units—the Stockton and Lockatong Formations of the Newark Supergroup and the Palisade Diabase. The near-vertical jointing patterns in the Palisade Diabase impart the characteristic columns associated with the cliff face. The tunnel was constructed using drill-and-blast methods simultaneously from each portal and in each direction from five 8-foot by 16-foot construction shafts.

Approximately 70% of the tunnel was unlined, meaning the rock was exposed and unsupported. The remaining sections of the tunnel were lined with brick arches and ashlar (stone) masonry sidewalls at abandoned construction shaft locations, fault zones, and other areas of poor rock mass conditions. The tunnel varied in width from about 26 feet wide at lined sections to 32 feet at unlined areas and ranged in height from 20 to 24 feet.

Project Design Elements

At the station cavern, the tunnel was enlarged from about 27 feet wide and 22 feet high to 65 feet wide and 34 feet high. Rock excavation was conducted using smooth-wall blasting techniques, with excavation sequences and maximum opening widths and round lengths specifically utilized to maintain opening stability. Rock support consisted of rock dowels with lengths that varied between 10 feet and 18 feet. The station cavern excavation was divided into three top headings and a bench for ease of excavation and ground support installation (Figure 3.59). Initial ground support shotcrete and smoothing shotcrete were placed using robotic means.

A waterproofing and drainage system was installed to minimize seepage into the station and to provide a drainage path for groundwater behind the final concrete lining, which consisted of 24 inches of reinforced cast-in-place concrete.

For the running tunnel, excavation was required to either widen or increase the height of the tunnel section for project requirements. When rock impinged within project tunnel section limits in the unlined sections, it was removed by smooth-wall blasting. Rock support in those areas included 10-foot-long rock dowels with a rock surface protection mesh when a cast-in-place liner was to be installed. However, in areas of good rock conditions, a rock surface protection net was used and no concrete liner was placed. For the brick-lined areas, excavation was completed using mechanical means with ground support installed after excavation (Figure 3.60).

Because of the availability of the tunnel below, the elevator/access shaft could be constructed using the raise bore and slash method (Ott and Jacobs 2003). Benefits of the raise bore included reduction of noise, dust, and vibrations at the station head-house area and elimination of the need to haul excavated material on the crowded and busy local streets. Additionally, it reduced vibration levels when the shaft was excavated due to the free face created by the raise bore. The raise was 8.25 feet in diameter followed by enlargement to 40 feet in diameter using drill-and-blast methods.

Rock support in the shaft excavation consisted of 10-foot rock dowels and a rock surface protection mesh. The shaft penetrated the station cavern excavation, creating a 40-foot-diameter opening within the 65-foot-wide cavern excavation (Figures 3.59 and 3.61).

The Weehawken Tunnel/Bergenline Avenue station has reestablished passenger rail service as originally intended more than 130 years ago (Figures 3.62 and 3.63).

Figure 3.60 | **Support at Brick-Lined Section**

Figure 3.61 | **Concrete Lined Shaft at Cavern Intersection**

Figure 3.62 | **New Bergenline Avenue Station at Track Level**

Figure 3.63 | **New East Portal of the Weehawken Tunnel**

FIRST ST. CLAIR TUNNEL— Michigan/Ontario

The St. Clair Tunnel under the St. Clair River between Port Huron, Michigan, and Sarnia, Ontario, was opened in September 1891. At 20 feet in diameter and 6,028 feet long in soft ground below a wide river, this tunnel would represent a remarkable engineering achievement even today (Figure 3.64).

Tunnel Construction

Joseph Hobson, the chief engineer of the St. Clair Tunnel Company, a subsidiary of the Grand Trunk Railway, took on the task of engineering and constructing the tunnel. Undaunted by abandoned attempts a few years earlier to a tunnel in similar ground conditions under the nearby Detroit River, Hobson proceeded in 1885 with geotechnical investigations comprising 11 boreholes across the river using 6-inch pipe driven down to the rock. The results of these borings indicated poor conditions for tunneling, but Hobson decided to sink small shafts on each side of the river, some 60 feet below the water surface, and to progress small drifts along the line of the tunnel. These headings advanced 20 feet in one case and 200 feet in the other before they were abandoned in July 1887 because of water and gas infiltration and quicksand conditions.

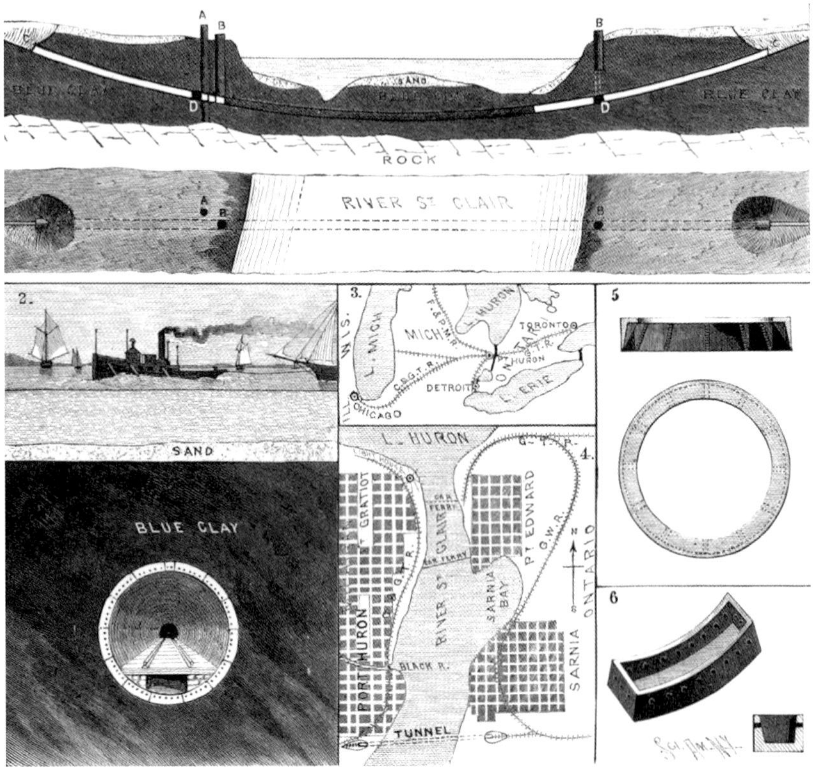

Tunnel Challenges

In spite of these failures, in the spring of 1888, Hobson and his managers decided to start the full-sized tunnel. A total of 110 borings were made at 20-feet centers along the alignment to establish the top of the clay layer that overlies the bedrock. The tunnel would be located within this clay layer to provide protection from the river above and from the gas in the rock below. Work was begun by constructing 23-foot-diameter brick-lined shafts by the sinking caisson method on each shoreline. The American shaft became stuck at about 30 feet deep and refused to sink further. Attempts to extend the shaft down were abandoned after reaching about 60 feet deep because of cracking in the lining. The Canadian shaft fared better in the sinking operation, reaching a depth of about 100 feet. However, at that point, settlement started to occur around the shaft and the blue clay began rising in the bottom of the shaft. The shaft began to lose its circular form and had to be backfilled to prevent collapse.

Incredibly, Hobson and his managers were not defeated and moved the construction operations to the portal areas, creating wide battered cuts. After large-scale slope failures, these cuts were made stable enough to proceed with the tunnel itself. Hobson had decided to use the recently developed shield method of tunneling, but he applied it at an unprecedented scale with two shields being manufactured at almost 21 feet 7 inches external diameter and 16 feet long (Figure 3.65). The shield faces were divided into 12 compartments, and each shield was propelled by 24 hydraulic rams with a total thrust capacity of 3,000 tons. The shields were equipped with crank-operated rotating arms to erect the cast-iron tunnel lining segments (Figure 3.66).

Figure 3.64 | **Tunnel Under the St. Clair River**

Figure 3.65 | **Tunneling Process for St. Clair Tunnel**

The tunnel shields were launched in July and September 1889 on the American and Canadian sides, respectively. The advance rates averaged 10 feet per day. For the under-river section, it was necessary to construct air locks within bulkheads near the shorelines and to apply compressed air within the tunnel. Pressures of up to 40 psi were used occasionally when pockets of gravel or quicksand were encountered to keep up with air loss that caused the river to "boil like a geyser" (*Engineering News* 1890). On these occasions it was always possible to plaster the face with clay to hold back the water long enough to get past the porous seam. Experience in working in compressed air was very limited, and decompression times were only determined by the time required to avoid the "benders, where a man's knees would wobble under him and he would bleed from the nose, mouth and ears" (Anonymous 1890). Haulage was by muck cars pulled by mules, since horses were not able to tolerate the compressed air.

Tunnel Progress

In August 1890, the two shields met under the river, on line within ¼ inch, and Hobson was able to be the first man to pass through the tunnel. The tunnel was opened to rail traffic in September 1891. The total cost of the work was $2.6 million, and approximately 700 workers were employed during its construction (Figure 3.67).

Figure 3.66 | **St. Clair Tunnel Shield**

Figure 3.67 | **Port Huron Portal, Circa 1907**

SECOND ST. CLAIR TUNNEL— Michigan/Ontario

Prior to the construction of the first tunnel, all freight was barged across the river. It was a slow, expensive process, which in winter was often impossible. After more than 100 years of service, the original tunnel was still structurally sound, with its bolted, cast-iron segmental lining still in excellent condition. But with a diameter of 20 feet, the tunnel was too small to accommodate the double-stacked container cars and other advances in the railroad's equipment that had developed over the years. The tunnel's bolted, cast-iron segmental lining is still in excellent condition today. The Canadian National Railway was again ferrying freight across the river. In 1991, a feasibility study identified several options to alleviate the shipping logjam at the St. Clair River, including

- Enlarging the existing tunnel,
- Building a bridge,
- Building an immersed-tube crossing, and
- Building either a deep-cover or shallow-cover parallel tunnel.

Recent advancements in tunneling technology led to the decision that a shallow-cover parallel tunnel would be the fastest and most economical solution.

Figure 3.68 | **TBM "Excalibore"**

Tunnel Construction

One hundred one years after the original tunnel was opened, Canadian National Railway reestablished the St. Clair Tunnel Company to oversee the design and construction of a new international railway tunnel, with Chief Engineer Duncan MacLennan at the helm. At 27.54 feet in internal diameter, the new tunnel would soon put an end to the costly and often dangerous task of floating freight across the swift St. Clair River. The St. Clair Tunnel Company set a time frame of 53 months for the design and construction of the project. Knowing that this timetable would be impossible to meet with conventional contracting methods, the St. Clair Tunnel Company developed a compressed schedule.

In March of 1992, final design for the new tunnel was begun. A technical committee was formed to consult on the most desirable design features for the tunnel boring machine (TBM). The panel consisted of experts from tunneling and engineering companies around the world. Although there were reportedly very few things that all of the experts agreed on, a consensus was reached, and St. Clair Tunnel Company delivered the specifications and a contract to Lovat Tunnel Equipment to build the TBM. With a diameter of 31.30 feet, it would be the largest pressurized face TBM ever employed in North America (Figure 3.68). In addition to the TBM, the St. Clair Tunnel Company also chose to procure the precast concrete segmental lining directly from a joint venture of experienced specialty precasting firms. Several of the other major, long lead-time items were also purchased directly (owner procured) and supplied to the contractor. It was believed that by using this Negotiated Compressed Schedule contracting method, the new tunnel could be opened 12 months earlier than would otherwise have been possible.

In early October 1992, tenders for the tunnel construction were submitted by four teams. Because of last-minute permit problems, however, actual construction was delayed until March 1, 1993. It is rumored that President Clinton signed off on the last permit while getting a haircut aboard Air Force One.

Tunnel Challenges

Not unlike the first St. Clair Tunnel, the second tunnel also faced many challenges. The geological challenges of the areas had not gone away. And in the 100-plus years since the first tunnel was constructed, the southern portion of Sarnia had grown into a sizable industrial center. Four hundred feet from the tunnel portal, the massive TBM passed a mere 18 inches below the main sanitary sewer for the City of Sarnia. Immediately after passing the sewer, the tunnel entered the Imperial Oil Company refinery property. After a short run beneath an open grassy area, the TBM passed under a three-story research laboratory. The separation between the laboratory and the tunnel was so small that the tunneling machine could actually be heard operating from the basement of the building. Because of the sensitive work taking place in the building, there was serious concern over what effect any ground movement might have. Again, constant monitoring was undertaken, and the TBM passed without any measurable heave or settlement—a true testament to the advances in tunneling in the mid-1990s (Figure 3.69).

Figure 3.69 | **Tunnel Section of Second St. Clair Tunnel**

Tunnel Progress

By this early stage in the project, the excavation was already advancing at an average rate 3.5 times faster than the original bore. A risk mitigation measure imposed by the railroad required the tunnel contractor to excavate below the river on a 24/7 basis. Over the remaining 70% of the tunnel excavation, the TBM was advanced at the incredible average rate of 57 feet per day—nearly six times that of the first St. Clair Tunnel. The best-day advance rate of 94 feet would have taken the miners and mules of 1890 two weeks to accomplish (Figure 3.70).

The Second St. Clair Tunnel claims several noteworthy accomplishments. Despite delays due to permitting and other issues associated with the creation of a new international border crossing, the railroad was still able to open the tunnel in early 1995—proof of the value of the Negotiated Compressed Schedule contracting model employed on the project. The fact that production rates such as those attained over the final 70% of the drive could be accomplished and sustained using precast concrete segments in a 30-foot-diameter tunnel should be recognized as evidence of both the quality of the machinery and the dedication of the people who worked with it. Finally, and perhaps most importantly, the demonstrated capability to efficiently construct a large-diameter, near-surface tunnel beneath sensitive structures while inducing minimal ground movements proved promising for the future of the earth pressure balance method of tunneling. The fact that the level of ground control demonstrated on the St. Clair River Project was achievable opened up new opportunities for tunneling in congested, urban areas. Several subsequent projects moved from the planning stage to actual construction based solely on the proven success of the Second St. Clair Tunnel.

Figure 3.70 | **Hole-Through of Second St. Clair Tunnel**

GREAT NORTHERN RAILWAY TUNNEL— Seattle, Washington

Great Northern Railway Tunnel No. 17, also known as both the Great Northern Tunnel and the Seattle Tunnel, was one of the earliest railroad tunnels constructed in the Pacific Northwest and was part of the northernmost United States transcontinental railroad corridors that extended from Seattle, Washington, to Saint Paul, Minnesota. The tunnel was constructed to relieve train-induced congestion and blockage of traffic to the waterfront in downtown Seattle and opened in 1905 (population of 150,000 at the time). The tunnel is 5,141 feet long, and about half of its length is beneath 4th Avenue with a soil cover ranging from 20 feet at the south portal to about 110 feet near midpoint at Spring Street. The north portal of the tunnel is between Virginia and Stewart Streets, and from this point, it runs southeast adjacent to Pike Street Market until it reaches 4th Avenue, under which it turns on a 4-degree right curve. The remaining part of the tunnel follows under 4th Avenue, turning south at Jefferson Street on another 4-degree right curve. The south portal is about two blocks beyond this curve, between Washington and Main Streets, and within less than two blocks of the new passenger station. Photographs taken during construction on the north and south portals are presented as Figures 3.71 and 3.72, respectively.

The tunnel and northwest rail line were subsequently incorporated into the Burlington Northern Railway in 1970, which then became part of the Burlington Northern Santa Fe (BNSF) Railway in 1996.

Tunnel Construction

The finished double-track tunnel was constructed as a 30-foot-wide by 28-foot-high horseshoe shape, making it the largest cross-section tunnel in the United States at the time. The tunnel was initially supported with timber and subsequently lined with concrete. A concrete slab was poured in the invert to prevent deterioration of the underlying soils and settlement of the tracks.

Excavation was through a wide range of soils, including sands and gravels, blue clay, and hard glacial till. Localized large water inflows were noted on a construction and geology profile. In the vicinity of 4th Avenue and Marion Street, the tunnel encountered a "prehistoric forest," with one tree measuring more than 3 feet in diameter embedded in the blue clay. The wood decomposed rapidly upon being exposed to air (*Railway Gazette* 1904).

Tunnel construction was accomplished by 1,000 men working continuously in three shifts from both the north and south portals. The tunnel was constructed using the multiple-drift or stacked-drift method consisting of seven small timber-lined 10- to 20-foot rectangular to trapezoidal drifts excavated around the perimeter of the horseshoe-shaped tunnel opening, with the timber support and concrete forms braced off the remaining core soils, as shown in Figure 3.73.

These drifts combined to form a 40-foot-wide by 32-foot-high excavated opening. The cast-in-place concrete walls were 4.5 feet thick at the base, 3.5 feet thick at springline, and 3.5 feet thick across the arch, forming a 30-foot-wide by 28-foot-high horseshoe-shaped tunnel, as shown on Figure 3.74. No construction records are available for this tunnel, although photos and drawings show the multiple-drift method with a large crew using shovels, picks, and wheelbarrows. Based on the reported schedule, tunneling began from the north portal with about 300 workers on April 4, 1903, with another 700 workers added to start from the south portal and to provide multiple shifts about two months later. When the two top headings met near the middle on October 26, 1904, the surveying proved to be accurate to within ¼ inch. Once the timber lining was completed, the 20-foot-wide soil core was excavated with a large electric-powered shovel, and the muck was removed with small rail-mounted muck cars with electric locomotives and electric-powered belt conveyors. The tunnel was completed, with the final concrete lining in place, just two

Figure 3.71 | **North Portal of Great Northern Tunnel with Timber for Tunnel Support**

months later. Electric hoists, concrete mixers, and ventilation fans were also used and drew power from the new Snoqualmie Falls Hydroelectric Plant, making this "one of the most complete applications of electricity to tunnel driving that has ever been made…" (*Railway Gazette* 1904). The tunnel was constructed without the use of compressed air, diesel- or gasoline-powered equipment, a shield, or any advanced dewatering methods.

Construction Impacts

City reports and a claim by the then new Seattle Public Library (Hussey et al. 1915) indicate that settlements of a few inches up to 3 feet occurred along the alignment over the few years following construction. Up to 4 inches of settlement was measured, resulting in tilting and cracking of the newly built Seattle Public Library Building and the nearby Lincoln Hotel, both of which were subsequently demolished and rebuilt. The library was located above the tunnel east of 4th Avenue, between Madison and Spring Streets and the hotel was located at the northwest corner of 1st and Pike. Surface settlement was documented to cause some damage to several other buildings, pavement, and utilities located above the tunnel. The subsidence likely occurred in two phases, with initial settlements occurring during construction, and long-term

Figure 3.72 | **South Portal**

Figure 3.73 | **Top Heading Excavation, Circa 1903**

settlements occurring due to rotting and compression of the roof support timbers and the wood blocking used to support the soil. To minimize the continued long-term settlements, an 80-foot deep, 4-foot by 6-foot shaft was excavated down to the crown of the tunnel, and an adit was excavated in both directions along most of the tunnel length to provide access to remove timber and fill the resulting cavity with concrete (Hussey et al. 1915).

Long-Term Performance, Upgrades, and Clearance Improvements

Over the last 110 years, very little damage has occurred or maintenance has been required for the concrete tunnel lining. Several tunnels have been constructed over or under the Great Northern Tunnel, including the Elliott Bay Interceptor Sewer Tunnel in the 1960s and the downtown Seattle bus tunnels in 1986. Construction of the twin 21-foot outside-diameter bus tunnels just 5 feet below the Seattle Tunnel invert required that the railroad tunnel be underpinned with jet grout columns in 1985, which may have been the first use of jet grouting with tunneling in the United States. For underpinning the railroad tunnel, 126 jet grout columns, each about 3 feet in diameter, were constructed from inside the railroad tunnel, as shown in Figure 3.75. These formed guide walls on either side of the future twin transit tunnels beneath the railroad tunnel invert slab, thus limiting settlements to less than 0.2 inch. Several major buildings, including the Rainier Tower and Benaroya Hall, have also been built over the tunnel with foundation elements such as piers, piles, shoring walls, and tiebacks extending down beside the tunnel. Although the Great Northern Tunnel was assessed before and after each major building and tunnel project, no indications of construction-related damage were observed.

The tunnel has also experienced three magnitude 6.5 to 7.2 earthquakes with no apparent structural damage, while the overlying 50-year-old double-deck Alaskan Way Viaduct suffered major damage during the 2001 earthquake.

In the early 1990s, an analysis was performed to assess possible means for improving clearances for double-stack container cars and tri-level automobile carriers. The very thick unreinforced concrete lining was deemed suitable for up to 9-inch-deep notches near the quarter arches to accommodate the larger freight cars. A roadheader was used to cut the notches, as can be seen in the 2005 photograph from the south portal shown in Figure 3.76.

Summary

In addition to being the widest and tallest tunnel in the United States when constructed, the Great Northern Tunnel was also one of the first tunnels constructed using electricity as a primary power source for excavation, muck hauling, ventilation, and lighting. After 110 years of use and only minor rehabilitation, including minor notching of the quarter arches for double-stack container car clearance, the Great Northern Tunnel continues to be a vital link in the BNSF system.

Figure 3.74 | **Nearly Completed Concrete-Lined Tunnel with Formwork in Distance**

Figure 3.75 | **Jet Grouting Operations**

Figure 3.76 | **Notch Cut by Roadheader in Great Northern Tunnel**

CASCADE TUNNEL— Stevens Pass, Washington

The Cascade Tunnel (also known as BNSF Railway Tunnel No. 15) beneath Stevens Pass, Washington, is the second longest railroad tunnel in North America and longest in the United States at 7.8 miles (41,183 feet). This single-track tangent tunnel was opened in 1929 by the Great Northern Railway and is currently part of the BNSF Pacific Division, Scenic Subdivision (Line Segment 37), located between the Berne (to the east) and Scenic (to the west) sidings. This area of northwestern Washington is shown in Figure 3.77.

Crossing the Cascade Range

A route through the northern part of the Cascade Range was long desired by the "Empire Builder" James J. Hill (Anderson 1987). Since founding the Great Northern Railway in 1889, Hill had wanted a connection to the Pacific Coast to open trade with the Orient. In 1892, the first line was built over Stevens Pass, operating on a slow and arduous route involving switchbacks to a summit at an elevation of 4,059 feet. Hill realized his dream in 1896 when a transportation agreement was negotiated with Nippon Yusen Kaisha, the largest steamship line in the Pacific, which brought goods from Asia to Seattle. Eight years after opening the route, in 1900, a 2.63-mile tunnel was constructed, reducing the summit elevation by 677 feet to 3,382 feet, reducing the maximum grade from 4% to 2.2%, shortening the distance by 9 miles, and reducing the running time by two hours. Studies were initiated in 1912 to further reduce the summit elevation and improve the operating economies. After interruption by World War I, the investigations were resumed in earnest in 1925 to construct a long tunnel and electrify the mountain grade (Mears 1929).

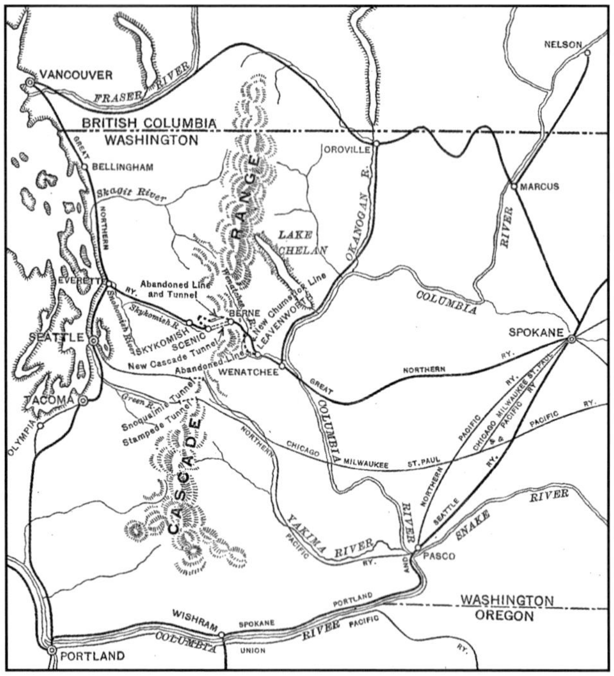

Surveying began in 1925, and by the end of that year, an acceptable route involving a 7.8-mile tunnel and 2 miles of new grade was established, eliminating 8 miles from the original route. Maximum curvature on the route was reduced from 10 degrees to 6 degrees, and the summit elevation was reduced by 501 feet to an elevation of 2,881 feet.

Layout of the tunnel is schematically depicted in Figure 3.78. It is lined with unreinforced concrete, averaging 19 to 24 inches thick. The concrete liner is horseshoe shaped with a design width of 16 feet and a clear height of 20 feet 10 inches above top of rail, with top of rail designed to be 2 feet above the invert. Large recesses (8-foot by 8-foot by 9-foot-high arch) in the liner are located approximately every 1,200 to 2,400 feet along the south tunnel wall, comprising 21 refuge bays. Eleven smaller (8 feet wide by 7 feet high by 1 foot 8 inches deep) recesses, called manbays or manhole recesses, are also located along the south tunnel wall. Finally, two "construction passing track recesses" are present in the south wall, one toward each end of the tunnel. Both are 14 feet high by 2 feet 2 inches deep; the eastern recess is 377 feet long and the western recess is 262 feet long.

The track lies on a uniform downgrade to the west at 1.5%. Drainage of the tunnel occurs through cross-drains that discharge at 14 locations into a mostly unlined drainage tunnel, called Pioneer Tunnel. Pioneer Tunnel was constructed parallel to and offset by 66 feet (centerline to centerline) to the south from the railroad tunnel. It was originally used to facilitate the drill-and-blast mining of the main tunnel bore from multiple faces. More details about the Pioneer Tunnel are discussed in the following subsection.

The Cascade Tunnel originally accommodated only electric-powered engines via an 11,000-volt catenary system suspended from the tunnel crown (Bauhof 1989). Figure 3.79 shows an

Figure 3.77 | **Cascade Tunnel Vicinity Map**

Figure 3.78 | **Tunnel Sketch**

electric-powered train entering the east portal. Electrification of this section of railroad extended from Wenatchee to Skykomish, Washington. The high-voltage transmission lines are embedded in the north tunnel wall. Electrical transformer bays (17 at 2,400-foot spacing) and pull boxes (69 at 600-foot spacing) are located in the north tunnel wall, across from the refuge bays. Pull boxes for telephone services (69 at 600-foot spacing) are embedded in the south tunnel wall.

In 1956, the catenary lines were removed and a ventilation system was installed when the railroad transitioned from electric to diesel locomotives. The ventilation system consists of two large fans and a door or "gate" to seal off the east portal, allowing a 20-to-40-minute flushing cycle.

With the introduction of larger freight containers, tight tunnel dimensions restricted the passage of double-stack container traffic. In 1990, the tunnel liner was enlarged with notches to accommodate passage for the largest double-stack container cars (20 feet 3 inches high by 8 feet 6 inches wide) through the tunnel. The notches can be seen in a photo of the west portal, Figure 3.80. Portions of the tunnel liner required additional support to accommodate up to 10 inches of concrete notching (measured perpendicular to the tunnel face). These remedial measures included grouting to fill voids behind the concrete liner with a neat cement or sand–cement mixture and installing resin-grouted rock dowels in the tunnel crown and sidewalls. The notches were designed to allow 2 inches of vertical clearance and 5 inches of horizontal clearance (20 feet 5 inches by 9 feet 4 inches) on each side of a double-stack container car. This design clearance is tight and relies on strict track maintenance with respect to cross-level as well as vertical and horizontal track alignment.

Figure 3.80 | **West Portal of Cascade Tunnel Showing Notches**

Figure 3.79 | **Electrification of Cascade Route**

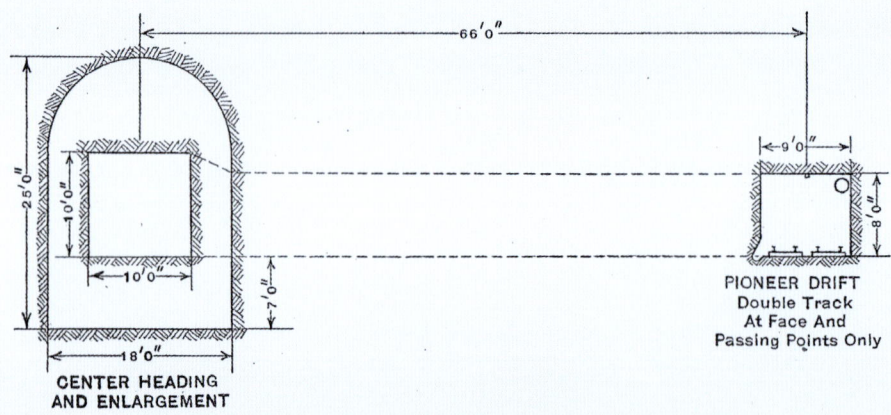

Figure 3.81 | **Cross Section of Pioneer Tunnel and Main Tunnel**

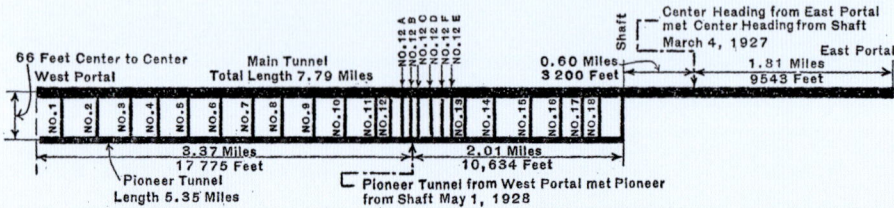

Figure 3.82 | **Plan for Main Tunnel, Pioneer Tunnel, and Crosscuts**

Figure 3.83 | **East Portal of Cascade Tunnel**

Pioneer Tunnel

The 8-foot-high by 9-foot-wide Pioneer Tunnel is offset 66 feet south and runs parallel to the Cascade Tunnel, as depicted in the cross section shown in Figure 3.81. Its headwall is located approximately 15 feet east of the west portal of the railroad tunnel. Its floor level is 7 feet above the final railroad tunnel floor to conform to the floor elevation of the 10-foot by 10-foot center heading of the railroad tunnel. However, east of Crosscut 5, the floor was gradually lowered to the same elevation as the railroad tunnel floor. Pioneer Tunnel does not extend along the entire railroad tunnel length; rather, its eastern end terminates near the Mill Creek shaft. Pioneer Tunnel was constructed in conjunction with the railroad tunnel to accelerate completion of Cascade Tunnel by opening up multiple excavation faces through 29 cross passages (crosscuts). A plan of the tunnel presented in Figure 3.82 shows this arrangement of crosscuts. The 8-foot-high by 9-foot-wide crosscuts were constructed at a 45-degree angle between the two tunnels. Eleven of the crosscuts have been sealed with concrete. Eighteen remain accessible but are closed by removable 2-foot by 4-foot steel doors to facilitate the ventilation flushing cycle.

Following construction, Pioneer Tunnel's central purpose shifted to drainage of the railroad tunnel. Fourteen drains (cross drains) discharge into Pioneer Tunnel, typically through 24-inch-diameter pipes running at a 45-degree angle between the tunnels (Mears 1929). All drainage from Pioneer Tunnel discharges at the west portal.

The current condition of Pioneer Tunnel is highly variable, which is assessed annually with inspections by BNSF staff. Much of the tunnel has remained unlined since its completion. Where supported by timber, some of the timber sets have deteriorated or collapsed. In several locations, the Pioneer Tunnel crown has experienced partial or total collapse. As a result, several of the drains have become blocked and are no longer functioning.

Several instances of repair in Pioneer Tunnel have been undertaken, including clearing the invert of fallen debris, cleaning out the drains, and re-supporting the arch with new timber sets or cast-in-place concrete. In 2008, a collapse occurred in Pioneer Tunnel near Crosscut 5 that blocked the drainage system, necessitating repairs. An additional rehabilitation program was undertaken in 2013, which involved replacement of numerous timber support elements.

Tunnel Length

The original constructed tunnel length was 41,152 feet, which is also the tunnel length displayed on the portal plaque above the west portal (Figure 3.80). After the transition to diesel locomotives in the 1950s, a fan house and portal gate were constructed at the east portal. The fans and guillotine gate added 16 feet to the tunnel length. In the 1970s, a more modern fiberglass sliding gate was constructed, which added another 15 feet, for a total increase of 31 feet to the original tunnel length. Presently, the tunnel length is 41,183 feet, as displayed on the portal plaque above the east portal (Figure 3.83).

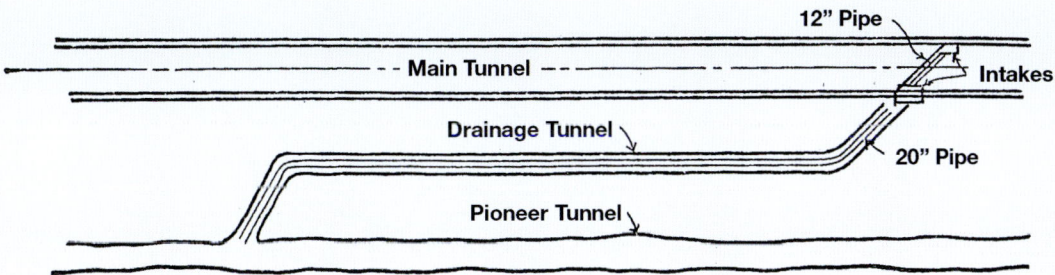

Figure 3.84 | **General Plan Showing Cross Drainage**

Ground Conditions

There was one deviation from the original excavation plan. At the west portal, exceedingly soft ground was encountered, requiring heavy timbering. Progress was so slow that in April 1926, only 100 feet was advanced on the center heading. Indications were that this condition would persist for 800 to 1,000 feet. Advantage was taken of the Tye River canyon located about 2,000 feet from the west portal. An incline 251 feet long was constructed on a 30-degree slope, which permitted operations in the Pioneer to resume 2,279 feet past the west portal. It was several months before excavation from the west portal of Pioneer caught up to the inclined adit.

Most of the intact rock encountered was altered granite of a fine texture, described as quartz diorite. In general, this material was excellent for drilling and excavation, but delays were frequent because of soft seams. Occasionally, rock in broken condition was encountered, aggravated by the presence of thick seams of graphite. This occurred at the Mill Creek shaft and in several short sections west of the shaft, requiring that the ground be timbered.

Construction involved excavation of 934,600 cy (cubic yards) of rock—839,700 cy from the main tunnel and 94,900 cy from Pioneer.

The eastern half of the tunnel was relatively dry. However, starting at a point about 3.85 miles from the east portal to 2.38 miles, excavation was met with substantial water inflows. Water flows up to 10,000 gpm (gallons per minute) were reached and successfully diverted into Pioneer Tunnel through 13 cross-drains constructed at 1,500-foot intervals on a 0.5% grade, as shown in Figure 3.84. Figure 3.85 shows the water inflow that occurred in the Pioneer Tunnel. Water inflows eventually subsided to less than half the maximum amount.

Construction Method

Because of the extreme length of the tunnel and the short time period allotted to its construction, plans were made to excavate from an intermediate point to provide more than two working faces. A view of the west portal during construction is shown in Figure 3.86. This necessitated the

Figure 3.85 | **Pioneer Tunnel Water Inflow**

Figure 3.86 | **West Portal During Construction**

construction of the Mill Creek shaft in the lowest valley crossing the tunnel alignment, located approximately one-third of the tunnel length from the east portal. With the peak of Cowboy Mountain at an elevation of 5,710 feet, the ground surface at the shaft was nearly 2,400 feet lower at an elevation of 3,307 feet, necessitating a shaft depth of only 622 feet to reach the tunnel invert.

The tunnel bore was generally 18 feet wide by 25 feet high in hard rock and designed with 1-foot-thick concrete walls and a 2-foot-thick crown throughout (*ASCE Transactions* 1932). Figure 3.87 shows the main tunnel dimensions. An initial tunnel heading of 10 feet by 10 feet was excavated and was located 7 feet above the final invert. Where soft ground, excessively broken rock, or rock with soft seams was encountered, the bore was enlarged to 22 feet wide by 27.5 feet high to accommodate timbering. Timbers sets of 12 inches by 12 inches were used when required by the character of the ground. Figure 3.88 shows a timbered section of the Cascade Tunnel near Crosscut 6.

The western section between Mill Creek shaft and the west portal was twice as long as the eastern section. A "pioneer tunnel" was driven from the west portal to the shaft with crosscuts at 45 degrees to the main tunnel at 1,500-foot intervals to provide multiple excavation headings to further reduce the construction duration. This also provided a haulage way for muck and supplies, as well as a route for air lines and water discharge pipes. The Pioneer Tunnel consisted of an 8-foot-high by 9-foot-wide heading constructed parallel to but offset 66 feet (center to center) south of the main tunnel and was located 7 feet above the final invert of the main tunnel. Figure 3.81 shows the arrangement of the main and Pioneer tunnels in cross section. The Pioneer Tunnel was driven from its west portal

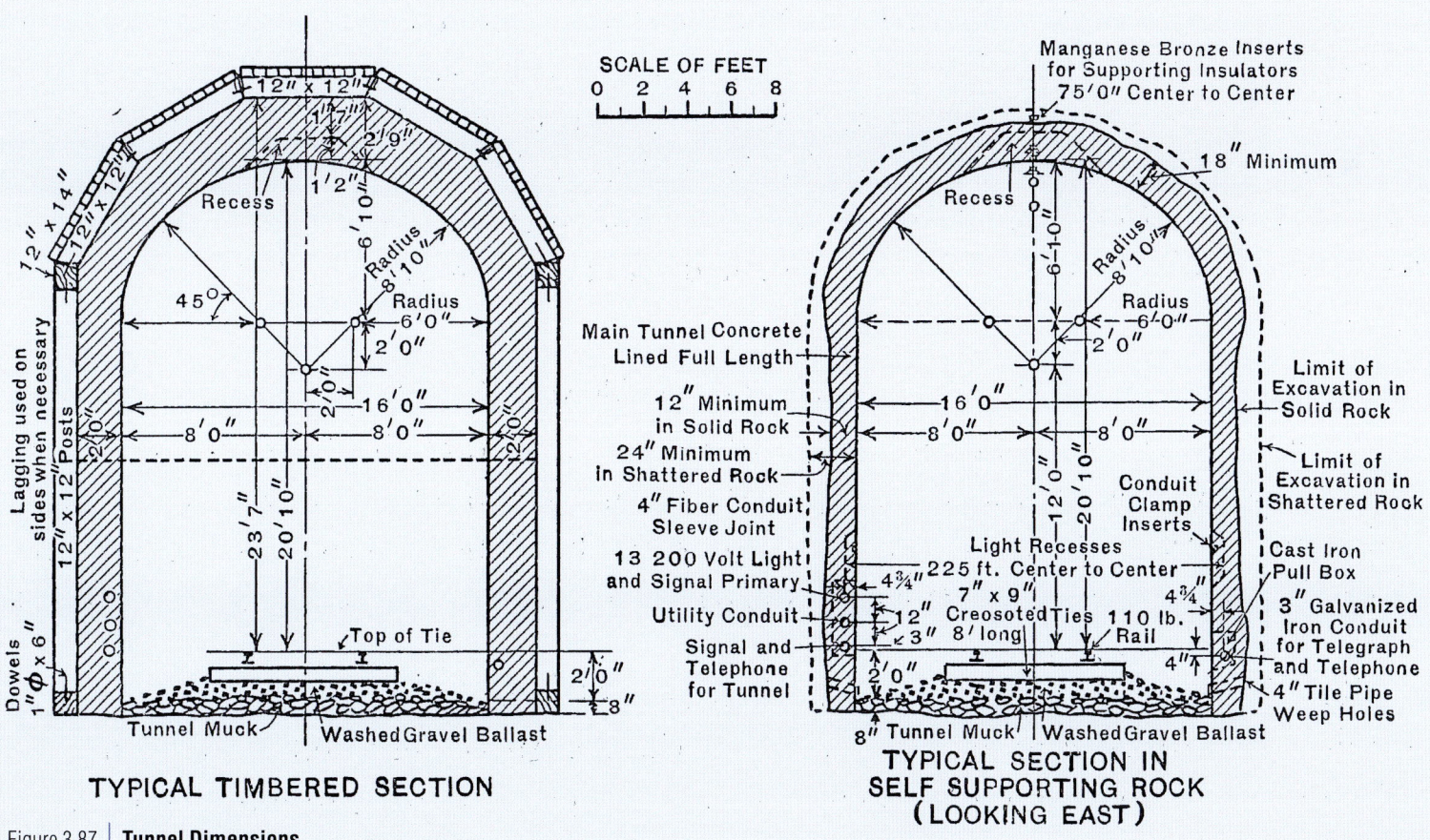

Figure 3.87 | Tunnel Dimensions

and from the Mill Creek shaft. Initially established at 1,500-foot intervals, crosscuts to the main tunnel were constructed at 45 degrees. From these crosscuts, a 10-foot by 10-foot center heading was advanced in the main tunnel. Later on, when considerable groundwater quantities were encountered, crosscuts were added at more frequent intervals. In addition, the invert of the Pioneer Tunnel was dropped to correspond to the final invert of the main tunnel to facilitate water diversion from the main tunnel into the Pioneer. Using this method of multiple headings allowed a workforce of up to 1,793 men to be involved simultaneously in tunneling operations: 767 at the west portal and Pioneer, 590 at the Mill Creek shaft, 411 at the east portal, and 25 at headquarters.

Concreting was started 1½ years after tunneling began, concurrent with tunnel enlargement. The last concrete was poured only 16 days after tunnel enlargement was completed. The total amount of concrete placed was 262,562 cy. Because of the great distances involved, in-tunnel mixing plants were used with movable forms (37.5 feet long) that were designed not to interfere with excavation operations. The mixing plant was 9 feet above the invert, allowing sufficient clearance below for mucking equipment. The plant produced 16 cy per hour, and forms could be safely moved 15 to 20 hours after placement, allowing completion of up to 400 feet of tunnel each week.

Construction Progress

Notice to Proceed (NTP) for a construction contract with a three-year duration was awarded to A. Guthrie & Company on November 25, 1925. Four days later, clearing was begun for the west portal camp. At 19 days from NTP, the approach cut for Pioneer Tunnel began, with tunneling at the west portal of the Pioneer Tunnel beginning 33 days from NTP. At 65 days from NTP, the Mill Creek shaft was begun. At 34 days from NTP, the approach cut for the east portal began, and tunneling of the east portal began at 74 days from NTP. Thus, within 74 days from NTP, work had begun at all three major points of entry.

Mill Creek shaft was completed in August 1926. Three headings spread out from the shaft—the Pioneer and main tunnel heading west and a main heading to the east. The two Pioneer headings met 2½ years later on May 1, 1928. Upon project completion, the Mill Creek shaft was sealed with concrete to a point 10 feet above the tunnel crown, then backfilled with broken rock to the surface.

The following average advance rates were accomplished: west Pioneer tunnel at 800 feet per month, west main tunnel heading at 800 feet per month, Mill Creek shaft at 115 feet per month, main tunnel heading east from the shaft at 500 feet per month, main tunnel heading west from the east portal at 500 feet per month, and tunnel enlargement at 900 feet per month (ENR 1926, 1927, 1928, 1929). Several world records for the rate of tunnel advance were established during this construction period. To accomplish this progress, work went on continuously day and night, including Sundays and holidays. Not a single shift was lost during the entire construction period.

The East Portal–Mill Creek shaft center headings met on March 24, 1927, and enlargement was completed on May 31, 1928. The East Portal–Mill Creek shaft enlargement was completed on November 20, 1928.

Summary

As the northernmost rail route in the United States across the Cascade Mountains, the Cascade Tunnel is an important east-west link for Burlington Northern Railway.

Figure 3.88 | **Timbered Section of Tunnel**

EAST SIDE ACCESS TUNNELS— New York

The East Side Access (ESA) project in New York is the first major expansion of the commuter rail network in New York in more than 100 years. When complete, it will provide transportation for 160,000 Long Island Rail Road passengers a day from Queens to a new terminal located on the east side of Manhattan directly beneath the existing historic Grand Central Terminal. In addition to improving access to the east side of Manhattan for Long Island residents, it will also help reduce overcrowding in Penn Station and the subways that connect Penn Station with the east side of Manhattan (Figure 3.89).

Improving commuter rail access between Long Island and the east side of Midtown Manhattan started in the late 1960s and early 1970s with the construction of the 63rd Street Tunnel between Second Avenue in Manhattan and Northern Boulevard in Queens. It was constructed as a four-track bi-level tunnel with the New York City Transit using the upper level for the F Train service and the lower level reserved for ESA. Various construction methods were used including immersed-tube tunnels under the East River, cut-and-cover used to construct the Queens Approach Tunnels, and a tunnel boring machine (TBM) to drive the tunnels on the Manhattan side from the end of the immersed tube to a location near Second Avenue. At that time the concept for ESA was to enter the lower level of the existing Grand Central Terminal. This would have required significant underpinning of structures along Park Avenue to expand the existing Metro North Tunnel that runs under Park Avenue.

No further work was undertaken on ESA following completion of the 63rd Street Tunnel until the 1990s when the Metropolitan Transportation Authority completed a Major Investment Study that recommended construction of the project as it is today.

Figure 3.89 | **Map of East Side Access Alignment in New York**

Although the additional route miles being added to the Long Island Rail Road network is only 3.5 miles, approximately 10 miles of tunnel has been constructed for this new facility. A variety of tunneling techniques have been used to create the new real estate beneath Manhattan and Queens into which the railroad will be installed. TBM tunneling, drill-and-blast, as well as sequential excavation methods were used, in some cases debuting in New York over the course of the project's life.

Constructing in an Urban Environment

Likely the most significant challenge faced in constructing the ESA is its location. In Queens, the bulk of the work occurs beneath and within the busiest railroad interlocking in the United States that handles more than 800 passenger trains per day heading in and out of Penn Station. In Manhattan, the alignment takes ESA along Park Avenue, beneath Metro North's existing cut-and-cover structure but also beneath some of the most expensive real estate in the world, including structures such as the Waldorf Astoria Hotel, JPMorgan Chase world headquarters, the MetLife (formerly Pan Am) Building, and the 100-year-old Grand Central Terminal itself.

Figure 3.90 | Station Cavern Under Midtown Manhattan

Drill-and-Blast

Drill-and-blast has been used extensively in the New York area for tunneling in the Manhattan Schist, but its use on the ESA faced significant challenges. Over time, property owners became better informed regarding the potential and perceived impacts that vibrations induced by blasting may have on their structures as well as the deterioration in workplace environmental quality caused by the noise and vibration of blasting operations. Another challenge was managing the fumes, dust, and pressure waves caused by blasting while not affecting excavation production cycles. The ventilation system was therefore designed to extract air from the underground works to one location, drawing the fume and dust-laden air through water curtains and filters to remove dust and fumes. Finally, the bulk of the blasting would occur beneath Grand Central Terminal, which handles 260,000 Metro North passengers daily and is one of the most visited buildings in the United States, averaging 750,000 people per day.

Commencing in 2009 and completed in 2013, drill-and-blast excavation methods were used to construct five vertical and four inclined escalator shafts, and for 40% of the excavation of 800,000 cy (cubic yards) of rock to create the station caverns, cross passages, and other spaces beneath Grand Central Terminal (Figure 3.90). More than 2,600 individual blasts were detonated with no delays, cancellations, or damage to Metro North Railroad's operations, Grand Central Terminal itself, or other properties.

Figure 3.91 | **TBM Mining Through Concrete Plug**

Roadheaders

Although discounted during the planning phase, roadheader technology had advanced sufficiently by the early 2000s such that two Sandvik MT750 roadheaders were used to excavate rocks in the 12–14 kpsi range for the station cavern top headings beneath Grand Central Terminal. With limited (35 to 45 feet) cover to Metro North Railroad's operational suburban level, the ability to excavate the 60-foot-wide top heading arch profile using nonexplosive methods resulted in the elimination of approximately 2,000 blast events.

Tunnel Boring Machines

The excavation of the tunnels beneath Manhattan and Sunnyside Yard in Queens was undertaken using TBMs. In Manhattan, hard-rock TBMs excavated 40,000 linear feet of tunnel in the Manhattan Schist. In Queens, 10,500 linear feet of tunnels were excavated through glacial till and beneath the water table using two slurry-faced TBMs, the first application of such technology in New York and purportedly only the fifth time slurry TBMs had been used in the United States.

Two 22-ft-diameter TBMs were used in Manhattan, a Robbins main beam TBM and a SELI double-shield TBM. Both TBMs were equipped with 17-inch disc cutters and achieved advance rates of more than 100 feet per day through the schist. Rock bolts, mesh, mine straps, and steel arch ribs were installed behind the TBMs for initial rock support. A waterproofed cast-in-place permanent lining was installed when mining was complete. Eight tunnels were mined using the two TBMs with one of the major challenges being the relaunch of the TBMs because there was no space for a reception shaft at 38th Street. A technique was borrowed from Chicago's Tunnel and Reservoir Plan project whereby once the TBM had completed its initial run, it was pulled back through the tunnel and a concrete plug placed ahead of the TBM that then provided a surface for the grippers to thrust off as the TBM commenced the next run (Figure 3.91). This technique was successfully used on six occasions, saving 30,000 cy of excavation that would have been needed to unearth separate launch caverns.

In Queens, two 22.5-ft-diameter Herrenknecht mixshields (Figure 3.92) were used to excavate four tunnels totaling 10,500 linear feet through glacial till containing abrasive sands and boulders. In some locations, the underlying gneissic rocks thrust up into the tunnel envelope, requiring the TBMs to be able to handle a full face of rock. These tunnels were excavated beneath Amtrak's modern-day storage yard, Sunnyside Yard, and the mainline tracks that handle 800 Amtrak and Long Island Rail Road passenger trains a day in and out of Penn Station in New York. The ability to manage ground movement when mining beneath 60-mph line speed tracks with less than one diameter of cover was a critical factor and for which the slurry TBMs proved to the be the ideal tool. Settlements at track level never exceeded ¼ inch and at no time were train operations suspended or affected by the mining operations. Another first for New York was the use of left- and right-hand tapered precast segmental linings as the permanent lining behind the TBMs.

Figure 3.92 | **Slurry TBM Ready for Launch**

Sequential Excavation Method

The most challenging tunnel portion of the project is the 120-foot long Northern Boulevard crossing that links the Manhattan and Queens sections. Crossing beneath Northern Boulevard, one of the busiest thoroughfares in the Queens area, the 60-foot-wide tunnel also passes beneath a five-track subway box as well as elevated subway lines. The geology in this location is glacial till over rock with a near-surface groundwater table.

Before tunneling could start, extensive ground treatment was undertaken to provide protection to the overlying structures, including compaction beneath the subway box. To cut off the groundwater and stabilize the perimeter of the tunnel, a horizontal ground freeze was used to provide a nominal 6-foot-thick frozen arch. Forty-three holes were drilled for the freeze, together with heat pipes and temperature monitoring holes between the frozen arch and existing subway structure to manage and monitor the spread of the freeze. Once a median temperature of 5°F was achieved in the arch, excavation of the tunnel started using multiple drifts in a three-over-three configuration (Figure 3.93). Excavation was undertaken using a small roadheader and hoe rams with lattice girders, mesh, and shotcrete installed against the frozen ground as initial support. Once mining was complete, a 30-inch-thick waterproof permanent lining was installed using shotcrete techniques and the freeze turned off. Compensation grouting was undertaken through the compaction grouting holes to control movement of the elevated structure and subway box as the freeze thawed.

Figure 3.93 | Northern Boulevard Tunnel Overall View

REFERENCES

AAR (Association of American Railroads). 1951. *American Railroads: Their Growth and Development*. Washington, DC: AAR.

AAR (Association of American Railroads). 2014. U.S. freight railroad statistics. July 14. Washington, DC: AAR.

AAR (Association of American Railroads). n.d. Slide 1. Victory by rail: America's railroads during wartime. http://archive.freightrailworks.org/wartime#slide-1 (accessed Dec. 16, 2015).

Ambrose, S.E. 2001. *Nothing Like in the World: The Men Who Built the Transcontinental Railroad 1863–1869*. New York: Touchstone/Simon & Schuster.

American-Rails.com. 2015a. The American railroads: A long and storied history. www.american-rails.com/ (accessed Dec. 16, 2015).

American-Rails.com. 2015b. Railroads in the Civil War. www.American-rails.com/railroads-in-the-civil-war.html (accessed Dec. 16, 2015).

Anderson, E. 1987. *Rails Across the Cascades*, 4th ed. Wenatchee WA: World Publishing.

Anonymous. 1890. St. Clair Tunnel: Success of a great engineering undertaking. In *The Toronto Mail*, Sept. 9. Toronto: Mail Printing Company.

ASCE Transactions. 1932. The Eight-Mile Cascade Tunnel, Great Northern Railway, A Symposium. American Society of Civil Engineers, Paper No. 1809. *ASCE Transactions* 96:915–1004.

Bauhof, F.C. 1989. Construction History of the Cascade Railroad Tunnel. In *Engineering Geology in Washington*, Vol. II. Edited by R.C. Galster. Washington Division of Geology and Earth Resources Bulletin 78. pp. 729–742.

Best, G.M. 1976. Story of the Golden Spike at Lang Station. Golden Spike Centennial Souvenir Program. www.scvhistory.com/scvhistory/golden-spike-centennial-best.htm.

Bickel, J.O., Kuesel, T.R., and King, E.H. 1996. *Tunnel Engineering Handbook*, 2nd ed. New York: Chapman & Hall.

Chinese Historical Society of America. 1969. A history of the Chinese in California: A syllabus. http://cprr.org/Museum/Chinese_Syllabus.html (accessed Dec. 16, 2015).

Cudahy, B.J. 2002. *Rails Under the Mighty Hudson*, 2nd ed. New York: Fordham University Press.

Drinker, H. 1893. *Tunneling, Explosive Compounds, and Rock Drills*, 3rd ed. New York: John Wiley & Sons.

Engineering News. 1890. The St. Clair Tunnel. *Engineering News*, November 8.

ENR (*Engineering News-Record*). 1926. Methods and progress in driving the 8-mile Great Northern Tunnel in the Cascades. *Engineering News-Record* 97(22):858–863.

ENR (*Engineering News-Record*). 1927. Progress on the Cascade Tunnel, Great Northern Railway. *Engineering News-Record* 99(6):224–225.

ENR (*Engineering News-Record*). 1928. Progress on the Cascade Tunnel, Great Northern Railway. *Engineering News-Record* (Oct. 11):555.

ENR (*Engineering News-Record*). 1929. Driving the Second Cascade Tunnel. *Engineering News-Record* (Feb. 28):334–338.

Federal Railroad Administration. 2015. Freight rail today. https://www.fra.dot.gov/Page/P0362 (accessed Dec. 16, 2015).

Gallamore, R.E., and Meyer, J.R. 2014. *American Railroads—Decline and Renaissance in the Twentieth Century*. Cambridge, MA: Harvard University Press.

Gilliss, J.R. 1872. Tunnels of the Pacific Railroad. *ASCE Transactions* 1(13):155–171.

HAER (Historic American Engineering Record). 1993a. No. PA-515, Philadelphia Railroad, Allegheny Tunnel, Cambria County, Pennsylvania. Washington, DC: National Park Service.

HAER (Historic American Engineering Record). 1993b. No. PA-516, Philadelphia Railroad, Gallitzin Tunnel, Cambria County, Pennsylvania. Washington, DC: National Park Service.

HAER (Historic American Engineering Record). 1993c. No. PA-520, Philadelphia & Reading Railroad, Black Rock Tunnel, Chester County, Pennsylvania. Washington, DC: National Park Service.

HAER (Historic American Engineering Record). 1993d. No. PA-539, Philadelphia & Reading Railroad, Flat Rock Tunnel, Montgomery County, Pennsylvania. Washington, DC: National Park Service.

Hallberg, M.C. 2009. *Railroads in North America—Some Historical Facts and an Introduction to an Electronic Database of North American Railroads and Their Evolution*. University Park, PA: Pennsylvania State University.

Hansen, B. 2010. Completing the loop: The Tehachapi Pass Railroad. *Civil Engineering* 80(4):38–39.

Harrison, N. 1995. Driving the new St. Clair River Tunnel. *Tunnels & Tunneling International* 27(1):17–20.

Hidy, R.W., Hidy, M.E., Scott, R.V., and Hofsummer, D.L. 1988. *The Great Northern Railway: A History*. Boston: Harvard Business School Press.

Hussey, E.B., Ober, R.H., and Blackwell, J.D. 1915. *Report Upon Cause of Settlement of the Seattle Public Library Building and Site, Seattle, WA*. Prepared for H.W. Findley, Assist. Corp Council for the City of Seattle.

McMechen, E.C. 1927. *The Moffat Tunnel of Colorado—An Epic of Empire*. Denver: Wahlgreen Publishing.

Mears, F. 1929. Conquering the Cascades: Guthrie & Co. *Railway and Marine News* 26(1): Section A.

Middleton, W.D., Smerk, G.M., and Diehl, R.L. 2007. *Encyclopedia of North American Railroads*. Bloomington, IN: University of Indiana Press.

Nicolls, W.J. 1897. *The Railway Builder: A Handbook for Estimating the Cost of American Railway Construction and Equipment*, 5th ed. Philadelphia: J.B. Lippincott.

NPS (National Park Service). 2015. Allegheny Portage Railroad: Staple Bend Tunnel. www.nps.gov/alpo/learn/historyculture/staplebend.htm

Ott, K., and Jacobs, L. 2003. Design and construction of the Weehawken Tunnel and Bergenline Avenue shaft. In *Rapid Excavation and Tunneling Conference: 2003 Proceedings*. Littleton, CO: SME.

Parker, H.W., Robinson, R.A., Godlewski, P.M., Hultman, W.A., and Guardia, R.J. 2001. Tunnel rehabilitation in North America. In *Progress in Tunneling After 2000: Proceedings of the AITES-ITA 2001 Congress*. Bologna: AITES-ITA.

Proctor, R., and White, T. 1946. *Rock Tunneling with Steel Supports*. Youngstown, OH: Youngstown Printing.

Railway Gazette. 1904. Great Northern terminal improvement at Seattle. *Railway Gazette* 37(21):510–512.

Roby, J., Sandell, T., Kocab, J., and Lindbergh, L. 2008. The current state of disc cutter design and development directions. In *North American Tunneling: 2008 Proceedings*. Edited by M. Roach, M. Kritzer, D. Ofiara, and B. Townsend. Littleton, CO: SME.

Rodrigue, J.P. 2015. *Geography of Transport Systems 1830 to 2012*. New York: Hofstra University.

Sandstrom, G.E. 1963. *Tunnels*. New York: Holt, Rinehart, and Winston.

Shannon & Wilson. 1993. *Black Rock and Flat Rock Tunnels Exploration Summary Report and 30% Design Submittal*. Submitted to the Consolidated Rail Corporation (CONRAIL).

Signor, J.R. 1983. *Tehachapi—Southern Pacific-Santa Fe*. San Marino, CA: Golden West Books.

Signor, J.R. 1985. *Donner Pass—Southern Pacific's Sierra Crossing*. San Marino, CA: Golden West Books.

Skinner, E.H. 1974. *The Flathead Tunnel—A Geologic, Operations, and Ground Support Study, Burlington Northern Railroad, Salish Mountains, Montana*. U.S. Bureau of Mines Information Circular 8662. Washington, DC: U.S. Government Printing Office.

University of Richmond. 2015. Rapid development of railways and canals in America. www.historyengine.richmond.edu/episodes/view/5173 (accessed December 2015).

Vogel, R.M. 2012. *Smithsonian Institution National Museum of History and Technology Bulletin 240*. In *The Project Gutenberg EBook of Tunnel Engineering. A Museum Treatment*. Paper 41, pg. 201–240.

White, M.W. 2004. Economics of Time Zones. Philadelphia, PA: Wharton School, University of Pennsylvania http://web.archive.org/web/20060911152247/http://bpp.wharton.upenn.edu/mawhite/Papers/TimeZones.pdf.

Wickham, G.E., Tiedeman, H.K., and Skinner, E.H. 1972. Support determinations based on geologic predictions. *Proceedings, North American Rapid Excavation and Tunneling Conference*. New York: American Institute of Mining, Metallurgical, and Petroleum Engineers. pp. 43–64.

Wikipedia contributors. 2015a. Baltimore and Ohio Railroad locomotives. In *Wikipedia, The Free Encyclopedia*. https://en.wikipedia.org/wiki/Baltimore_and_Ohio_Railroad_locomotives (accessed Dec. 16, 2015).

Wikipedia contributors. 2015b. First Transcontinental Railroad. In *Wikipedia, The Free Encyclopedia*. https://en.wikipedia.org/wiki/First_Transcontinental_Railroad (accessed Dec. 16, 2015).

Wikipedia contributors. 2015c. Stourbridge Lion. In *Wikipedia, The Free Encyclopedia*. https://en.wikipedia.org/wiki/Stourbridge_Lion (accessed Dec. 16, 2015).

Wikipedia contributors. 2015d. Electric locomotive. In *Wikipedia, The Free Encyclopedia*.https://en.wikipedia.org/wiki/Electric_locomotive (accessed Dec. 16, 2015).

4 TRANSIT TUNNELS

As a resident or tourist, we cannot imagine navigating through America's great cities without riding the subways.

In fact, many people take a city's subway system into account before affixing the label of "great," simply because a subway system provides fast and efficient travel from point A to point B. Imagine having to take a taxi the full length of Manhattan from Battery Park to Harlem. Or between any number of the thousands of business and tourist destinations in New York. The luster of the Big Apple would quickly tarnish for the millions that traverse this great city on a daily basis without the ability to travel readily from place to place among the city's skyscrapers. The same is true for any major metropolitan area with economic aspirations of growth in part made possible with the efficiency of a transit system constructed underground in tunnels. Development of modern urban mass transit systems is only possible because of tunneling.

As the human population continues to increase, so does the demand for more efficient use of the fixed space within our cities. The ability to move large numbers of people to places of employment, shopping, and recreation will play a large role in determining which cities prosper and which experience non-sustainable growth and inevitable urban blight. Our older, larger cities were necessarily the first to face this eventuality. Boston, New York, Philadelphia, and Chicago began their subway systems early in the twentieth century. They were the logical starting places for transit tunnels, as they were the largest metropolitan areas of their era. During this same period, cities in the western United States still enjoyed the benefit of smaller populations and abundant open land. Those areas were eventually compelled to go underground with mass transit to maintain growth. Decisive civic leadership in places such as Buffalo and Pittsburgh moved proactively to construct subway systems before the need overwhelmed their cities, providing for both controlled growth and revitalization of downtown business districts. The City of Seattle and surrounding areas delayed their development of transit and are now expanding their mass transit system as Sound Transit is both building and planning new underground transit in emerging areas, where municipal planners expect growth. Los Angeles (Figure 4.1), with its legendary freeways, is expanding its mass transit system, which is only 25 years old, with extensive tunnels to meet the historic transportation challenges of that city.

Many great cities within the United States are forever joined to their subway systems for continued growth and survival. As these cities continue to grow, so do most of their transit facilities. History has shown that going underground early and often is the key to sustainable economic prosperity.

Douglas (1963) described the plight of London in the mid-nineteenth century:

Toward the middle of the last century London was dying—slowly, painfully, and with a great deal of protest. No physician had to be called in to diagnose the trouble; it was all too apparent to those who lived there for, wherever they went, they encountered the great thrombosis of traffic which clogged the highways that were the veins and arteries carrying the city's blood.

And so it was at the end of the nineteenth century for the great commercial centers of the United States as well—Boston, New York City, Philadelphia, and Chicago. The only solution was to move pedestrian traffic underground.

Figure 4.1 | **Transit Tunnel, Los Angeles Metro Red Line**

Previous Page | **Metro Station in Washington, D.C.**

CHAPTER FOUR | TRANSIT TUNNELS

127

In the early 1900s in the large cities of Boston and New York, the congested conditions demanded an underground solution, and once started, the initial transit tunnels were rapidly expanded. Transit tunnels have been in use for more than 100 years in those cities. Other cities were slower to adopt underground transit, such as Philadelphia and then Chicago in the late 1930s. These pre–World War II transit systems were constructed by the cities largely without federal government funding. After World War II, mass transit projects were funded by the U.S. government. However, mass transit with tunnels was not universally considered as preferred or needed at that time. Through the enactment of the Federal-Aid Highway Act of 1956, new highways were considered to be in direct competition with transit systems. Transit had to compete with the alternative of more highways in studies that evaluated multiple modes of transportation of individual people driving automobiles on highways, buses on highways or dedicated guideways, and electrified rail systems. The story of how the transit system came into existence played out differently in each city, in some cases with many years or decades of delay resulting in incompletion or no transit system being constructed (for example, Cincinnati).

Over a century ago, as the United States was struggling to manage massive congestion on city streets, Boston and New York City, in a sense, raced to put in place an underground transit system like those already constructed in London and Paris. Boston was first in 1897, followed by New York City in 1904. Other eastern U.S. cities achieved some underground transit when they put street cars in tunnels, as was the case in Philadelphia in 1928. Underground transit lines were constructed in Chicago in the 1930s (opened in 1943). Due to the outbreak of World War II, further work on the Chicago system was not pursued and further underground expansion has not taken place.

With substantial political support in the 1960s, several underground transit systems were initiated, the major ones being in San Francisco (1972); Washington, D.C. (1976); Atlanta (1979); Baltimore (1983); Buffalo and Pittsburgh (both in 1985); Seattle (1990, bus tunnel); and Los Angeles (1990). All have a remarkable story to tell.

ART IN TRANSIT

Transit is more than just the tunnels. Those who designed and built these public works have practiced the *art of tunneling* and recognize the place art has in transit. Experience of transit professionals (APTA 2013) has shown that key features of aesthetics, function, and durability are critical to the success of a transit project. High-quality public art and design improve the appearance and safety of a facility, add vibrancy to public spaces, and make patrons feel welcome, often resulting in higher usage of the facility. As riders descend to station platforms, site-specific integrated artworks connect users to the community, to humanity and beauty offering a transit experience of contemplation and joy.

Some artwork examples have been included in the sections that follow. This artwork graces the completed transit structures that, in great contrast, were built with heavy construction equipment and huge tunnel boring machines.

WHAT ARE TRANSIT TUNNELS?

Transit tunnels by definition carry people, typically in electric-powered vehicles with steel wheels on steel rails or in rubber-tired vehicles on fixed guideways. Electrification is either by power from an overhead line typically associated with *light-rail* transit or third-rail electrification with *heavy-rail* transit. Regardless, the tunnels are largely the same and vary in internal size ranging from about 17 to 20 feet in diameter.

Several terms are used in the United States and also around the world to refer to essentially the same types of mass transit. In the United States, transit tunnels are often called *subways*, such as the New York City Subway, whereas in London, these tunnels are called the *Underground* or the *Tube*. In other U.S. cities, the transit system agency name has evolved to be the name of the system, such as *BART*, which refers to the Bay Area Rapid Transit system in the San Francisco Bay Area. A more global name has evolved in some of the world's major cities where the term *Metro* is used, such as the Washington (D.C.) Metro, Atlanta Metro, Los Angeles Metro, and Paris (France) Metro.

Transportation that does not involve individual automobiles or trucks on highways is typically termed *mass transit*, or in some instances *rapid transit*. For most transit systems in the United States, it means electrified rail lines on dedicated rights-of-way without the interference of surface street traffic. Tunnels are required to achieve the dedicated rights-of-way in cities where elevated (overhead) transit lines are not practical, or more typically, not socially acceptable in the center of major cities for new systems.

In this book, the simplified term *transit* is used to designate all forms of transportation that transport many individual people in a public system. Historically, *transit* does not include transportation via highways or by heavy-rail commuter trains.

Figure 4.2 | **Boston, Massachusetts**

BOSTON, MASSACHUSETTS

Greater Boston has the oldest subway system in the United States, with operations beginning in 1897. Since then, the population of the metropolitan area has grown from approximately 700,000 people comprising approximately four communities to more than 4,000,000 in 2014, encompassing 176 cities and towns (Figure 4.2). To accommodate this large growth, the original Boston Elevated Railway Company (BERY), which included the subways, underwent numerous changes and expansion. In 1947, due to economic issues and a desire to maintain public transportation at a reasonable price, the Commonwealth of Massachusetts purchased the BERY stock and formed the Metropolitan Transit Authority (MTA), a public corporation. In 1968, the MTA was again transformed into the Massachusetts Bay Transportation Authority (MBTA), which absorbed the Eastern Massachusetts Street Railway Company and several commuter rail services, greatly expanding the MBTA system. Currently, the MBTA is the fifth largest mass transit system in the country (Figure 4.3).

Between 1870 and 1890, the city of Boston and its suburbs grew rapidly with extensive horse-drawn trolleys and, in the 1890s, electrified trolleys operating on the major city streets. This rapid growth led to legislation in 1891 to establish "a commission to promote rapid transit for the city of Boston and its suburbs" (Commonwealth of Massachusetts 1891). In 1892, the commission reported to the legislature about the combining of numerous railroad terminals and, most importantly, a subway and elevated lines to connect the downtown area with the suburbs. In 1893, the legislature established a Board of Subway Commissioners who recommended creating the BERY and the Boston Transit Commission to design and construct the Tremont Street Tunnel (Commonwealth of Massachusetts 1894). The BERY would construct and operate elevated lines connecting Boston's North End via Atlantic Avenue to Roxbury and Forest Hills south of the city.

The original Tremont Street Subway is located near Park Street on the Boston Common and continues to a portal where the rail line running at street level goes underground at the Public Garden on Boylston Street. The cut-and-cover tunnel involved excavation of a large trench, construction of the tunnel structure, and then backfilling above the new construction to reinstate the surface. Tremont Street Subway was constructed using steel support sets and

CHAPTER FOUR | TRANSIT TUNNELS 129

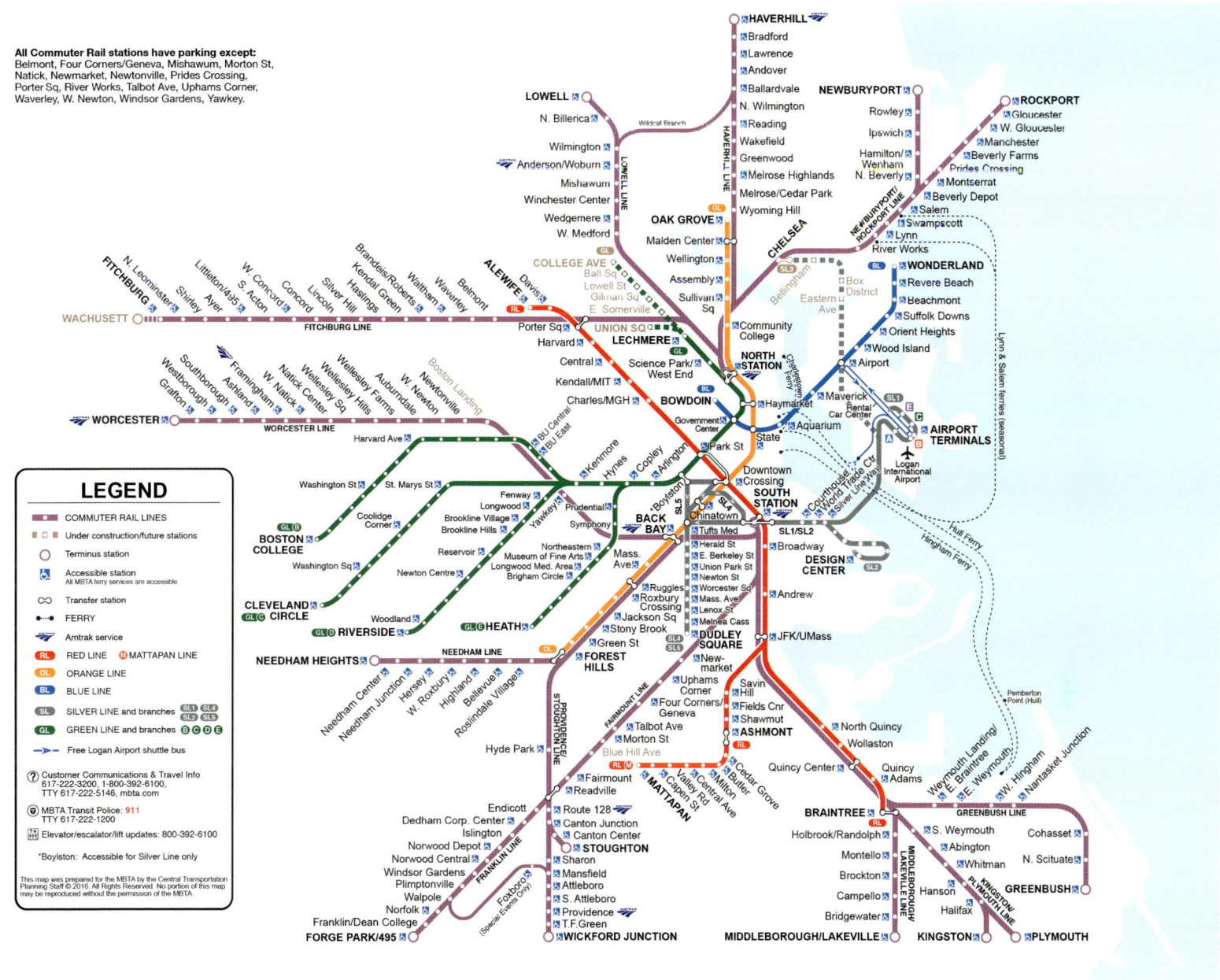

Figure 4.3 | **Boston MBTA System Map**

had a final lining of brick jacked-arch construction with waterproofing provided by Roman concrete. Roman concrete *(opus caementicium)* is a mortar based on a hydraulic setting cement obtained from natural material (typically limestone), which differs from portland cement that is manufactured using gypsum and lime as binders. This section opened for traffic on September 1, 1897, with an additional section connecting the main line to a southern portal at Pleasant Street in the South End (Figure 4.4). The second section of the Tremont Street Tunnel had an underground flyover to separate revenue traffic going to South Boston from the revenue traffic going to Roxbury (Figures 4.5 and 4.6). The second section of the Tremont Street Tunnel was opened on September 3, 1898 (BTC 1899). The opening of the Tremont Street Tunnel removed the streetcars from Tremont Street and the downtown area of Boston.

Figure 4.4 | **Opening Day, September 1, 1897**

Figure 4.5 | **Tremont Street Tunnel, Single Bore**

As the city grew, demand for an expanded underground transit system intensified. The next major accomplishment for the Boston Transit Commission was excavation of the 2.1-mile-long East Boston Tunnel at a depth of more than 90 feet in primarily Boston blue clay under Boston Harbor using open-shield mining and compressed air. The use of compressed air for tunneling was a somewhat new technology of the time and this was one of the first applications in subway construction in the United States. Compressed air maintained stability of the tunnel face and prevented water from flowing and the clay from squeezing into the tunnel being excavated. At this point in the history of tunneling, this tunnel would have been considered impractical to construct by any other means. Tunneling took only about one year and was a significant improvement in underground rail transportation because it connected the Tremont Street Tunnel with the East Boston Tunnel at Scollay Square Station (now Government Center), continuing the line to the downtown Bowdoin Loop in Bowdoin Square. Originally, the East Boston Line had trolley service. In the 1920s, the East Boston Tunnel was converted for accepting heavy-rail transit vehicles by enlarging the tunnel (Figure 4.7). Today this tunnel is the rail transit link to Boston's Logan International Airport.

During the same period, the Washington Street Tunnel was being constructed of similar jacked arch construction (Figure 4.8) as used for the first underground transit lines under Tremont Street. This tunnel was to join the elevated lines in the South End at the Pleasant Street portal and through downtown Boston to North Station (BTC 1910). The Washington Street Tunnel allowed for easy passenger connections between the growing South End and North End of Boston, and it also allowed easy passenger access to the south and north railroad stations. The Washington Street Tunnel was completed in 1910. This line connected the East Boston Tunnel at State Street and provided a future connection to the Dorchester Tunnel at Summer/Winter Street Stations. At the same time, the Tremont Street Tunnel was extended to Kenmore Square, as a twin-cell cast-in-place concrete tunnel, and was completed in March of 1912 (BTC 1913). The line was further extended to Commonwealth Avenue and Audubon Circle in 1930 (BTC 1932).

The Ashmont–Harvard Square Tunnel was a cut-and-cover tunnel from the portal at Dorchester Avenue in South Boston to Park Street. This section of tunnel was constructed using traditional cut-and-cover technology with cast-in-place concrete and was completed in May of 1912. The tunnel, now at Park Street on the Red Line and Summer on the Orange Line, allowed for connections from Dorchester to Forest Hills and North Station.

Figure 4.6 | **Flyover at Pleasant Street Portal**

In 1969, a short (½-mile) underground turnout and tunnel was constructed connecting the Highland Branch of the Boston and Albany Railroad right-of-way, providing service to the communities west of Boston to Route 128 (BTC 1913). In the 1930s until the beginning of World War II, the Huntington Avenue Branch of the Tremont Street Tunnel connected a light-rail line from Copley Square Station southerly to the portal at Northeastern University. The total length of the Tremont Street Tunnel and its subsequent construction (Green Line) is approximately 4.2 miles.

The Beacon Hill Tunnel was constructed between 1909 and 1912 (BTC 1913) and was a twin-cell, circular bored tunnel constructed of cast-in-place concrete. This tunnel was necessary to avoid disruption to the historic Beacon Hill neighborhood and the Massachusetts State House, which are above the tunnel. The transit-line crossing of the Charles River to Cambridge was on the newly constructed Longfellow Bridge. The Cambridge Tunnel was constructed to continue the line between the Charles River and Harvard Square in Cambridge (Figure 4.9). The tunnel was cut-and-cover, cast-in-place concrete construction with a single and double cell that was opened for service in March of 1912. The east portal is at Main Street and the west portal is at the Brattle Street yards at Harvard Square. Overall, the Ashmont–Harvard Square rapid transit line is approximately 7 miles long (MBTA 2016b).

There was little expansion of the subway system from 1940 to 1965, which was due to extreme financial losses by the operator, the Boston Elevated Railway Company. This monetary deficit was primarily due to a combination of the original agreements in 1897 for setting fares and the Great Depression of the 1930s. The Commonwealth of Massachusetts purchased the BERY bonds in 1947 and formed the MTA to create a regional government-owned transportation company. The MTA was enlarged in 1968 to incorporate commuter rail systems in eastern Massachusetts and became the MBTA, and remained as a government-owned transportation authority.

Between 1965 and 1967, the line names were changed to reflect an easier identification system:
- The Ashmont–Harvard Square Line became the Red Line.
- The Forest Hills–Everett Line became the Orange Line.
- The Tremont Street Tunnel and all of its extensions became the Green Line.
- The East Boston Tunnel became the Blue Line.

With public funding available, the MBTA began to expand the existing tunnel system, commencing with the expansion of the Orange Line north to Melrose, identified as the Haymarket North Project. This project created better access to downtown for Charlestown, Somerville, Medford, and Melrose. The majority of the Orange Line is at grade with the only tunnel section being the crossing for the Charles River. This project added approximately 1 mile to the tunnel system by using an immersed-tube tunnel (Figure 4.10) to cross the Charles River to the Bunker Hill portal in Charlestown (Keville 1987).

At the same time as the Haymarket North Project was under way, the first section of the Southwest Corridor Project commenced and the cut-and-cover, cast-in-place concrete South Cove (now Tufts Medical Center) Station was constructed and completed in 1971.

In 1976, the continuation of the Southwest Corridor Project began construction of a tunnel and "boat" section transit way, which was shared with Amtrak, commuter rails, and the MBTA Orange Line, and was completed in 1987 (Figure 4.11). The existing elevated line to Forest Hills that was built in 1909 had deteriorated and required replacement. The project removed the elevated rail line, which was founded on an embankment, and depressed the Orange Line and Amtrak railroad lines into a boat- or U-shaped, depressed section. The elevated portion of the Southwest Corridor in Jamaica Plain was a further barrier to development, and the removal of the elevated sections has made upscale development possible in that part of the city. A 4.7-mile tunnel with a 53-acre linear park was created over the depressed section in the Back Bay and South End.

Figure 4.7 | **East Boston Tunnel Portal, Atlantic Station, 2016**

Figure 4.8 | **Jacked Arch Construction for Roof of Washington Street Tunnel**

Figure 4.9 | **Harvard Square Station, Cambridge Tunnel, 1912**

Figure 4.10 | Haymarket North Project Immersed-Tube Section, 1968

In 1980, MBTA extended the Red Line north from Harvard Square to Alewife Circle in Cambridge. The 3.2-mile extension has a large garage at its northwest terminus and provides stops in Somerville and Cambridge, which greatly reduced commuter time into the city without driving a car. Construction for the tunnel was by cut-and-cover methods for the section extending from Porter Square to Alewife. The Porter Square Station was a drill-and-blast rock cavern and is the deepest station in the MBTA system. A tunnel boring machine that was equipped to bore through hard-rock conditions was used to tunnel between Harvard Square and Porter Square in Cambridge. For the design of the Harvard Square Station, it was the first time in North America where slurry walls were used as part of the permanent structure. The exterior wall on the left in Figure 4.12 is a precast panel mounted on the slurry wall. The Red Line tunnel with the Alewife extension now had a total length of 8.7 miles (MBTA 2016a).

In 2004, MBTA opened Phase I of the new Silver Line connecting the South End and Roxbury to downtown (Figure 4.13). This was possible as a result of moving the Orange Line ¼ mile to the west because the community requested an alternate service on Washington Street, the site of the old Orange Line.

Figure 4.11 | **Back Bay Orange Line and Amtrak Station, Southwest Corridor**

The history of this modern-day transportation improvement is complex. As a result of the relocated Orange Line to the Southwest Corridor, the stations shifted for most of the line approximately 1 mile to the west. This posed a problem for the MBTA in that the South End and part of Roxbury no longer had easy access to public transportation. The MBTA reviewed the available options, including providing bus or light rail on the surface and constructing another tunnel on the old Orange Line alignment. Based on the success of combined tunnel and at-grade street operations with dual-mode operating buses in Seattle, it was decided that the most economical method to provide transportation in this area was to adopt a diesel/electric bus (trackless trolley) option using dedicated bus lanes on the surface. To connect the new bus line to downtown, the decision was made to create a tunnel from South Station to the Seaport District in South Boston, with an eventual tunnel link between Chinatown and South Station. The Chinatown Tunnel has not yet been constructed. However, there is an easy passenger connection at Chinatown Station on the Orange Line to connect to South Station via Downtown Crossing. The new line is called the Silver Line and in recent times has been extended to East Boston and Logan Airport via the Ted Williams Tunnel. The Silver Line has greatly enhanced the development of the Seaport District of South Boston.

During the subsequent Phase II of the Silver Line extension, a 1.5-mile-long tunnel connecting South Station to South Boston was constructed. Phase II of the Silver Line consisted of a connector tunnel from South Station to South Boston at the Convention Center stop (MBTA 2016b).

Figure 4.12 | **Harvard Square Station, 2010**

This phase was particularly challenging. The tunnel was to be constructed using a combination of cut-and-cover similar to the earlier phase. However, it was now combined with tunneling by the sequential excavation method (SEM) and an immersed-tube tunnel at the crossing of Fort Point Channel. With SEM, the tunnel excavation was completed by a carefully designed sequence of well-defined smaller excavations and supported at each stage within this sequence. Adding an immersed-tube tunnel involved excavating an underwater trench within the channel and submerging a prefabricated twin-tunnel unit into position and then backfilling.

The cut-and-cover sections were traditional slurry wall construction typically used in South Boston. The most difficult portion of the project was the mining of the SEM section under the Russia Wharf Buildings, which are listed on the National Register of Historic Places, and have building foundations supported by a maze of timber piles. At this location, the historic structure foundations were underpinned along with ground stabilization using ground freezing to support the building throughout mined tunnel construction (Figure 4.14). *Ground freezing* is a construction process where pipes are installed in the ground through which a chilled brine liquid is circulated. As salt (sodium chloride) saturated water, this brine is chilled to below −27°F. This freezes the soft ground to a rock-like mass to

Figure 4.13 | **Silver Line, South Station Bus Platform**

improve ground stability for mining the tunnel. The system of ground freezing was maintained until tunneling was completed.

The crossing of the Fort Point Channel was performed using three concrete immersed-tube tunnel sections (Figure 4.15). The landside bulkheads of the immersed-tube tunnels were constructed using slurry wall techniques.

Future Expansion

By 2016, the only expansion currently being contemplated was on the Green Line, extending its service from Lechmere in Cambridge to Tufts University in Medford. This extension is at grade and is being built on existing railroad right-of-way. The MBTA has completed the preliminary engineering of the final section of the Bus Rapid Transit (BRT) Silver Line from the South Station transportation hub (Red Line), through Chinatown Station (Orange Line) and Boylston Street Station (Green Line), to connect all of Boston's downtown lines and to provide a one-seat ride from the city's Roxbury section via local streets to connect with the rapid transit lines. At this time, however, the ridership projections do not justify the capital investment. The MBTA has also completed additional studies over the past 15 years, including discussions of a railroad link between South Station and North Station. But these dialogues have been preliminary in nature, and with funding in question, it is unlikely such a link will be built in the foreseeable future. The near future work in the Boston transit tunnels will focus on the rehabilitation of the existing facilities and upgrading of electrical and mechanical systems. Most of the system's rolling stock is more than 30 years old and is in the process of being replaced as funding allows.

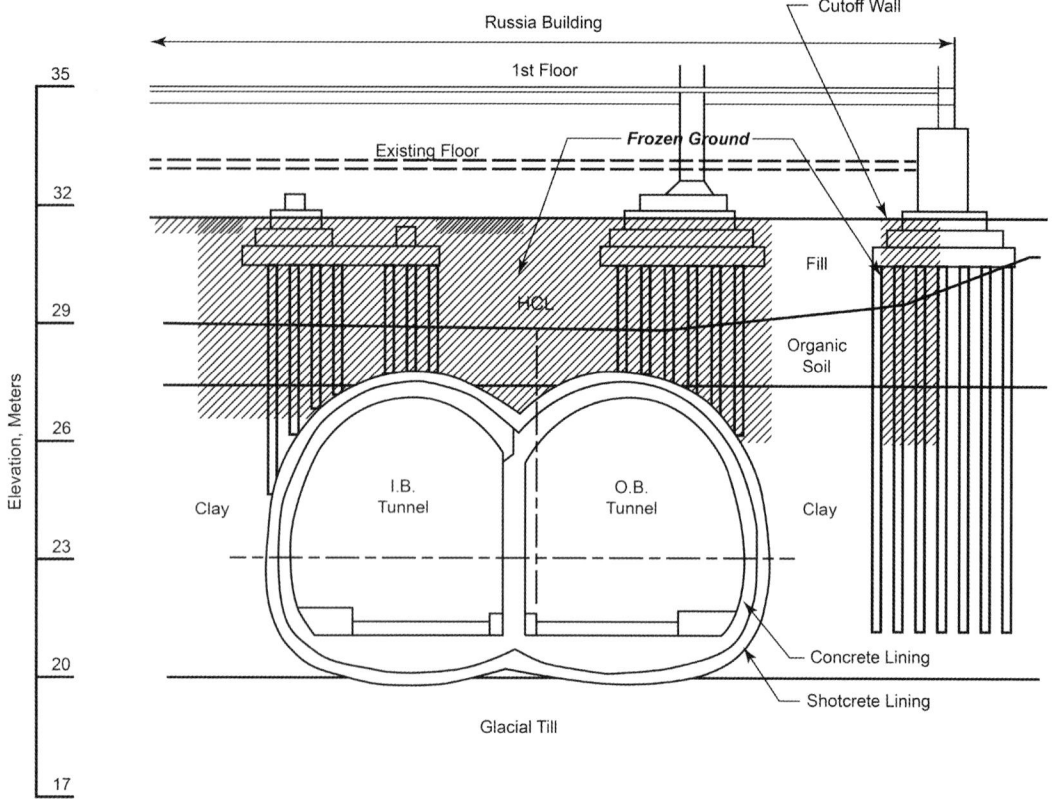

Figure 4.14 | **Ground Freezing at Russia Wharf**

Summary

Transit tunnels constructed over the past 100 years in and around Boston proper have contributed to major advancements in the tunneling industry, including methods of mined (SEM) construction using ground stabilization and ground freezing, immersed-tube tunnels, structural building underpinning, and permanent slurry wall construction for cut-and-cover tunnel and station construction. The challenges for MBTA in developing future underground transportation are to satisfy local concerns and environmental impacts, to limit the cost of construction, and to maintain for their existing systems. The MBTA currently has limited local funding, so the immediate focus will be on aging infrastructure and upgrading to current standards as funding allows.

CHAPTER FOUR | TRANSIT TUNNELS | 135

Figure 4.15 | **Immersed-Tube Tunnel Sections for Crossing Fort Point Channel**

136 THE HISTORY OF TUNNELING IN THE UNITED STATES

Figure 4.16 | **New York City, Lower Manhattan**

NEW YORK CITY

New York City would not be the transportation hub it is today without the miles of tunnels that create its subway system. The island of Manhattan, one of the five boroughs (counties) comprising all of the city, can be seen in Figure 4.16 as an island bounded by rivers. An area of only 23 square miles, Manhattan today has a population density of more than 70,000 people per square mile.

New York City did not start out with subway tunnels, but the need for them was clearly indicated before the end of the nineteenth century. The dense population that was growing because of massive immigration from many parts of the world was clogging streets and making travel within the city slow or practically impossible at peak times. As can be seen in Figure 4.17, the solution before subways was elevated trains and streets full of trolleys and horse-drawn wagons. Later as automobiles came to the city streets, congestion got worse. The New York Subway and the tunnels that create it was the solution to the ever-growing transportation needs and eventually displaced the trolleys and most of the elevated railways. No longer powered by steam but electric traction power, some lines of the New York transit system still operate on elevated tracks.

CHAPTER FOUR | TRANSIT TUNNELS | 137

Today, the 8 million people who live in the boroughs of New York City are connected with the more than 20 million people in the metropolitan area by yet more tunnels under the rivers. As shown in Figure 4.18, creating this massive transit system with 659 miles of mainline track (of which about 60%, or about 560 miles, is underground constructed by all types of tunneling methods) required more than using the best available tunnel construction technology of the time. New York City in many ways led tunneling technology change and set standards for years to come.

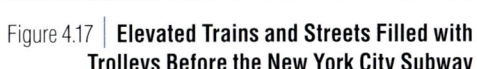

Figure 4.17 | **Elevated Trains and Streets Filled with Trolleys Before the New York City Subway**

Tunnels Making History

The history of tunneling in the New York City metropolitan area includes many first-time technological events, including

- The first use of a circular shield in the United States and the first tunnel in the world where the shield was propelled by hydraulic jacks,
- The first recorded use of compressed air in a tunnel in the world,
- The first use of an immersed-tube tunnel for mass transit in the United States, and
- The first use of a heavy-rail electric traction system for mass transit in the United States.

138 THE HISTORY OF TUNNELING IN THE UNITED STATES

Figure 4.18 | New York City Subway

However, the genesis of urban mass transit tunneling in the United States really begins in London—first with the completion of the Thames Tunnel in 1843, followed by the opening of the Metropolitan District Railway between Paddington and Farringdon in 1863. These tunnels represented great milestone achievements on a world scale. The Thames Tunnel, designed by Marc Brunel, was by its nature a subaqueous (under the river) tunnel mined using an innovative steel shield. In history, this method became "shield tunneling," where the shield permits erection of a tunnel lining and provides some support of the ground at the tunnel face and some protection to the workers within the shield. This method was a technological achievement, but initially it came at a great financial and human cost. Regardless, the Thames Tunnel proved that large subaqueous tunnels were feasible, which would permit transit tunnels to cross rivers instead of via a bridge. The Metropolitan District Railway demonstrated that trains could be used in transit tunnels. However, the system used steam-driven, coal-fired locomotives, so the air belowground became noxious and difficult to breathe. The tunnel was opened up to the atmosphere whenever possible to ventilate and improve air quality underground. The problem was eliminated when the underground rail line was eventually electrified.

First Transit Tunnel

In 1869, the first transit tunnel in New York City, a pioneering effort to put transit underground in the crowded city, was constructed in secret. Alfred Ely Beach, the publisher of *Scientific American*, built his pneumatic tube subway covertly against the wishes of William "Boss" Tweed, the corrupt political leader of New York City, who refused to grant him a construction permit.

The pneumatically powered subway under Broadway from Warren Street to Murray Street was only 312 feet long, with fans creating sufficient force to push a car holding 20 people and move it along the entire length of the tunnel (Figure 4.19). The tunnel was constructed though dry sand using an 8-foot-diameter circular shield, and it was the first tunnel to use hydraulic jacks to propel the shield forward. All of this was performed without the knowledge of the authorities. The tunnel was open to the public in 1870 but closed soon afterward. At the time, pneumatic propulsion methods had limited application for longer tunnels and therefore proved to be a poor choice for train propulsion when compared to steam, cable, or electric propulsion systems.

Shields and Compressed Air Make Subaqueous Tunnels a Reality

One of the greatest challenges of the time for tunneling through soil conditions, particularly when crossing under major rivers, was handling the groundwater at the exposed tunnel face. In subaqueous tunnels, groundwater pressures could be in excess of 50 psi (more than three atmospheres above normal air pressure). One effective method to control groundwater during tunneling was to pressurize the tunnel using compressed air.

The use of compressed air for "excavating, sinking, and mining" was patented in 1830 by Lord Cochrane. Copperthwaite (1906) states that the first use of compressed air in underground work was in 1839 by Jacques Triger who used it for sinking a mining shaft. Glossop (1976) lists 20 applications of compressed air from 1841 to 1869, primarily for constructing bridge piers. The use of compressed air in subaqueous transit tunneling was pioneered in the United States in 1879 by DeWitt Clinton Haskin, an experienced railroader with no hands-on experience as a tunneler. He was influenced by the successful use of compressed air during the construction of a bridge over the Mississippi River in St. Louis by James E. Eads. The Brooklyn Bridge caissons were also constructed using compressed air during the same period.

The first use of compressed-air tunneling for a railroad tunnel in the United States began under the Hudson River in 1879 when Haskin attempted to drive a tunnel under the Hudson River with compressed air but without the use

Marc Isambard Brunel (1769–1849)

As the second-born son to a wealthy farmer in France, it would have been customary for Brunel to enter the priesthood. However in 1786, he became a naval cadet and served on several voyages to the West Indies, returning to France in 1792. As a Royalist sympathizer, he was forced to leave his native France for the United States during the French Revolution in 1793, leaving behind an Englishwoman, Sophia Kingdom, who would eventually become his wife. Having become an American citizen while in the States, Brunel was appointed chief engineer for New York City in 1796 and designed multiple docks and commercial buildings for the city, although no official records exist today.

In 1799, Brunel returned to England and was reunited with Sophia. They married and subsequently raised three children, the youngest of which, Isambard Kingdom Brunel, would become a prominent engineer in his own right. In 1818, Brunel patented a tunneling shield that was used for the successful construction of the Thames Tunnel following several significant flooding events. With participation of Marc Brunel's son, Isambard Kingdom Brunel, the tunnel was opened in 1843 and received more than one million visitors in the first three months of service.

The "tunneling shield" patented over a century and a half ago by Marc Brunel progressively evolved over many years to make transit tunnels feasible to construct in all geologic conditions in major cities in the United States, as well as throughout the world.

Figure 4.19 | **Alfred Beach Pneumatic Tube Subway**

of a shield, with disastrous results. The death rate for miners at the time was more than 25%, and many more miners suffered illness (although they did not necessarily die) from "caisson disease." Later known as decompression sickness or the bends, caisson disease would occur if the workers returned to free air at the surface too rapidly from the pressurized air within the tunnel. After many re-starts where work had ended primarily due to lack of funds, the Hudson and Manhattan tunnels were eventually completed in 1908 by William G. McAdoo and Charles M. Jacobs using Ernest Moir's innovative medical lock to help minimize the occurrence of caisson disease. Figure 4.20 shows a typical arrangement of air locks for compressed-air tunneling. In the photo, the muck lock can be seen in the bottom of the tunnel with rail tracks, and the manlocks and emergency locks in the upper part of the tunnel. Other practical construction engineering functions had to be adapted for work under compressed air, such as surveying as shown in Figure 4.21, where the tunnel alignment survey had to be carried through the compressed-air lock.

Meanwhile in London, Peter Barlow and James Henry Greathead developed a tunneling technique in similar ground conditions by using a combination of circular shields, hydraulic jacks, cast-iron tunnel liners, and compressed air with some success. The first successful subaqueous crossing using both compressed air and a shield occurred during Greathead's mining of the City and South London Tunnel under the Thames in London (1886–1890).

The first compressed-air, shield-driven tunnel in the United States was Joseph Hobson's St. Clair River Tunnel that was constructed under the raging St. Clair River between Port Huron, Michigan, and Sarnia in Ontario, Canada, from 1888 to 1891. The successful development of soft-ground tunneling using a shield in conjunction with compressed air under rivers with high groundwater pressures enabled more tunnel routes to be viable and opened the "first door" to urban mass transit tunneling as we know it today.

Figure 4.22 shows a cast-iron tunnel lining built by this method after completing the cleanup and just before installing the final cast-in-place concrete lining. Figure 4.23 displays the typical design for a New York transit tunnel—in particular, the Vesey Street Tunnel, Lexington Line.

Electric Traction Power

Another innovation or "second door" for mass transit was opened by Frank Sprague with the introduction of large-scale electric traction power. The prime movers of urban transit in the United States in the late nineteenth century were steam engines and horses. Neither were suitable for mass transit in densely populated urban areas. Steam engines were used primarily for elevated railroads where unconfined ventilation and exhaust at the surface were possible. However, columns supporting the elevated structures interfered with roadway traffic, blocked out the sun, and reduced real estate values along the alignment. Steam engines were also notorious for belching out hot ash onto the pedestrians below. Electrification of railroads began with initial attempts to electrify the locomotives on the elevated rails.

Frank Sprague, born in Connecticut and raised in Massachusetts, graduated from the U.S. Naval Academy in 1878 with a passion for the new field of electricity and was particularly interested in the applications of electrical systems to transportation. After resigning from the navy, he first worked with Thomas Edison in 1884 before starting his own company, the Sprague Electric Railway and Motor Company. Sprague's initial triumph was the electrification of the Richmond, Virginia, passenger railway. Important visitors to Richmond in 1888, who would later have a major impact on electrification throughout the United States, included Henry Whitney of Boston. Whitney was the driving force behind the early Boston subway system. He envisioned underground electric trolleys when he visited Richmond. Sprague's company offices were in New York City. Consequently, his first tests of an electric traction system in the late 1880s were for an elevated railway in the city (Sprague 1916).

Figure 4.20 | **Compressed-Air Tunneling Locks**

Figure 4.21 | **Compressed-Air Tunnel Surveying Through a Manlock**

Figure 4.22 | **Tunnel During Construction Showing Segmental Cast-Iron Tunnel Lining**

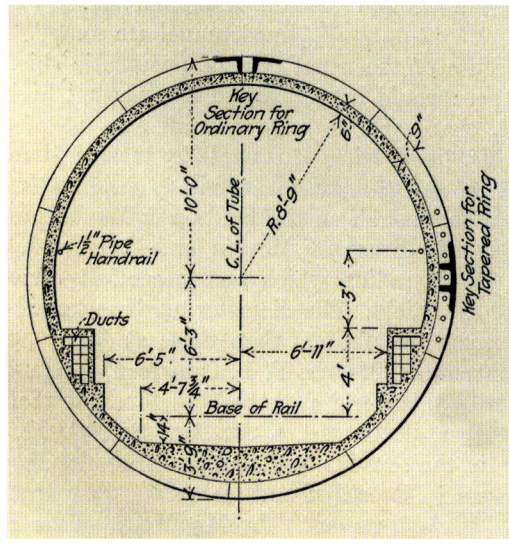

Figure 4.23 | **Typical Design for a New York Subway Tunnel**

William Barclay Parsons (1859–1932)

Parsons was a prominent American civil engineer who founded one of the largest American civil engineering firms, Parsons Brinckerhoff. He earned a bachelor's degree from Columbia College in 1879, and a second from Columbia's School of Mines in 1882.

Parsons served in the New York National Guard during the Spanish-American War and led the legendary "Fighting Engineers" of the Eleventh Engineers Regiment of the First Army during World War II. He was also one of the founders of the Society of American Military Engineers.

As a consultant and member of many boards, Parsons held many advisory roles abroad and in the United States, including chairman of the Chicago Transit Commission and transportation advisor to the cities of San Francisco, Toronto, and Detroit. He also contributed to the design of subway systems in Boston and Philadelphia.

Parsons authored several books, including *Engineers and Engineering in the Renaissance*, two technical manuals, and numerous professional papers and articles. Among his many achievements, he became chief engineer of the New York Rapid Transit Commission at the age of 35 and oversaw construction of the first New York City Subway, the Interborough Rapid Transit subway line.

Sprague followed these advances of the traction system with the successful electrification of the South Side Elevated Railroad in Chicago using what is considered by many to be Sprague's most significant mass transit innovation: cars with a multiple-unit system of train control. Without Sprague's invention, the electrification of railways and the early success of underground railroads in Boston, New York, and elsewhere would not have been possible.

Rapid Transit System

In 1894, the Board of Rapid Transit Railroad Commissioners instructed their chief engineer, William Barclay Parsons, to travel to Great Britain and Europe to study rapid transit systems and to bring back recommendations for a system in New York City. On November 20 of that same year, Parsons submitted his report that was simply titled "Report on Rapid Transit in Foreign Cities." Parsons had visited three cities in Great Britain—London, Glasgow, and Liverpool—and two cities on the continent, Paris and Berlin. He also reviewed American practice, including the Baltimore Belt Railroad electric traction locomotives that powered trains passing through an 8,350-foot-long cut-and-cover tunnel section beneath Howard Street in downtown Baltimore, Maryland; and the electrification of Chicago's elevated Intramural Railway. The success of the Intramural Railway in 1894 prompted the elevated Metropolitan South Side Elevated Railroad to engage the services of Frank Sprague and go electric in 1895.

On March 24, 1900, the official ground breaking took place with thousands of citizens attending and led by city dignitaries including Parsons who was in charge of the work. *The Race Underground* provides a modern-day summary of the history of the Boston and New York transit systems (Most 2014). For the sake of simplicity and economy, Parsons selected the cut-and-cover method to construct much of the subways under the city streets. Excavations needed to be only deep enough to get below the many utilities that were already in the city streets as

Figure 4.24 | **Typical Cut-and-Cover Construction of New York City Subway in Early 1900s**

well as the many to be constructed in coming years. Figure 4.24 shows typical cut-and-cover construction in the early 1900s. Other parts of the system required tunneling in hard-rock conditions as well as in soil requiring tunnel shields (Figure 4.25).

The Rapid Transit Subway, 21 miles long, opened to service on October 29, 1904. This was the beginning of the transit construction that continues today. Soon after the opening of the first line, major projects extended the subway to the east side of Manhattan, and to the boroughs of Queens and Brooklyn. Some were constructed on bridges, but many were in subaqueous tunnels using shields working under compressed air.

Legacy of Subaqueous Tunnels in New York

The successful completion of the initial phase of the New York City Subway system (Interborough Rapid Transit, or IRT)—initially financed primarily by August Belmont Jr. (a wealthy financier whose name is also the namesake for the famous horseracing venue, the Belmont Park Race Track), with William Barclay Parsons as its first chief engineer led to a demand from the citizens of New York City that subways be extended into the boroughs beyond Manhattan. This demand led to use of compressed air and tunnel shields in soft ground, mixed face, and rock, and a three-decades-long renaissance in the method of tunneling to construct rapid transit subaqueous tunnels:

- Joralemon Street Tunnel, twin tubes, 1903–1906
- Pennsylvania Railroad East River Tunnels, four tubes, 1904–1909
- Pennsylvania Railroad Hudson River Tunnels, twin tubes, 1905–1906
- Steinway Tunnel, twin tubes, 1905–1907
- Clark Street Tunnel, twin tubes, 1914–1919
- Montague Street Tunnel, twin tubes, 1914–1920
- East 60th Street Tunnel, twin tubes, 1916–1919
- East 14th Street Tunnel, twin tubes, 1916–1921
- East 53rd Street Tunnel, twin tubes, 1927–1930
- Fulton Street Tunnel, twin tubes, 1927–1932
- Jackson Avenue Tunnel, twin tubes, 1928–1931
- East 161st Street Tunnel, three bores, 1928–1932
- Rutgers–Jay Street Tunnel, twin tubes, 1930–1933

Figure 4.25 | **Tunneling in Difficult Ground Conditions for First New York Subways**

After construction of the many subaqueous transit tunnels, there was little subway expansion during the Great Depression, World War II years, and many years after spanning the 1930s through the 1960s. Most tunnel work involved constructing highway tunnels, for example, Queens–Midtown Tunnel, Brooklyn-Battery Tunnel (now Hugh Carey Tunnel), and the third tube of Lincoln Tunnel.

During those years of no transit expansion, subway lines were made more efficient with the creation of a free connection between adjacent lines. Most of the subway tunnel work was to lengthen platforms to accommodate longer trains with tighter headways. Short tunnels of a few thousand feet were excavated by drill-and-blast to create express tracks. Overall, this was a period of time for New York City rapid transit construction that involved making "service improvements" but not constructing "new routes."

THE HISTORY OF TUNNELING IN THE UNITED STATES

Modern Day Transit Tunnels in New York

New York continues to expand its transit system with major projects now under design and construction that were originally conceived decades ago, and in some cases, a century ago. These projects include the 63rd Street connection, an extension of the No. 7 Line from Times Square to 34th Street, the Second Avenue Subway, and expansion and upgrades of existing stations to improve capacity, convenience, and accessibility for the disabled. In lower Manhattan, much of that work has involved rebuilding the transit lines under the buildings after the destruction of the World Trade Center towers in 2001.

Second Avenue Subway

Expansion of the subway system in New York was envisioned soon after the time of constructing the original lines that opened in 1904. Two world wars, the Depression, and a focus on constructing tunnels for highways resulted in new transit lines not starting until many decades later. The general public dissatisfaction with the elevated railways had been manifest by the Second and Third Avenue elevated lines (Second Avenue El and Third Avenue El) being demolished in 1942 and 1955, respectively (Wikipedia 2016), but without plans in place to be replaced with an underground transit tunnel. It was not until the 1970s that construction of another transit line on the East Side of Manhattan was started (but not finished). At that time, other major cities is the United States, including San Francisco and Washington, D.C., were close to opening new transit systems with extensive tunnels, so another subway line in New York fit with the resurgence of subway construction in the United States. Construction on the Second Avenue Subway started in 1972 but was stopped in 1975 as a result of the City's fiscal crisis. Other concurrent work that did continue, however, included the 63rd Street tunnels in Manhattan that connected to immersed-tube tunnels constructed in the East River between Manhattan and Queens.

Figure 4.26 | **Roadway Decking for Cut-and-Cover Subway Construction**

Figure 4.27 | **Excavation Below Decked-Over Street, Installing Pipe Struts for Second Avenue Subway**

Figure 4.28 | **Excavation Showing Connection to TBM-Excavated Tunnels for the 96th Street Station**

Figure 4.29 | **Subway Station with Walls Partially Concreted Showing Surface Access Opening in Street Deck**

Design began in 2002 and construction started in 2007 on the Second Avenue Subway, a new two-track subway line that will run from 125th Street and Park Avenue at the north, east along 125th Street to Second Avenue and south along Second Avenue to the Financial District in lower Manhattan, with 16 new stations and 8.5 miles of track to Hanover Square. The total cost of this work is estimated to exceed $17 billion. The project has been broken into four construction phases. The initial Phase 1 operating segment includes three new stations (96th, 86th, and 72nd Street Stations) with a connection to the existing Broadway line at the 63rd Street Station at Lexington Avenue. Rehabilitation of the 63rd Street Station is also included in the scope of Phase 1, which opened January 1, 2017.

Cut-and-Cover Station Construction. Modern construction and its effect on the city has dramatically changed from a time more than a century ago when the first transit tunnels were constructed. In contrast to conditions shown 100 years ago in Figure 4.24, when the street was fully open to construction, major street excavations that were required to construct the Second Avenue Subway are now decked over to permit traffic flow and local services. An example is construction of the 96th Street Station under Second Avenue by cut-and-cover, as shown in the sequence of Figures 4.26 through 4.29. In Figure 4.26, the excavation in the Second Avenue roadway is decked over. Access to the underground works is through limited openings in the deck. Decking over the excavations addresses strong public concerns for impacts to traffic and business along Second Avenue. What was an unavoidable and tacitly acceptable construction impact from transit construction 100 years ago is not acceptable today.

Below the deck in Figure 4.27, excavation is taking place without interfering with the street traffic above. The figure shows pipe struts that laterally support the excavation walls being installed under the utilities hanging from the roadway decking. When the excavation had proceeded to full depth, as shown in Figure 4.28, a "launch box" was created where the TBM was assembled and hard-rock tunneling commenced to connect to the next station.

Figure 4.30 | **72nd Street Station Under Construction**

Figure 4.31 | **TBM-Excavated Tunnel and Interface with Drill-and-Blast Excavated Turnout Cavern**

This figure shows the station fully excavated with the start of the circular TBM-excavated tunnels in the distance. Figure 4.29 shows a later stage of station construction with the station walls partially concreted.

Mined Cavern Station Construction. The tunnels were first mined between 91st Street and 63rd Street using a hard-rock TBM. The stations at 86th and 72nd Streets were then constructed as large mined rock caverns. Figure 4.30 is an underground station that has been fully excavated but shown before installation of the waterproofing and final concrete lining. In the far distance of the figure, the circular outline of the TBM-excavated tunnels can be seen. Figure 4.31 shows the detail of the work being done by the drill-and-blast method of rock excavation.

Figure 4.32 shows concrete formwork for constructing the final lining of a station arch for the Second Avenue Subway. Groundwater leaks are common in the original subways as well as in those built in more recent times. Leaky, wet tunnels are undesirable and result in eventual deterioration of the tunnel structure. For modern work, constructing dry tunnels, cut-and-cover stations, and station caverns is a priority as well as a condition for creating a modern subway. To do this, the underground structures are fully enveloped with a waterproof membrane (PVC, or polyvinyl chloride) between the rock and the cast-in-place concrete lining. In Figure 4.32 the yellow waterproof membrane can be seen on the wall as the custom formwork is being put in place to construct the concrete lining.

Figure 4.32 | **Complex Station Cavern Formwork Showing Yellow Waterproof Membrane in Arch Before Final Concreting**

No. 7 Line Extension

The Metropolitan Transportation Authority's No. 7 Line Extension project extends the No. 7 IRT subway line from its historic original terminus at Times Square to the Jacob Javits Center on the far west side of midtown Manhattan. The No. 7 Line Extension project is a component of Manhattan's West Side Hudson Yards Redevelopment Project—an ongoing effort to revitalize the area surrounding the Long Island Rail Road's West Side Yard. This $2.1 billion project was completed and open for service in 2015. The No. 7 Extension project used the first double-shielded tunnel boring machines (TBMs) to excavate transit tunnels under New York City.

The original New York subway station entrances are now considered quaint and historic, like the one shown in Figure 4.33 from a historic photo of more than 100 years ago compared to today. Today, new subway entrances often make an architectural statement and, where space is available in the crowded city, are larger, such as the one shown in Figure 4.34 for an entrance to the No. 7 Line Extension in Midtown Manhattan at 34th Street. Underground, the path to the subway station platforms matches the elegance of the entrances with smooth design lines, as shown in Figure 4.35.

Station Upgrades and Interconnectivity

Many stations are currently undergoing renovation or upgrades to meet modern requirements better, as shown for the Fulton Street Transit Center in Figure 4.36. Linking five subway stations and several transit lines, this project simplifies complex underground connections that have evolved over the decades. It serves commuters traveling from the five boroughs and the New Jersey Waterfront. The Fulton Center advertises itself as being key to re-energizing lower Manhattan in the continuing recovery from the 9/11 destruction of the World Trade Center and provides accessibility expressed as "Whether you are going to or coming from Fulton Center, it's easy to get from point A to point B. With nine MTA subway lines and five subway stations, more than 300,000 individuals have easy access to Fulton Center. Connecting subway lines transport people from Grand Central in 10 minutes, and from Penn Station or Times Square in 15 minutes. The Dey Street Concourse will enable effortless accessibility to the World Trade Center" (Fulton Center 2016).

Figure 4.33 | **Original New York City Subway Entrance at 72nd Street Station, Early 20th Century and Today**

Figure 4.34 | **No. 7 Line Extension Subway Entrance at 34th Street**

Figure 4.35 | Underground Concourse in New 34th Street Station on No. 7 Line Extension

CHAPTER FOUR | TRANSIT TUNNELS | 149

Figure 4.36 | **Fulton Street Transit Center**

Figure 4.37 | **View of San Francisco from Twin Peaks**

SAN FRANCISCO, CALIFORNIA

San Francisco is a world-renowned scenic tourist destination and the financial and cultural center of northern California (Figure 4.37). The city is, in effect, the transportation hub of the greater San Francisco Bay Area that includes to the east across the bay the cities of Oakland and Berkeley, and to the south along the peninsula the "Silicon Valley" cities of Palo Alto, Cupertino, Santa Clara, and San Jose. Tunnels have been important in the development of the region's transportation system since the early part of the twentieth century when streetcar lines for the San Francisco Municipal Railway (Muni) were extended west and relocated underground, and they continue to be essential with the recent expansion and construction of the Central Subway through downtown San Francisco. The outstanding achievement where transit tunnels have played a key role is the creation of the Bay Area Rapid Transit (BART) system that opened in 1972 with the goal of a full-circle rapid transit line around the bay.

Bay Area Rapid Transit System

BART is a heavy-rail transit system that serves the Bay Area, linking San Francisco to Alameda and Contra Costa Counties to the east and northern San Mateo County to the south. Currently operating 44 stations and 104 miles of twin track, 37 miles of which is in subway, the system averaged 421,000 riders per day and 127 million trips in 2015 and continues to grow.

BART was touted in a 1966 issue of *Civil Engineering*-ASCE magazine as the most advanced rapid transit system in the world (Godfrey 1966) and became the first in a series of "modern" heavy-rail transit systems in the United States that were to follow, including the Washington Metro (1976) and the Metropolitan Atlanta Rapid Transit Authority in Georgia (1979). These transit systems were also designed to the same vision of rider comfort specifically aimed at luring suburban commuters away from their vehicles.

Initial awareness of the need for new transit for the Bay Area began in the late 1940s in the face of rapid region-wide urban and suburban growth following World War II and the realization that more highways and buses alone were not going to meet the region's needs. An extensive Key System had served Berkeley, Oakland, and San Leandro with ferry services, followed by streetcars, and then finally bus services, that were heading to and from San Francisco from 1903 into the late 1950s. This was one of the many rail systems in the United States that met its demise directly or indirectly in the hands of National City Lines and its owners—major corporations in the oil and truck industries—who together saw a better future in buses and automobiles than rail transportation.

Planning for BART began in earnest in 1952 with the formation of the San Francisco Bay Area Rapid Transit Commission to evaluate rail transit for the nine counties bordering the bay. The grand vision of a state-of-the art rail network circling the bay and extending north to Marin County succumbed to political and financial forces that began with San Mateo County opting out, Santa Clara County voters choosing highways instead, and Marin County being forced to withdraw over the challenges of adding rail to the Golden Gate Bridge and the increased cost of their share with fewer counties on board to share the cost.

Figure 4.38 | **Initial BART System and Extensions Showing Tunnel Locations**

Construction began in 1964 on the initial system, shown on a geographic map of the region in Figure 4.38, covering three counties (San Francisco, Contra Costa, and Alameda) with 34 stations linked by 75 miles of twin track (21.5 miles of which were tunneled in rock and soils), 3.6 miles in an immersed tube under the bay, 25 miles of aerial structure, and the remainder at grade. Tunnel construction methods included cut-and-cover for the Oakland and Berkeley sections, drill-and-blast through the Berkeley Hills, and hand mining with shields.

Shield-Driven Tunnels in Soft Ground

Underground utilities pose a major complication for construction of new underground transit in the city streets. As can be seen in Figure 4.39 (Kuesel 1972), the utilities under Market Street in downtown San Francisco demanded that BART be constructed by tunneling wherever possible. Extensive and aging, utilities of all types (water, sewer, gas, electric, communications) can be seen in the figure, which was prepared by the designers in 1965. Cut-and-cover station excavations on Market Street had to deal with all utilities by relocating them out of the way, or where possible, preserving in place. Avoiding the complexity and cost of utility work was a good reason to bore a tunnel between the stations on Market Street, which required shield tunneling in soft ground.

To control settlements in the saturated soft ground, specifications required all tunnels through San Francisco, Berkeley, and Oakland to be driven using a shield, and where compressible soils were encountered, with compressed air. Several contractors opted for closed-faced tunneling machines like the MEMCO (Mining Equipment Manufacturing Company) cutting wheel machine shown in Figure 4.40 to drive just over half of the tunnels, with the remainder hand-mined using conventional shields. Two thirds of the tunneled excavation employed compressed-air methods. The tunnel shield shown in Figure 4.40 is closed face (Thon and Amos 1968), but it is not the same as the more modern closed "pressurized face" tunnel boring machines (TBMs) that evolved decades later to replace the need for using compressed air to make tunneling feasible. Project-wide average tunneling rates ranged from 40 feet per day for the machine-driven tunnels and down to 25 feet per day for the hand-mined shield headings (Peterson and Frobenius 1971).

A standardized, six-segment plus key welded steel lining (Figure 4.41) was used for all of the driven tunnels and was the first use for a transit tunnel in the United States. Cast-iron segments had been used previously. Fabricated by Kaiser Steel in Napa, California, the 16.5-foot inside-diameter bolted rings were adopted for their flexibility and ductility under earthquake conditions. Lead wool caulk tamped into caulking grooves along all joints provided watertightness. Pea gravel or cement injected through threaded ports was used to fill the annular void outside the lining after ring erection and advancement of the tunnel shield. Other notable aspects of the BART system are discussed in the following subsections.

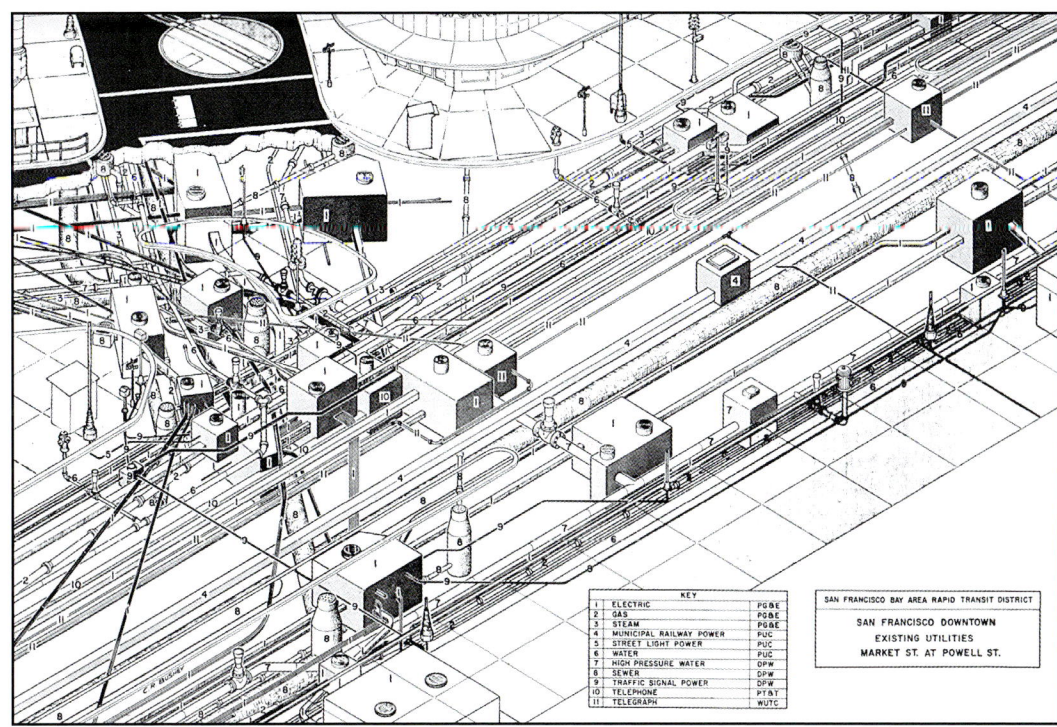

Figure 4.39 | **Utilities Under Market Street in San Francisco**

Figure 4.40 | **MEMCO Closed-Face Cutting Wheel Machine**

Figure 4.41 | **Fabricated Steel Segmental Tunnel Lining**

Market Street Tunnels and Stations

Four underground stations were included along Market Street, the heart of San Francisco, with stacked platforms and a common mezzanine shared by BART and Muni, as shown in Figure 4.42. The stacked platform configuration required construction of four soft-ground tunnels, two upper and two lower, as shown in Figure 4.43 (Kuesel 1972). Soldier pile–tremie concrete methods that had been recently successfully applied in San Francisco by the Ben C. Gerwick Company saw extensive use along Market Street as the preferred watertight wall system for controlling movements and supporting the saturated soft and compressible soils at the edge of San Francisco Bay.

The Transbay Tube

The Transbay Tube, at 3.6 miles long and 135 feet below the bay's surface, was the longest and deepest immersed-tube tunnel in the world at the time. Seismic joints designed to accommodate 6 inches of transverse displacement and 6 inches of longitudinal movements are installed where the tube meets the ventilation structures (caisson) and land-based tunnels at the shore. Since 2012, 40 years after first put into service, the Transbay Tube has been undergoing a seismic retrofit to bolster the joints and stiffen the shell and to address liquefaction risks in the surrounding backfill soils. Figure 4.44 shows two of the 58 steel shells for the Transbay Tube fabricated at Bethlehem Shipyards at Pier 70 in San Francisco. Ranging in length from 273 to 366 feet and weighing up to 14,000 tons, the 3/8-inch-thick steel single-shell tubes were launched into the bay where interior concrete was poured and then towed to the prepared trench at the bottom of San Francisco Bay. Each tube was lowered using 500 tons of rock placed on the overhead ballast pockets and finally backfilled with soil (Murphy and Tanner 1966). Figure 4.45 shows the tubes being lowered from barges to the bottom of the bay. Today, in the face of a steadily increasing regional population now nearing double what it was in the 1960s when BART was designed, the Transbay Tube finds itself challenged by its own success as a rapid route across the bay. Studies investigating ways to increase passenger throughput and a second transit crossing of the bay are underway (Bay Area Council Economic Institute 2016.)

Berkeley Hills Tunnels Through an Active Fault

The 3.2-mile-long Berkeley Hills Tunnels were driven by drill-and-blast methods through the active Hayward Fault, one of the major seismic hazards in the region. Anticipating heavy ground (high loads on tunnel supports) and recognizing the fault's creep and potential for future movement, the twin tunnels were oversized by 1 foot to 17.5 foot inside diameter and supported using steel sets with invert struts followed by a cast-in-place circular reinforced concrete lining. Rails were mounted on timber ties embedded in the concrete invert to allow adjustment for the horizontal creep. Since construction, the fault has averaged about 0.16 inch of right lateral creep per year, resulting in just over 7 inches of total offset to date, and the rails have been successfully reset to maintain reliable service. In response to the steadily diminishing vehicle clearance envelope in each tunnel, the localized fractures now visible in the reinforced concrete linings resulting from fault

Figure 4.42 | **Market Street Looking Toward Ferry Building Illustrating BART and Muni Transit Lines Operating Underground**

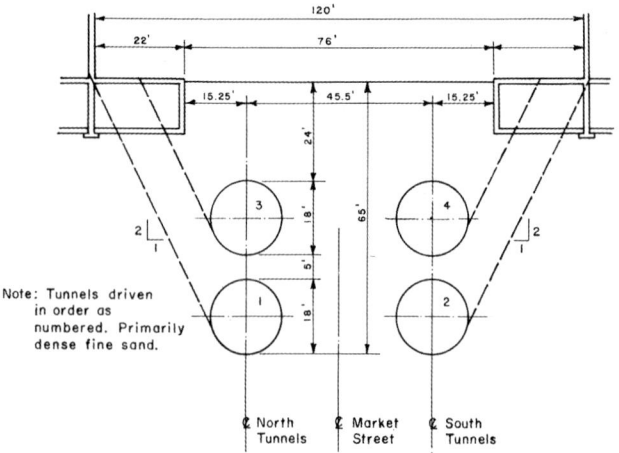

Figure 4.43 | **Market Street Tunnels for BART and Muni**

CHAPTER FOUR | TRANSIT TUNNELS | 155

Figure 4.44 | **Steel Shells for the BART Transbay Tube at Bethlehem Shipyards**

movement and the potential for significantly larger, sudden displacements during an earthquake on the Hayward Fault, BART has begun seismic retrofit studies to evaluate options for enlarging or replacing the existing tunnels where they cross the Hayward Fault.

Seismic Design of Underground Structures

Major advancements in the state-of-practice of seismic design for underground structures were introduced and put into practice on BART. The core principles of the deformation-based design methods for tunnels were pioneered by Kuesel and the BART design team. The seismic design of tunnels needed to be different from that of aboveground structures, such as building and bridges, because belowground structures react quite differently during an earthquake. During earthquake shaking, aboveground structures react depending on many factors of weight (mass),

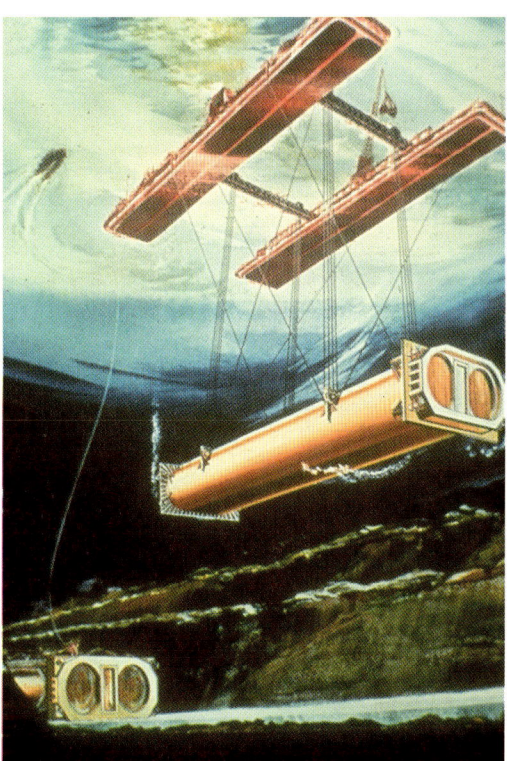

Figure 4.45 | **Lowering of Transbay Tubes to the Bottom of San Francisco Bay**

Thomas R. Kuesel (1926–2010)

Kuesel was a recognized authority on tunnel and bridge engineering and former partner at Parsons Brinckerhoff. During a career spanning more than four decades, Kuesel contributed to the design of more than 130 bridges and 140 tunnels in 36 states and on six continents worldwide. He directed the design of 20 miles of subway, 25 miles of aerial structures, two hard-rock tunnels, and a 3.6-mile immersed-tube tunnel under San Francisco Bay to link San Francisco and Oakland. He was also involved in subway projects in Boston, New York, Baltimore, Washington, D.C., Pittsburgh, and Los Angeles. He was instrumental in 1976 as the Engineer of Record for first use in the United States of permanent, one-pass, precast concrete tunnel linings that enabled their prevalent use today. Kuesel was co-editor of the *Tunnel Engineering Handbook*, first published in 1982, which is a comprehensive guide to the design and construction of tunnels. Kuesel was elected to the National Academy of Engineering in 1977.

156 THE HISTORY OF TUNNELING IN THE UNITED STATES

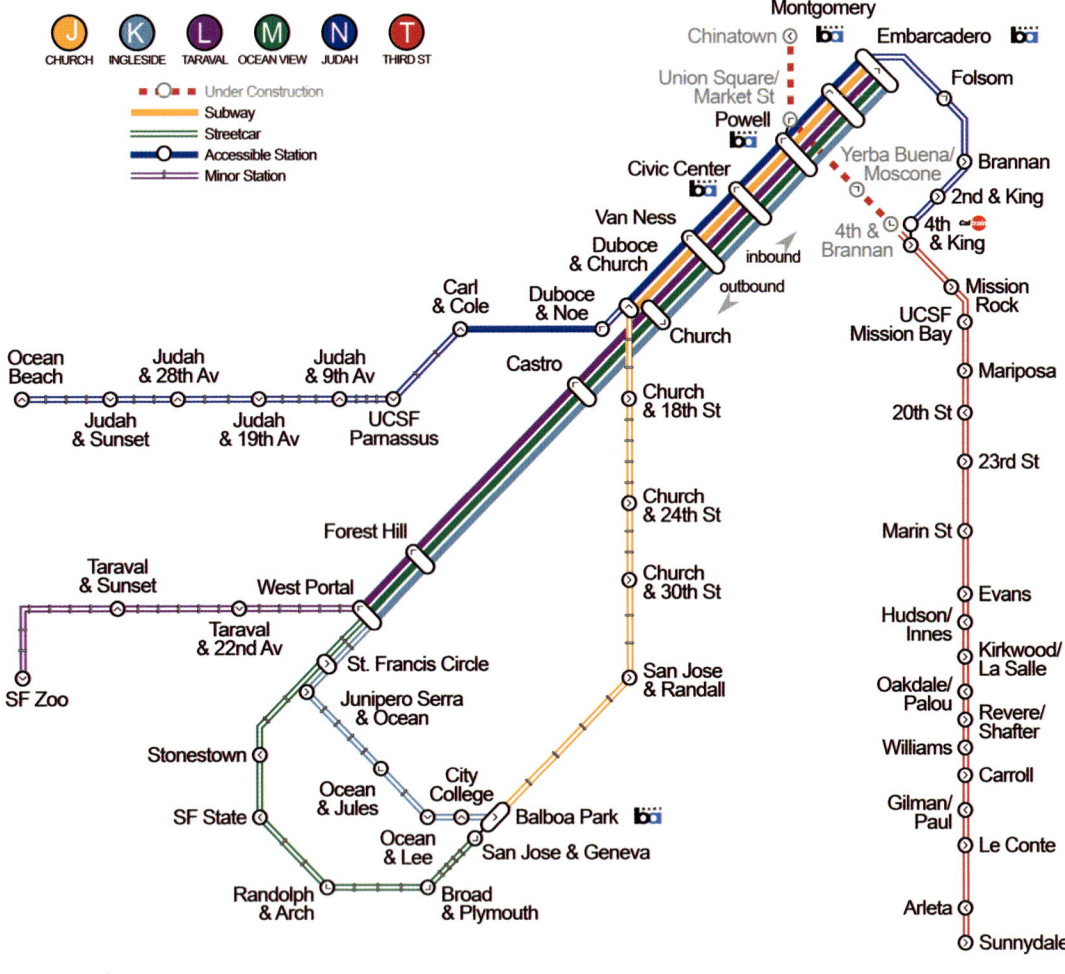

Figure 4.46 | San Francisco Muni System Map, 2016

height, and rigidity of the structure. The structure movements are unrestrained above the ground, and when movement is too great, the structure will be damaged or, in the extreme, fail and collapse. For the right conditions, the earthquake motions can result in larger (amplified) movements of the structure. Tunnels, on the other hand, have the natural advantage of being restrained by the surrounding ground, which limits movement. While aboveground structures have no external lateral support beyond being fixed to their foundations, tunnels can be considered to have complete and redundant lateral support by the ground surrounding the tunnel. In engineering terms, this is called *soil–structure interaction*.

The now accepted principles of soil structure interaction were employed where underground structures are designed to move with the deformations imposed on them by an earthquake and to have ductility (the ability to deform without breaking) instead of strength. A significant test came with the 6.9 magnitude Loma Prieta earthquake on October 17, 1989. The BART Transbay Tube was virtually undamaged, as well as other tunnels, and only closed for post-earthquake inspections. Despite all BART transit facilities and tunnels performing well during that event, seismic upgrades are in process to ensure that the rapid transit system will be able to sustain a significantly greater earthquake in future years and maintain critical life-line transportation services to the public.

BART Today and Tomorrow

In the ensuing years since BART first opened, it has been extended to San Mateo and Santa Clara Counties. Twenty years after the first transit trains rolled into service, construction resumed with extensions to the east (Pittsburg/Bay Point and Dublin/Pleasanton) and one to the south (Colma) that together added five more stations and 23.4 miles of double track. The San Francisco International Airport (SFO) extension came into service in 2003, adding 8.7 miles of new railway and four stations at South San Francisco, San Bruno, SFO, and Millbrae; the last station connects with Caltrain, the heavy-rail commuter line operating on the San Francisco Peninsula. Much of the alignment north of the airport is in cut-and-cover tunnel. As of early 2016, construction was nearing completion on the Warm Springs and Barryessa extensions into Silicon Valley (with anticipated completion by 2016 and 2018, respectively). BART Silicon Valley Phase 2 will complete the long-awaited original vision of reaching San Jose by tunneling through downtown San Jose and adding four more stations, including one adjacent to the Caltrain Diridon Station, which will provide a tie to the California high-speed rail system now under design.

CHAPTER FOUR | TRANSIT TUNNELS

San Francisco Municipal Railway and the Central Subway

San Francisco's long and colorful history of transit includes extensive use of streetcars and the famous cable cars that even now continue to operate on the hills of downtown San Francisco. Early in the twentieth century, Muni started on a large building program that resulted in construction of tunnels. In 1914, the Stockton Street tunnel opened under Nob Hill, allowing streetcars from downtown to go to North Beach and the new Marina District. Alignment of the Central Subway under construction (2016) follows Stockton Street 90 feet under this heavily trafficked roadway tunnel. In 1918, the 2.7-mile-long Twin Peaks Tunnel opened, making the southwestern quarter of the city available for development, and in 1928, the 0.8-mile-long Sunset Tunnel opened, bringing streetcars to the Sunset District. The Twin Peaks tunnels were subsequently connected to the new Muni tunnels under Market Street when the BART and Muni tunnels opened in 1972. The famous San Francisco Cable Cars only operate on the surface on a 3-foot-6-inch track gauge driven by a continuously moving 1.25-inch-diameter wire rope that follows the tracks beneath the street surface. Three lines of the original system remain totaling 10.7 miles all powered from the Cable Car Barn on Russian Hill, remarkably, by a 510 horsepower electric motor. Figure 4.46 shows the Muni system, which is being expanded by the addition of approximately 2 miles of transit tunnel under the heart of San Francisco and scheduled to open in 2019.

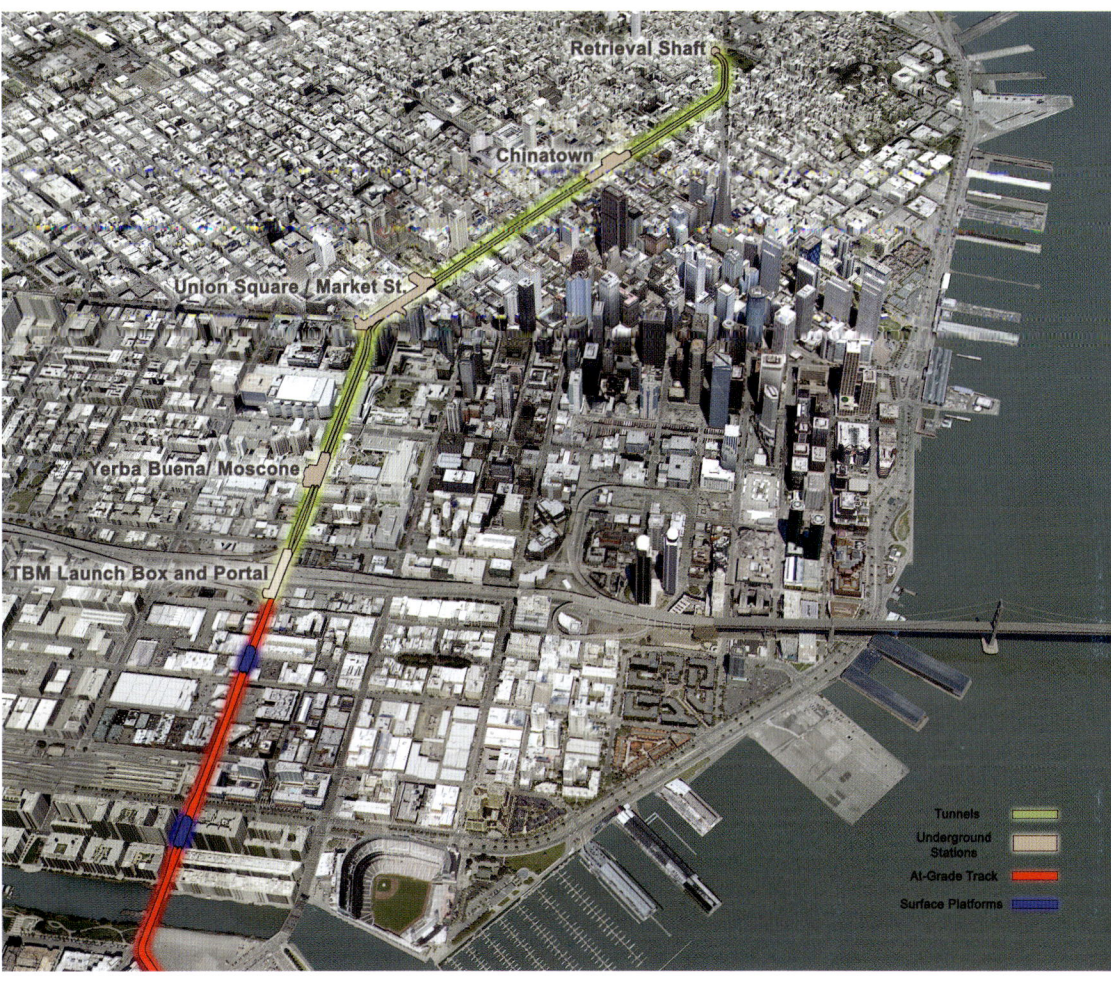

Figure 4.47 | **Muni Central Subway Alignment**

Expansion of Muni Today—The Central Subway

Forty years after the opening of the BART system, the San Francisco Municipal Transportation Agency embarked on construction of its largest underground transit project with the start of the Central Subway. This is the second phase of the Third Street light-rail line. Phase 1 of the line went into service in 2007 with 5.4 miles of twin track and 18 platforms along Third Street. It runs through the burgeoning Mission Bay and University of California–San Francisco medical office developments and extends south to Visitacion Valley near the southern border of the city of San Francisco. Phase 2, shown in Figure 4.47, extends the line 1.7 miles northward under Market Street and beyond to Chinatown with three underground stations at Yerba Buena/Moscone, Union Square/Market Street, and Chinatown.

Two new 20.7-foot-diameter Robbins earth pressure balance machines (EPBMs) were assembled and launched from a 400-foot-long launch box excavated beneath Fourth Street and supported using soldier pile–tremie

Figure 4.48 | **Low-Headroom Clamshell Digging SPTC Panels for the Launch Box**

concrete (SPTC) diaphragm walls, much like that used for BART many years earlier. Custom-built low-headroom clamshells, as shown in Figure 4.48, were required where the SPTC walls of the launch box crossed under aerial freeway structures.

Two-thirds of the tunnels were driven through saturated, dense, clayey sand and stiff clays with the middle third of the alignment encountering Franciscan bedrock under Nob Hill, a highly fractured and variably weathered formation of sandstone and siltstone. Figure 4.49 shows "Mom Chung," the first of the two EPBMs to be launched, and her hybrid cutterhead chosen by the contractor and fabricated for the mixed soil and rock.

Measures taken to protect the many commercial and residential structures that border the tunnel alignment along Fourth and Stockton Streets and Columbus Avenue included continuously read settlement prisms on all adjacent buildings and compensation grouting pipes at six locations where the tunnels crossed under or very near buildings. One of the more challenging and closely watched aspects of the project was the crossing of the Market Street Subway with its four tunnels used by BART and the Muni. Figure 4.50 shows the Central Subway tunnels where they crossed 11.8 feet

Figure 4.49 | **"Mom Chung," One of Two EPBMs Employed for the Tunnels**

CHAPTER FOUR | TRANSIT TUNNELS 159

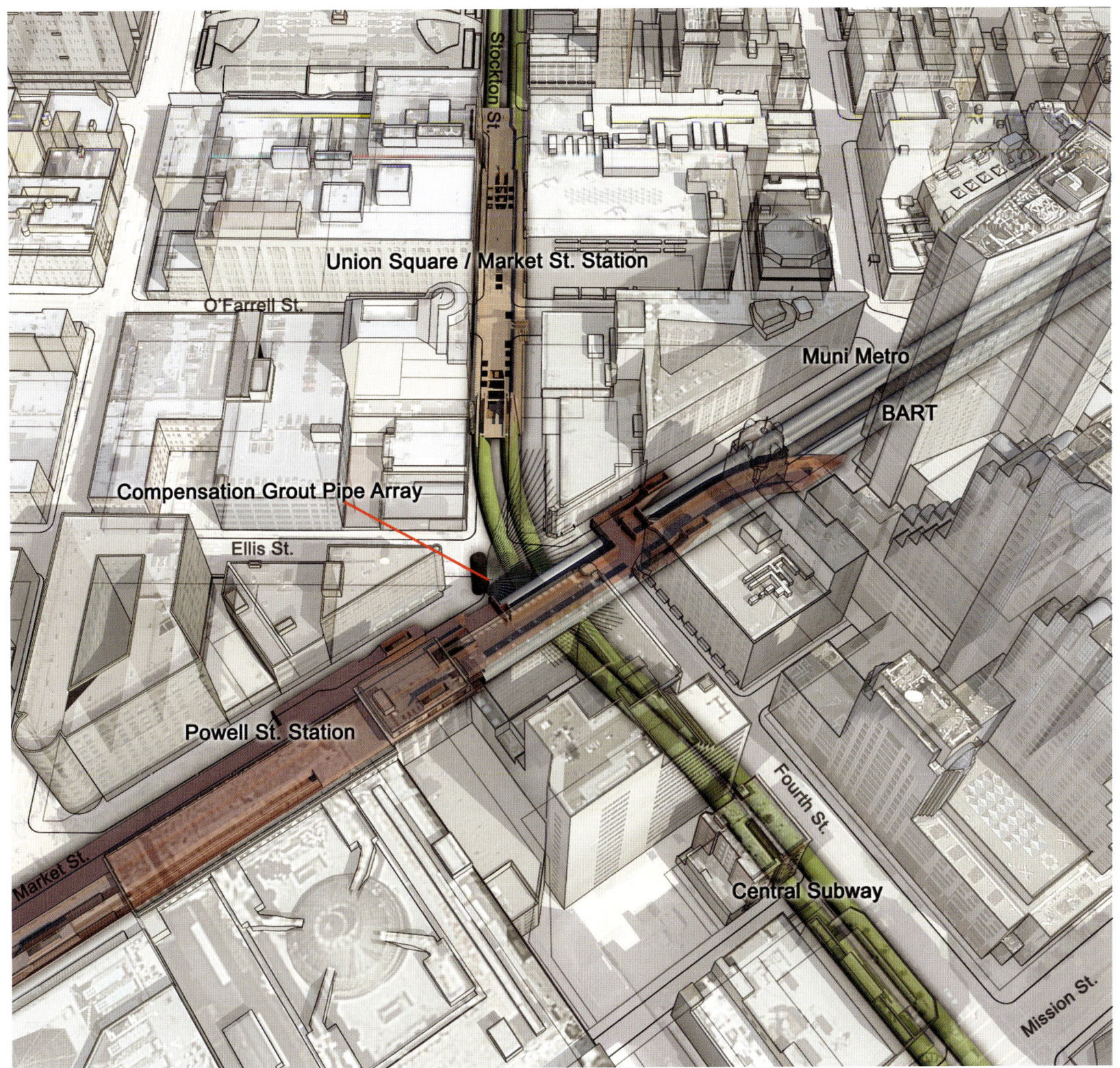

Figure 4.50 | **Muni Central Subway Crossing Under Existing Market Street Subway Tunnels**

Figure 4.51 | Ellis Street Grouting Shaft at the Crossing Under BART

below the BART tunnels. Compensation grout pipes placed from a 95-foot-deep shaft at Ellis and Stockton Streets, shown in Figure 4.51, served as a contingency measure in the event adverse settlement trends arose. The grouting array placed under the BART tunnels can be seen in Figure 4.52. Ultimately, no active compensation grouting was required and settlements of the two BART tunnels were minimal, 0.1 to 0.2 inches, well within the tolerances imposed by BART (Leong et al. 2015).

The tunnels were driven beyond the Chinatown Station to a retrieval shaft in North Beach laying the groundwork for an eventual Phase 3 that is hoped will someday extend Muni service to North Beach and Fisherman's Wharf. Figure 4.53 shows the second TBM, "Big Alma," arriving in the retrieval shaft one week after Mom Chung's arrival.

The final works for the Central Subway that are now underway are the three underground stations and one surface platform to be followed by the rail track and systems. The cut-and-cover Union Square/Market Street Station, which will provide concourse level access to the Powell Street BART Station, utilizes a combination of inclined and vertical 4-ft diameter tangent and secant piles and jet grouting. The platform and crossover caverns for the Chinatown Station are being mined using sequential excavation methods with an off-street access shaft to preserve traffic flow along Stockton Street in the heart of Chinatown. When the $1.6 billion Central Subway opens to the public in 2019, it will serve 43,700 travelers daily.

CHAPTER FOUR | TRANSIT TUNNELS

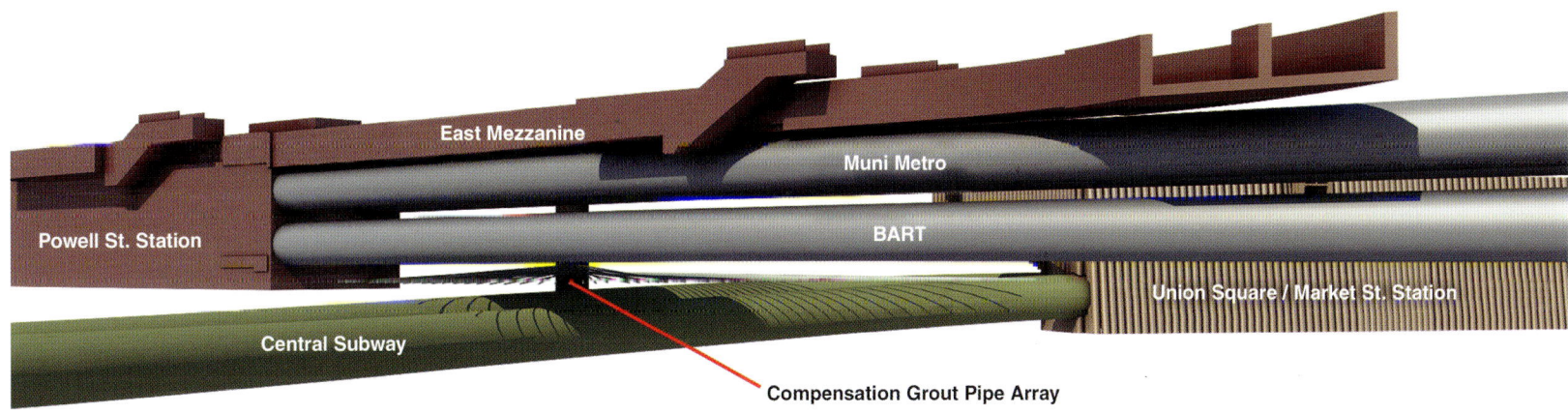

Figure 4.52 | **Compensation Grout Pipe Array Placed from the Ellis Street Grouting Shaft**

Figure 4.53 | **"Big Alma" Hole-Through in North Beach Area**

Figure 4.54 | **Atlanta Skyline at Night from Buckhead, North of Downtown**

ATLANTA, GEORGIA

Transit tunnels in downtown Atlanta, Georgia, are a key part of the Metropolitan Atlanta Rapid Transit Authority's 48-mile rail transit system. Known as MARTA, construction of this rail transit system started in 1975 and the first commercial rail service began in 1979. Underground transit lines were constructed by cut-and-cover, shield tunneling, and hard-rock mining.

Shown at night in Figure 4.54, Atlanta is Georgia's state capital with a population of 420,000 residents in the midst of the large metropolitan Atlanta area having a population of 4.5 million. MARTA operates the combined rail transit and bus system in principally three counties (Fulton, Clayton, and DeKalb) with a combined population of about 1.6 million. The transit system has a direct rail connection to Hartsfield–Jackson Atlanta International Airport, the busiest airport in the United States on the basis of total passenger traffic. Combined bus and rail train ridership of the MARTA system is 432,000 people each day along principally north–south and east–west lines, shown in Figure 4.55, that pass through transit tunnels that connect to Five Points Station, the transit system hub in downtown Atlanta.

CHAPTER FOUR | TRANSIT TUNNELS

Tunnels and Transit Station Cavern

Cut-and-cover construction was required as transitions from at-grade track to the start of tunneling or between stations. Traditional shield tunneling with compressed air was used for construction of a short section of tunnel under Broad Street for about six blocks (1,000 feet) in mixed-face conditions (rock and soil), with the central two blocks of tunneling being the most difficult. The tunnel shields started in highly weathered rock, then passed through a zone of mixed-face conditions, and eventually into saturated soil. All tunneling was located below the groundwater table, which created unstable soils at the face of the shield. To create safe and stable tunneling conditions, tunneling was accomplished using compressed air. Where tunneling involved hard rock in the lower part of the tunnels, a smaller tunnel was excavated by drilling and blasting in free air and was known as a pilot invert drift (small-diameter tunnel in the bottom of the larger tunnel). Despite all the modern advances in tunneling up to that time, the MARTA tunnel work was as difficult as the construction required in tunnels for transit and rail systems in New York City in the early 1900s.

In great contrast to the poor geologic conditions for the tunnel described above, unusually favorable rock conditions did exist in downtown Atlanta in the form of massive gneiss (a hard metamorphic rock) with very widely spaced joints. This geologic condition provided the basis for an economical rock reinforcement design for the Peachtree Center Station. This station was constructed 120 feet beneath downtown Atlanta in a 60-foot-wide cavern having a 600-foot-long center platform. In this situation, the rock was used as a structural material and a stable cavern was created using rock dowels and rock bolts (rather than structural steel and reinforced concrete), which became the permanent structural system. Although the cavern roof has a thin concrete lining, the reinforced rock creates a safe and stable underground transit station cavern. The architects and civil engineers left the rugged gneiss rock walls exposed, as can be seen in Figure 4.56. Figure 4.57 shows the station in operation.

Figure 4.55 | **MARTA System Map**

THE HISTORY OF TUNNELING IN THE UNITED STATES

Figure 4.56 | **Natural Rock Walls of Cavern for MARTA's Peachtree Center Station**

Figure 4.57 | **MARTA's Peachtree Center Station**

Future Transit in Atlanta

MARTA was conceived in the 1960s to serve the five counties comprising the Atlanta metropolitan area. Disagreements on paying sales taxes to fund the system resulted in participation from only three counties. Expansion of the system has been advocated many times over the years with little progress. Atlanta was the site of the 1996 Olympics and transit was shown to be essential to getting thousands of people from distant parking lots to the intense Olympic activities in downtown Atlanta. Despite the dramatic demonstration of the value of transit, no extensive growth of the transit system has taken place.

Only recently has the funding base increased when, in 2014, Clayton County agreed to be a part of the system. Proposals for expansion of the MARTA system have been received positively by the public. Expanded transit brings economic growth and jobs with the projection of bringing billions of dollars into the local economy and tens of thousands of jobs. Atlanta's international competitiveness will also be increased by transit system improvements as public transportation is greatly valued by global firms for deciding on investing in new facilities where the workforce can use public transit. Where needed, tunnels will likely be required to make possible a transit line that otherwise would be environmentally unacceptable or physically not constructible through the typically hilly region of metropolitan Atlanta.

CHAPTER FOUR | TRANSIT TUNNELS

Figure 4.58 | **Baltimore, Maryland**

BALTIMORE, MARYLAND

Founded in 1729, Baltimore is the largest city in Maryland and the second largest seaport in the mid-Atlantic (Figure 4.58). The city's population is 620,000 within the Baltimore metropolitan area population of 2.8 million. Baltimore is home to the world-renowned Johns Hopkins Hospital.

The Baltimore Metro, originally called the Baltimore Region Rapid Transit System, operates along a 15.5-mile-long route between downtown Baltimore and Owings Mills in the suburbs northwest of the city. The Metro system consists of 14 stations, including underground stations at the historic Lexington Market, Charles Center, and Johns Hopkins University.

The Baltimore Metro subway went into revenue service in 1983 and was the fourth new rapid transit system construction in the United States since World War II. It was designed to bring a planned intermodal transit system to the Baltimore metropolitan area, projected at the time to have a population greater than 4 million. The system was initially planned as a regional transit system to provide extensive downtown and suburban coverage with six radial lines and a downtown hub, similar to the planned transit system for Washington, D.C. Because of funding limitations, the system was reduced to 28 miles extending from downtown Baltimore northwest to Owings Mills and south to Anne Arundel County.

Figure 4.59 | **Baltimore Metro and Light-Rail Lines**

The northwest section of transit line would alleviate traffic along the most congested travel corridor in the city, and the extension south would link the city and northwest suburbs to Baltimore–Washington International Thurgood Marshall Airport, or BWI (then known as Friendship Airport). The northwest line was constructed in part along an existing rail corridor of the Western Maryland Railway, and part of this line was also built in conjunction with the Northwest Expressway, I-795. The south extension to the airport was eliminated because of community opposition. A northeast extension of the Metro to Johns Hopkins University was later added to the original Metro line, thereby linking the northwest suburbs with the university. Additional transit service in Baltimore has been established through the 30-mile Baltimore Central Light Rail Line, which runs primarily at grade from the northern suburbs of Hunt Valley and Timonium south through the city to BWI, similar to alignments within the original Metro system plan. The Baltimore Metro and light-rail lines are shown in Figure 4.59.

Baltimore Metro Tunnels

Baltimore Metro consists of the northwest line, often referred to as Section A, and the northeast extension, also known as Section C. Section A extends from Charles Center to Reisterstown Plaza and consists of 2.95 miles of mined tunnel, 1.25 miles of cut-and-cover tunnels, 0.95 mile of surface alignment, and 2.35 miles of aerial structure (Desai 1979). Section C extends from Charles Center to Johns Hopkins Hospital and consists of 1.6 miles of mined tunnel. All underground stations were built using cut-and-cover techniques. Section A was opened in 1983 and Section C in 1994. An extension of the northwest line from Reisterstown Plaza to Owings Mills was constructed using aerial and at-grade segments (Section B) and opened in 1987.

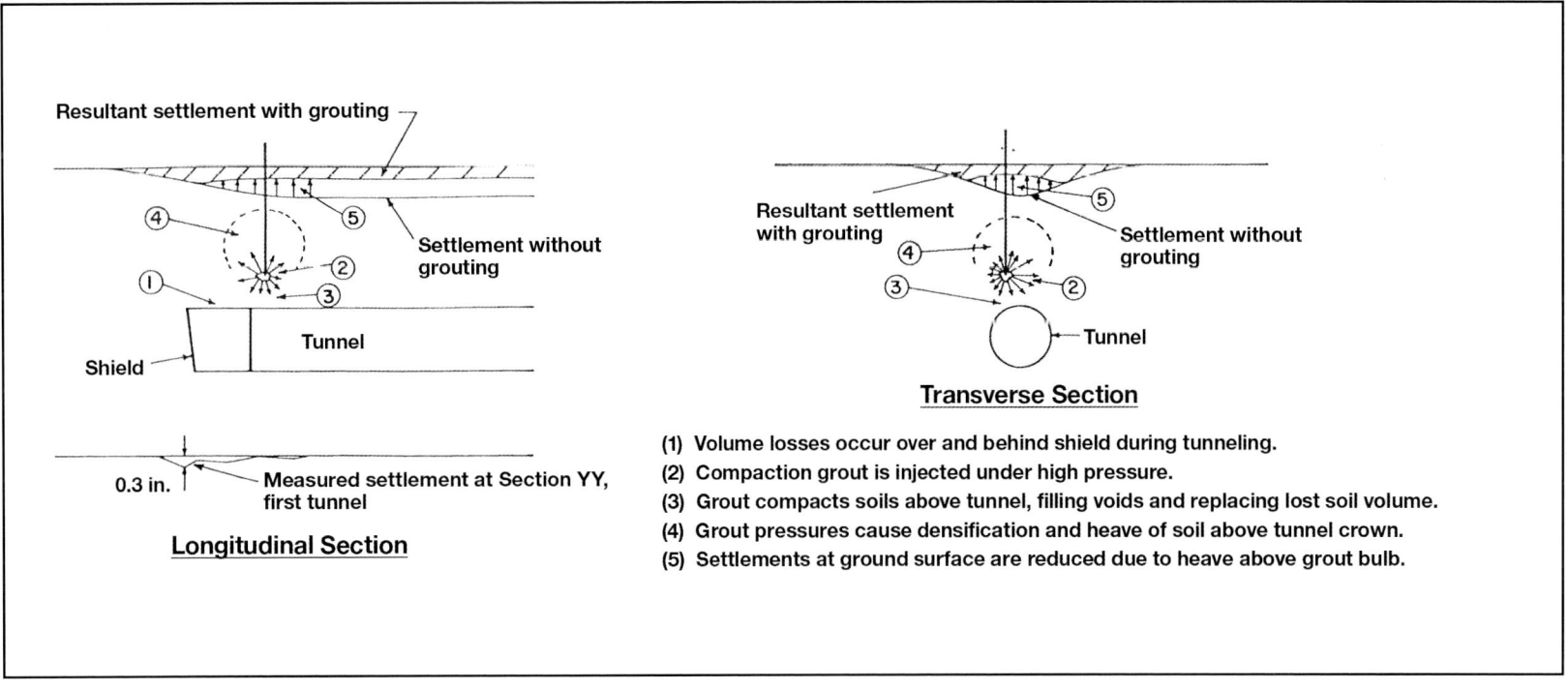

Figure 4.60 | **Schematic of Compaction Grout Program to Mitigate Tunneling Settlement**

The first 7.5-mile segment to be built, Section A, connected the northwest downtown to the Baltimore County line. The tunnels included hard rock, mixed face, and soft ground. The hard-rock tunnels were mined using full-face excavation with steel sets for initial support, with a subsequent cast-in-place concrete final liner. Rock excavation was by drill-and-blast. In areas where decomposed rock was encountered, the tunneling was changed to a top heading and bench method, whereby the tunnel was excavated in stages.

Both soft-ground and mixed-face tunnels were mined with open-face tunnel shields, and dewatering and compressed air for groundwater control. For the Lexington Market Tunnels in Section A, one tunnel was supported with steel liner plates and the other with precast gasketed concrete segments. The use of concrete segments on the one tunnel was a change in the design as part of a federally funded special program to explore the use of precast segments due to the rising cost of metals and the successful use of segmental precast linings in Europe (Foster and Butler 1981). Implemented as a demonstration project, this was the first use of precast gasketed and bolted concrete segments in the United States for a transit-sized tunnel.

A new ground improvement technique called "compaction grouting" was developed on the Baltimore transit tunnel construction. Ground settlements resulting from tunneling were mitigated by the first use of compaction grouting on Section A (Figure 4.60). Compaction grout pipes were installed above the tunnel alignment within the zone of anticipated settlements. Just after the tunneling shields passed, bulbs of grout were created by injecting low-slump grout at a depth of 5 to 10 feet above the tunnel to densify the soils overlying the tunnel and prevent ground losses and settlements from propagating to the ground surface. The ground settlement attributable to tunneling was demonstrated to be significantly reduced, as can be seen in Figure 4.60. In the case of this first use in Baltimore, ground settlement was diminished to an acceptable level that protected adjacent buildings.

Additional compaction grouting was performed under the footings of adjacent buildings as needed to mitigate settlements (ENR 1978; Baker et al. 1981). This new approach to settlement control (that is, densifying the soils over the tunnel after tunneling but before the settlement propagated to the surface) was occasionally referred to later as the "Baltimore Method" for mitigation of tunneling settlement compaction grouting.

Section A also included the use of chemical permeation grouting for support of two existing railroad tunnels that the Metro mined underneath. One existing tunnel was an early twentieth-century concrete with stone facing and brick arch Amtrak tunnel along the northeast corridor (Desai and Ku 1981). The vertical clearance between the Metro tunnels and the Amtrak tunnel was approximately 13 feet. The second tunnel was the nineteenth-century brick arch Baltimore and Ohio (B&O) railroad tunnel (now CSX Railroad) under Howard Street. The protection scheme for the B&O tunnel included permeation grouting of the soils around and underneath the existing tunnel and installation of a steel arch within the brick tunnel. This allowed the Metro tunnels to be mined using an open-face shield with compressed air beneath the B&O tunnel having a minimum clearance of 7 feet between the existing railroad tunnel and the crown of the Metro tunnels.

Section C tunnels were mined in mixed ground consisting of Cretaceous sediments, residual soil, and weathered rock using tunneling shields, excavators, and compressed air (Figure 4.61). Although the approach was similar to the work on Section A, the depth below groundwater was much higher on Section C. To limit the required supporting pressures for using compressed air, the alignment depth for Section C was established to be as shallow as possible, thereby reducing the cost of construction for the tunnels.

Several areas along the Section C tunnels encountered zones where gasoline was present that had contaminated the ground. The initial encounter resulted in a shutdown of the project for almost a year while the excavation equipment was retrofitted to be explosion proof and safe for working in such conditions. Additional contaminated areas consisting of dry cleaning solvents required all workers to wear respirators within the tunnels (Smith et al. 1993). Working conditions within the tunnels in the contaminated areas are shown in Figure 4.62.

Figure 4.61 | **Leading Edge of Tunnel Shield Used for Baltimore Metro Section C**

The use of tunneling shields and compressed air for tunnel face support resulted in several challenges. Compressed air that escaped through pervious sediments migrated up through the groundwater and carried gasoline vapors into nearby buildings, resulting in a significant vapor monitoring and extraction effort and a thorough communication exercise to inform the public of the unfortunate ongoing events (Edwards and Merrill 1995). The compressed air also proved inadequate to support the face in some areas, resulting in raveling conditions at the tunnel heading, the need for temporary face support, excessive lost ground, and formation of sinkholes at the surface (Figure 4.63).

Today's Tunneling Technology and Future Tunnels in Baltimore

The existing Baltimore transit tunnels were completed before major technological improvements in tunneling equipment. Current tunneling methods can successfully address most of the difficult construction conditions of poor and contaminated soils that were encountered on Section C. These conditions can now be addressed through the use of modern pressurized-face tunnel boring machines (TBMs). Modern TBMs can provide adequate face support without the need for continuous compressed-air operations, and the explosion hazard of petroleum-contaminated soils is avoided because onboard TBM electrical equipment is regularly designed to be explosion proof when gassy or potentially gassy conditions are anticipated.

The transit system within the Baltimore region has expanded since the completion of the Metro through the construction of the Central Light Rail Line. The light-rail line runs north–south (see Figure 4.59) and provides additional connectivity to the northern suburbs and transit service to the major league baseball and football stadiums and the regional airport (BWI). The light-rail line is entirely at grade or on aerial structures. Both the Central Light Rail and Metro have stations at Lexington Market, allowing riders to transfer between the systems. Additional studies have been undertaken to expand the Baltimore transit network with additional light-rail lines, including possible tunnel sections in the downtown area, but none have progressed to fruition. The most recent scheme for a 14.6-mile light-rail line (called the Red Line) included two tunnel sections (Cooks Lane and Downtown Tunnels) that were in the advanced stage of design when funding for the line was cancelled in 2015.

Figure 4.62 | **Use of Respirators for Tunneling in Contaminated Soil Zones on Baltimore Metro Section C**

Figure 4.63 | **Face Stability Issues Encountered Despite Use of Compressed Air on Baltimore Section C**

Figure 4.64 | Buffalo, New York

BUFFALO, NEW YORK

Buffalo is a mid-sized city within a metropolitan area population of more than 1 million. The city experienced rapid growth into the suburbs during the 1950s and 1960s when the state announced an ambitious plan to build a new campus for the University at Buffalo in Amherst with a transit connection to the existing campus in North Buffalo (Figure 4.64).

After breaking ground in late 1978, the Buffalo Metro Rail first carried passengers in 1985. An earlier plan with a surface guideway comprised of predominantly overhead structures created strong community opposition that resulted in the 6.5-mile Metro Rail project with only 1.5 miles on the surface in a pedestrian mall downtown and in tunnels outside the downtown area to the South Campus of the University at Buffalo. The underground section includes 3.5 miles of twin tunnels, 1.5 miles of cut-and-cover, four tunneled stations, four cut-and-cover stations, one tunneled crossover, one tunneled pocket track with crossover, and two tunneled turnouts that will enable future extensions. The rail system is shown in Figure 4.65, and key elements of the underground works are shown in Figure 4.66.

CHAPTER FOUR | TRANSIT TUNNELS

Fortunately, advantage could be taken of the favorable ground conditions, as the massive dolomite rock underlying the city proved to be ideal for tunneling. Twin tunnels were bored with four open-face hard-rock tunnel boring machines (TBMs). The four tunneled stations were constructed by drill-and-blast enlargements of the TBM bore, as were the crossovers and pocket track (Figure 4.67). The tunneled stations have inclined escalator shafts between the platform tunnels that were blasted in a delicate sequence to protect the narrow rock pillars as well as the buildings and utilities above (Figure 4.68).

A significant undertaking in its day, this system has served the city of Buffalo successfully over the last three decades and is partly responsible for recent growth in the medical campus and the conversion of many older commercial buildings to residential along the alignment (Figures 4.69 and 4.70).

172 THE HISTORY OF TUNNELING IN THE UNITED STATES

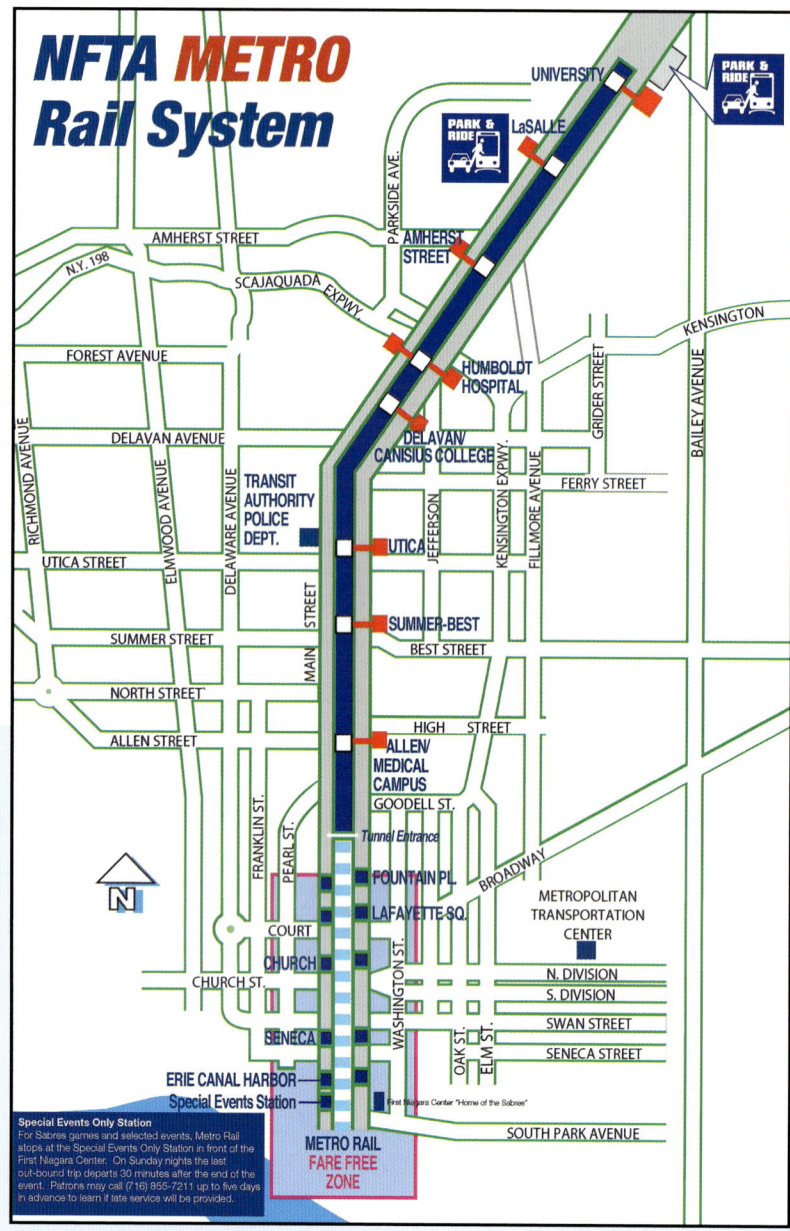

Figure 4.65 | **Buffalo Metro Rail System**

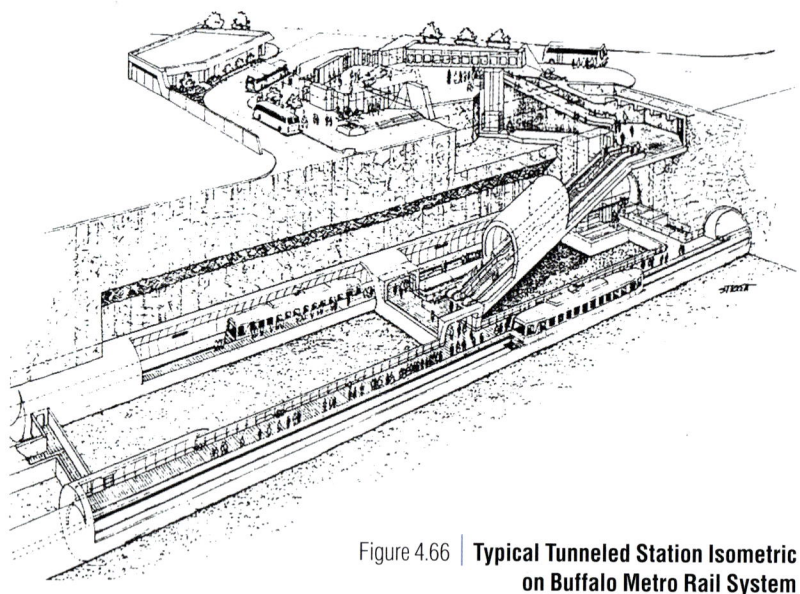

Figure 4.66 | **Typical Tunneled Station Isometric on Buffalo Metro Rail System**

Figure 4.67 | **Drill Jumbo Used for Drilling Blast Holes for the Platform Tunnel Enlargement of TBM Bore**

Figure 4.68 | **Station Platform Tunnel (foreground) at Cross Passage with Inclined Tunnel for Escalator (center)**

Figure 4.69 | **Downtown Buffalo**

Figure 4.70 | **Buffalo Metro Rail**

174 THE HISTORY OF TUNNELING IN THE UNITED STATES

Figure 4.71 | **Pittsburgh, Pennsylvania**

CHAPTER FOUR | TRANSIT TUNNELS | 175

PITTSBURGH, PENNSYLVANIA

Pittsburgh, Pennsylvania, known as the "City of Bridges," is located in the southwestern portion of the state and creates an established compact and vibrant urban center, as well as being a cultural and recreational destination. Pittsburgh, the city, has a modest population of 305,700 and an area of 58 square miles, including some in the waters of the adjacent Monongahela and Allegheny Rivers that bound the city, as can be seen in Figure 4.71. However, the city is the heart of the much larger metropolitan area of 5,343 square miles with a population of about 2.4 million.

History—From Street Cars to Light Rail and Subways

The Pittsburgh light rail (commonly known as "the T") is a 26.2-mile system. It functions as a subway in downtown Pittsburgh and largely as an at-grade light-rail service in the suburbs south of the city. The system is owned and operated by the Port Authority of Allegheny County (Port Authority). It is the successor system to the streetcar network (locally termed *trolleys*) formerly operated by Pittsburgh Railways, the oldest portions of which date back to 1903. The streetcar network in Pittsburgh was interconnected with those of other streetcar companies in adjacent counties. The Pittsburgh light-rail line is a remnant from the city's streetcar days and is one of only three light-rail systems in the United States that continues to use Pennsylvania Trolley (broad) gauge rail (5 feet, 2½ inches) on its lines instead of standard gauge (4 feet, 8½ inches) (Wikipedia 2016).

In October 1981, the Port Authority began construction on its first "modern" light-rail service, the T. The work included extensive cut-and-cover subway construction in downtown Pittsburgh. Creating the light-rail transit line to the south involved adapting existing bridges and tunnels, and constructing the Mount Lebanon Tunnel

176 THE HISTORY OF TUNNELING IN THE UNITED STATES

Figure 4.72 | **Light-Rail System Map, Downtown Pittsburgh**

to avoid major impacts to this community south of downtown Pittsburgh. The T started operation to the south in 1985 and to downtown Pittsburgh in 1987. In 2012, the underground system was expanded when the North Shore Connector tunnel under the Allegheny River was opened to service, connecting downtown Pittsburgh and the North Shore community (Port Authority of Allegheny County 2016). The downtown Pittsburgh light-rail system map is shown in Figure 4.72.

Downtown Pittsburgh Subway

Creating the new system involved using an old trolley route to connect downtown Pittsburgh to the South Hills Village area. This former street car line was reconstructed (completely double tracked) and routed from the South Hills Junction (Port Authority Station) through the Mount Washington transit tunnel. The 3,500-foot-long Mount Washington Transit Tunnel is an outstanding example of continued adaptation of an existing tunnel for modern-day transit. Constructed by drill-and-blast methods and opened in 1904, it was used only by trolleys for decades. The tunnel was upgraded to allow for joint use by bus and trolley traffic in 1973, and then in 1985 was refit for joint use by light-rail transit and buses. The line emerges from the tunnel at a newly constructed station at Station Square (shopping center) before crossing the Monongahela River on the Panhandle Bridge (Figure 4.73). At the end of the bridge, the line leads into a newly built, 1.1-mile-long downtown subway (cut-and-cover tunnel) with four stations (Figure 4.74). Upon completion of the subway, all former streetcar lines were removed from the surface streets of downtown Pittsburgh, vastly improving the chronic traffic congestion.

Mount Lebanon Tunnel

In 1979, the U.S. Department of Transportation's Urban Mass Transportation Administration approved a research grant to demonstrate the cost-effectiveness of innovative domestic and international design and construction technologies for tunneling. The grant was to develop alternative designs for a tunneled portion of the Port Authority's South Hills Corridor Light Rail Rehabilitation Program. The Dormont/Mount Lebanon Tunnel project consisted of a 2,480-foot-long twin-track tunnel in sedimentary rock and 2,363 feet of cut-and-cover, open cut, and at-grade transit line with stations at each portal (Cavin et al. 1985).

The project is notable in that it was part of the early introduction and adaptation of European tunneling methods to the United States, specifically the new Austrian tunneling method (NATM). The project was bid with two alternatives, one labeled "traditional

Figure 4.73 | **Panhandle Bridge Across the Monongahela River**

design" of initial ground support with rock dowels or steel ribs followed by a cast-in-place concrete final lining. The NATM alternative design consisted of rock bolts, light steel ribs, and reinforced shotcrete final lining. The NATM alternative was successfully bid and constructed. In this alternative, the contracting approach had flexibility to install tunnel ground support as ground conditions dictated. The project demonstrated that this approach, which involved both the flexible contracting approach for payment and the tunneling methods, could be bid and constructed in the United States. In the years since this project began, NATM terminology has evolved to be popularly referred to as the *sequential excavation method* (SEM).

The North Shore Connector

The North Shore Connector (NSC) light-rail transit project, extended the Port Authority's light rail 1.2 miles from the downtown business district to the north shore of the Allegheny River. Opened in 2012, this tunnel provides direct access under the river to the recently developed PNC Park and Heinz Field professional sports arenas, new business and residential developments, museums, and potential future extension of the light-rail transit to a major hospital and the Pittsburgh International Airport (Figure 4.75).

The project area bedrock is overlain by about 50 to 70 feet of soil on land and 20 to 30 feet in the river. A generalized subsurface profile along the project alignment in Figure 4.76 shows the sequence of uppermost fills, alluvium, and fluvioglacial deposits (sand, gravel, and silty clay material) that lie above the bedrock.

Based on the difficult soil and rock conditions anticipated, a 22-foot diameter, pressurized-face slurry TBM was selected for tunneling (Figure 4.77). Construction included mining twin tunnels, each 2,240 feet long, and 1,282 feet of cut-and-cover tunnel (Roy et al. 2009).

Figure 4.74 | **Construction of Steel City Plaza Station, Downtown Pittsburgh, 1987**

Figure 4.75 | **Aerial View of the North Shore Connector Project in Pittsburgh**

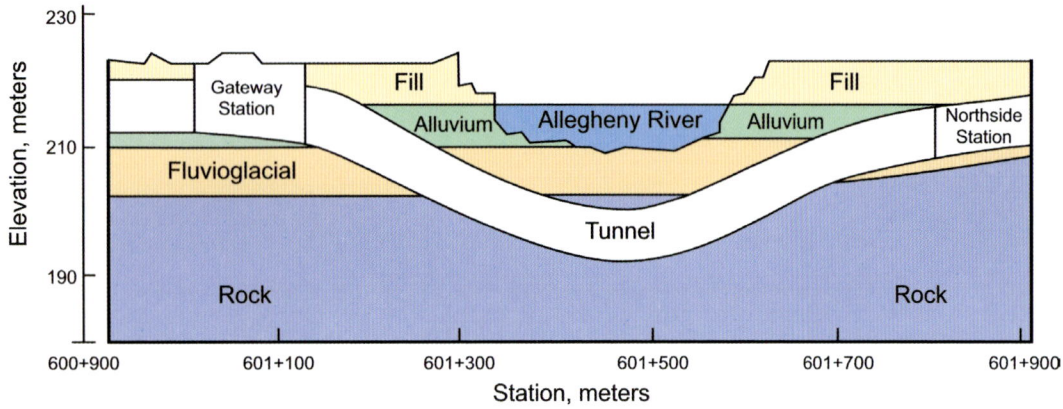

Figure 4.76 | **Schematic View of Geological Profile**

The tunnel alignment provides a direct connection to the existing Gateway Center Station, built during construction of the downtown Pittsburgh Subway. The vertical and horizontal alignment for the tunnels were constrained by river depth, minimum tunnel cover needs, maximum tunnel grades, and the southern tunnel endpoint at the existing Gateway Center Station. To avoid major obstructions, the horizontal tunnel alignment across the river follows an S-curve having a tight radius of only 600 feet. The limited distance between the tunnel endpoints pushed tunnel grades up to a steep average maximum of 7%.

There were several construction challenges on this project, including mixed-face tunneling through cohesive soils and rock beneath the Allegheny River, shallow tunnel cover, alignment through a narrow city street, steep grades, tight curves, numerous obstructions, and the underpinning and relocation of foundations for an overhead highway structure. Adjacent to the tunnel alignment are a historic structure on shallow-spread footing foundations, a high-rise building on deep foundations, and, over the tunnel alignment, a newly constructed building requiring the tunnel to pass directly beneath the building between its deep foundation piles. Figure 4.78 shows these site conditions in a schematic section along Stanwix Street.

A seven-segment, universal tapered, precast concrete lining was used for construction of the tunnel. Geometry of the segments was such that the taper allowed the lining to be rotated and erected to fit the tortuous TBM alignment that combined vertical and horizontal curves. The tunnel segments are 12-inch-thick, 4-foot-long, steel-cage-reinforced concrete. To maintain a watertight concrete liner, a gasket made of EPDM (ethylene propylene diene monomer) was bonded to the perimeter of each segment, which were then bolted together to form a ring. The completed tunnel showing the lining segments and bolt pockets for connecting the tunnel segments is shown in Figure 4.79.

Along Stanwix Street, the tunnel alignment is adjacent to Penn Avenue Place (see Figure 4.78), an eight-story historic office building constructed in the early 1900s and founded on shallow-spread footing foundations. Since geotechnical analysis indicated that underpinning the foundations could have resulted in even larger movement than the tunneling operation would potentially cause, soil improvement by jet grouting was chosen as the best alternative to prevent damage to the building. Solid jet grout blocks were constructed around both tunnels along Stanwix Street prior to tunneling. Given the close spacing of the tunnels to each other in Stanwix Street, ground stabilization by jet grouting was also required between the tunnels to protect the first tunnel lining constructed from damage during tunneling for the second tunnel. The close proximity of the two tunnels can be seen in Figure 4.80 at the reception shaft.

The TBM met its first major obstacle during tunneling beneath the Equitable Resources Building, a six-story steel-frame structure founded on steel pile

Figure 4.77 | **Herrenknecht Mixshield TBM Used for the North Shore Connector**

foundations extending to the rock below. The plan was to allow one of the bored tunnels to pass safely between the pile foundations. The potential for lateral movement of the building piles was evaluated, and the building was then monitored for movement during the tunnel boring operation directly below. The TBM passed successfully within 3 feet of the building foundation piles without causing damage to the building. Overall, tunneling was completed with no adverse effects on the community or nearby structures.

As the TBM reached the bedrock beneath the river bank at a 7% grade and with the lower portion of the cutting face boring into the harder material, there was some concern regarding the ability of the TBM to cut into the rock at this soil–rock interface. However, this proved to not be a major problem and the tunnel was successfully completed. The TBM is shown in Figure 4.81 "holing through" at the end of the tunnel drive.

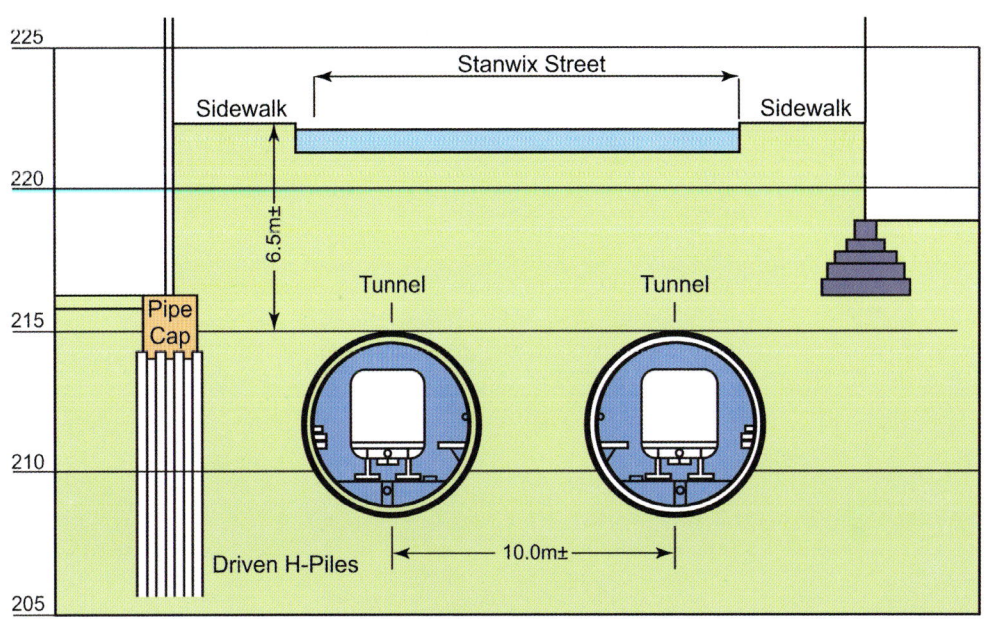

Figure 4.78 | **Stanwix Street Tunnel Alignment and Adjacent Building Foundations**

Figure 4.79 | **Segmental Precast Concrete Tunnel Lining**

180 THE HISTORY OF TUNNELING IN THE UNITED STATES

Figure 4.80 | **Completed Tunnels Beneath Stanwix Street at Reception Shaft**

Transit Tunnels and Pittsburgh

More than a century ago, a tunnel was constructed under Mount Washington to create trolley transit lines to the south of downtown Pittsburgh. Trolley service in the city continued until the 1960s and 1970s when trolleys started to be replaced by rubber-tired buses. This was also the time when Pittsburgh was experiencing the end of steel production—the closing of the steel mill and the massive loss of jobs. As these major societal changes were taking place, tunnels were key to recovery and urban transformation of downtown Pittsburgh with the opening of the subway in 1987. At the same time, the Mount Lebanon Tunnel extended transit underground to the south of downtown Pittsburgh much like the Mount Washington Transit Tunnel had done for trolleys years before. Twenty-five years later, a modern tunnel boring machine constructed a tunnel under the river for a transit line extension that opened in 2012. Tunnels clearly have been an essential part of creating and improving Pittsburgh's transit system over time and helped craft the revitalized Pittsburgh of today.

Figure 4.81 | Face of the TBM Reaching Its Final Location, Completing the Tunnel Drive

WASHINGTON, D.C.

The Washington Metro serves the Washington, D.C., metropolitan area, including the nearby suburbs of Maryland and Virginia. As a major tourist destination (Figure 4.82), mass transit is an important component of the local transportation system. Over the past four decades, the Washington Metro has been instrumental in the growth and quality-of-life improvement for this metropolitan area.

Defining the Metro

In the post–World War II environment of Washington, D.C., many political forces were engaged in the process of shaping transportation systems and facilities within and around the nation's capital. In 1952, the U.S. Congress passed the National Capital Planning Act mandating preparation of plans for movement of people and goods in the region. At that time, transit within tunnels was not universally considered as a preferred or needed option. An underground rail transit system in Washington had to compete with many political interests advocating new, well-funded highway projects that resulted from the Federal-Aid Highway Act of 1956. During the late 1950s and the 1960s, other political forces were working to change the city from the image of a crime-ridden, decaying city with the more affluent population fleeing to the suburbs and the inner city becoming more racially segregated and generally poorer. Development and new highways were often in conflict with planning an attractive and livable city.

A signature event for modern transit in Washington was the creation of the National Capital Transportation Agency in 1960 with a goal to develop a rapid rail system. In January 1967, the National Capital Transportation Agency was replaced with the Washington Metropolitan Area Transit Authority (WMATA). WMATA had representation from the counties of Montgomery and Prince George's in Maryland; the counties of Arlington and Fairfax in Virginia; the Virginia cities

CHAPTER FOUR | TRANSIT TUNNELS 183

Figure 4.82 | **Washington, D.C.**

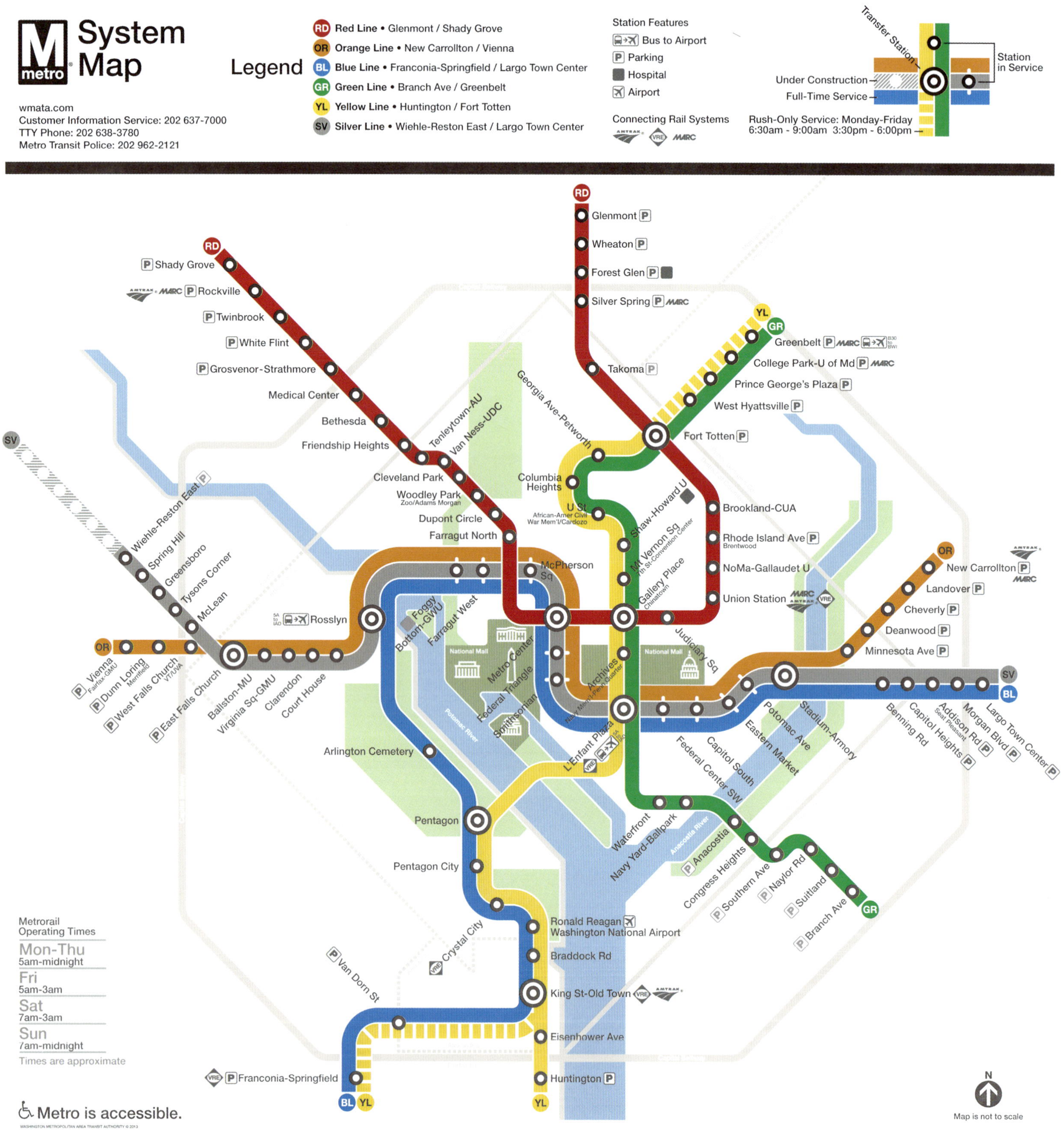

Figure 4.83 | **Washington Metro System in 2015**

of Alexandria, Falls Church, and Fairfax; and the District of Columbia. It took another three years for WMATA to officially obtain funding authority and issue the first construction contract at the end of 1969.

Ground was broken on December 9, 1969, on the first section of the Washington Metro in the heart of the city at Judiciary Square, a cut-and-cover station and line tunnels. This was a momentous event achieved after nearly two decades of political wrangling and study involving the U.S. Congress, the neighboring states of Maryland and Virginia, and the District of Columbia.

Under WMATA, this transit system with many miles of tunnels and complex stations came into being. Of necessity in the inner city, much of this transit system was built underground. The Washington Metro was originally designed to be a 103-mile-long and 83-station system with several lines (Red, Orange, Yellow, Green, and Blue), or spokes, extending from downtown Washington to the suburbs.

To great public acclaim, after a little more than six years of construction, a section of the Red Line (Rhode Island Avenue to Farragut North Stations) opened with free rides on March 27, 1976, and for revenue service on March 29, 1976.

Of the original 103-mile system, approximately half is underground. Comprising the underground portion, 19 miles are rock tunnels, 12 miles are soft-ground mined tunnels, and the rest are cut-and-cover construction. A total of 48 stations are underground, 11 of which were excavated in rock, 1 was excavated in soil, and 36 were constructed by cut-and-cover methods. WMATA divided the underground work into 62 tunnel contracts: 15 in rock, 16 in soil, and 31 by cut-and-cover (Alldredge 1974).

Construction methods were greatly influenced by the local geology. Consequently, some portions of the Washington Metro system are entirely in soil, some areas are completely in rock, and other portions were excavated through both geological conditions. The Potomac and Anacostia Rivers and Rock Creek, among other tributary streams and creeks, greatly influenced the construction methods. Tunnels had to be constructed under rivers, while others had to be constructed at substantial depth in the steep gorge of Rock Creek that runs through the city.

Several extensions were undertaken soon after the completion of the original 103-mile system in 2001. These extensions included the Blue Line extension to Largo, Maryland, an in-fill station at New York Avenue (renamed NoMa-Gallaudet U in 2012), station improvements at Rosslyn Station, and the new Silver Line that extends Metro service through Tysons Corner in Virginia to Dulles International Airport, as shown in Figure 4.83. Today the Metro is the second busiest transit system in the United States (after New York City's transit system).

Tunneling Technology Changes on the Washington Metro

During the nearly 50 years of construction on the Washington Metro system, technology improvements and lessons learned from the global tunneling industry were implemented. For soil tunnels, these improvements have included the change from a "two-pass" tunnel lining having a first pass of temporary support followed by a second pass of traditional cast-in-place concrete permanent lining to a "one-pass" tunnel lining consisting of segmental precast concrete. Soft-ground tunnels were originally excavated using open-face tunnel shields but changed over time. By the late 1980s, tunneling using closed, pressurized-face tunnel boring machines (TBMs) permitted tunneling in a wider range of soil conditions with much less risk of damaging overlying utilities or structures, greater safety for tunnel construction personnel, and without the need to dewater the soils.

Ground improvement techniques that made tunneling possible in weak or very wet soils at the start of Metro construction were largely limited to cement grouting or ground freezing. This changed over time as technologies

Ralph Brazelton Peck (1912–2008)

Ralph Peck was a world-recognized educator and consultant in foundation engineering, soil mechanics, and tunneling. His work on construction of the Chicago Subway starting in the late 1930s gave him insights to the practical side of tunnel design and construction that he applied throughout his many years of consulting and teaching as a professor of civil engineering at the University of Illinois at Urbana-Champaign.

Peck co-authored the book *Soil Mechanics in Engineering Practice* in 1948 with Karl Terzaghi that became the seminal reference for what today is geotechnical engineering. His paper "Deep Excavations and Tunneling in Soft Ground" in 1969 set the stage for modern design of tunnels. Peck was awarded the National Medal of Science in 1975 "for his development of the science and art of subsurface engineering, combining the contributions of the sciences of geology and soil mechanics with the practical art of foundation design."

As a consulting engineer, he provided advice on more than a 1,000 projects, including tunnels, dams, foundations, and mines worldwide. His contribution to transit tunnels was unparalleled as a tunneling consultant, among many projects, for the San Francisco BART, the Washington D.C. Metro, and the Los Angeles Metro.

---

evolved for chemical grouting that permitted use in fine-grained soils, jet grouting (replacement of soil by grout), compaction grouting, and compensation grouting where grouting is undertaken as tunneling takes place.

For tunnels in rock, the work started when tunneling technologies were transitioning from the traditional method of drill-and-blast excavation with rock support using structural steel (steel sets) to more modern methods. The first rock tunnel running north from Dupont Circle Station was a single double-track tunnel excavated by drill-and-blast methods. Later, tunnel contractors used hard-rock TBMs between and through the stations. Drill-and-blast excavation continued out of necessity to excavate station caverns in rock, as well as many smaller-size excavations such as for cross passages between tunnels.

Groundwater leaks were a major problem during early construction. Dry tunnels without leaks became possible when polyvinyl chloride (PVC) membranes were placed between the initial and final linings of rock tunnels (Sauer et al. 1987). It was part of the early introduction and adaptation of European tunneling methods and initial use in the United States of the new Austrian tunneling method (NATM), also known in North America as the sequential excavation method (or SEM; Heflin 1985). This was the first time in the United States where the owner fully accepted the method and used it for subsequent projects. This approach to tunneling integrated the several techniques that had evolved globally over the years and was used to successfully excavate tunnels in rock and, for the first time in the United States, in soil (Heflin et al. 1991).

The precedent for geotechnical instrumentation and monitoring for tunneling that is undertaken worldwide today was tacitly set by tunnel research work commenced on the first projects of the Washington Metro. Metro engaged the University of Illinois at Urbana-Champaign Department of Civil Engineering to conduct field research on soft-ground tunneling, rock tunneling, and cut-and-cover excavations. The lessons learned for instrumentation and procedures for tunneling were published as *Methods for Geotechnical Observations and Instrumentation in Tunneling* (Cording et al. 1975).

Washington Metro engaged a board of experts to peer review and provide advice on all aspects of design and construction. The Washington Metro Tunnel Board of Consultants (Ralph B. Peck, Don U. Deere, and A.A. Mathews) became the forerunner for what is today a common practice within the underground industry.

Transit Tunnels in Soft-Ground Conditions

Tunnels excavated in soft-ground conditions, basically tunnels in soil, were in the downtown area, and points south and east of the city, including the crossing under the Anacostia River. The initial portions of the Red Line downtown were constructed mostly by the cut-and-cover method. Only one of the initial construction contracts was tunneled. Later tunnel construction for the Yellow and Green Lines used soft-ground tunneling methods that improved over time.

Methods and equipment for soft-ground tunneling evolved with increasing levels of sophistication over the 30 years that the Metro was being built out. The great demand for tunnel construction that the Metro created, along with the concurrent global demand for tunnels in urban areas, can be said to be a factor motivating the technological improvements for tunneling.

On the Red Line downtown, the first tunnel was in soil and was constructed with an open-face shield with a two-pass tunnel lining. From a historical perspective, the project started using a soil tunneling method that had not changed much in 100 years. Where conditions permitted, a few more tunnels in soil were constructed with open-face shields. An example of this type of shield is shown in Figure 4.84, where the shield is being assembled before lowering in the shaft to start tunneling. In the background can be seen a building in the Watergate complex,

Figure 4.84 | **Open-Face Tunnel Shield During Assembly**

Figure 4.85 | **Large Excavator Used with Open-Face Digger Shield**

where the infamous Watergate scandal started with arrests on June 17, 1972. This picture was taken weeks earlier in May. As political "dirty tricks" were taking place aboveground, the difficult and dirty job of tunnel construction continued underground.

In parallel with the initial tunneling, soft-ground tunneling methods and equipment were evolving to use more sophisticated tunnel shields with a rotating cutterhead and a variety of tunnel lining types. For many years, hard-rock TBMs with powerful rotating cutterheads had been in use. Thus, it was somewhat of a transference of technology to add a rotating cutterhead to a soft-ground tunnel shield. The ultimate step in technological change for soft-ground tunneling was the pressurized-face TBM with a one-pass segmental precast concrete tunnel lining. An entirely different tunneling technology that did not use a tunnel shield, the SEM, also came to be used for tunneling in soft ground.

Open-Face Tunnel Shields and Two-Pass Tunnel Linings

An open-face shield, also known as a "digger shield," consists of a circular steel shield with a hydraulic excavator somewhat like a backhoe (Figure 4.85). Use of a mechanical digger in a tunnel shield was in itself a technology change, where before, tunnel workers excavated by hand with shovels. Figure 4.84 is an example of such a tunnel shield used on the Metro to construct the tunnels from downtown at Foggy Bottom Station to the rock tunnels under the Potomac River (Blue Line). Digger shields were also used on the Mid-City E Route (Green Line) from the U Street Station north to Fort Totten.

The first soft-ground tunnel (2,200 feet of twin tunnels on the Red Line) crossed under Lafayette Park, across Pennsylvania Avenue from the White House. Tunneling started in soil and was expected to encounter a mixed-face condition with weak weathered rock in the invert (bottom of the tunnel). The weathered rock was the reason for employing the massive digger bucket (Figure 4.85) for tunnel excavation, which had been used to successfully construct a tunnel in weak sedimentary rock in California. For both tunnel drives, the rock was found to be much harder than expected and could not be excavated by the gigantic digger mounted in the shield. The tunnels were completed by hand mining from the other end of the project using steel ribs and liner plate for ground support.

Figure 4.86 | **Steel Ribs and Timber Lagging Initial Tunnel Lining**

Figure 4.87 | **Final Cast-in-Place Concrete Lining of Two-Pass Tunnel Lining System**

Many lessons were learned on this first WMATA soft-ground tunnel project, given that the first tunnel drive had excessive "lost ground" (over-excavation of soil) and substantial surface settlement (Hansmire and Cording 1972, 1985). Lost ground in soft-ground tunnels leads to settlement and potential damage to structures and utilities. Changes in shield design for the second tunnel bore greatly reduced ground losses and surface settlement. The detailed field measurements from a research test section in Lafayette Park contributed greatly to understanding the nature of lost ground with open-faced shield tunneling. The terminology for distinguishing lost ground as *face loss, shield loss, or tail loss* was developed on this project.

A two-pass cast-in-place concrete lining was typically specified, although alternatives using cast-iron segments were considered but rarely used because of cost. Figure 4.86 shows a typical initial tunnel lining of steel ribs and wood lagging that was erected ring by ring within the tunnel shield as tunneling took place ("first pass"). In Figure 4.87 the final cast-in-place concrete tunnel lining can be seen at the time before concrete is placed in the next segment of tunnel ("second pass" of the two-pass tunnel lining).

In the 1970s, construction of one rail segment consisted of the Pentagon Route (twin 1,500 linear feet, Yellow Line) and Branch Route (twin 2,900 linear feet, Green Line) Tunnels that run south out of the L'Enfant Plaza Station for a total of 8,800 linear feet. Two custom-built TBMs with articulated shields and telescoping joints combined with a hydraulically powered claw excavator were used to bore the tunnels.

In addition, an 8,300-linear-foot single-bore tunnel (Blue, Orange, and Silver Lines) was constructed through soft-ground conditions between L'Enfant Plaza and the Federal Center SW Metro Station, under the Penn Central Railroad, with the lowest tunnel invert approximately 55 feet below the ground surface. The contractor was able to select the final lining from three alternatives: cast-in-place concrete, segmental steel, or segmented cast iron. Because of various economic and supply factors, both segmental steel and segmental cast-iron liners were used. This was perhaps the last use of cast-iron tunnel linings for transit tunnels in the United States.

A cast-in-place concrete final tunnel lining was typical for soft-ground tunnels constructed in the 1970s. Grouting was used to seal leaks that were observed after applying the concrete. The ability to completely seal leaks was difficult because they could appear later with seasonal rainfall and any long-term shrinkage of the concrete as the system was put into operation. For cut-and-cover tunnel structures, the exteriors were covered with bentonite panels for waterproofing. Bentonite panels are like a heavy cardboard filled with bentonite, a very impervious naturally occurring clay. When the panels come in contact with water, the bentonite clay would swell with the intention that this would create a waterproof barrier. Bentonite panels and grouting were not as waterproof as intended, and continuing groundwater infiltration problems have plagued the Washington Metro tunnels constructed using these methods. Later tunnel projects introduced an impermeable PVC membrane between the initial and final lining.

Beginning in 1983, WMATA started specifying final linings consisting of a single, one-pass, bolted precast concrete segment having synthetic rubber gaskets to seal against groundwater as an alternative to fabricated steel tunnel liners on the Mid-City E Route (Green Line; Hart 1989).

Much of the early soft-ground tunnel construction included widespread use of cementitious and chemical grouting to provide ground improvement. For the Pentagon Route and Branch Route Tunnels that ran south out of the L'Enfant Plaza Station, grouting was used in advance of tunneling to stabilize granular soils below existing bridge foundations and prevent the development of running conditions in the crown of the tunnel headings. In the 1990s on the Mid-City E Route, extensive grouting was completed for the Park Road Tunnels (between the Columbia Heights and Georgia Avenue–Petworth Stations). Grouting for ground improvement in tunneling was tested in the early years of Metro construction (1971) using sodium silicate grout on a portion of the Red Line to protect the adjacent Treasury Building and underground Secret Service pistol range (Figure 4.88). This was the first major use of chemical grouting as a means for ground improvement for tunneling.

Shield Tunneling with Rotating Cutterheads

For this technological change in soil tunneling, the tunnel shield has a rotating cutterhead for excavation of the soil. A rotating cutterhead is shown in Figure 4.89 from inside the tunnel shield at the tunnel face. Marks from the cutterhead teeth can be seen in the clayey soils. It is only possible to use this type of "wheeled excavator" in firm soils without groundwater. The geologic conditions were right for its use in a section of soft-ground tunnel that runs near the Capitol building on the Yellow Line.

Tunneling below the groundwater table in soil that cannot be dewatered has always been a challenge that was initially overcome by the use of compressed air and dewatering. Keeping the soil and water from coming into the tunnel shield at the front where the digging takes place made the difference between success and disaster. Shown in Figure 4.90 is a tunnel shield with a rotating cutterhead where the openings in the front are adjustable and are designed to be closed to prevent water-bearing soil from rushing into the tunnel. In Figure 4.91, the shield with rotating cutterhead has completed tunneling, and the one-pass segmental concrete tunnel lining can be seen behind the shield. The congested conditions in the tunnel during constructon are shown in Figure 4.92.

Tunneling, by its nature, excavates soil and creates a hole in the ground supported by the tunnel lining. The excavated soil, in tunneling terms called "tunnel muck," has to be taken out of the tunnel. Tunnel muck is traditionally brought out of the tunnel in muck cars that transport the excavated soil back to the shaft where tunneling started. In Figure 4.93, tunneling has just started and the TBM can be seen in the upper part of the photo starting to tunnel out of the shaft. The tunnel muck will be brought out of the tunnel by the muck cars pulled by the small locomotive. Muck removal by conveyors was used on later projects.

Figure 4.88 | **Drilling for Sodium Silicate Grouting in Lawn Next to U.S. Treasury Building**

Figure 4.89 | **Soil Tunneling Using Open-Wheel Rotating Cutterhead**

Figure 4.90 | **TBM Entering Excavation for Metro Station**

Figure 4.91 | **TBM and Precast Concrete Segmental Tunnel Lining**

Figure 4.92 | **Inside TBM Tunnel Under Construction**

Figure 4.93 | **Tunneling Set Up at Shaft for Removal of Tunnel Muck**

Pressurized-Face TBMs and One-Pass Tunnel Linings

In the 1980s and 1990s, Metro was extending the Green Line southeast to Branch Avenue and northeast to Greenbelt (Mergelsberg et al. 1987). This included a crossing below the Anacostia River and required extensive tunneling through the southeast quadrant of Washington, D.C. The timing of this work corresponded to major improvements in tunneling technology and the transition to the use of pressurized-face TBMs. *Pressurized face* means the cutterhead is enclosed in a chamber at the front of the tunnel shield, which permits maintaining pressure on the tunnel face being excavated and keeps the remaining ground surrounding the TBM from entering the excavation. The excavated material is removed by an enclosed screw conveyor (like an auger used in agriculture to move grain) or as a slurry and pumped in a pipe to the ground surface. Respectively, these are known as earth pressure balance tunnel boring machines (EPBMs) and slurry TBMs.

The Anacostia River Tunnel (Green Line) is a 2,500-foot-long twin tunnel, including a 1,500-foot-long section under the river, 500-foot sections on each shore, an emergency access shaft on the south shore, and a deep fan/pumping station on the north shore. This was the first time a one-pass precast concrete tunnel lining and pressurized-face TBM were used for a transit-sized tunnel in the United States. This eventually became the industry standard.

The segmental precast concrete tunnel lining is shown in Figure 4.94 assembled as a test ring on the ground surface to demonstrate that the lining segments could be bolted together and meet quality tolerance requirements. This is a rare example of the ring being assembled horizontally as if it were in the ground. It was found on later projects to be not necessary to assemble the ring horizontally. Assembly of a few tunnel rings vertically on the ground is much safer and sufficient to demonstrate that the segments can be bolted together satisfactorily, and is the practice today. Figure 4.95 shows the completed tunnel lining after tunneling and before transit station concreting. The tunnel lining has a "waffled" look that is created by the recesses to form pockets to permit segment-to-segment and ring-to-ring connections with bolts. This design follows how cast-iron and fabricated steel linings were shaped to accommodate connection bolts between the segments. Segmental precast concrete tunnel linings later evolved to eliminate the deep bolt pockets, which in turn made segments much more efficient and less costly to fabricate.

CHAPTER FOUR | TRANSIT TUNNELS | 191

Figure 4.94 | **Mock-Up of Segmental Tunnel Lining Prior to Construction of Anacostia River Crossing**

Figure 4.95 | **Segmental Precast Concrete Lining Showing Deep Bolt Pockets**

The pressurized-face TBM is shown in Figure 4.96 in the manufacturer's shop in Japan before shipping to the United States. This is an earth pressure balance type of TBM. At the time, the terminology for this type of tunneling equipment was being formed in the industry and the manufacturer's term for the machine type was slime shield. Functionally, it was an EPBM type. Figure 4.97 shows the TBM cutterhead as it has just completed tunneling and come through the cast-in-place-concrete slurry wall support of the 100-foot-deep shaft.

The Anacostia River tunnel has played a key role in the redevelopment of southeast Washington, D.C. In 1963, the U.S. Navy returned substantial land along the Anacostia River to the General Services Administration. The land had been used for manufacture of naval ships and weapons (gun barrels) of an earlier era. After environmental cleanup, a new headquarters for the Department of Transportation was constructed over the Metro tunnels that connect to the crossing under the river. In 2005, remaining lands were turned over to developers. One notable development was construction of Nationals Park (home of the Washington Nationals baseball team) which opened in 2008. Specifically to meet needs of the ballpark, a second entrance to the nearby Navy Yard Metro Station was constructed. Today, the area is full of new residential and commercial buildings with a growing number of restaurants and recreational activities. The Metro was clearly an essential element in the creation of this new neighborhood that was unimaginable when the TBM cutterhead was pulled out of the ground 30 years ago, as shown in Figure 4.98. A dismantled portion of the TBM can be seen in front of an old brick power plant remaining in the historic Washington Navy Yard. Without the anchor of Metro access, made possible by tunneling, it is likely that quite a different development would have evolved, or none at all.

Figure 4.96 | **TBM Used for the Anacostia River Tunnel Crossing**

Figure 4.97 | **TBM Breaking into Reception Shaft at Washington Navy Yard After Completion of Anacostia River Tunnel Drive**

Construction continued to extend the line by tunneling. The Congress Heights Station and Line (Green Line) Project included 10,000 linear feet of bored and cut-and-cover twin tunnels, five emergency access and fan shafts, a cut-and-cover double crossover, a hand-mined drainage pumping station, and a traction power substation. The alignment, contained entirely in tunnel, posed significant environmental challenges in preserving rare northern magnolia bogs. A combination of tunneling methods were used, including an EPBM in addition to the NATM.

NATM/SEM Tunneling in Soft Ground

The contract for the Fort Totten Station and tunnels in 1988 was the only contract out of three offered by WMATA where the contractor chose to utilize NATM. It was the first use of NATM in soft ground (sands and clays) in the United States. This project included 958 linear feet of twin tunnels, in addition to a portion of the station excavation located west of the Fort Totten Station. Specified initial support included shotcrete applied in three stages, with welded wire fabric and lattice girders.

The Dulles Corridor Metrorail Project extends Metro service from West Falls Church Station through Tysons Corner to Wiehle Avenue Station (Phase I) and to Dulles International Airport, terminating at a station beyond the airport in Loudoun County, Va. (Phase II). SEM tunneling (as the terminology evolved) was employed for a 1,700-linear-foot section through Tysons Corner in Phase I of the project. SEM was used, including a single- and double-grouted steel pipe arch canopy pre-support because of shallow cover and soft ground (Figure 4.99).

Rock Tunnels

The WMATA system also required extensive tunneling through rock conditions. Rock tunnels extending over 19 miles of the Washington Metro system were constructed for the Red Line extending from Dupont Circle Station north to the Rockville Station, approaches to and under the Potomac River between Foggy Bottom and Rosslyn Stations, and the north end of the Red Line between Wheaton and Glenmont Stations.

Early rock tunnel projects that started in 1970 included construction between Dupont Circle and Rock Creek Park running north under Connecticut Avenue (Red Line) and crossing under the Potomac River (Blue, Orange, and Silver Lines) to Rosslyn Station. These projects were constructed using drill-and-blast methods with grouted rock bolts, shotcrete, and structural steel support (ribs). For the Rock Creek Tunnel, construction started with an initial and final lining design of shotcrete and rock dowels. Geologic conditions were such that steel supports were required for much of the tunnel, which required substantial amounts of shotcrete to fully encapsulate and form the composite steel rib and shotcrete final lining. The result was a very large overrun of steel support ribs and shotcrete quantities, and underrun of rock bolts, with huge additional cost to Metro. This type of lining using shotcrete and rock bolts fell out of favor and was not considered to be an acceptable technique for tunnel lining construction until more than a decade later when NATM/SEM techniques were used. Otherwise, final tunnel linings were constructed of reinforced cast-in-place concrete.

Another subway segment that carries the Blue, Orange, and Silver Lines under the Potomac River to Rosslyn Station includes two 6,200-foot-long single-track tunnels, three ventilation shafts, and a mid-river pumping station. The project included 4,200 feet of hard-rock tunneling under the river, 600 feet of mixed-face tunneling, and 1,400 feet of tunneling in saturated sands and clays. Control of water during construction was a primary concern in both the soft-ground and rock tunnels.

Figure 4.98 | **Removal of TBM at Washington Navy Yard**

Figure 4.99 | **Steel Pipe Arch Canopy Pre-Support Construction**

In the 1980s and early 1990s, the Wheaton Station cavern and Glenmont Route tunnels (Red Line) were constructed by SEM in rock and utilized a continuous PVC membrane waterproofing system. The project included 14,450 linear feet of single-track tunnel with a cover of 120 to 200 feet. The Wheaton Station included a 600-foot-long station chamber plus a 200-foot-long rail crossover chamber that is 50 feet wide and 35 feet tall. These tunnels and station represented the early use of SEM for a transit system in the United States and the first use of SEM by WMATA.

Rosslyn Station is an important underground hub station in Arlington, Virginia, that serves the Orange, Blue, and Silver Lines. The station was constructed in the early 1970s and was opened in 1977. As part of a new residential and commercial development, an access improvement project was undertaken in 2009 (Figure 4.100) and completed in 2013 to accommodate expanded station capacity associated with the new development. This work included a new station entrance and a new track-level mezzanine that expands the passenger capacity of the station. Construction included a new vertical elevator and stair shaft entrance from the ground surface, connected by a 35-foot-wide by 40-foot-high mezzanine constructed using SEM techniques leading toward the existing station.

Figure 4.100 | **Cavern Construction for Rosslyn Station Improvements**

Immersed-Tube Tunnels

Not all tunnels for WMATA were constructed by tunneling but by another method called an "immersed-tube tunnel" (sometimes called "sunken-tube tunnel") as shown in Figure 4.101. This image shows the steel shell of the immersed-tube tunnel being launched from the steel fabrication yard. This type of construction was needed to cross under the Washington Channel in soft soils and now forms part of the Yellow Line.

Figure 4.101 | **Launching of Immersed-Tube Tunnel Section for Crossing Under Washington Channel**

Figure 4.102 | **Iconic Underground Metro Station Showing Vaulted Roof and Absence of Interior Columns**

Figure 4.103 | **Glenmont Station**

Station Caverns in Rock

Design standards for the underground space were determined to be of critical importance from the early planning stages of the Washington Metro. Harry Weese & Associates was engaged as the project architect and established the iconic design of the underground stations with the vaulted ceiling as shown in Figure 4.102. Many of the underground stations look the same whether the station was constructed in soft ground or rock. The simple elegance of the design and the attention to detail were carried throughout the whole system, including the aerial and at-grade sections of the rail guideways.

Overall, the program for Washington Metro involved the construction of 87 stations, 11 of which were constructed as caverns by mining in solid rock. Each structure was more than 600 feet long and was wide enough to accommodate two trains and one or two passenger platforms (a single center platform or two side platforms). Structures were also high enough to attain a stable arching action without the need for interior columns (Daugherty 1981).

Early plans for the Washington Metro defined the interior configuration of the underground cut-and-cover stations to be large, free span, coffered vaults (vaulted arch) approximately 30 feet high and more than 50 feet wide so they would be harmonized with the monumental architecture of the nation's capital (Daugherty 1981). The mined Dupont Circle Station was faithfully designed to look the same as the cut-and-cover stations with cast-in-place concrete. The interior station roof was architectural, since the rock (of poor quality in this location) had to be supported separately by extensive structural steel during station excavation.

The geometry of subsequent rock cavern stations were adjusted to be more constructible by incorporating the hard-rock TBM tunneling for the running tunnels as they passed through the station as a first stage of cavern excavation. Glenmont Station on the Red Line is an example (Figure 4.103). The interior of the stations have a different shaped ceiling constructed

CHAPTER FOUR | TRANSIT TUNNELS | 195

Figure 4.104 | **Dupont Circle Station near the End of Construction**

of precast concrete elements, while the architectural integrity was maintained with an appearance similar to the deep-coffered cast-in-place concrete design. Later stations were designed according to the dual-chamber concept, where the station has two separate train rooms rather than one, with the passengers entering and exiting from facilities excavated from the intermediate pillar of rock.

Each station had site-specific geologic details to be accommodated to achieve a constructible design. Dupont Circle Station is shown in Figure 4.104 near the end of construction before installing the side platforms, and Figure 4.105 shows the Gallery Place/Chinatown Station in service.

196 THE HISTORY OF TUNNELING IN THE UNITED STATES

Figure 4.105 | **Gallery Place/Chinatown Station in Service**

Conclusion

When the latest expansion of the system to Dulles International Airport is opened, the Washington Metro will serve 99 stations and operate on 129 miles of track on six interconnecting lines. Metro provides a critical transportation link to a population of approximately 6 million within a 1,500-square-mile jurisdiction and has allowed economic growth to expand to all corners of the region. In the 1950s when the system was first conceived, most jobs were centered in downtown Washington, and most of the workforce commuted by bus or car. Today, transit-oriented development has increased residential, commercial, and government facilities near most of the stations increasing the importance of the Metro system as a critical transportation link for the region.

CHAPTER FOUR | TRANSIT TUNNELS

Figure 4.106 | Los Angeles, California

LOS ANGELES, CALIFORNIA

Los Angeles, the City of Angels, is also known as the Entertainment Capital of the World. The greater metropolitan area is home to more than 14 million people living between the Santa Monica and San Gabriel Mountains to the north and east, and the Pacific Ocean to the west (Figure 4.106). Los Angeles County has a population of about 10 million, within which the City of Los Angeles is the second most populous city in the United States (4 million), but only ranks twelfth in land area, accounting for its urban density. Today, Los Angeles is one of the largest metropolitan economies in the world. Even though leisure and hospitality are leading industries in Los Angeles, the fashion industry, health services, and aerospace/technology are also strong economic drivers. The area is also home to the largest port complex in the Western Hemisphere. A vibrant economy coupled with 75 miles of beaches and world-famous attractions brought over 45 million visitors to the area in 2015.

Whether having experienced it in person or seen it in the movies, most Americans are familiar with the infamous traffic jams that have plagued the Los Angeles area for decades. Los Angeles was the last of the United States' mega-cities to take mass transit underground, not opening its first modern subway line until 1990. Even so, the Los Angeles region has been working on various forms of public transit for more than 140 years.

Figure 4.107 | **Hollywood Line Subway Terminal Building Placard**

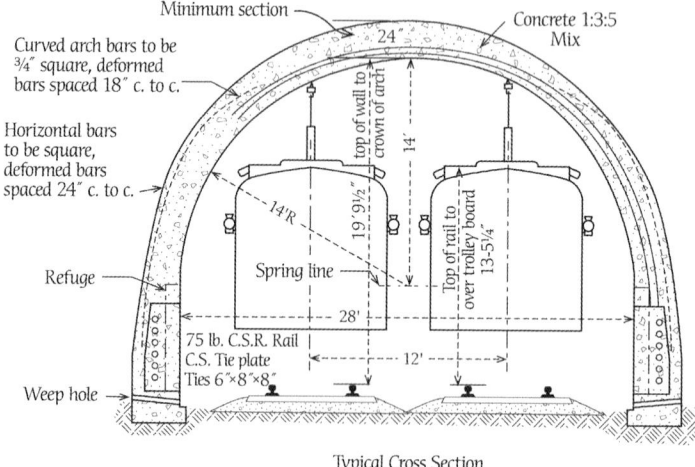

Figure 4.108 | **Pacific Electric Tunnel Cross Section**

History—1873 to the 1960s

In July 1873, the Los Angeles City Council chartered the Main Street Railroad Company to install rail lines with horse-drawn cars. Although never actually constructed, this action is said to have been the springboard for many future ventures in mass transit for the region. From 1874 through 1887, as many as 11 additional independent ventures were chartered to provide horse-drawn and cable-car service in the downtown region.

In 1902, the first electric rail transit line in the area opened, transporting passengers between Los Angeles and Long Beach. By 1911, eight regional carriers merged into the Pacific Electric Railroad Company, controlling the majority of passenger rail traffic in the Los Angeles Basin. Pacific Electric maintained a standard rail gauge (4 feet 8½ inches) that allowed the company to compete with steam-powered rail lines for both passengers and freight. In late 1925, Pacific Electric opened its first underground subway line (Figures 4.107 through 4.109). Dubbed the "Hollywood Line," the 28-foot-wide Pacific Electric Tunnel ran 4,325 from Fourth and Hill Streets to a portal near Glendale and Beverly Boulevards (William-Ross 2008).

By 1933, increasing popularity of automobiles was depleting ridership on both rail and bus lines in Los Angeles although buses fared much better because of their versatility in changing routes with demand. The rail lines enjoyed a brief spike in ridership due to gas rationing during World War II. When the war ended, so did gas rationing, signaling the end of commuter rail transportation in Los Angeles. In 1955, the Hollywood Line was converted to carry bus traffic only and was eventually abandoned. The connection of the Hollywood Line to the downtown terminal remains forsaken but has not been completely demolished. Foundations of modern buildings have cut through sections of the tunnel. As an ironic sign of progress, a new subway line, under construction downtown since 2014, will demolish the portion remaining under South Flower Street.

Although bus transport continued to thrive in the area since the mid-1950s, it would be many decades before subways would return to Los Angeles. Within a few years the remaining private bus and rail systems were eventually consolidated into the newly established Los Angeles Metropolitan Transit Authority (LAMTA), completing the transformation of public transportation in the region from private to public ownership.

State legislation enacted in 1964 created a new transportation agency—the Southern California Rapid Transit District, or SCRTD. One of this new agency's directives was to design and construct a rail transit system for Los Angeles County. The greatest achievement of the SCRTD was to secure federal financing for the future Metro Rail subway projects.

Numerous theories abound as to why the development of underground transit in Los Angeles lagged so far behind other major U.S. cities. Stories of automobile and tire manufacturers along with oil companies influencing local politicians to build roads instead of rail are numerous and often retold without factual basis. The truth behind this lack of underground initiative is likely twofold. First, it is well known that Los Angelinos do love their cars. Trips to the nearby beaches and mountains are readily enjoyed in the comfort of one's own vehicle. Second, the landscape and layout of the region did not translate well into a mass transit market. Both Manhattan and downtown Boston have natural barriers to land travel, and Los Angeles is less constrained. Downtown Los Angeles is land accessible from all directions. With an area of 4,058 square miles, Los Angeles County is composed of urban and suburban sprawl. This growth eventually covered what was once cheap and plentiful land, stretching to the north, south, and east with new towns and communities popping up everywhere. By the early 1960s, rail and streetcar lines had disappeared, and Los Angeles was transformed into the quintessential car-friendly American City.

CHAPTER FOUR | TRANSIT TUNNELS

Figure 4.109 | **Hollywood Line Portal near Glendale Boulevard on Opening Day in 1925 (inset) and No Longer in Use in 2015**

1970s to Present and Creation of the Modern Transit System

Finally in the 1970s, the county decided to return to rail-based public transport and to build a network of subway and light-rail lines. In 1976, California Assembly Bill No. 1246 created another Southern California transportation agency—the Los Angeles County Transportation Commission (LACTC). This commission was charged with managing all public transit in Los Angeles County, starting a movement underground that continues to this day. Another act by the state legislature in 1993 provided for the merger of SCRTD and LACTC, creating the Los Angeles County Metropolitan Transportation Authority, or LACMTA (now called Metro) that exists to this day, operating, maintaining, constructing, and expanding the Los Angeles subway system.

Design and construction of the regional rail system started in 1985. The Metro Blue Line opened in July 1990. This 22.2-mile-long light-rail line has a half-mile section of cut-and-cover tunnels connecting to 7th Street/Metro Center Station in downtown Los Angeles. The present-day and under-construction Metro rail transit system is shown in Figure 4.110. When current subway projects are completed, Metro will have a combined total of approximately 32 miles of light-rail and heavy-rail underground transit.

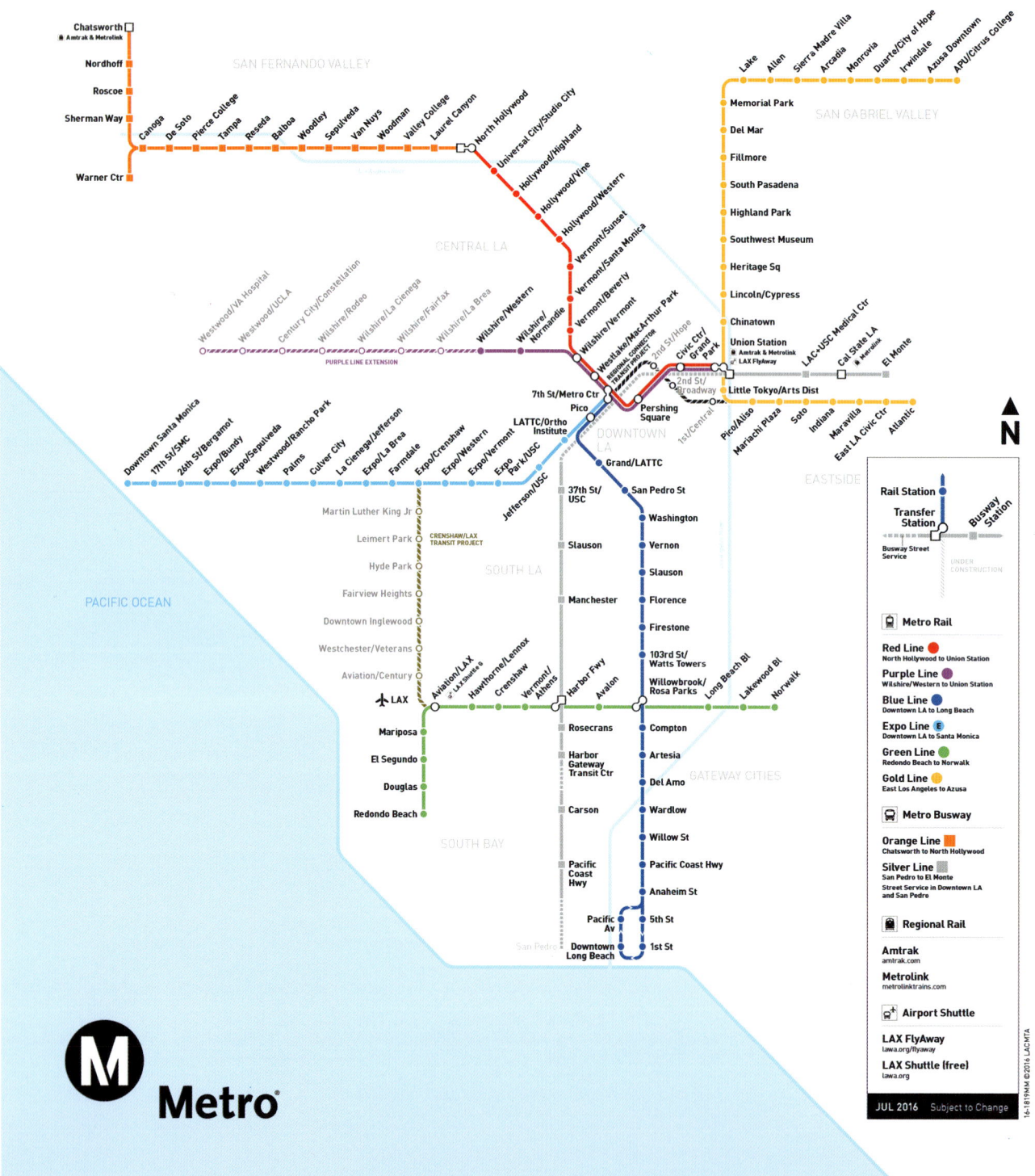

Figure 4.110 | Los Angeles Metro Transit System, Current Operating Routes and Sections Under Construction

Los Angeles Subway Tunnels Made Safe from Earthquakes and Gases

Los Angeles Metro tunnels have involved major technical challenges that, when combined, no other major city in the world has had to confront. With success in meeting those challenges have come notable firsts in the tunneling industry and advancement of the state of the art in tunnel engineering.

Earthquake Design Codes and Design Methods for Tunnels

Los Angeles is situated in a seismically active zone and has a history of damaging earthquakes. Metro developed the first seismic design code and a complementary design method for subway tunnels to withstand the seismic effects of earthquakes. This work built upon the prior seismic design work done for the San Francisco Bay Area Rapid Transit (BART) system. Two levels of design are based on the statistical probability of earthquake occurrence. For the more routine and less intense case, called the Operating Design Earthquake, the subways are designed to withstand the largest earthquake expected in 150 years with none to minimal damage and the ability to maintain operations. During the most intense case, called the Maximum Design Earthquake, the subways are designed to withstand the 2,500-year earthquake, sustain repairable damage, and maintain life safety, or, in other words, not collapse. In statistical terms, Operating and Maximum Design Earthquakes have a 4% and 50% chance of occurrence in 100 years, respectively. For the Regional Connector project, the Maximum Design Earthquake corresponds to a magnitude 7.0 event.

In January 1994, the Los Angeles area experienced the Northridge earthquake with a magnitude over 6.4. The ground motion downtown (Red Line, Segment 1) was recorded at 0.3g, which corresponds to the 150-year earthquake. The system sustained no discernable damage and returned to operation once power was restored. The design criteria and methods of design have been adapted by other transit systems throughout the world.

Tunnels Crossing Faults

Geologic fault movement in Los Angeles can range from inches to several feet. Design philosophy for fault crossings recognizes that it is not possible to prevent damage in a strong earthquake with fault movement. The Red Line Segment 3 rock tunnels were constructed with hard-rock tunnel boring machines (TBMs), but where the tunnels cross the Hollywood Fault, each tunnel was oversized by drill-and-blast excavation. With Maximum Design Earthquake fault movement of about 6 feet, each tunnel will be of sufficient size to permit rail realignment and operations after repairs.

Solutions for Safe Tunneling Through Gassy Environments

The Los Angeles Basin has been a prolific hydrocarbon (oil and gas) producing region. Methane is the main production-related gas, coexisting with hydrogen sulfide. Methane is explosive, and hydrogen sulfide, while also explosive, can be deadly at lower concentrations. The gas hazard was made abundantly clear on March 24, 1985, when a methane gas explosion and fire occurred at the Ross Dress for Less store in Los Angeles near the proposed subway alignment. This event played a part in delaying new subway construction, as federal legislation was enacted that for many years prevented construction of transit tunnels along a portion of Wilshire Boulevard.

Construction of tunnels in these "gassy" conditions is made safe by imposing special working restrictions on mechanical and electrical systems and enforcing stringent ventilation requirements. For subway operations, redundant measures are taken to practically eliminate the risk of gas intrusion.

Figure 4.111 | **Installation of HDPE Water/Gas Barrier for Subway Station with Detail of HDPE Membrane (inset)**

To prevent hazardous gas and water intrusion in the operating systems, cut-and-cover tunnel and station structures are enveloped with a continuous water/gas barrier system of hydrocarbon-resistant high-density polyethylene (HDPE) sheeting. In Figure 4.111, the white membrane is being installed for station construction. In the inset photo, a business card gives scale to the white membrane and the black preformed waterstop. The HDPE is bicolored, white over black, so that if damaged during installation, black color on the back side shows where to repair. Elsewhere in the United States and the world, tunnels are made watertight using a polyvinyl chloride (PVC) membrane. However, PVC is not a barrier to methane gas; therefore, HDPE was required for Los Angeles. Permanent gas monitoring equipment automatically reports to Metro operations should gas be detected, and fans are activated to purge the gas. The underground ventilation system is specially designed to maintain air circulation at all times, even during a complete regional electrical power outage. These extraordinary measures to operate the subway in the gassy geologic conditions of Los Angeles, including after major earthquakes, are unprecedented in the world.

Metro Red Line Tunnels

Design and construction of this all-underground 17.4-mile-long heavy-rail transit line for the Metro Red Line took place in three segments. The first segment was 4.4 miles of tunnels with four stations and opened in 1993. Tunnels were constructed with open-face shields such as the one shown in Figure 4.112. Tunnels have two-pass tunnel linings (described in a later section). A typical cut-and-cover section of transit tunnel is shown in Figure 4.113. Figure 4.114 shows a cut-and-cover tunnel and the crossover tracks before entering the underground station in the distance. Rail crossovers are critical to operating the rail service efficiently because they permit trains to switch over to the other track to bypass track maintenance or an operating problem in one tunnel.

Red Line construction continued at a rapid pace for years, but not without controversy. The Red Line was originally designed to be part of a much longer route that would have extended the underground transit line well to the west of the downtown area. From 1995 to 1997, the feasibility of continuing tunneling to the west on Wilshire Boulevard was considered by using slurry-face TBMs. However, that plan for continuing the subway to the west was tabled in 1997 when continued funding (through Measure C) was not approved by voters.

The second segment of the Red Line north and west of downtown Los Angeles to Hollywood, to the famous intersection of Hollywood and Vine, was opened to service in 1999, adding five underground stations. Tunneling was expedited by a construction shaft, as shown in Figure 4.115, from which four tunnel drives were launched. This was done to reduce construction delays by allowing the stations to be constructed separately from the tunneling.

The last segment of the Red Line began construction in 1995 and was opened to service on July 24, 2000. This segment takes passengers north from Hollywood to the San Fernando Valley, terminating in North Hollywood. Crossing under the Santa Monica Mountains, the distance between the Hollywood/Highland and Universal City Stations is 3.25 miles, making it the longest station-to-station ride on the entire system.

Figure 4.112 | **Open-Face Digger Shield**

Figure 4.113 | **Los Angeles Metro Cut-and-Cover Transit Tunnel**

Metro Gold Line Eastside Extension

Tunneling technology advanced as rapidly as the ability to navigate the crowded streets of Los Angeles declined. In 2007, a 1.4-mile tunnel section of the 6.6-mile Gold Line under the Boyle Heights section of Los Angeles was completed. The two TBMs used to excavate these tunnels, "Suzie" and "Marj," are shown in Figure 4.116. One TBM is shown in Figure 4.117 as it tunneled into the excavation for Soto Station, one of the two underground stations on this extension. This completed tunnel section is shown in Figure 4.118. Light-rail transit cars are powered by the overhead conductor anchored at the top of the tunnel.

Successful completion of these tunnels demonstrated to local officials that developments in pressurized-face tunneling technology could allow TBMs to more safely excavate through the often unforgiving soft-ground geology of the Los Angeles Basin. Planners at LACMTA now had the technological means they needed to safely move westward through the county. Behind public pressure to relieve congestion on surface streets, the previous forbidding federal law was repealed. Ironically, the same politician that introduced the legislation to stop the westward expansion was the same politician behind the repeal.

The first Los Angeles Metro heavy-rail subway tunnel linings were "two-pass" lining systems—typically an initial lining of unbolted, expanded, precast concrete segments (or steel ribs with timber lagging) and a final cast-in-place concrete lining with HDPE membrane in between. Two-pass tunnel linings are not often used for TBM-excavated tunnels anymore, having been replaced with "one-pass" bolted and gasketed segmental precast concrete tunnel linings.

Figure 4.114 | **Los Angeles Metro Subway Tunnels Transition into Subway Station**

Los Angeles transit tunnel design and construction has spanned the great technological developments of the industry of recent decades. Pressurized-face tunneling and one-pass precast concrete segmental linings were pivotal changes that were quickly adapted to Metro subway construction.

Transit tunneling in Los Angeles of the late 1980s and into the 1990s was dominated by the use of open-face mechanized shields (Figure 4.112). Although tunneling with this equipment was economical, several projects required extensive soil pretreatment by grouting to prevent excessive settlement and damage to utilities and buildings. Red Line tunnels in soil were completed by the year 2000 using these traditional open-face tunnel shields. Globally, however, the tunneling technology of pressurized-face TBMs was evolving. The first use of a pressurized-face TBM for a transit tunnel in the United States was on the Washington, D.C., Metro's Anacostia River Crossing Tunnel in the late 1980s (see Figure 4.96). Los Angeles Metro required the use of pressurized face tunneling for the first time on the Gold Line Eastside Extension project. This project was constructed with virtually no ground settlement above the tunnels. Pressurized-face tunneling has now become standard practice in most urban soft-ground tunneling projects. All subsequent subway tunneling in soil in progress or planned in Los Angeles uses or will use pressurized-face tunneling techniques.

The initial motivation to use pressurized-face TBMs was to minimize ground loss and related surface settlement and risk of damage to utilities and structures. Inherent with these machines, tunneling can take place below the groundwater table without having to first lower the groundwater by dewatering.

To complement a pressurized-face TBM, the tunnel lining also has to be watertight and gastight. Precast concrete segmental tunnel linings had been used for many years globally as a replacement for the more costly cast-iron segments used for soil tunnels starting more than a hundred years ago. Erected within the TBM, the precast lining is installed ring by ring as tunnel excavation takes place.

Los Angeles, however, has the special conditions related to hydrocarbons. Given the risks of methane and hydrogen sulfide gas leakage into the tunnel, the use of a double-gasketed tunnel liner (one-pass system) for use with the pressurized-face TBMs was adopted as a Metro standard. Seismic conditions led to the design of a convex-to-convex shape on radial joints, principally to flex during earthquakes so that the dual gaskets remain sealed to prevent gas intrusion.

To prove the gasket system, Metro undertook a lengthy laboratory testing program conducted at the University of Illinois. The testing program evaluated the structural capacity of the segments under seismic and ground loads, and included gas leakage testing through gaskets and gasket material testing. To test segments under cyclic loading to simulate an earthquake, segments were rotated under working loads. No damage to gaskets under such cyclic loading was

Figure 4.115 | **Construction Shaft on Metro Red Line Tunnels Expedites Tunneling**

206 THE HISTORY OF TUNNELING IN THE UNITED STATES

Figure 4.116 | **Metro Gold Line Eastside Extension Earth Pressure Balance TBMs**

Figure 4.117 | **TBM Entering Excavation for Soto Station, Metro Gold Line Eastside Extension**

observed. The double gaskets on the edges of the precast concrete tunnel lining segments can be seen in Figure 4.119. A typical ring of tunnel lining, 5 feet wide, is made up of six individual segments, as shown in Figure 4.120.

Major Expansion of the Los Angeles Metro with Tunnels

In November 2008, voters in Los Angeles County approved Measure R that added a half-cent sales tax for 30 years to pay for new transportation projects and improvements. This tax is being used to fund billions of transportation projects, including major subway tunnels. Figure 4.110 shows the Los Angeles Metro System and tunnels under construction. The tunnel projects (Purple Line, Regional Connector, and Crenshaw/LAX [Los Angeles International Airport]) and the technical challenges are presented in the following section.

Purple Line

The Purple Line extends heavy rail 9 miles west along Wilshire Boulevard and adds seven new stations. This rail extension has evolved from at one time being known as the "Subway to the Sea," then the "Westside Subway Extension," and finally officially named the "Westside Purple Line Extension" (Figure 4.121). The existing Red Line running between Wilshire/Western and Union Station was renamed as well. Construction of the 3.9-mile-long Segment 1 of the new Purple Line broke ground on November 7, 2014.

The full Purple Line Extension was originally planned for a 2035 opening. However, strong funding and future public transit demands are expected to accelerate construction. The entire length of the Purple Line will be underground. Major challenges for the design and construction involve tunneling in gassy ground and "tar sands" in the vicinity of the La Brea Tar Pits, earthquakes, and geologic faults.

Regional Connector Transit Corridor Project

When it opens in 2022, the Regional Connector Line will provide service from the Little Tokyo area through the Financial District to the existing Metro Center Station. Completely underground, this 1.9-mile extension will knit together the Blue, Gold, Expo, Red, and Purple Lines (Figure 4.122). The project will provide a one-seat, one-fare ride for commuters from Azusa to Long Beach and from East Los Angeles to Santa Monica without the need to transfer between rail lines for major east/west and north/south trips. Almost every type of underground construction is required for this project, including soft-ground tunneling, deep excavations on major city streets, cut-and-cover station excavations up to 100 feet deep, and a 58-foot-wide by 288-foot-long mined cavern for a rail crossover that permits trains to cross to the other track when one line is shut down for any reason. The crossover cavern is approximately the size of three professional basketball courts.

Figure 4.118 | **Los Angeles Metro Tunnel Constructed with Segmental Precast Concrete Tunnel Lining**

Figure 4.119 | **Precast Concrete Tunnel Lining Segments with Double Gaskets on Joints**

Figure 4.120 | **One Ring of Segmental Precast Concrete Tunnel Lining**

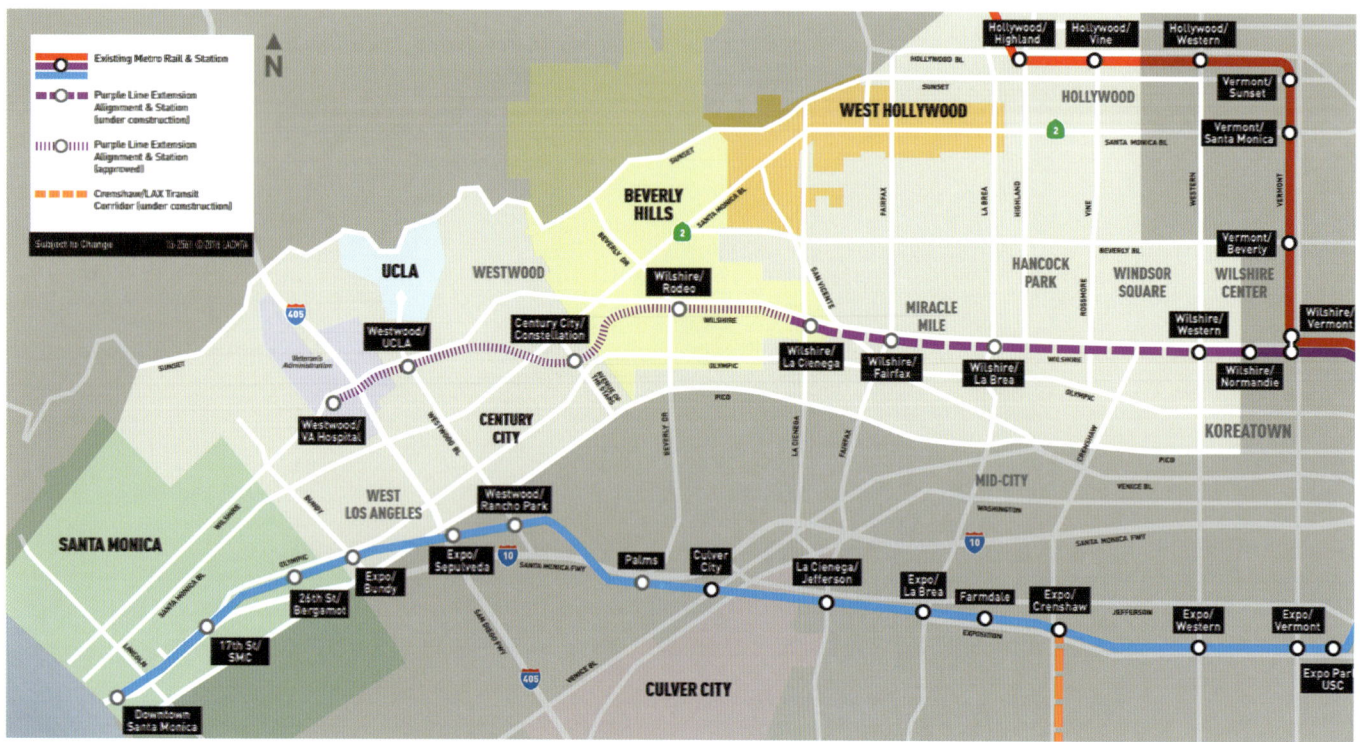

Figure 4.121 | **Los Angeles Metro Purple Line Alignment**

CHAPTER FOUR | TRANSIT TUNNELS | 209

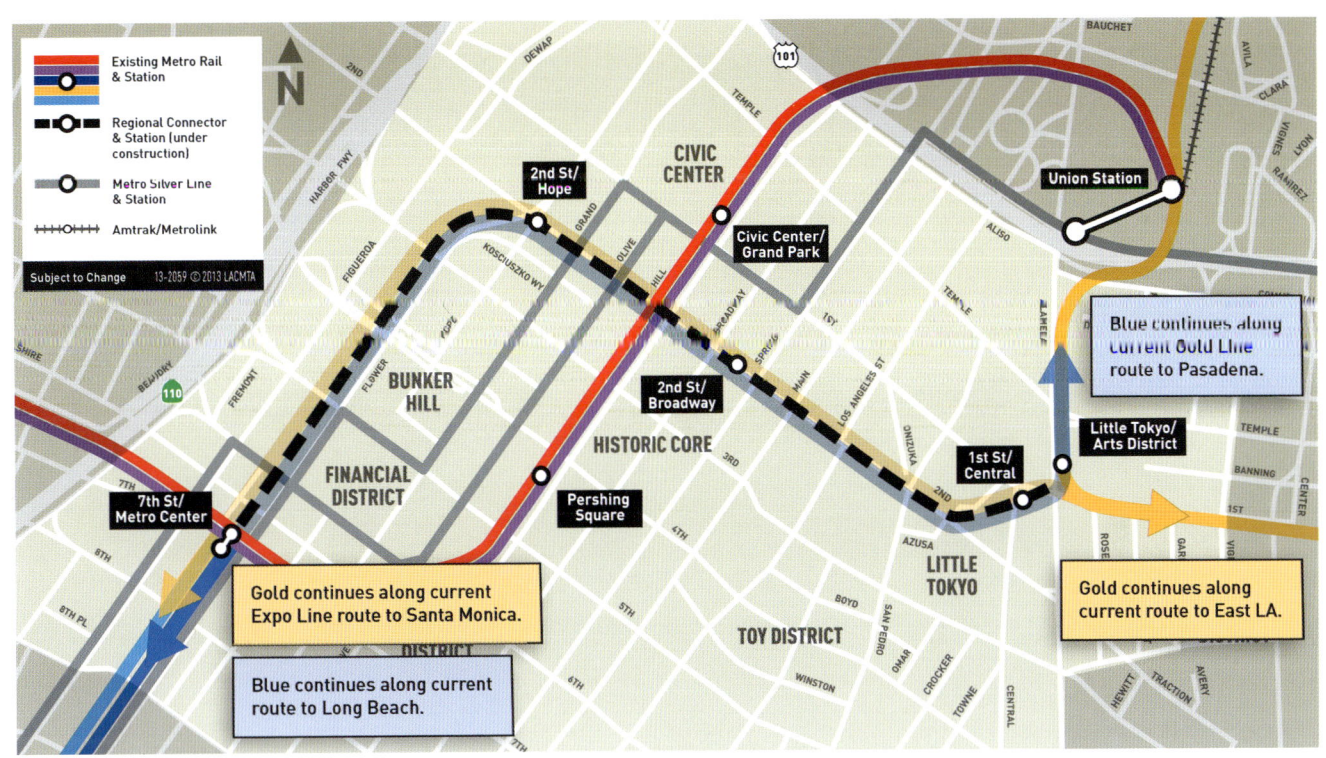

Figure 4.122 | Regional Connector Alignment Through the Heart of Downtown Los Angeles

Figure 4.123 | TBM Being Rebuilt for Regional Connector

210 THE HISTORY OF TUNNELING IN THE UNITED STATES

Figure 4.124 | Front View of TBM Being Rebuilt for the Regional Connector Project

Figure 4.125 | Los Angeles Metro Crenshaw/LAX Transit Project Stations

CHAPTER FOUR | TRANSIT TUNNELS | 211

Figure 4.126 | **Hollywood and Vine Station Platform**

The 21.5-foot-diameter TBM to be used to excavate the Regional Connector tunnels is shown in a shop being refurbished for use in Figures 4.123 and 4.124. The color scheme for this cutterhead was selected because the Regional Connector ties the Gold and Blue Lines together. This particular TBM has a special history of use. When new, it first built tunnels for the Metro Eastside Extension and is the machine with the cutterhead painted blue in Figure 4.116. Subsequently, the TBM was used to construct subway tunnels in Seattle for Sound Transit and is shown before and after tunneling in Figures 4.143 and 4.144 in the Seattle section. Multiple reuse of these complex and costly TBMs is exemplary of how tunnels are becoming more economical to construct when the machines are reusable.

Crenshaw/LAX Project

Breaking ground in January 2014, the Crenshaw/LAX Transit Project is an 8.5-mile light-rail line with eight stations (Figure 4.125). The rail line will run between the Expo Line on Exposition Boulevard and the Metro Green Line, with 2.1 miles underground. The centerpiece of this project is the 1.2-mile twin-tunnel segments (Exposition Boulevard to 48th Street/59th Place to 67th Street/LAX south runways) with all remaining areas at grade or elevated. The Crenshaw/LAX Transit Project will serve the Crenshaw District, Inglewood, and Westchester areas.

Stations

Subway stations in Los Angeles are unique from those in other major American cities. In Los Angeles, each existing station boasts unique architecture and finishes, depicting the vibrant culture or social theme of its neighborhood. The Hollywood and Vine Station, for example, displays the area's rich history in motion pictures with a film-reel-lined platform ceiling and old-time Hollywood movie camera (Figures 4.126 and 4.127). Figure 4.128 shows the Civic Center Station.

Figure 4.127 | Hollywood and Vine Station and Vintage Movie Camera

Figure 4.128 | **Mezzanine Level of the Civic Center Station**

From its beginning, art has been a part of Metro's subways. Examples are shown in Figure 4.129 in existing subways. This site-specific, integrated public art is an enhancement to the appearance of the subways and is regarded as making patrons feel welcome and inclined to use the facility. Metro manages engagement of artists on a location-by-location basis. For the current expansion of the subways, art is embedded from the beginning within the final architectural and structural design, not added on later. This process allows for strategic placement of art in the stations and station entrances in a way that balances placement, lighting, and overall visual effects.

The Los Angeles Metro system is in the infancy stage when compared to the transit systems of New York, Boston, and Washington, D.C. The clean, modern, and distinctively decorative finishes make it a new standard for mass transit in the United States. A major transportation bond measure (Measure M) was approved by the Los Angeles County voters in the November 2016 elections. This will provide $120 billion in transportation infrastructure funding that includes expanding rail and rapid transit with miles of subway tunnels.

CHAPTER FOUR | TRANSIT TUNNELS | 215

Universal City/Studio City Station

Wilshire/Western Station

Soto Station

Hollywood/Highland Station

Figure 4.129 | **Public Art in Metro Station**

Figure 4.130 | **Seattle, Washington**

SEATTLE, WASHINGTON

Seattle is a beautiful coastal city situated on Puget Sound (Figure 4.130). Hilly and surrounded by water, Seattle has transportation challenges. It is a popular place, home to the legendary firms of Microsoft, Amazon, and The Boeing Company, and has been growing in population. In 2010, Seattle's population was 608,660 and the Seattle metropolitan area has more than 3.5 million inhabitants, making it the 15th largest metro area in the United States.

Sound Transit is Seattle's transit tunnel system, which started in 1983 with the approval of a plan for a 1.3-mile transit tunnel under downtown. The goal of the initial project was to create a new downtown right-of-way for special "dual-mode" buses and, eventually, rail transit. This work was undertaken by the Municipality of Metropolitan Seattle, commonly known as "Metro," and was eventually taken over by the multi-county agency Sound Transit. Construction started in 1987 and the tunnel opened in 1990. In 2009, Sound Transit began light-rail service on 15.8 miles of double track from downtown Seattle to Sea-Tac (Seattle–Tacoma) International Airport. A key link in this rail line to the airport is the tunnels excavated by an earth pressure balance tunnel boring machine (EPBM) through Beacon Hill and its deep underground station mined in glacial soils.

Sound Transit completed construction of the University Link tunnels, 3.5 miles of underground light-rail alignment from downtown Seattle to the University of Washington, which opened for revenue service on March 19, 2016. Also under construction are the Northgate Link tunnels, 4.3 miles of light rail, mostly underground, from the

University of Washington to a point just south of the Northgate Station, which will open in 2021 (see Figure 4.131).

The Downtown Seattle Transit Project

Downtown Seattle is a hilly city, bounded by water on the east and west sides that makes transportation difficult. Since its founding in 1851, the Seattle landscape has been shaped by regrading to smooth out and completely remove hills and by construction of city streets and highways. Regardless of these major changes and creation of a working road network, there was a need for a transit system that would not rely on often congested roadways and provide an efficient way to get between the downtown area and the suburban communities. An underground transit system would be ideal but costly, and it was not until the early 1980s that a scheme emerged that gained both political and community support. What resulted was a tunnel under 3rd Avenue that became the keystone event for the start of modern transit in Seattle.

Creation of a regional transit system, however, required more than just one downtown tunnel. The problem remained that connections to the suburban communities, such as Bellevue and Kirkland east of Seattle, would require riders to transfer from a bus to the underground transit vehicle. The political leaders in those and other cities found this condition to be unacceptable. Therefore, to achieve a one-seat ride, the compromise was to use "dual-mode" electric-diesel buses operating on electric power in the tunnel under 3rd Avenue and using diesel engines on the surface. This removed the diesel buses from downtown streets and did not require the suburban riders to transfer to another mode of travel. Only dual-mode buses operated in the tunnels for several years before the light-rail system was constructed.

Known at the time as the Downtown Seattle Transit Project (DSTP), it entailed the boring of two parallel tunnels 60 feet beneath city streets and construction of five distinct transit stations by cut-and-cover methods. Construction started in 1987, and the

Figure 4.131 | **Sound Transit Light Rail and Proposed Extensions, 2009**

218 THE HISTORY OF TUNNELING IN THE UNITED STATES

1.3-mile-long twin tunnels opened for bus service on schedule in 1990 (Critchfield and MacDonald 1989).

The tunnels were constructed with an open-face tunnel shield using a backhoe type "digger" for excavation, as shown in Figure 4.132. Figure 4.133 shows the view after the tunnels and station were completed and made ready for rail service. Public art is well established in Sound Transit stations, like the large wall mural in Westlake Station shown in Figure 4.134.

The system was designed to accommodate future use of light-rail trains, in combination with buses, in the same facility. In 2009, operation of both light-rail vehicles and buses started, as shown in Figure 4.135.

Start of Light-Rail Transit in Tunnels

In November 1996, voters approved a Regional Transit System Plan called *Sound Move* that incorporated elements of commuter rail, light rail, and express bus service into a comprehensive high-capacity regional

Figure 4.132 | **Digger Shield Used for Constructing Seattle Bus Tunnel**

Figure 4.133 | **Completed Bus Tunnel Connections to Station** transit system. Thus, 15 years after the start of the DSTP, the tunnel was closed on September 24, 2005, to be retrofitted for use by light rail (in addition to buses) as part of Sound Transit's plan to build the extensive light-rail system. Although rails were included when the tunnel was first built, they were not insulated adequately to prevent stray electric current from the trains from corroding nearby utility lines. Given that correcting this situation required removal of the original rails, Sound Transit also took this opportunity to lower the tunnel roadway to allow level train boarding. In 2009, Sound Transit began light-rail service on 15.8 miles of double track from downtown Seattle to Sea-Tac Airport, which includes the Beacon Hill Tunnels and Station.

Beacon Hill Project

The Beacon Hill Project was completed as part of Sound Transit's Link light rail connecting downtown Seattle with Sea-Tac Airport. One would be hard pressed to find another project as complex as this one, given its configuration, the variety of issues to address, and the required construction techniques (Akai et al. 2007).

The Beacon Hill Project consisted of 4,300 feet of twin tunnels that pass east and west through the 300-foot-high glacially sculpted Beacon Hill (Figure 4.136). The tunnels were constructed primarily in very hard soils placed by glaciers called "glacial till." These glacial soils are found throughout the region and vary greatly in composition from a mixture of clay, sand, and boulders to irregular strata of very dense silt, sand, and gravel (see geologic profile, Figure 4.136). The project includes a 476-foot-long, 165-foot-deep mined station with twin platform tunnels constructed by the sequential excavation method (SEM).

This deep "binocular" station, meaning that it was built as two tunnels rather than one cavern, has been mined through some of the most challenging soft-ground conditions in the United States and was successfully completed in mid-2007. Some of the station tunnels are the largest soft-ground tunnels excavated by SEM in North America. At 165 feet deep, the station includes platform, concourse, cross adit, and emergency ventilation tunnels, together with station egress and ventilation shafts, as shown in Figure 4.137. Various excavation sequences were used for the different tunnels, including the single-heading, single-sidewall drifts (a *drift* is one of several stages for excavating the whole tunnel), and the twin-sidewall drifts for the impressive 45-foot-wide by 42-foot-high concourse cross adits (an *adit* is a smaller or short tunnel excavated from a larger tunnel) (Figures 4.138 and 4.139). Excavation for the station involved removal of approximately 60,000 cubic yards of material, and the station consists of a variety of geometries and cross sections.

To assess the risks associated with subsurface conditions and the proposed method of construction, an exploratory shaft was constructed at the location of the Beacon Hill Station. The test shaft, being a unique North American

Figure 4.134 | **Public Art in Westlake Station**

Figure 4.135 | **Light Rail and Dual-Mode Bus Transit Operating in Westlake Station**

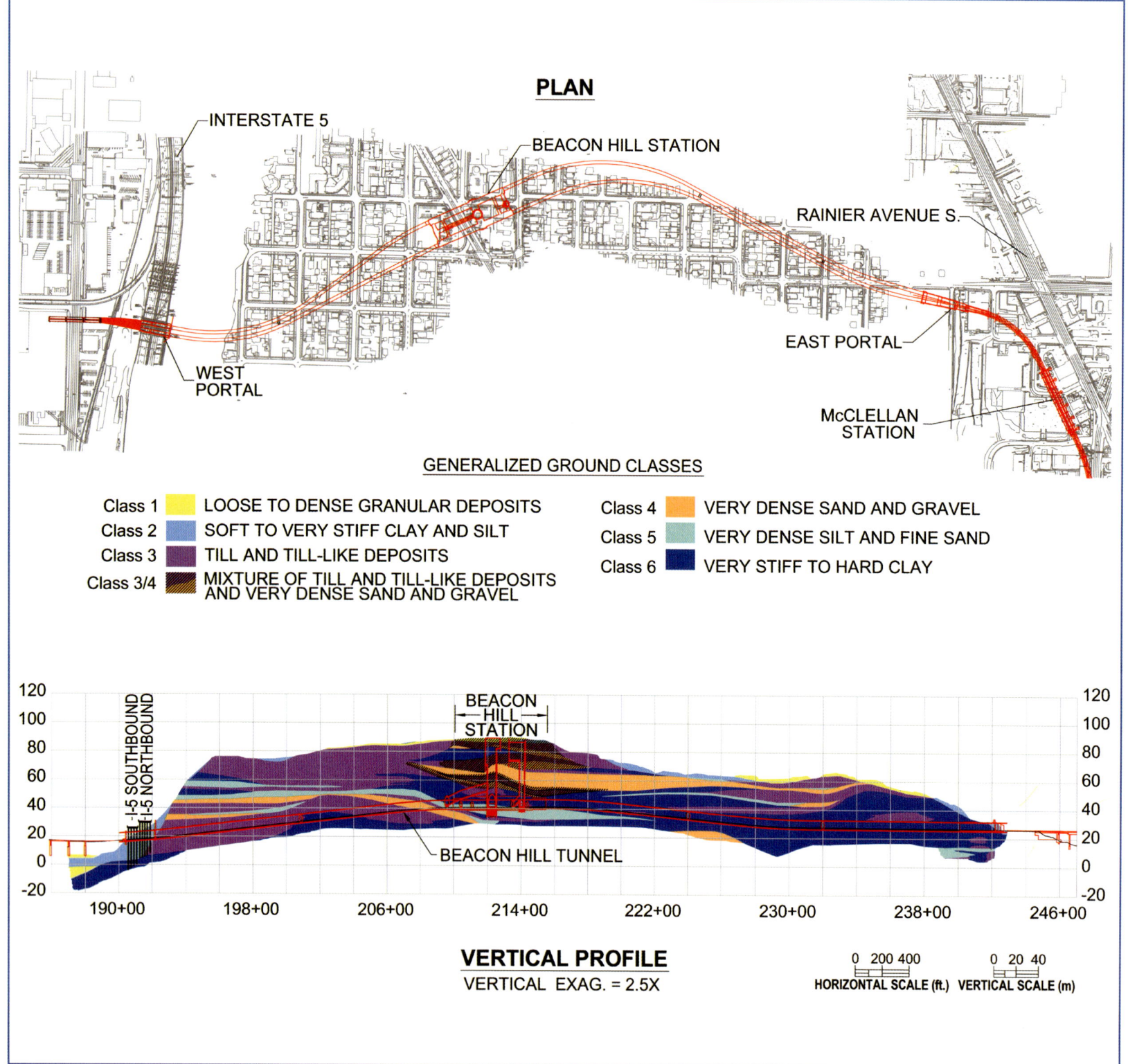

Figure 4.136 | Plan and Geologic Profile of the Beacon Hill Tunnel

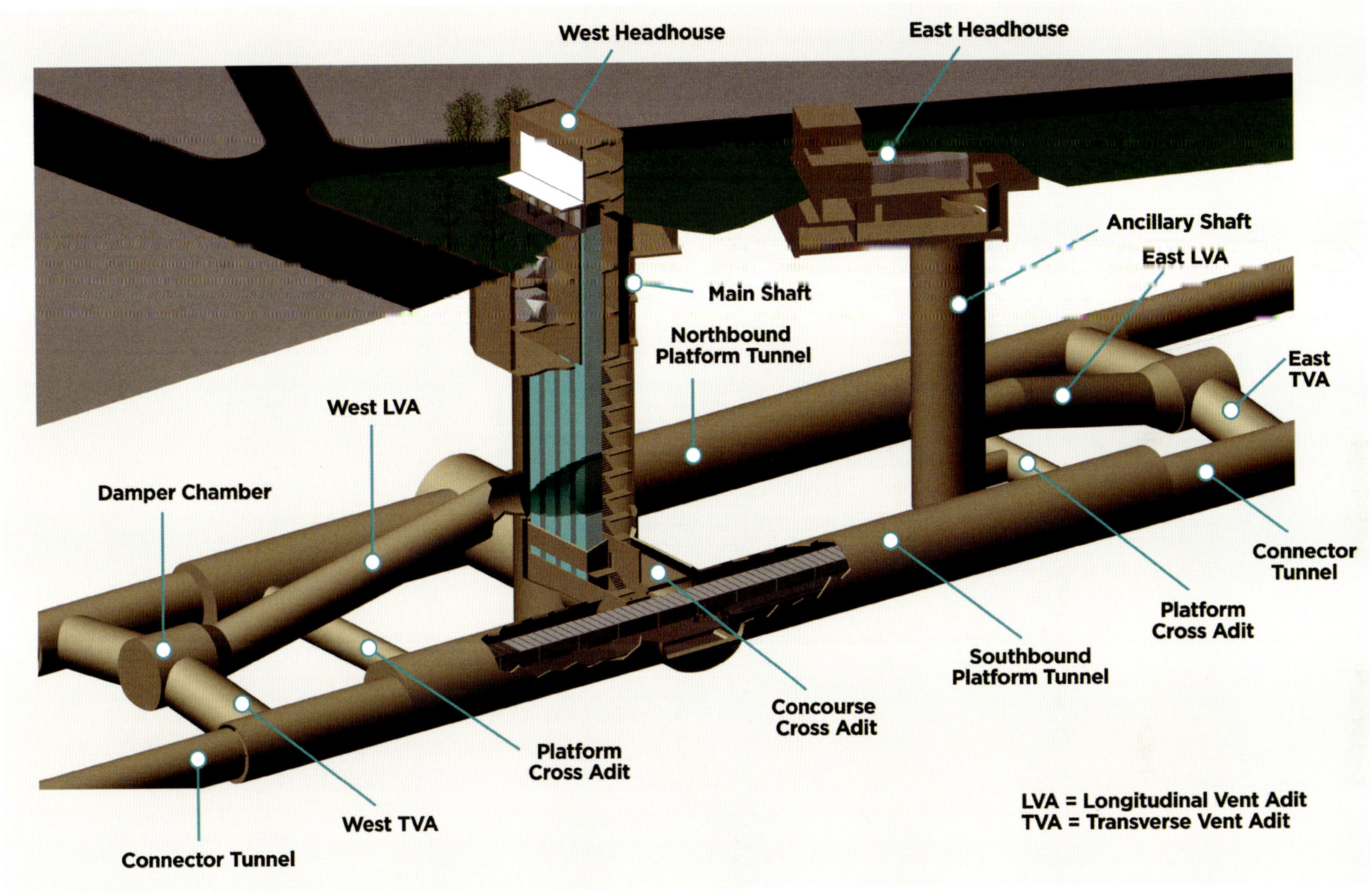

CHAPTER FOUR | TRANSIT TUNNELS | 221

Figure 4.137 | **Beacon Hill Station Configuration**

Figure 4.138 | **Concourse Cross Adit— Dual Sidewall Drifts, Top Heading**

Figure 4.139 | **Concourse Cross Adit—Dual Sidewall Drifts**

application for a vertical pilot bore, allowed ground conditions and stability to be assessed with respect to constructability of the shaft and station using SEM. This resulted in revisions to the base design concepts and construction sequences of the main and ancillary shafts (Tattersall et al. 2004). Additional soil borings required by the contract documents and borings conducted for instrumentation, dewatering wells, and geologic probe holes drilled prior to construction redefined the understanding of the subsurface conditions at the east end of the station, which resulted in an 88-foot shift of the entire station to the west.

Stability of the excavation during construction was managed with various types of ground improvements, including jet grouting, vacuum-assisted deep dewatering wells from the surface, pipe arch canopies (a row of grouted pipes installed just outside the tunnel excavation ahead of the excavation face [Figure 4.138]), grouted pipe and rebar spiling installed ahead of the excavation face, chemical grouting, and localized vacuum dewatering (Varley et al. 2007).

The twin running tunnels were constructed using a 21-foot-diameter EPBM that was assembled on the approach embankment at the west portal, under the Interstate 5 freeway viaduct. The 18-foot-10-inch internal-diameter tunnels were lined with bolted, gasketed, precast concrete segments. As a mitigation for anticipated subsurface conditions, the EPBM was required to be continuously operated in closed-face mode at a minimum face pressure of 50 psi at the invert (bottom) of the forward chamber of the machine and to be adjusted to the actual conditions encountered. In-service operations for Sound Transit's light-rail service through the Beacon Hill Station and tunnels began in 2009. Figure 4.140 shows the Beacon Hill Station at platform level on opening day with a train approaching the station.

Tunnels Expand Light-Rail Service

In November 2008, voters approved an extensive program of transportation projects to be implemented over the 15-year time period from 2009 to 2023. This plan, when fully implemented, will add 34 miles of light rail extending north from the University of Washington through Northgate and on to Lynnwood; to the east from downtown Seattle through Mercer Island and Bellevue to Redmond's Overlake Transit Center, and to the south from Sea-Tac Airport through the Kent Des Moines Road area and on to Highline Community College and Redondo/Star Lake. This expansion of the regional light-rail transit was driven by increased traffic congestion and the realization that there is no way to build sufficient highway infrastructure to relieve existing traffic and accommodate future

CHAPTER FOUR | TRANSIT TUNNELS

Figure 4.140 | **Beacon Hill Station—Platform Level on Opening Day in 2009**

growth in the region. Tunnels are critical elements of this expansion that involve the University Link Extension and the following underground extensions:

- **The Northgate Link Extension** is a 4.3-mile extension of light rail that runs north from the University of Washington Station, going under campus via twin-bored tunnels to be constructed by EPBMs, to an underground station at NE 45th Street (U District Station), continuing to an underground station at NE 65th Street (Roosevelt Station), and via tunnel, retained cut and elevated sections to Northgate. This extension project is in the final design and construction stages. Revenue service is expected to begin in September 2021.

- **The East Link Extension** expands light rail to East King County via Interstate 90 from downtown Seattle to the Overlake Transit Center area of Redmond, with several new stations. Revenue service to the Redmond Technology Center is forecast for early 2023. The route includes a single one-half mile tunnel, approximately 30 feet high and 35 feet wide, constructed in downtown Bellevue by using SEM.

THE HISTORY OF TUNNELING IN THE UNITED STATES

Figure 4.141 | **Map of the University Link Extension Tunnels**

University Link Tunnels

This 3.5-mile extension of the initial light-rail transit system is located entirely underground and starts from a non-revenue stub tunnel constructed as part of the DSTP at the north end of the existing system in downtown Seattle, and ends at the University of Washington campus near Husky Stadium (Figure 4.141). The tunnel and geologic profile are shown in Figure 4.142. Twin-bored tunnels were constructed in two contract segments. The first segment originated south from Husky Stadium, passing under State Route 520 and the Lake Washington Ship Canal, and then to the Capitol Hill underground station. The second segment commenced at Capitol Hill Station and proceeded south under the Interstate 5 freeway to the stub tunnel. Both underground center platform stations were constructed by cut-and-cover; the Capitol Hill Station and the University of Washington Station are approximately 65 and 100 feet deep, respectively. Revenue service began on March 19, 2016.

Tunneling with EPBMs

Consisting of twin 11,400-foot-long segmentally lined tunnels, the first segment was excavated with 21.5-foot-diameter EPBMs (Burdick et al. 2013). The twin tunnels are connected by 16 cross passages excavated using SEM. The project had to plan for hyperbaric interventions (working in compressed air) above 4.5 bar (about 65 psi) and had to perform SEM on multiple cross passages from within an active EPBM tunnel.

From a previous tunnel project for the Los Angeles Metro Gold Line Eastside Extension, the contractor had two EPBMs available. Conveniently, the internal tunnel diameter for the Seattle project was identical to that of the Los Angeles tunnels, and although the required EPBM operating pressures for Los Angeles were lower, there was potential for a remanufacture to meet Seattle conditions. When first used, the new EPBMs shown in Figure 4.116 were named "Marj" and "Suzie." After refurbishment for this Sound Transit project, they were renamed "Talgo" and

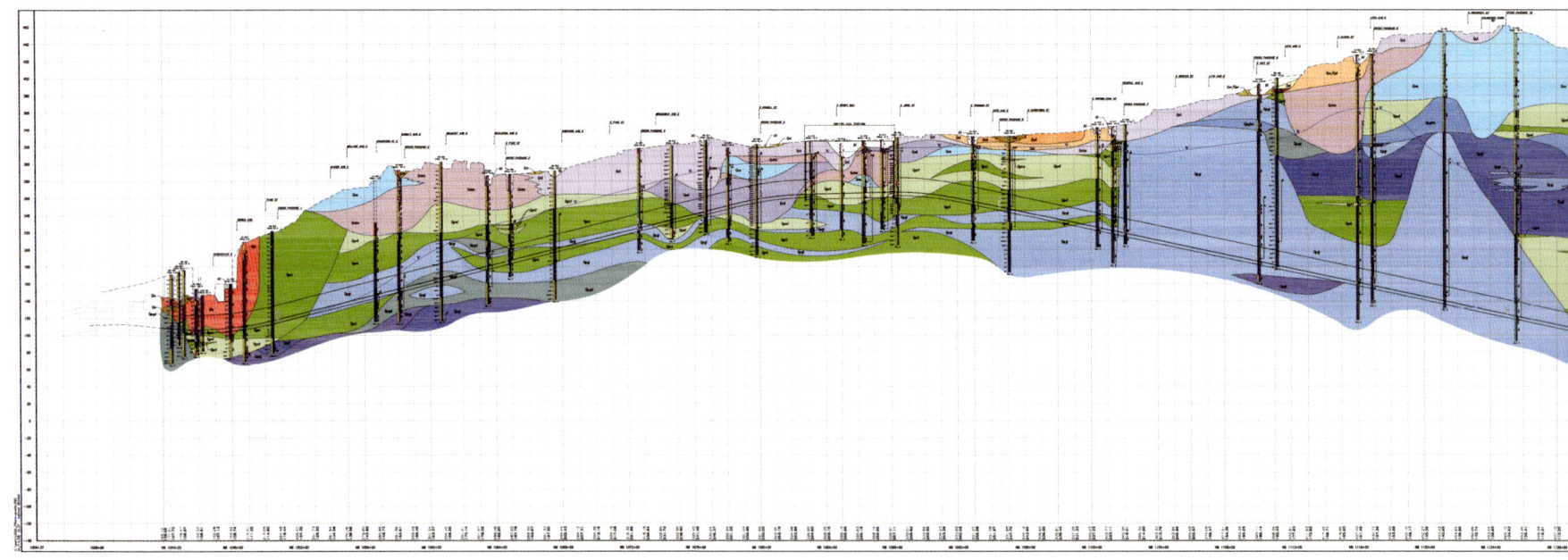

Figure 4.142 | **Profile of the University Link Extension Tunnels**

"Balto" and are shown in Figure 4.143. These two EPBMs successfully completed the two bores and are shown in the receiving shaft of the Capitol Hill Station in Figure 4.144. Exemplary of efficient reuse of these modern tunnel boring machines, one of these EPBMs was refurbished again in early 2016 and transported back to Los Angeles to excavate the tunnels for the Los Angeles Metro Regional Connector Project.

University Link is the second segment of tunnels that consists of 3,800 feet of twin tunnels, with five cross passages, two low-point sump alcoves, and a station box excavation approximately 560 feet long, 74 feet wide, and 75 feet deep for the future Capitol Hill Station. Tunneling included two crossings under Interstate 5 using one EPBM for both tunnel drives (Frank et al. 2007). The station shell serves as the receiving shaft for completion of tunnels driven from the north in the adjacent construction contract, and for the launch of the EPBM for the two tunnel drives to the south. Figure 4.145 is a 2012 aerial view of the station excavation and is representative of the scale of construction for new transit stations. In the left inset, the false tunnel used to launch the EPBM to excavate one of the two tunnels south toward downtown Seattle can be seen with a ventilation duct to the street level. At the other end of the station excavation, the two EPBMs in Figure 4.145 can be seen temporarily stored after completing tunnel drives from the north.

Cross Passages

Construction of cross passages was done by hand largely by tunnel workers, as shown in Figure 4.146. A large number of cross passages (16) on the first segment made it fairly obvious that the project could not be completed within the allotted time if cross-passage work did not commence until after the completion of EPBM mining. To meet the schedule deadline, the contractor undertook serious planning to simultaneously construct cross passages as the tunnel was still being constructed with the EPBM. As an added difficulty, a minimum of two cross passages would have to be excavated simultaneously while several others would be undergoing waterproofing installation and final lining construction concurrently.

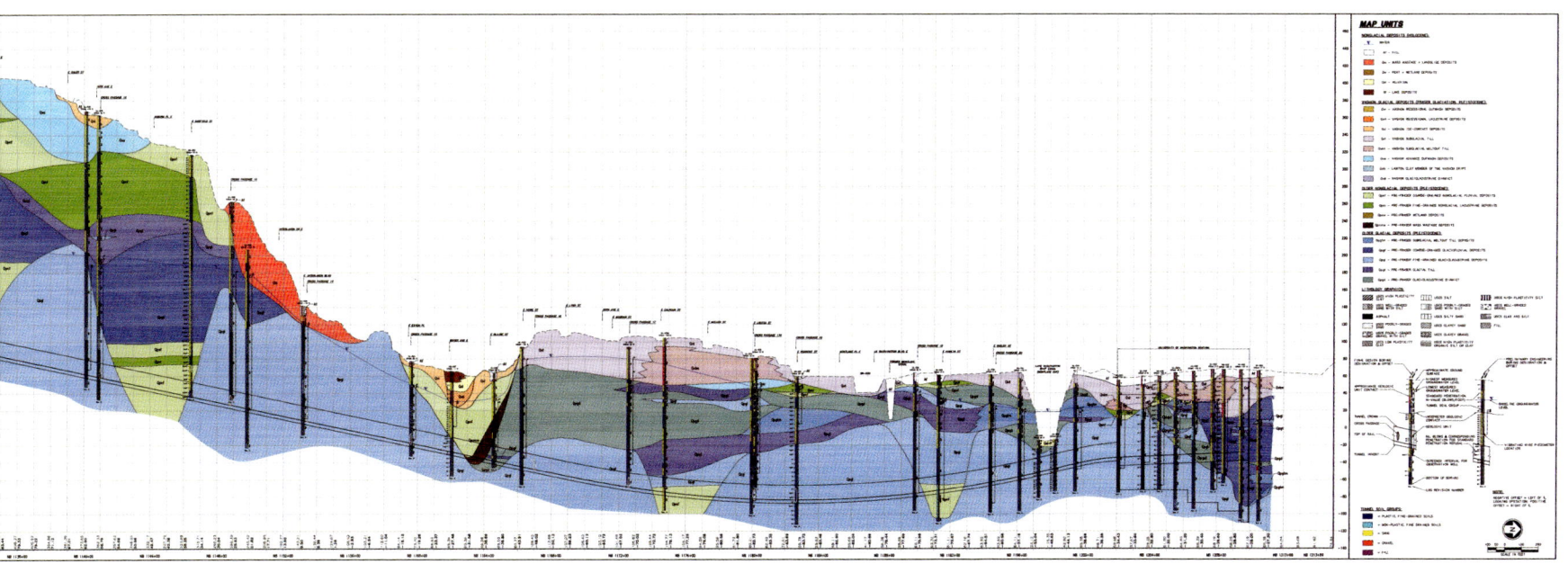

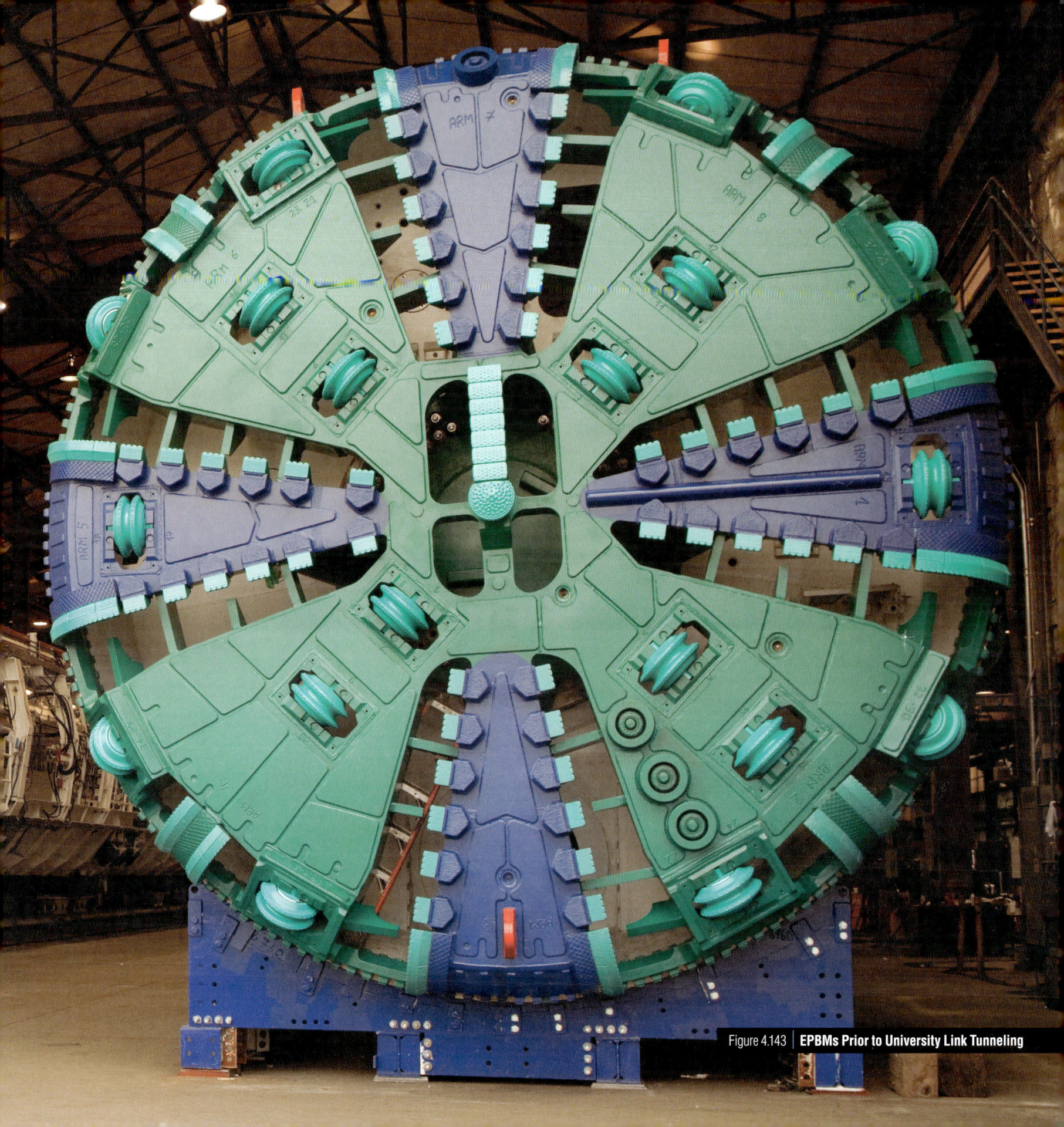

Figure 4.143 | EPBMs Prior to University Link Tunneling

THE HISTORY OF TUNNELING IN THE UNITED STATES

The Future of Transit Tunnels in Seattle

Seattle did not enter the world of underground transit until 1990, almost 100 years after the first underground transit system was opened to service in the United States. However, Seattle is catching up. Expansion of the system is well under way with long tunnels that will be built north of downtown Seattle along alignments that pass under established residential neighborhoods and will minimize construction disruption. To the east of downtown Seattle, a tunnel will avoid extensive disruption to the business center of the city of Bellevue. Sound Transit 3 was approved by voters in 2016 to expand Sound Transit with $54 billion of funding spent over 25 years that will add 62 miles of light rail and will include a new transit tunnel to the northwest Seattle neighborhood of Ballard. Seattle may be one of the later U.S. cites to build an extensive underground rail transit system, but it has inadvertently benefited by the great technological improvements in the tunneling industry. Modern EPBM tunneling equipment and the interactive design and construction approach of SEM, coupled with improved contracting practices, make tunnels more feasible to construct with less cost and risk compared to earlier times.

Figure 4.144 | **EPBMs After University Link Tunneling and Ready for Dismantling**

Figure 4.145 | **Aerial View of Excavation for Capitol Hill Station, 2012**

CHAPTER FOUR | TRANSIT TUNNELS 229

THE FUTURE OF TRANSIT TUNNELS IN THE UNITED STATES

Mass transit systems are major capital investments, taking years, if not decades, to plan and construct. Many of our larger cities have longed for modern, underground mass transit. For some, the political will to expend great sums of money has been a roadblock. But eventually, gridlock usually outweighs the financial barriers as the compelling need for an underground transit system is demonstrated. What many citizens often fail to realize is that even though these systems cost many millions to billions of dollars to design and construct, they are super-charged economic engines of fiscal growth for the immediate communities and the local region. Not only does the actual construction create hundreds of jobs, neighborhoods adjacent to and near the new transit stations thrive. Businesses relocate near mass transit stations so that their employees can easily get to and from work. Retail benefits from the ready supply of consumers passing through the area. As the dream of improved mobility becomes reality, real estate values climb with the high demand for property close to transit stations. The vast potential for profit and new tax revenue streams is always on the minds of civic leaders and developers alike. In Washington, D.C., and Los Angeles, for example, subway lines have been designed to serve aging parts of these cities, with leaders knowing that mass transit systems are exactly the economic stimulus needed to help redevelop a neighborhood.

Transit tunnels will continue to be essential to the future growth and sustainable development of the major cities in the United States.

Figure 4.146 | **Sequential Excavation of a Cross Passage for the University Link Alignment**

REFERENCES

Akai, S., Murray, M., Redmond, S., Sage, R., Shetty, R., Skalla, G., and Varley, Z. 2007 Construction of the C710 Beacon Hill Station using SEM in Seattle. In *Rapid Excavation and Tunneling Conference 2007 Proceedings*. Littleton, CO: SME. pp. 943–963.

Alldredge, W.S. 1974. Washington Metropolitan Area Transit Authority's experience with contractual relationships in tunneling contracts. In *Proceedings 1974 Rapid Excavation and Tunneling Conference*, Vol. 2. Littleton, CO: SME. pp. 1137–1162.

APTA (American Public Transportation Association). 2013. Best practices for integrating art into capital projects. APTA SUDS-UD-RP-007-13. www.apta.com/resources/standards/Documents/APTA-SUDS-UD-RP-007-13.pdf.

Baker, W., MacPherson, H., and Cording, E. 1981. *Compaction Grouting to Limit Ground Movements: Instrumented Case History Evaluation of the Bolton Hill Tunnels*. Report No. UMTA-MD-06-0036-81-1. Washington, DC: U.S. Department of Transportation, Urban Mass Transportation Administration.

Bay Area Council Economic Institute. 2016. The case for a second Transbay transit crossing. www.bayareaeconomy.org/wp-content/uploads/2016/02/BACEI_TransbayCrossing_Feb2016b.pdf.

BTC (Boston Transit Commission). 1899. *Fifth Annual Report of the Boston Transit Commission, for the Year Ending August 15, 1899*. Boston: Rockwell and Churchill.

BTC (Boston Transit Commission). 1910. *Sixteenth Annual Report of the Boston Transit Commission, for the Year Ending June 30, 1910*. Boston: Chapple Publishing.

BTC (Boston Transit Commission). 1913. *Nineteenth Annual Report of the Boston Transit Commission, for the Year Ending June 30, 1913*. Boston: Rockwell and Churchill.

BTC (Boston Transit Commission). 1932. *Report of the Transit Department for the Year Ending December 31, 1931*. Boston: Boston Printing Department.

Burdick, M., Krulc, M., McLane, R., and Brandt, J. 2013. University Link Light Rail TBM tunnel UWS to CHS contract U220: Case history. In *Rapid Excavation and Tunneling Conference 2013 Proceedings*. Englewood, CO: SME. pp. 987–1013.

Cavin, B.P., Rhodes, G.W., and Mussger, F.K. 1985. NATM provides improved design and construction methods for U.S. tunnel projects. In *Proceedings 1985 Rapid Excavation and Tunneling Conference*, Vol. 2. Littleton, CO: SME. pp. 645–644.

Commonwealth of Massachusetts. 1891. General Laws of the Commonwealth of Massachusetts, Chapter 365 Acts of 1891.

Commonwealth of Massachusetts. 1894. General Laws of the acts of the Commonwealth of Massachusetts, Chapter 584 Acts of 1894.

Copperthwaite, W.C. 1906. *Tunnel Shields and the Use of Compressed Air in Subaqueous Works*. New York: Van Nostrand.

Cording, E.J., Hendron Jr., A.J., MacPherson, H.H., Hansmire, W.H., et al. 1975. *Methods for Geotechnical Observations and Instrumentation in Tunneling*. Report to the National Science Foundation by the Department of Civil Engineering, University of Illinois at Urbana-Champaign. No. UILU-ENG-75-2022, December.

Critchfield, J.W., and MacDonald, J.F. 1989. Seattle bus tunnel construction. In *Rapid Excavation and Tunneling Conference 1989 Proceedings*. Littleton, CO: SME. pp. 341–359.

Daugherty, C.W. 1981. Metrorail's dual chamber rock tunnel station—Two can be simpler than one. In *Proceedings 1981 Rapid Excavation and Tunneling Conference*. Littleton, CO: SME. pp. 1186–1205.

Desai, D.B. 1979. North American subway system—Baltimore Rapid Transit Project. In *1979 Rapid Excavation and Tunneling Conference Proceedings*, Vol. 2. New York: American Institute of Mining, Metallurgical, and Petroleum Engineers. pp. 1545–1566.

Desai, D.B., and Ku, C.C. 1981. Mixed face tunneling in urban setting. In *1981 Rapid Excavation and Tunneling Conference Proceedings*, Vol. 1. New York: American Institute of Mining, Metallurgical, and Petroleum Engineers. pp. 351–382.

Douglas, H. 1963. *The Underground Story*. London: Robert Hale.

Edwards, C.A., and Merrill, K.D. 1995. Baltimore Metro northeast extension. In *1995 Rapid Excavation and Tunneling Conference Proceedings*. Littleton, CO: SME. pp. 93–110.

Engineering News-Record. 1978. Grouting limits settlement as transit bores advance. *Engineering News-Record* 201(11):22–23.

Foster, E.L., and Butler, G.L. 1981. Some observations on the use of precast segmented concrete liner in urban tunnels. In *1981 Rapid Excavation and Tunneling Conference Proceedings*. New York: American Institute of Mining, Metallurgical, and Petroleum Engineers. pp. 498–518.

Frank, G. DiPonio, M., and Cowles, M. 2007. Construction of the University Link Light Rail Tunnel U230 in Seattle, WA, case history. In *Rapid Excavation and Tunneling Conference 2013 Proceedings*. Littleton, CO: SME. pp. 953–966.

Fulton Center. 2016. About Fulton Center: Unparalleled accessibility. www.fultoncenternyc.com/ (accessed June 8, 2016).

Glossop, R. 1976. The invention of and early use of compressed air to exclude water from shafts and tunnels during construction. *Geotechnique* 26(2):253–280.

Godfrey, K.A. 1966. Rapid transit renaissance. *Civil Engineering-ASCE* 36(12).

Hansmire, W.H., and Cording, E.J. 1972. "Performance of a soft ground tunnel on the Washington Metro. In *Proceedings, North American Rapid Excavation and Tunneling Conference,* Vol. 1. New York: AIME. pp. 371–389.

Hansmire, W.H., and Cording, E.J. 1985. Soil tunnel test section: Case history summary. Paper No. 20129. *Journal of the Geotechnical Division, ASCE* 111(11):1301–1320.

Hart, A.J.R. 1989. Conventional precast concrete bolted liners and their use in the Washington Metro. In *Proceedings 1989 Rapid Excavation and Tunneling Conference*. Littleton, CO: SME. pp. 360–368.

Heflin, L.H. 1985. WMATA Use of the new Austrian tunneling method (NATM) for lining and support. In *Proceedings 1985 Rapid Excavation and Tunneling Conference*, Vol. 1. Littleton, CO: SME. pp. 381–391.

Heflin, L.H., Wagner, H., and Donde, P. 1991. U.S. approach to soft ground NATM. In *Proceedings 1991 Rapid Excavation and Tunneling Conference*. Littleton, CO: SME. pp. 141–155.

Keville, F.M. 1987. *The Building of a Transit Line: The Massachusetts Bay Transportation Authority's Haymarket North Extension Project.*. Boston: MTBA.

Kuesel, T.R. 1972. Soft Ground Tunnels for the BART Project. In *Proceedings, 1st North American Rapid Excavation and Tunneling Conference, Chicago*. New York: American Institute of Mining, Metallurgical and Petroleum Engineers.

Leong, M.W., Wilson, S., Fowler, M., and Tricamo, A. 2015. Central Subway tunnels: Success under San Francisco. In *Rapid Excavation and Tunneling Conference 2015 Proceedings*. Englewood, CO: SME.

MBTA (Massachusetts Bay Transportation Authority). 2016a. *MBTA Transit History—Orange Line*. Boston: MBTA.

MBTA (Massachusetts Bay Transportation Authority). 2016b. *MBTA Transit History—Silver Line.* Boston: MBTA.

Mergelsberg, W.A., Leech, W.D., and Viner Jr., R.L. 1987. Washington Metropolitan Area Transit Authority underground projects. In *Proceedings, 1987 Rapid Excavation and Tunneling Conference*, Vol. 2. Littleton, CO: SME. pp. 1112–1133.

Most, D. 2014. *The Race Underground.* New York: St. Martin's Press.

Murphy, G.J., and Tanner, D.N. 1966. The BART trans-bay tube. *Civil Engineering-ASCE* 36(12).

Peterson, E., and Frobenius, P. 1971. Soft-ground tunneling technology on the BART project. *Civil Engineering-ASCE* 41(10).

Port Authority of Allegheny County. 2016. Light Rail facts page. www.pittsburghtransit.info/lrt.html.

Roy, P.A., Boscardin, M.D, and Miller, A.J. 2009. Pittsburgh North Shore Connector tunnel project overview and update. In *Hong Kong Tunneling Conference 2009, Proceedings*. Hong Kong: Institute of Materials, Minerals, and Mining.

Sauer, G., and Garrett Jr., V.K. 1987. Achieving a dry tunnel. In *Proceedings 1987 Rapid Excavation and Tunneling Conference*, Vol. 1. Littleton, CO: SME. pp. 461–478.

Smith, R.E., Eisold, E.D., and Schrad, M. 1993. Compressed air tunneling through a contaminated zone. In *1993 Rapid Excavation and Tunneling Conference Proceedings*. Littleton, CO: SME. pp. 735–750.

Sprague, F.J. 1916. *The Growth of Electric Railways.* New York: American Electric Railway Association.

Tattersall, C., Gregor, T., and Lehnen, M. 2004. Design and impact of the Beacon Hill exploratory test shaft. In *North American Tunneling 2004: Proceedings of the North American Tunneling Conference 2004.* Leiden: A.A. Balkema.

Thon, J.G., and Amos, M.J. 1968. Soft-ground tunnels for BART. *Civil Engineering-ASCE* 38(6):52–55.

Varley, Z., Martin, R., Robinson, R., Schmall, P., and Parmantier, D. 2007. Beacon Hill Station dewatering wells and jet grouting program. In *Rapid Excavation and Tunneling Conference 2007 Proceedings*. Littleton, CO: SME. pp. 346–359.

Wikipedia contributors. 2016. Pittsburgh light rail. In *Wikipedia, The Free Encyclopedia.* https://en.wikipedia.org/wiki/Pittsburgh_Light_Rail (accessed Jan. 25, 2016).

William-Ross, L. 2008. LAistory: The 1925 "Hollywood Subway." http://laist.com/2008/07/12/laistory_the_19.php.

5 HIGHWAY TUNNELS

The history of highway tunnels in the United States is inextricably related to the history of automobiles and the roads on which they traveled.

Predating automobiles were horse-drawn carriages, and later, rail transit vehicles powered by horses, then electricity. Occasionally, passageways were created across difficult terrain such as waterways or hillsides. For instance, the Washington and LaSalle Street Tunnels were constructed in Chicago to provide access across the Chicago River in the 1860s and 1870s, respectively (Figure 5.1). The Washington Street Tunnel is one of the first vehicular tunnels in America (Richardson and Mayo 1975). The first successful American gasoline-powered automobile was designed in 1893 with commercial sales starting in 1896. By 1900, several companies were manufacturing automobiles, and in the next decade Ford introduced the Model T and General Motors was founded. There was a tremendous demand for automobiles in the United States because of its vast area and widely spaced population centers. By 1913, there were close to a half million cars on the roads. It quickly became obvious that government involvement was needed to organize and build an adequate road system for this new and very popular mode of transportation. The Federal Aid Road and Federal Highway Acts were passed in 1916 and 1921, respectively. During the Great Depression of the 1930s, President Franklin D. Roosevelt initiated a program to construct a network of "superhighways" to provide work for people struggling through the poor economic conditions of the time. The Federal-Aid Highway Act of 1938 legislated studying the feasibility of building road networks that would stretch across the United States. World War II interrupted this initiative, but near the end of the war, the Federal-Aid Highway Act of 1944 authorized construction of 40,000 miles of new national highways or "interstates." As automotive routes were improved and new ones built, various lengths and sizes of highway tunnels were necessary. Some of the highway tunnels were simply conversions from existing railroad tunnels. Others were far more formidable and unprecedented engineering undertakings.

ANATOMY OF A HIGHWAY TUNNEL

Early vehicular roadways consisted of unpaved routes converted from pedestrian footpaths and horse and buggy travel ways. No safety and traffic control measures such as lane striping, stoplights, and signage existed. Early highway tunnels tended to be fairly short and narrow, providing a minimum clearance for one or at most two vehicles to pass through. There was minimal structural support to the walls and ceilings. Often, the roadway surface consisted of dirt or bare rock with poor drainage. There was little to no lighting beyond what the vehicle had on board, and air quality relied on natural air flow. No traffic control devices or the like were in place. This was adequate for early versions of the comparatively primitive slow-moving vehicles. As the number of vehicles on the road dramatically increased, local, state, and federal agencies developed requirements and standards to be used for the roadway systems in the United States. The Federal Highway Administration, state Departments of Transportation, and municipal public works departments began taking responsibility for constructing and maintaining public roadways. Other groups were organized to participate in the development of design and construction guidelines and standards, such as the American Association of State Highway and

Figure 5.1 | **LaSalle Street Tunnel in Chicago**

Previous Page | **Mount Carmel Tunnel in Zion National Park**

Transportation Officials (AASHTO), American Road and Transportation Builders Association (ARTBA), American Society for Testing and Materials (ASTM), and Public Works Standards Inc. (PWSI) that publishes *Standard Specifications for Public Works Construction ("Greenbook")* every three years.

In modern times, highway tunnels require design features (Figure 5.2) such as adequate lane width, horizontal and vertical curvature limits, vertical overhead clearance, and traffic control signals and signage, while additional facilities are necessary to provide good visibility, clean air quality, and emergency access and egress. These become more important as the tunnel length increases beyond about 1,000 feet. During daylight hours, tunnel lighting must prepare the driver for entering into the "black hole" of a tunnel, rapid eye adjustment to tunnel conditions when entering, and then ambient brightness at the tunnel exit portal. Similar conditions need to be addressed at nighttime. The wall finishes on the tunnel such as bright paint, tiles, or special panels also affect reflectivity and brightness in the tunnel and should provide for minimum maintenance efforts.

Air quality needs to be controlled due the presence and accumulation of noxious vehicle exhausts, particularly during times of heavy traffic. Tunnels generally longer than 1,000 feet require some type of mechanical ventilation fan system. These fans are also used to provide fresh, clean air in the unlikely event of a vehicle fire or other emergency requiring motorists to leave their vehicles and exit the tunnel on foot. Closed-circuit televisions and other sensors are usually installed to monitor and assess conditions in the longer tunnels. Often the data collected with this equipment is communicated to a traffic control operations center for safe and continuous monitoring and incident response.

Other safety features of long tunnels often include a walkway for pedestrian evacuation, cross passages between twin tunnels or into a safe refuge area, emergency telephones, and fire suppression extinguishers and water sources for the fire department to use if necessary. All of these features require extra room in the tunnels as well as the necessary supporting electrical power, wiring, water pipes, and drainage. Structural and mining requirements often require tunnel shapes and sizes that provide extra room for some of these facilities (Figure 5.2). Additional equipment and rooms for operations and maintenance are often situated at the tunnel entrances in small portal buildings.

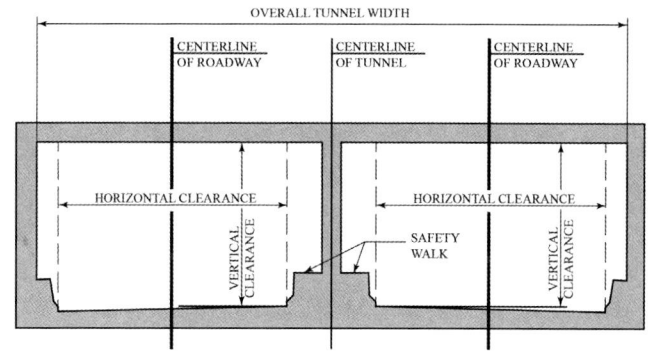

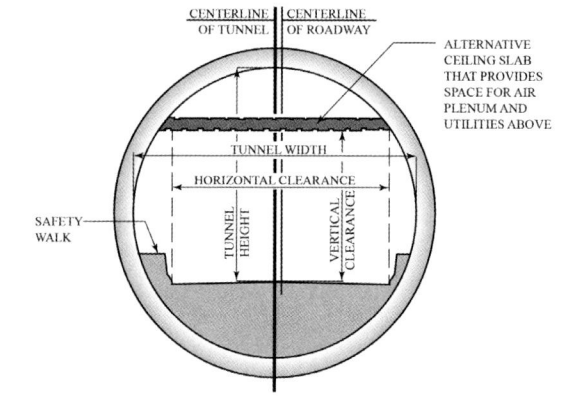

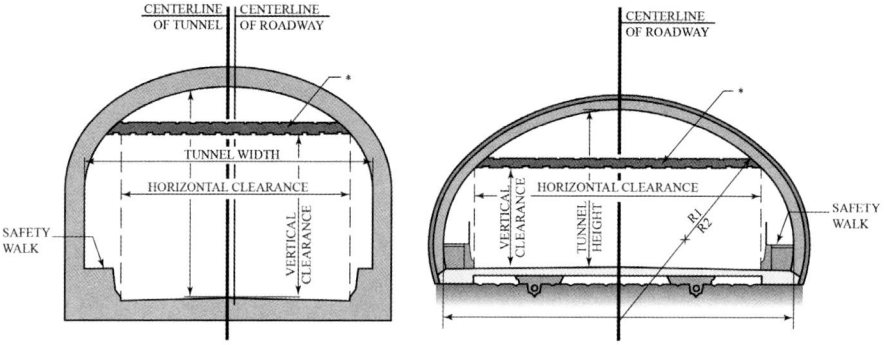

Figure 5.2 | **Typical Cross Section of Highway Tunnels**

CONNECTING AMERICA'S FIRST OFFICIAL ROADWAYS

Most of the early highway tunnels were short and constructed in rock. Natural ventilation was adequate. An early example is the Stockton Street Tunnel in San Francisco, California, under Nob Hill near Chinatown (Figure 5.3).

Figure 5.3 | **Stockton Street Tunnel in San Francisco at Nob Hill**

This tunnel opened on December 29, 1914 (Wallace 1949). The tunnel was constructed to reduce the road grade from a maximum of 18% to slightly more than 4% (Tilton 1915). The 50-foot-wide by 911-foot-long tunnel was carved out of highly fractured and decomposed Franciscan Complex clay shale. Timber supports were required during excavation, followed by a cast-in-place concrete final lining.

Much later, other tunnels such as the Broadway Tunnel (Figure 5.4) were constructed to penetrate the hills in San Francisco and to ease traffic congestion. The Broadway Tunnel was opened in 1952 and has 1,616-foot-long twin two-lane bores under Russian Hill (Gonzales-Hardy 2010).

At the beginning of the twentieth century, other cities in California such as Los Angeles were having similar challenges to efficient travel across steep terrain and traffic congestion that would be improved by the construction of tunnels. The unpaved and unlit Third Street Tunnel under Bunker Hill was opened in March of 1901. The relatively primitive, damp, and muddy conditions in this tunnel led the *Los Angeles Times* on March 27, 1901, to call it a "veritable stench in the nostrils of the public" (Richardson 2008). Soon thereafter, the tunnel was paved with appropriate drainage facilities and lighting was added. Despite this tunnel's inauspicious beginnings, traffic flow increased drastically

Figure 5.4 | Broadway Tunnel in San Francisco

and a second adjacent tunnel, the Second Street Tunnel, was constructed in 1924. Not only is this tunnel still used heavily for vehicular traffic, but it has become quite popular for the filming of movies such as *The Terminator* and *Independence Day*, car advertisements, fashion shows, and parties.

In another area of Los Angeles, the hills of Elysian Park were obstacles to efficient traffic flow between downtown Los Angeles and Pasadena across the Los Angeles River. In addition to other roadway facility improvements in the early 1930s, four tunnels were constructed along what eventually became the Arroyo Seco Parkway. The tunnels carried two lanes in each direction (northbound and southbound) as well as a sidewalk. From south to north, the tunnels are 755, 461, 130, and 405 feet long, respectively; 46.5 feet wide; and 28.3 feet high. Eventually, a parallel roadway was constructed for southbound traffic and the tunnels were converted exclusively to four lanes of northbound traffic.

The Wawona Tunnel provides access to Yosemite National Park in California. The tunnel was constructed through the granite bedrock that dominates the Sierra Nevada and iconic landmarks in the park such as Half Dome. The tunnel is two lanes wide and 4,233 feet long. Bidirectional traffic flows through it since it opened in 1933. Figure 5.5 shows some of the original construction details used to build the tunnel as well as the method of ventilation.

Similar development and expansions of the highway network in the western United States were taking place north of California. Highway 101 was constructed along the Pacific Ocean coastline extending into Oregon. Many places on the Oregon coast were virtually inaccessible in the early twentieth century. Small fishing villages existed as remote outposts, separated by rocky headlands and dense tree-covered hills. Several bridges and tunnels were required to provide vehicular access. One of these tunnels was the 427-foot-long Arch Cape Tunnel (Figure 5.6) that was blasted out of the hard basalt that formed the steep cliffs.

California and Oregon were not the only states facing physical obstacles to the expanding network of roadways within U.S. cities. In Pittsburgh, Pennsylvania, growth of the city in the 1920s began to spread to the South Hills that are south of the Monongahela River. Better access to downtown was needed, prompting the construction of the Liberty Tunnels through Mount Washington and the Liberty Bridge across the river. The twin tunnels are each two

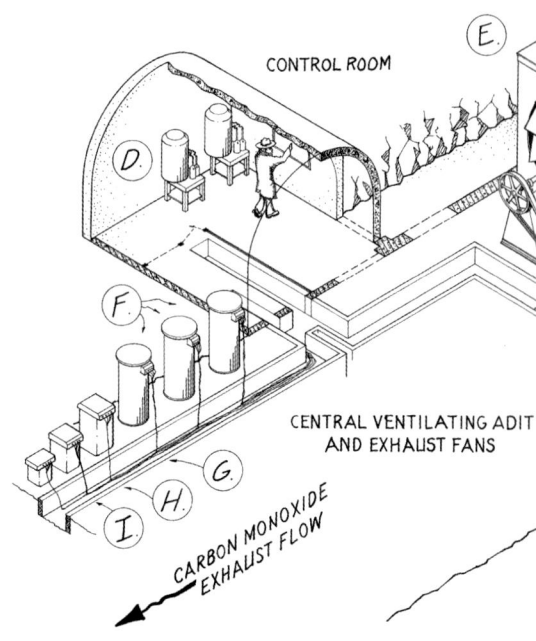

WAWONA TUNNEL CONSTRUCTION DETAILS

A Tunnel drilled & blasted to rough dimension
B Wood falsework constructed as single moveable unit ... (10-foot sections)
C Concreting machine, on temporary railroad, elevated mixture for pouring.
D Carbon monoxide detectors, located at tunnel crown, analyze increasing levels and activate ventilation fans.
E Three-8'x 0" diam. fans are automatically controlled to exhaust carbon monoxide.
F Three-50 Kw, 60 cycle transformers
G One- 20 Kw, constant current transformer
H One- 20 Kw, constant current transformer
I One-2,200 volts, 60 cycle transformer

CHAPTER FIVE | HIGHWAY TUNNELS

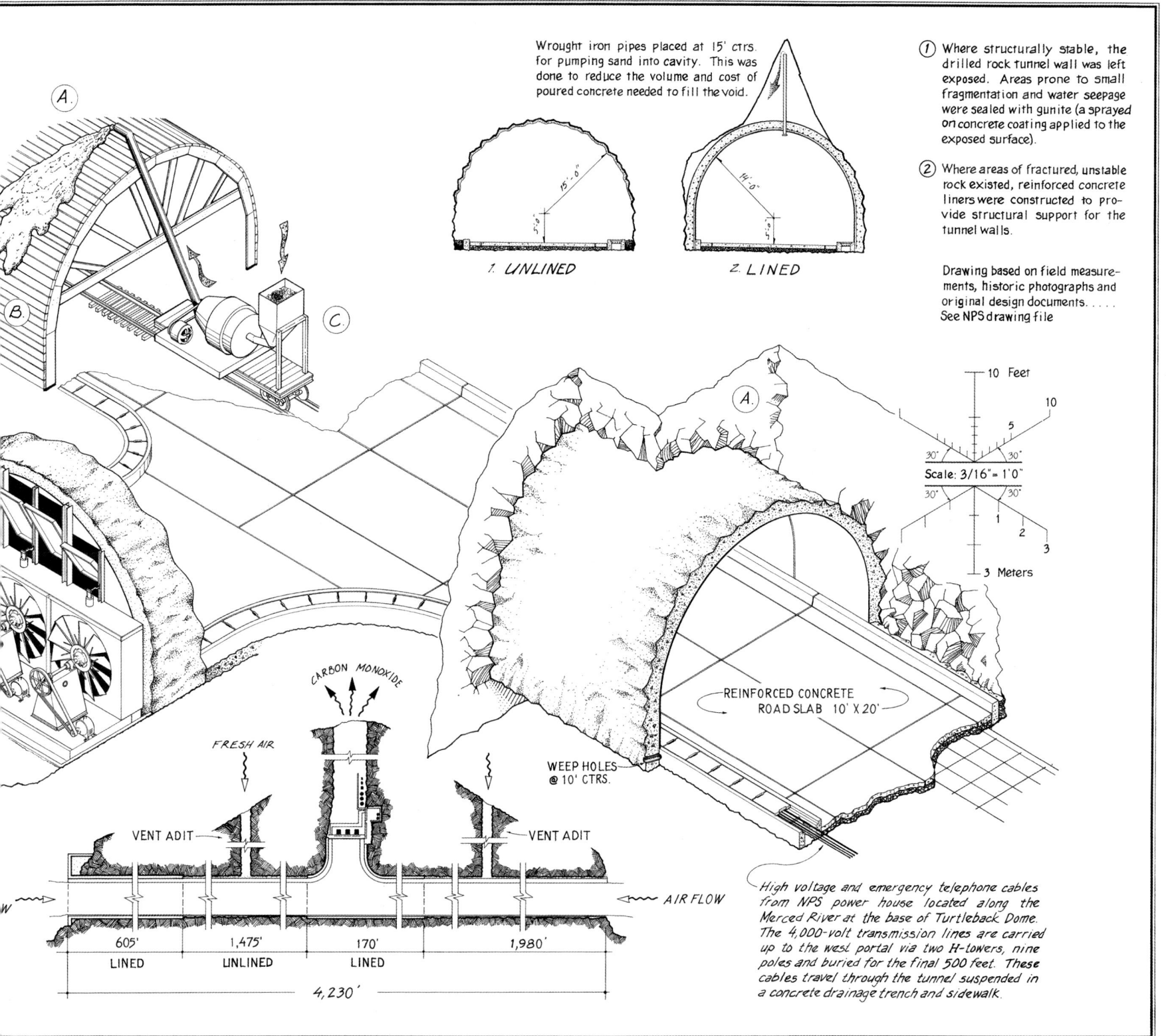

Figure 5.5 | **Ventilation System of the Wawona Tunnel in Yosemite National Park**

Figure 5.6 | Arch Cape Tunnel Along the Oregon Shoreline

lanes wide with a sidewalk (Figure 5.7) and 5,889 feet long from portal to portal. Eleven cross passages provided access/egress between the two tunnels. This was a fairly long vehicular tunnel for the times, and the lack of ventilation during high traffic-flow periods became a serious problem. In 1928, four 200-foot-deep vertical shafts and a fan plant at the top of the mountain were constructed to solve the car exhaust problem.

North of the Monongahela River, the Armstrong Tunnels were built in the late 1920s beneath Duquesne University hilltop campus in Pittsburgh's Bluff (Boyd's Hill) neighborhood. These twin tunnels are 1,298 feet long and 19.4 feet wide. One unusual feature is that the tunnels have a sharp curve between Locust Street and McAnulty Drive. Also, the tunnel walls are lined with Italian Renaissance stone and the portals are constructed with cut stone.

In New England, three notable tunnels were constructed to improve transportation conditions. In the town line between New Haven and Hamden, Connecticut, 1,200-foot-long twin two-lane tunnels were built for passage of the Route 15 Wilbur Cross Parkway. This tunnel, located in south-central Connecticut, opened in 1949 and was originally named West Rock Tunnel because it passes through West Rock Ridge, an intrusive volcanic fault block ridge composed of diabase. It was renamed Heroes Tunnel in 2003.

In Vernon, Connecticut, a short 108-foot-long tunnel was required under an embankment for the railroad between Manchester to Bolton Notch in the mid-1840s. This grade separation was needed to preserve access along Tunnel Road between Lake Street and Vernon Center. Vernon Tunnel is 14 feet wide and 16 feet high, consisting of 30 keystone arches made of sandstone blocks. Another name for this tunnel is the Keyhole Tunnel.

On the east side of Providence, Rhode Island, a tunnel originally constructed for trolleys in 1914 was repurposed for bus travel exclusively in 1948. This is likely the oldest example of rapid transit by bus in North America. The East Side Trolley Tunnel is 2,000 feet long under College Hill between South Main Street and Thayer Street. Using this route allows buses to avoid steep grades, slow traffic, and traffic signals. Emergency vehicles are also permitted access to the tunnel. The west end of the tunnel was constructed below the 1890s vintage Rhode Island School of Design, requiring extensive and careful underpinning and construction methods.

Figure 5.7 | **Liberty Tunnels in Pittsburgh**

In addition to the expanding roadway systems within cities, highways were being constructed between cities. One of the more ambitious undertakings at the time was the Lincoln Highway, the first transcontinental highway for automobiles across the United States (Weingroff 2011). This highway was dedicated in 1913 and went from New York City's Times Square to San Francisco's Lincoln Park, traversing several states and cities including New York, New Jersey, Pennsylvania, Ohio, Indiana, Illinois, Iowa, Nebraska, Colorado, Wyoming, Utah, Nevada, and California. The Lincoln Highway was the first national memorial to President Abraham Lincoln and it inspired the Federal Aid Highway Act of 1956 (discussed later in this chapter). The Lincoln Highway became known as "The Main Street Across America" and brought great prosperity to the hundreds of cities, towns, and villages along its path. To accomplish this feat, the highway adopted many routes and configurations along the way.

In Nevada on the eastern shore of Lake Tahoe, the Lincoln Highway was routed along a one-lane hanging bridge and rock wall adjacent to the Cave Rock area, considered sacred by the Washoe Tribe. However, this became a major access route to Lake Tahoe and one lane did not provide adequate capacity. So in 1931, a two-lane tunnel was constructed through Cave Rock. It is 153 feet long and unlined. This part of Lincoln Highway eventually became part of U.S. Route 50. At the time, the 1.1-mile-long Mount Carmel Tunnel in Zion National Park had just been constructed and the same tunnel crews were used for this construction. A second 410-foot-long two-lane bore was constructed through Cave Rock in 1957 (Figure 5.8) and is concrete lined.

Figure 5.8 | **Cave Rock Tunnel near Lake Tahoe in Nevada**

EARLY SUBAQUEOUS CROSSINGS

As tunnel engineers set their sights on higher mountains and wider rivers, more complex tunneling methods and specialized environmental control systems became necessary and possible, such as mechanical ventilation. One such tunnel to use this new technology between New York and New Jersey was the Holland Tunnel, and it was also subsequently used on the Lincoln Highway, with both achievements being quite remarkable undertakings. Prior to the opening of the Holland Tunnel in 1927, several thousand automobiles were crossing the Hudson River between New York and New Jersey via ferries every day, and the numbers were rising rapidly after the end of World War I. Rail tunnels had been successfully constructed across the river between terminals in New Jersey and Manhattan during the beginning of the twentieth century. So it was just a short time before similar methods were used to provide a vehicular crossing of the river (Figure 5.9). A bridge crossing would have required large amounts of land for the approaches, significant foundation construction down to bedrock, and major disruptions to ship traffic on the Hudson River.

The Hudson River Vehicular Tunnel (later renamed the Holland Tunnel) was a joint undertaking by the New Jersey Interstate Bridge and Tunnel Commission and the New York State Bridge and Tunnel Commission in 1920 (Lange 1993). These agencies were the precursors to the Port Authority of New York and New Jersey that is still in existence today. The tunnel consists of twin tubes with two 10-foot-wide lanes in each direction and 12.5 feet of clearance to the suspended ceiling and air plenum. Figure 5.10 provides an interesting comparison of the previously constructed Hudson River rail tunnels and the Holland Tunnel. The bores are 8,558 and 8,371 feet long in the westbound and eastbound directions, respectively, and nearly 100 feet below river level. Steel shields to protect the workers at the tunnel face and compressed air to hold back the groundwater flowing into the tunnel were used to carve the tunnels out of the river bottom soil.

CHAPTER FIVE | HIGHWAY TUNNELS | 243

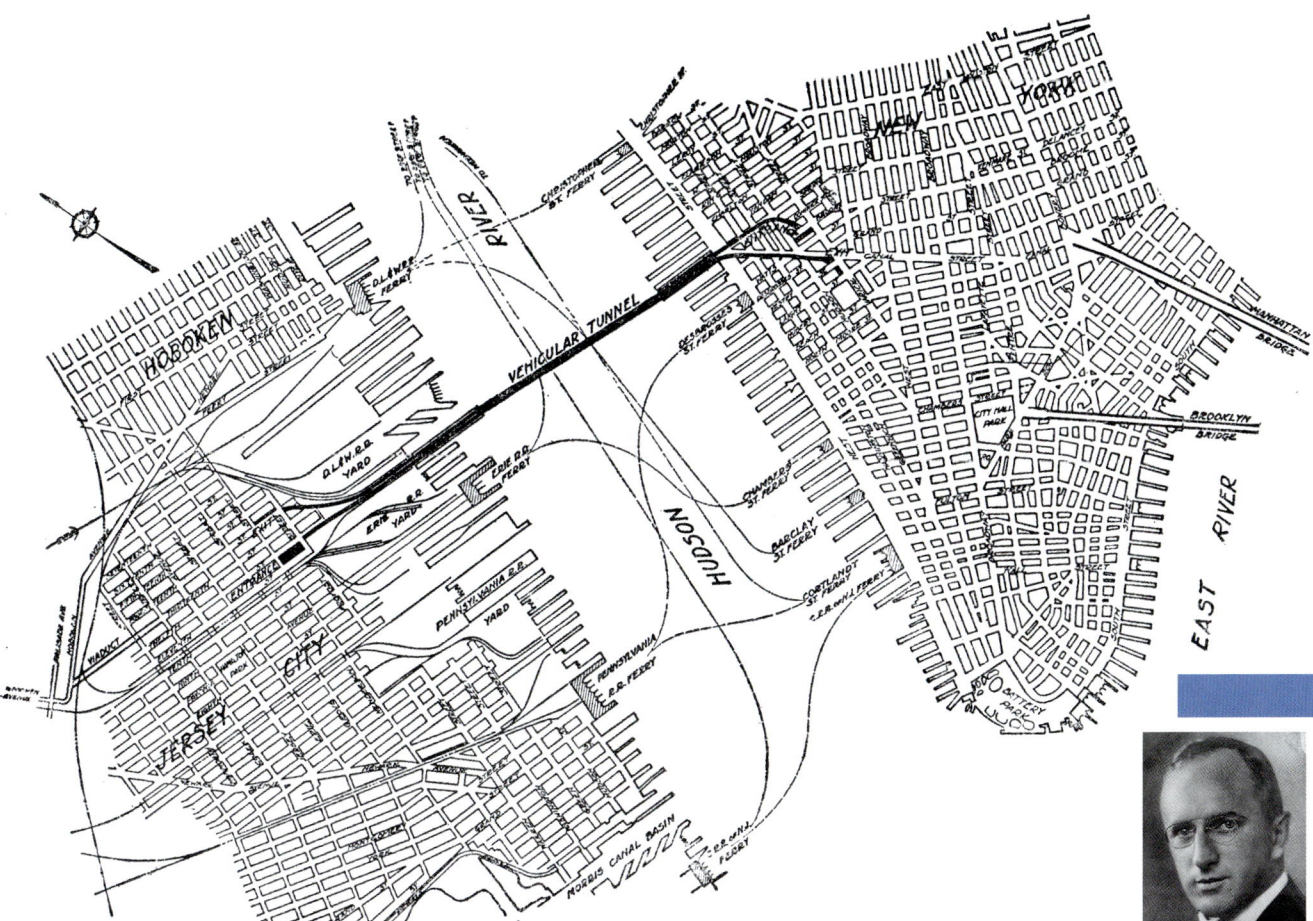

Figure 5.9 | **Location Map of Holland Tunnel in New York City**

Heavy cast-iron segments were bolted together in a 30-foot-diameter ring to support the excavation (Figure 5.11). Working under compressed air is like deep sea diving, requiring good physical condition of the workers and careful entry and exit procedures to avoid getting decompression sickness, also called the bends or caisson disease. The tunnel construction began in bedrock on each side of the river in huge pressurized steel caissons (Figure 5.12). Tunneling under the river took place from these caissons on each shore. The workers within the tunnel were called "sandhogs." After the tunnels were excavated and supported with a circular cross section (Figure 5.13), they were lined with concrete and ceramic tiles (Figure 5.14).

The requirements for ventilating the motor vehicle exhaust fumes and also for providing safety in the event of a fire in the Holland Tunnel were unprecedented. In previous long tunnels for railroads, ventilation was accomplished by blowing air longitudinally along the tunnel from one end to the other (longitudinal ventilation). This was not feasible for such a long vehicular tunnel. With the help of mining engineers, scientists at universities, and experience overseas, a new type of ventilation system was developed that blew air into the tunnel under and up through the roadway (supply air) with extraction openings in the ceiling above the roadway (exhaust air). This is called transverse ventilation. Eighty-four fans in four ventilation buildings were required to provide ventilation for the Holland Tunnel.

Clifford Milburn Holland (1883–1924)

Holland graduated from Harvard University with a bachelor of science degree in civil engineering in 1906. He served as assistant engineer during construction of the Joralemon Street Tunnel for the New York Rapid Transit Commission and oversaw construction of the Interborough Rapid Transit (IRT) subway line. He was subsequently the engineer-in-charge for the construction of the Clark Street, Montague Street, and 14th Street subway tunnels. Holland then served as chief engineer on the Hudson River Vehicular Tunnel project, also in New York City. Holland died of a heart attack at the age of 41, which was attributed to the long hours and stress caused by working in the compressed air of the tunnel. The tunnel was renamed in his honor by the New York State Bridge and Tunnel Commission and the New Jersey Interstate Bridge and Tunnel Commission on November 12, 1924.

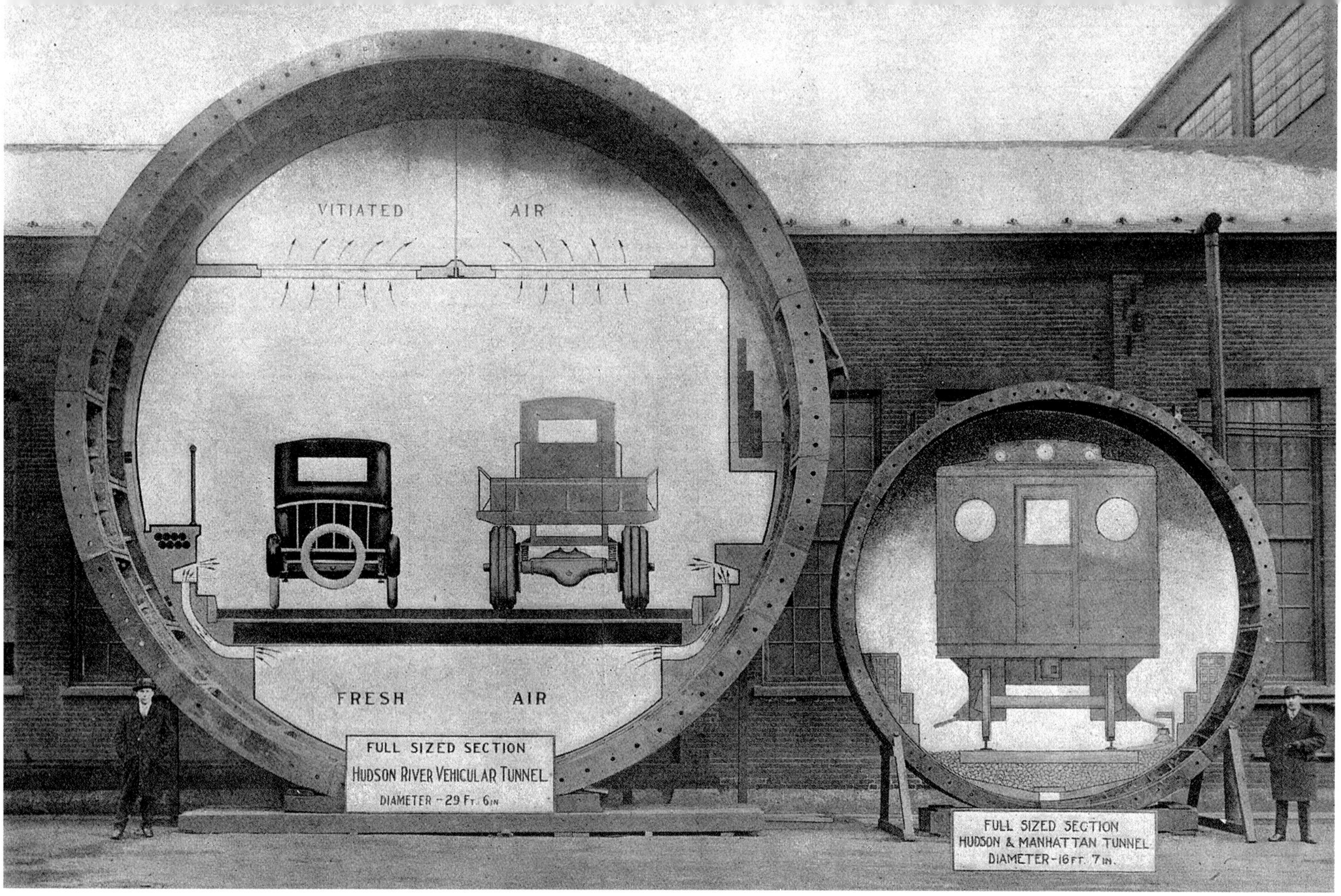

Figure 5.10 | **Comparison of the Holland Tunnel to a Typical Railroad Tunnel**

Figure 5.11 | **Installing Cast-Iron Tunnel Segments in the Holland Tunnel**

The Holland Tunnel was named after Clifford Holland, the first chief engineer of the project. Holland unfortunately passed away shortly before the tunnel drives from New Jersey and New York were joined together in 1924. Holland was succeeded by Milton Freeman who died four months later. Ole Singstad took over to oversee completion of the project in 1927. Singstad is also credited with pioneering the novel ventilation system used. The tunnel was designated a National Historic Civil and Mechanical Engineering Landmark in 1982 and a National Historic Landmark in 1993.

Not long after the Holland Tunnel opened, Singstad designed the Midtown Hudson Tunnel (later renamed the Lincoln Tunnel after Abraham Lincoln). Initially, two bores were constructed with the first bore opening in 1937 (Figure 5.15). The construction methods used for the Lincoln Tunnel (and the Holland Tunnel) were very similar and are depicted in Figure 5.16. The second bore opened in 1945 after delays due to World War II. A third bore was later constructed and was opened in 1957. These tunnels were almost identical to the Holland Tunnel in length, diameter, and physical characteristics (Figures 5.17 and 5.18).

In 1940, the Queens–Midtown Tunnel was opened linking midtown Manhattan with Queens under the East River. The Triborough Bridge and Tunnel Authority (TBTA) took ownership of the tunnel in 1946. While construction was very similar to the Holland Tunnel, the twin tubes of the Queens–Midtown Tunnel were one and a half foot wider

CHAPTER FIVE | HIGHWAY TUNNELS | 245

Figure 5.12 | **Excavation Within a Caisson**

Figure 5.13 | **Cross Section of the Holland Tunnel**

Ole Knutsen Singstad (1882–1969)

Singstad was a pioneering Norwegian-American civil engineer recognized for significant advances in tunnel ventilation and immersed-tube tunnel design. He became the chief engineer of the Holland Tunnel upon the death of Clifford Holland in 1924 and would become one of the premier highway tunnel builders in North America. One of his significant contributions and achievements is the development and design of the full transverse ventilation method for road tunnels. Fully transverse ventilation was pioneered during the construction of the Holland Tunnel, and most of the highway tunnels constructed in the twentieth century in North America were designed with transverse ventilation systems.

to accommodate the wider cars of the period (TBTA 2015). This tunnel was also designed by Ole Singstad and provides two lanes of traffic in each bore. Constructed just over 100 feet below the East River, the south tube is 6,272 feet long and the north tube is 6,414 feet long. While it took seven years to complete the Holland Tunnel, it only took four years to complete the Queens–Midtown Tunnel (www.nycroads.com 2015). With the Lincoln Tunnel, the Queens–Midtown Tunnel provided a continuous route from Long Island to New Jersey through Manhattan (Figures 5.19 through 5.21). The tunnel was rehabilitated in 2002.

Figure 5.14 | **Police Traffic Monitor in the Holland Tunnel**

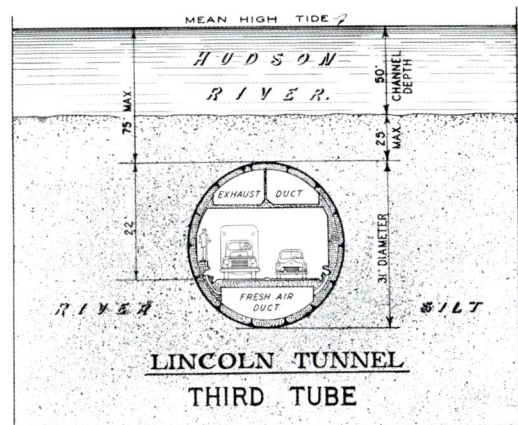

Figure 5.15 | **Cross Section Through the Lincoln Tunnel Below the Hudson River**

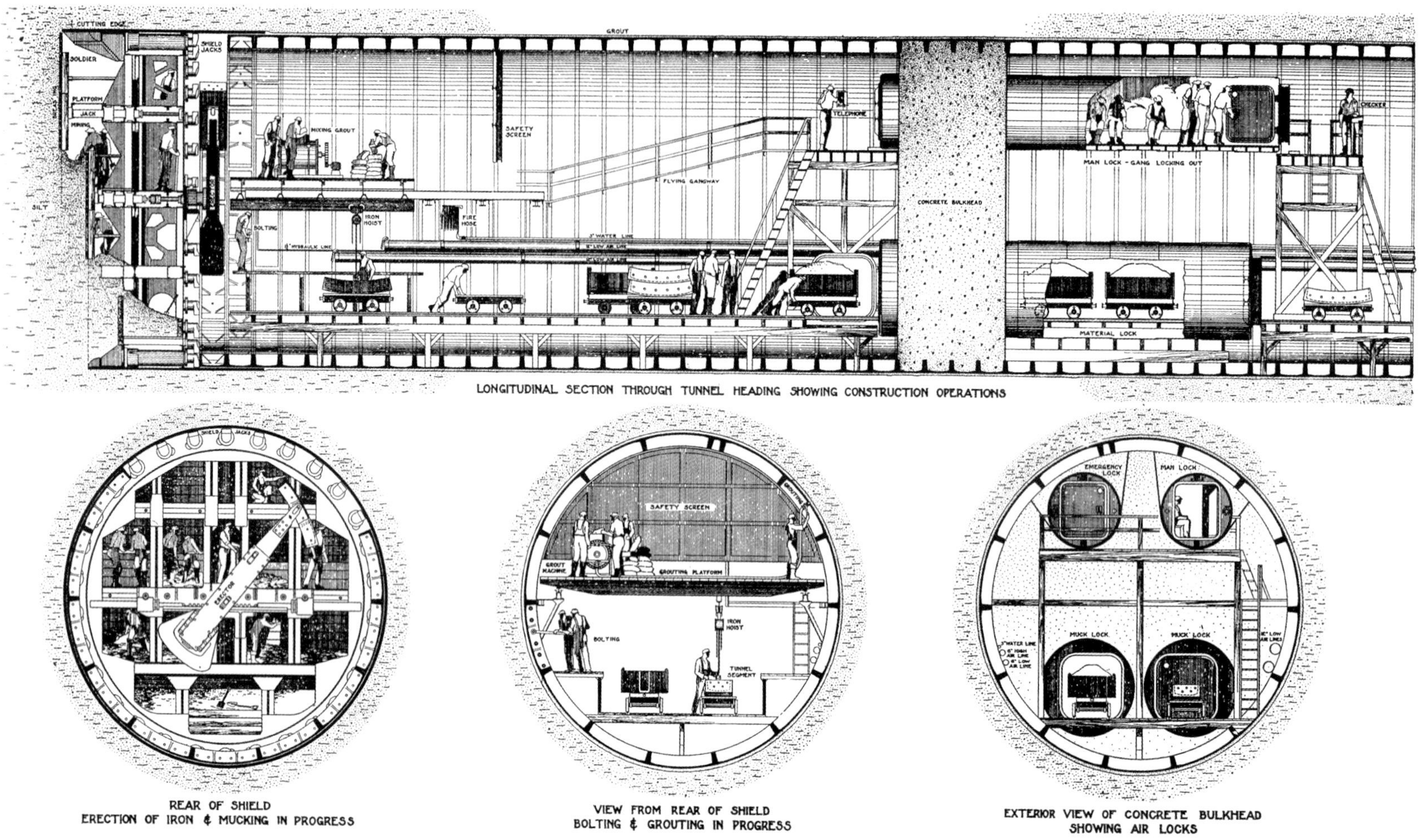

Figure 5.16 | **Construction Methods Used for the Lincoln Tunnel Third Bore**

Figure 5.17 | **Work in Progress on the Lincoln Tunnel**

The Brooklyn-Battery Tunnel (renamed the Hugh L. Carey Tunnel) is also a TBTA facility (now part of New York City's Metropolitan Transportation Authority Bridges and Tunnels) and opened in 1950 after delays due to World War II. At 9,117 feet long, it is the longest continuous vehicular tunnel in North America (TBTA 2015), connecting lower Manhattan with Brooklyn. It took 10 years to construct, and a huge parking structure was also constructed as part of this project in conjunction with the Manhattan tunnel portal facilities. Some people, including Robert Moses, preferred to build a bridge in lieu of a tunnel at this location, but that proposal was overwhelmingly opposed by many, including Eleanor Roosevelt.

The first tunnel under the Boston Harbor, the Sumner Tunnel, used mining and lining construction methods very similar to the ones used in New York. It had two-way traffic connecting downtown Boston with East Boston. The 29.5-foot-diameter initial lining consists of steel liner plates with 19.7 inches of cast-in-place concrete placed inside for the final lining. At road level, each lane is just over 11 feet wide with a 13.6-foot vertical clearance, and there is a 3.3-foot-wide walkway on the side of the tunnel (Russell 2002). The tunnel opened in 1934 and was named after William Sumner who fought in the War of 1812 and is best known for his role in the urban development of East Boston. In 1961, a parallel tunnel using the same construction methods

Figure 5.18 | **Entrance Portal to the Lincoln Tunnel**

Figure 5.19 | **Cross Section of the Queens–Midtown Tunnel Under the East River**

was opened to ease traffic to and from Logan Airport. The Sumner Tunnel was converted to one-way traffic with the newer Callahan Tunnel carrying traffic in the opposite direction. A third harbor crossing was completed as part of the "Big Dig" in 1995. The Big Dig is discussed later in this chapter.

While the tunnels in New York and Boston described previously utilized very similar mining and lining construction methods, quite different methods were being used in other places to cross rivers and estuaries in the same time period. These methods consisted of prefabricating long tunnel sections, floating them to the location like barges, dredging a trench, and then sinking the sections carefully into place and joining the sections end to end to form the new tunnel. This technique is known as the immersed-tube method. One such tunnel using this technique was the Posey Tube between Oakland and Alameda in California, which opened in 1928 to replace the Webster Street swing bridge across San Antonio Creek. The Posey Tube named after George Posey, Alameda County surveyor and chief engineer on the project, is the second-oldest underwater vehicular tunnel in the United States, preceded only by the Holland Tunnel. The Posey Tube is 4,436 feet long and 37 feet in diameter. It was the world's first concrete immersed tube and originally had one traffic lane in each direction. In 1963, a second tube was constructed called the Webster Street Tube that then allowed two lanes of traffic to flow in the same direction in each tunnel.

Another immersed-tube vehicular tunnel of this era is the Detroit–Windsor Tunnel between Detroit, Michigan, and Windsor, Ontario, which opened in 1930. This was the third subaqueous vehicular tunnel constructed in North America and the "first vehicular subway ever built between two nations," according to the Detroit–Windsor Tunnel website (dwtunnel.com). As early as 1870, Detroit citizens were greatly debating the relative merits of a bridge and a tunnel between Detroit and Windsor. The railroads favored a bridge, whereas shipping interests believed that a bridge structure would be hazardous to navigation, due to the exceedingly high masts of the sailing ships that forged the Detroit River at that time. The tunnel is 120 feet short of a mile at 5,160 feet. At its lowest point, the two-lane roadway is 75 feet below the Detroit River surface. Here too, Ole Singstad had significant involvement in the ventilation system design, working closely with a fellow Norwegian-American engineer named Soren Thoresen, the head design

Figure 5.20 | **Sandhog Crew Installing Tunnel Segments in the Queens–Midtown Tunnel**

Figure 5.21 | **Installing Cast-Iron Segments**

engineer for the project. Instead of using precast concrete sections, this immersed-tube tunnel was constructed using steel tubes. After constructing cut-and-cover tunnels on the land portions, shield-mined tunnels led to nine watertight steel tubes that were floated into place and sunk into a trench that had been dug in the river bottom. Cast-in-place concrete collars provide underwater seals between the steel tube sections. Twenty feet of stone was placed around and above the tunnel to hold it down and for protection from ship anchors.

Other places the immersed-tube tunneling method was used in the United States around this time period include the Bankhead Tunnel under the Mobile River in Alabama (opened in 1941), the Washburn Tunnel under the Houston Ship Channel in Texas (opened in 1950), and the Baltimore Harbor Tunnel under the Patapsco River in Baltimore, Maryland (opened in 1957). A unique aspect of the 1.45-mile-long Baltimore Harbor Tunnel along Interstate 895 is that it was constructed using 21 binocular twin-tube sections that were 69 feet wide by 300 feet long, providing four lanes of traffic within one structural section. The tunnel was an instant success. It eliminated 51 traffic signals for through traffic in downtown Baltimore, provided a cross-harbor route for local commuters, and diverted up to 40% of commercial vehicle traffic from local streets.

PENNSYLVANIA TURNPIKE TUNNELS

The tunnels previously discussed were somewhat isolated undertakings, scattered from coast to coast across the United States, and used a variety of tunnel construction methods. Engineers in Pennsylvania had their sights on a bigger mission—America's first "superhighway." Prior to 1940, east–west automobile routes through Pennsylvania's Allegheny Mountains were quite dangerous due to sharp curves and steep grades. In 1934, a superhighway was proposed to mitigate these poor driving conditions, utilizing the abandoned South Pennsylvania Railroad Company line between Carlisle and Irwin, Pennsylvania, over a distance of 160 miles (Figure 5.22). It was here that the Pennsylvania Turnpike was born (Pennsylvania Turnpike 2015). To make this possible, seven railroad tunnels were converted to two-way vehicular tunnels at Laurel Hill (4,541 feet long), Allegheny (6,070 feet long), Ray's Hill

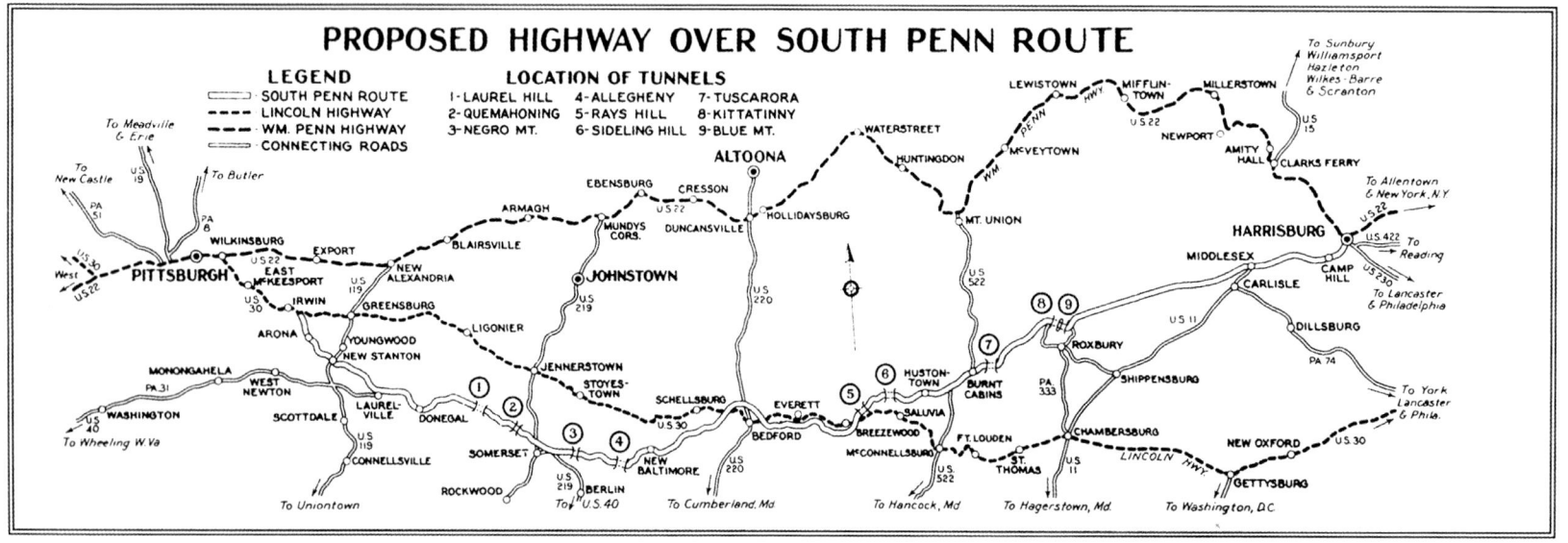

Figure 5.22 | Initial Route of the Pennsylvania Turnpike

Figure 5.23 | Allegheny Mountain Tunnel on the Pennsylvania Turnpike

Figure 5.24 | Inside of Allegheny Mountain Tunnel

Figure 5.25 | **Construction of the First Lehigh Tunnel on the Pennsylvania Turnpike**

Figure 5.26 | **Second Lehigh Tunnel on the Pennsylvania Turnpike**

(3,532 feet long), Sideling Hill (6,782 feet long), Tuscarora (5,326 feet long), Kittatinny (4,727 feet long), and Blue Mountain (4,339 feet long). Allegheny Tunnel, shown in Figure 5.23, was typical in appearance for these tunnels. As traffic volumes increased, parallel tunnels were eventually constructed at Blue Mountain, Kittatinny Mountain, Tuscarora Mountain, and Allegheny Mountain (Figure 5.24). In the 1960s, the Ray's Hill, Sideling Hill, and Laurel Hill tunnels were eventually bypassed. After World War II, the Pennsylvania Turnpike was extended eastward to Philadelphia and westward to the Ohio border. Further expansions in the 1950s connected the turnpike across the Delaware River to the New Jersey Turnpike and to Scranton, requiring an additional tunnel through Blue Mountain (Figure 5.25). This tunnel was named the Lehigh Tunnel. A second Lehigh Tunnel bore was constructed in the 1990s (Figure 5.26).

Also in Pennsylvania, the Squirrel Hill Tunnel was completed in 1953 to provide better access to Pittsburgh from the eastern suburbs. Previously, the William Penn and Lincoln Highways (Penn–Lincoln Parkway) were the only routes in this direction. Like many highway tunnels, the Squirrel Hill Tunnel connects to bridges at each end. The tunnel consists of twin arch-shaped, reinforced concrete bores that are 4,225 feet long and approximately 29 feet 4 inches wide. There are eight cross passages for safety and emergency egress. Up until this time, most highway tunnels were either naturally ventilated (no mechanical ventilation systems) or transversely ventilated (mechanical systems with air supply and exhaust plenums). The Squirrel Hill Tunnel uses a semi-transverse ventilation system that has an air supply plenum only above the roadway and uses the roadway area itself for exhaust flow. Consequently, some semi-transverse tunnel ventilation systems were developed to be reversible.

EARLY TUNNELS THROUGH RIDGES AND MOUNTAINS

Another conversion from railroad tunnel to vehicular tunnel warrants mention here albeit far from Pennsylvania. This tunnel is the Whittier Tunnel in Alaska and is very unique in that now both railroad trains and vehicles use it. The 2.5-mile-long Whittier Railroad Tunnel was constructed by the U.S. Army in 1941 so that the town of Whittier could be used as a deep water port during World War II. This railroad spur between Portage and Whittier became Alaska's main supply link for the war effort (ADOT 2015). After the war, the population and infrastructure of Whittier continued to grow and it became a federally run commercial port supporting freight ships, cruise lines, fishermen, and recreational boaters. To further improve access, the Alaska Railroad began offering a vehicle shuttle service between Portage and Whittier in the mid-1960s. Despite best efforts by the shuttle service, further improvements in vehicular access were deemed necessary in the 1990s (Figure 5.27). The tunnel was converted to dual use with trains and cars taking turns traveling through the tunnel. Modifications to the walls and floor of the tunnel were undertaken as well as improvements to the ventilation and other systems (Figures 5.28 and 5.29). The modified World War II vintage rail tunnel was opened to vehicular traffic on June 7, 2000, and was renamed the Anton Anderson Memorial Tunnel after the man who oversaw the construction of the original rail tunnel for the U.S. Army Corps of Engineers. This tunnel is the longest highway tunnel and the longest combined rail and highway tunnel in North America.

Figure 5.27 | **Alcove Mining in the Whittier Mixed-Use Tunnel in Alaska**

The Yerba Buena Island Tunnel, the world's largest single-bore transportation tunnel, originally carried rail and highway traffic (Figure 5.30). The San Francisco–Oakland Bay Bridge, along Interstate 80 in California, was opened in 1936 and at that time consisted of a complex truss and cantilever bridge and a double suspension bridge east and west of Yerba Buena Island, respectively (Figure 5.31). On Yerba Buena Island, traffic passes through a 540-foot-long, 76-foot-wide, and 58-foot-high tunnel excavated by drill-and-blast methods and a unique work sequence (Figures 5.32 through 5.35). The portals were constructed with cast-in-place reinforced concrete and designed in a manner consistent with the Art Deco trends popular in the day. The double-deck bridge and tunnel originally carried six lanes of bidirectional car traffic on the upper deck and three lanes of truck traffic and two commuter rail lines on the lower deck (Caltrans 2015). Train service across the bridge began on September 23, 1938, and ended in April 1958. In 1963, the Bay Bridge was reconfigured to have five westbound lanes on the upper deck and five eastbound lanes on the lower deck. Reconstructing the double-deck roadways within Yerba Buena Island Tunnel was a major engineering feat. The east span of the Bay Bridge was damaged in the Loma Prieta earthquake of 1989 and has since been replaced with a self-anchored suspension span.

Figure 5.28 | **Passing Pockets in the Whittier Tunnel**

Figure 5.29 | **Installing Prefabricated Road and Rail Invert Panels in the Whittier Tunnel**

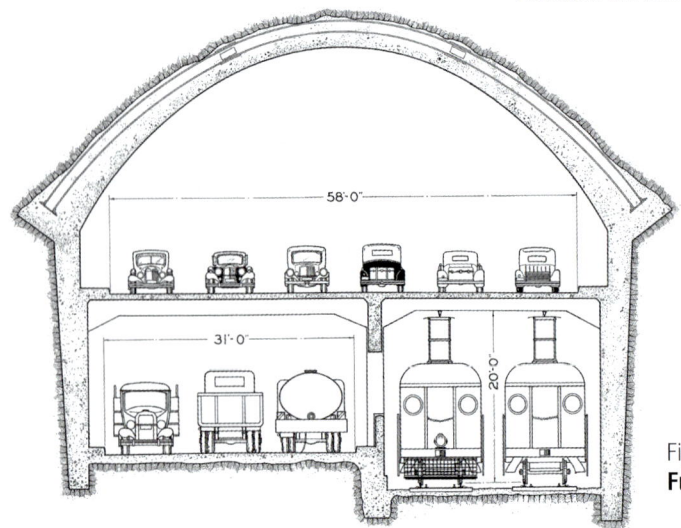

Figure 5.30 | **Yerba Buena Tunnel Full Cross Section—As Designed**

Figure 5.31 | San Francisco–Oakland Bay Bridge and Tunnel

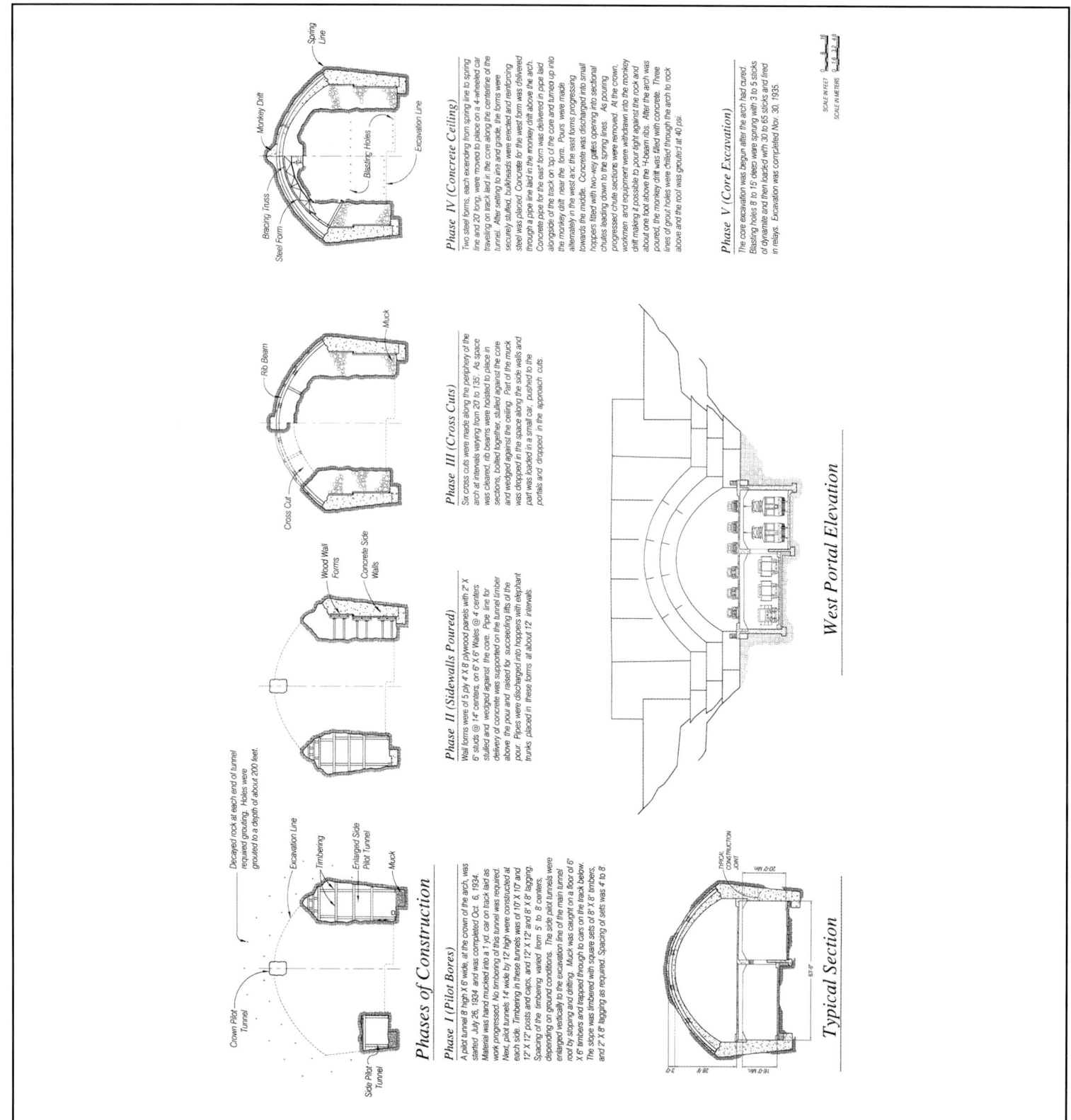

Figure 5.32 | **Yerba Buena Tunnel Excavation Sequence**

Also in the San Francisco area, two tunnels provide access between the Golden Gate Bridge and Marin County along U.S. Highway 101. Prior to construction of the highway in 1909, there were only Indian trails and primitive roads providing access to a ferry terminal in Sausalito. The first bore was finished in 1937, and the second in 1954. The 1,000-foot-long twin bores were named the Waldo Tunnel after the Waldo Point nearby. Each tunnel is four lanes wide. There are rainbows painted on the portals and some call it the Rainbow Tunnel. In 2015, efforts began to rename the tunnel the Robin Williams Tunnel after the actor, who lived nearby, committed suicide in 2014.

An early version of the Caldecott Tunnel in the Oakland–Berkeley Hills of California between Alameda and Contra Costa Counties was first contemplated in the mid-1800s and eventually opened in 1903 (Solon and McCosker 2014). This predecessor tunnel had several names, including the Kennedy Tunnel (Figure 5.36), the Inter-County Tunnel, and the Broadway Tunnel. It was merely 17 feet wide near timber supports, barely large enough to fit one horse-and-buggy carriage at a time. Illumination of the tunnel was done by lighting the end of a newspaper on fire to warn oncoming traffic. Eventually, twin bore tunnels were constructed about 500 feet lower and opened in 1937, and the earlier tunnel was closed in 1947. The new tunnels along State Highway 24 sparked extensive development of the East Bay of San Francisco, transforming the small farming community to a major suburb that continues to grow more each day. The first twin tunnels are each 3,610 feet long, 27 feet wide, and carry two lanes of traffic. In 1964, the third "reversible" tunnel was opened to ease traffic congestion, and in 2013, the fourth bore was also opened to ease the traffic congestion and to eliminate the labor-intensive need of deploying and redeploying traffic cones every day to reverse traffic through the third bore. The third and fourth bores are 3,771 and 3,389 feet long, respectively. The fourth bore has a modern appearance and functions using the latest lighting, ventilation, and communication technologies (Figure 5.37). While transverse ventilation with air supply and exhaust plenums in the suspended ceiling are used in the older three bores, longitudinal ventilation is used in the fourth bore with jet fans mounted and visible above the roadway. The tunnels are named in honor of a former Alameda County supervisor and president of the Joint Highway District, Thomas E. Caldecott.

Another bridge and tunnel combination of this era worth noting is the Lake Washington Floating Bridge and the Mount Baker Ridge Tunnel that opened in 1940. The Mount Baker bridge and tunnels were part of U.S. Highway 10 between Seattle and Mercer Island in Washington (ASCE 2007). The bridge was later renamed the Lacey V. Murrow Memorial Bridge. Prior to the construction of the bridge and tunnel, travelers had to drive around the 20-mile-long and up to 4-mile-wide Lake Washington. The 14-mile-long route around the lake was cut in half by the new route created with the 1.5-mile-long floating bridge and one-quarter-mile long tunnels through the 260-foot-high Mount Baker Ridge on the east shore of the lake. Each tunnel is two lanes wide and 1,466 feet long. The tunnel excavations were 23 feet high and nearly 29 feet wide. While similar vehicular tunnels of this size had been constructed in rock, these were the first to be constructed in soil consisting of blue clay. Wooden timbers were used to hold the excavation open until a 2-foot-thick concrete lining could be installed. Portal structures were constructed at the ends of the tunnels that were quite attractive. As discussed later, this

Figure 5.33 | **Yerba Buena Tunnel, East Portal Excavation, 1935**

Figure 5.34 | **Yerba Buena Tunnel, Interior Excavation**

Figure 5.35 | **Yerba Buena Tunnel, East Portal Construction**

route eventually became the route for Interstate 90 and another great feat of tunnel engineering consisting of a 63-foot-diameter soil tunnel for the adjacent Homer M. Hadley Memorial Bridge in 1989.

TUNNELS BUILT AFTER THE FEDERAL-AID HIGHWAY ACT OF 1956

By the 1950s, only a small portion of the Interstate Highway System legislated after World War II had been constructed. President Dwight D. Eisenhower had been very impressed with the advantages and mobility that highways in Europe provided during the war and came to appreciate the value that highways could provide to national defense. This notion led Eisenhower and Congress to create and pass the Federal-Aid Highway Act of 1956 that dramatically influenced U.S. community development around the automobile and the more extensive roadway network that ensued during the following decades. Similar legislation has been renewed in Congress year after year supporting local, state, and federal roadway construction, including the formation of the Federal Highway Administration in 1966. The Federal Highway Administration has been a significant source of funding for complex highway construction over the past several decades. Tunnels have continued to play a significant role in formulating and extending efficient highway routes across bodies of water and through mountains and other obstacles in the United States.

Figure 5.36 | **Kennedy Tunnel in Berkeley Hills, California, Circa 1903**

One of the first highway tunnels opened after the 1956 Highway Act was the Fort Pitt Tunnel that was constructed to improve access to downtown Pittsburgh from the South Hills and West End areas. The tunnel is named after Fort Pitt built in the 1700s by British colonists at the confluence of the Monongahela and Allegheny Rivers. The tunnel is 3,614 feet long and 28 feet wide and connects to the Fort Pitt Bridge over the Monongahela River. The tunnel was constructed using drill-and-blast methods and heavy steel supports. A concrete lining was placed after the tunnel was holed through.

The John H. Wilson Tunnel was also constructed in the 1950s on the island of Oahu in Hawaii. The tunnel is located on the Likelike Highway (Route 63) that connects Honolulu to Kaneohe. The twin bores are approximately 2,800 feet long and each carry two lanes of traffic through the Koolau Mountains. Wilson was the engineer who headed up the project. The tunnels were constructed through volcanic basalt. Highly weathered basalt and soil were also encountered during construction. Tunnel excavation was completed using "stacked drifts" starting with an exploratory drift at the top of the tunnel cross section, footing drifts in each lower corner, and then drifts in between to form a concrete arch to support the poor ground. Tunnel construction began in 1954 and was completed in 1960. The Wilson Tunnel and Likelike Highway were instrumental in transforming this area of Oahu from an agricultural community to

CHAPTER FIVE | HIGHWAY TUNNELS | 257

Figure 5.37 | **Fourth Bore of Caldecott Tunnel, California**

a bustling suburb of Honolulu. Hawaii Route 61, often called the Pali Highway, also connects downtown Honolulu with the windward side of Oahu. Toward the latter part of the twentieth century, Interstate H-3 was also constructed through the Koolau Mountains and will be discussed later.

Tunnels from Portsmouth to Norfolk, Virginia, were constructed to replace ferries crossing the Elizabeth River. The 3,350-foot-long Downtown Tunnel was opened in 1952. The 4,194-foot-long Midtown Tunnel was opened 10 years later. A second 3,813-foot-long parallel bore was added to the Downtown Tunnel and opened in 1987 (VDOT 2015). All three tunnels carry two lanes of traffic each and were constructed using the previously mentioned immersed-tube tunnel technique with prefabricated steel tube sections floated to the highway alignment and submerged on the river bottom. This was the same technique used by Ole Singstad for the Baltimore Harbor Tunnel. Other highway tunnels in the area used this same construction technique, including the Hampton Roads Tunnels that opened in 1957 and 1976 and the Chesapeake Bay Tunnels, opened in 1964 and 1999.

The Interstate 64 Hampton Roads Bridge-Tunnel is a combined 3½ miles long including 7,479-foot-long twin immersed-tube tunnels. Hampton Roads is the area where the James River flows into the Chesapeake Bay. The first immersed-tube tunnel was opened in 1957. A twin parallel tunnel using the same construction methods was opened in 1976. The tunnels are connected to Norfolk and Virginia Beach at the south end and to Hampton and Newport News at the north end via bridges and artificial portal approach islands. Each of the two roadways consists of two lanes.

In 1954, the Virginia General Assembly formed the Chesapeake Bay Ferry Commission, now the Chesapeake Bay Bridge-Tunnel Commission, and in 1956 began the operation of a ferry service across the Chesapeake Bay between Virginia's Eastern Shore and the mainland in Norfolk/Virginia Beach. As demand grew, the commission

Dwight D. Eisenhower (1890–1969)

In 1956, Eisenhower signed the bill that authorized the Interstate Highway System. Creation of the Interstate system was justified by the Federal-Aid Highway Act of 1956 as essential to national security during the Cold War. Eisenhower's goal to create improved highways was influenced by his involvement in the U.S. Army's 1919 Transcontinental Motor Convoy and his experiences in Europe during World War II, particularly the German Autobahn. He determined that an improved ability to move logistics throughout the country would prove beneficial for military operations and provide an engine for continued economic growth.

Figure 5.38 | **View of the Original Thimble Shoal Channel Tunnel**

identified the need for a fixed crossing and secured $200 million in revenue bonds to construct the link. Construction began in 1960 and after just 42 months, the facility was opened to the public on April 15, 1964, and ferry services were discontinued. Still considered the largest bridge-tunnel complex in the world, this landmark facility stretches 17.6 miles across the mouth of the Chesapeake Bay, which is one of the busiest navigable waterways in the United States.

This fixed link is comprised of more than 12 miles of low-level trestle, two 1-mile immersed-tube tunnels under Thimble Shoal and Chesapeake Channels, two high-level bridges, 2 miles of causeway, four human-made portal islands, and 5½ miles of approach roads (Figure 5.38). The immersed-tube tunnels consist of prefabricated composite structural steel and reinforced concrete tube elements, each 37 feet in diameter and approximately 300 feet long (Figure 5.39). The double-steel-form shell elements, including internal reinforcing steel cages, were fabricated in Orange, Texas, and towed approximately 1,700 miles to the outfitting yard in Norfolk, Va. The elements were then floated out and sunk into a dredged trench on the bay bottom, then covered with backfill and protective armor stone (Figures 5.40 and 5.41).

Following its opening in 1964, the Chesapeake Bay Bridge-Tunnel was selected as "One of the Seven Engineering Wonders of the Modern World" in a worldwide competition with more than 100 other major engineering projects.

Figure 5.39 | **Fabrication of the Immersed-Tube Tunnel Units**

Figure 5.40 | **Floating the Immersed-Tube Units into Position**

In 1990, the Virginia General Assembly enabled the commission to move forward with the Parallel Crossing Project, expanding the facility from a single lane in each direction to two lanes. The Parallel Crossing Project was comprised of the expansion of toll plazas, trestles, bridges, and roadways, but, because of financial constraints, did not include the expansion of the human-made islands or the inclusion of parallel tunnels under Thimble Shoal and Chesapeake Channels.

In 2013, the commission approved acceleration of the construction of the Parallel Thimble Shoal Channel Tunnel via design–build delivery, the first of two tunnels needed to complete the full parallel crossing. The commission adopted Virginia's Public–Private Transportation Act guidelines, and in 2016, a design–build contract for the parallel tunnel was awarded. A bored tunnel will be constructed using an earth pressure balance machine in lieu of an immersed-tube tunnel, which will result in reduced impact to maritime traffic associated with the Navy's North Atlantic Fleet as well as commercial marine traffic in and out of the Port of Virginia. This will be the first use of bored-tunnel technology in the Chesapeake Bay. Construction of the Parallel Thimble Shoal Tunnel is projected to begin in 2017.

Figure 5.41 | **Forming the Final Tunnel Liner Inside the Immersed Tube**

Design and construction of the Monitor-Merrimac Memorial Bridge-Tunnel was similar to the Chesapeake Bay Bridge-Tunnel, only over a shorter distance. This tunnel is named for the famous ships that fought in 1862 during the Civil War. It connects Newport News on the Virginia Peninsula to Suffolk in South Hampton Roads via Interstate 664. This 4.6-mile-long stretch of highway includes a four-lane tunnel that is 4,800 feet long, two human-made portal islands, and 3.2 miles of twin trestle. It opened in 1992. Fifteen 300-foot-long immersed tubes were used to build the tunnel. Each steel tube consisted of twin circular sections each having room for two traffic lanes and a walkway.

The Virginia Department of Transportation also owns and operates the 4,229-foot-long Big Walker Mountain Tunnel in Wytheville, Virginia, which opened in 1972, and the 5,412-foot-long East River Mountain Tunnel in Bluefield,

Figure 5.42 | **Entrance to Lytle Tunnel in Cincinnati**

Figure 5.43 | **Inside of Lytle Tunnel**

West Virginia, which opened in 1974. These two-lane twin tunnels are along the north–south Interstate 77 and were constructed to lessen the hazards along narrow and twisted mountainous roads especially dangerous during winter conditions.

The Lytle Tunnel carries Interstate 71 beneath historic Lytle Park in Cincinnati's Central Business District's southeast corner (Figure 5.42). This tunnel connects to Fort Washington Way (U.S. 50), is 1,099 feet long, and has six lanes including two southbound, three northbound, and an exit ramp to downtown (Figure 5.43). It was opened in 1969 and is the longest vehicular tunnel in Ohio.

The Lowry Hill Tunnel is a fine example of fitting a highway facility into an urban area with minimal environmental impacts. The tunnel was constructed in 1969 to carry Interstate 94 under historic and local properties and local streets in Minneapolis, Minnesota. The tunnel was mined out of bedrock and is 1,497 feet long. It has two lanes in each direction plus a shoulder.

A few blocks downriver from the Bankhead Tunnel in Mobile, Alabama, is the George Wallace Tunnel. This tunnel consists of two-lane twin immersed tubes and carries Interstate 10 underneath the Mobile River between downtown Mobile and Blakeley Island where it connects to the Jubilee Parkway, an elevated structure on Mobile Bay. The tunnel is 3,000 feet long and opened in 1973.

In Colorado, U.S. Route 6 is a treacherous two-lane road that tends to follow the twists and turns of the Colorado River. While those highway geometrics were convenient and economical for construction at the time, they do not meet modern highway design standards. To straighten roads along rivers but minimize negative environmental impacts, tunnels are sometimes required. An early example of this type of improvement is the No Name Tunnel constructed approximately 1 mile east of Glenwood Springs. Each of the twin bores carries two lanes of traffic and is 1,045 feet long.

The twin-bore Eisenhower–Johnson Memorial Tunnel, located on Interstate 70 approximately 60 miles west of Denver, Colorado, has the unique distinction of being the highest vehicular tunnel in the United States (CDOT 2015). It crosses the Continental Divide at an elevation of 11,155 feet (Figure 5.44). This area is called Loveland Pass and a highway under the pass had been contemplated for at least three decades prior to construction. The westbound bore was originally bidirectional and was opened in 1973. Originally called the Straight Creek Tunnel during construction, it was renamed the Eisenhower Memorial Tunnel to honor President Dwight D. Eisenhower just before it was opened. The second bore for eastbound traffic opened in 1979 and was named in honor of Edwin C. Johnson, a past governor and U.S. senator who had actively supported an interstate highway system across Colorado. Both tunnels are approximately 1.7 miles long, 40 feet wide, and 48 feet high with a ceiling just under 14 feet above the roadway and air supply and exhaust plenums above the ceiling. Despite an exploratory tunnel being driven in 1964 through the granite, gneiss, and schist bedrock, very severe and dangerous mining conditions occurred during construction where the rock was intensely faulted and unstable (CDOT 2015). Multiple small tunnel drifts were required to mine the full cross section in these areas. Another first occurred on this project when Janet Bonnema, an engineering technician, sued the project for denying her the right to work in the tunnel. Despite the superstitious feelings of the miners at that time, she won her suit and was one of the first, if not the first, woman in the United States to work in a civil works tunnel during construction.

West of the Eisenhower–Johnson Memorial Tunnel are also tunnels through Glenwood Canyon on Interstate 70. Two-lane U.S. Route 6 through Glenwood Canyon was the precursor to Interstate 70 and one of the most dangerous sections of highway in Colorado. But bringing the highway up to modern design standards and expanding it to four lanes in this very sensitive Colorado River Valley environment was no small undertaking. With great attention to

Figure 5.44 | Eisenhower–Johnson Memorial Tunnel on Interstate 70 in Colorado

environmental preservation, safety, and mobility, the new facilities required elevated roadway, including 40 bridges and viaducts stretching more than 6 miles between at-grade sections. The highway also has 15 miles of retaining walls and a 4,000-foot-long tunnel with bores for traffic in both directions (Figures 5.45 and 5.46). The retaining walls are secured with ground-anchored tiebacks and soil anchors, and the highway is paved with cast-in-place, post-tensioned pavement slabs cantilevered 6 feet beyond the retaining walls.

Located approximately 50 miles west of Vail, Colorado, the main Glenwood Canyon tunnels are called the Hanging Lake Tunnels named in honor of the natural bowl-shaped lake formed by mineral-spring-deposited limestone above the tunnels. There are actually four tunnels, two in each direction, joined by a four-story-high operations and maintenance facility about half way between the portals and tucked into the valley formed by the two ridges of rock that the tunnels were driven through. Motorists driving through the Hanging Lake Tunnels are unaware of this facility because the tunnels continue right through the building, which maintains a similar appearance to the tunnels connecting on either side. Semi-transverse ventilation is used in this tunnel with mechanical ventilation capable of either blowing or sucking air through the ceiling plenum and the tunnel themselves providing the companion duct for air supply or exhaust.

The 582-foot-long single-bore Reverse Curve Tunnel is approximately 2 miles east of the Hanging Lake Tunnels along Interstate 70. It has two westbound lanes and is ventilated naturally. Construction of this tunnel was completed in 1989. A testament to the improvement that resulted from this highway project was stated by Row et al. (2004):

Since the route through Glenwood Canyon was upgraded to an interstate, the number of annual crashes dropped nearly 40 percent, despite significant increases in traffic volume. In the 5 years prior to construction, 1975–1980 (excluding 1977), the average number of annual crashes was 106. Over the most recent 5-year period for which statistics are available, 1997–2001, the average number of annual crashes dropped to 67. This reduction is significant, considering that average daily traffic has more than doubled since the start of construction.

Figure 5.45 | Construction of the Glenwood Canyon Tunnel on Interstate 70

CHAPTER FIVE | HIGHWAY TUNNELS | 263

Figure 5.46 | **Completed Glenwood Canyon Tunnel Project**

The Glenwood Canyon project was the first of four similar highway tunnel projects on the U.S. Interstate system that required blasting through rock ridges. The other three are the Cumberland Gap Tunnel where Tennessee, Kentucky, and Virginia converge; the Bobby Hopper Tunnel in Arkansas; and the Trans-Koolau (Tetsuo Harano) Tunnel on the Island of Oahu in Hawaii.

Prior to the opening of the Cumberland Gap Tunnel in 1996, U.S. Highway 25E traversed a steep, winding, two- and three-lane route that passed over the Cumberland Gap in the Cumberland Gap National Historical Park (Abramson and Slakey 1990). Pioneers such as Dr. Thomas Walker and Daniel Boone first discovered the "Wilderness Trail," opening the western territory to settlers and pioneers. Cumberland Gap was known as the "Gateway to the West." With the opening of the tunnel, the Cumberland Gap and the Wilderness Trail were restored; and visitors are now able to gain the perspective of the early travelers. Modern technology has paved the way for restoration of the past. The historical nature of that route coupled with technical problems of roadway improvements in such difficult terrain led to relocating the highway into twin two-lane tunnels through Cumberland Mountain of the Appalachian Range.

Figure 5.47 | **Tennessee Portal of Cumberland Gap Tunnel Project**

Figure 5.48 | **Entrance to Bobby Hopper Tunnel in Arkansas**

Figure 5.49 | **Inside of Bobby Hopper Tunnel Showing Jet Fan Ventilation**

The Federal Highway Administration served as the design and construction manager for the project on behalf of the owner, the U.S. Department of the Interior, National Park Service. The project included 5 miles of new four-lane approaches to the tunnels and two interchanges, one for the Cumberland Gap park entrance and another with U.S. Route 58. Construction encompassed seven roadway bridges (four in Kentucky, three in Tennessee), a 200-foot railroad steel box girder bridge, two pedestrian bridges on hiking trails, and three parking areas. An abandoned railroad tunnel under old U.S. Route 25E was repaired that later housed electrical, telephone, cable, and water lines under the new U.S. Route 25E and U.S. Route 58 interchange. Vehicular crossovers at tunnel entrances were constructed to allow for two-way traffic through the tunnel if one tunnel had to be closed. The east tunnel portal is in Tennessee and the west tunnel portal is in Kentucky (Figure 5.47). It is one of only two mountain vehicular tunnels in the United States that cross a state line, the other being the East River Mountain Tunnel on Interstate 77 between Virginia and West Virginia. The Cumberland Gap Tunnels are 4,600 feet long.

Another project intended to eliminate a dangerous vehicular journey was the Bobby Hopper Tunnel project in Arkansas. Between Alma and Fayetteville, U.S. Route 71 was an extremely perilous stretch of road. To make the trip on this road safer and faster, Interstate 49 was created. At an elevation of 1,640 feet above sea level, twin parallel tunnels were mined, not bored, through the mountain in a horseshoe geometry. Blasting, drilling, and excavation removed native shale and sandstone rocks, slowly chipping to the desired width and length. The hollowed-out channel was lined with reinforced concrete, as were both openings. The finished total length is 1,595 feet, with a width of 38 feet and a height of 25 feet, as measured from the roadway to the top of the arch, allowing for two lanes of traffic and shoulder space on each portal (Figures 5.48 and 5.49).

The Interstate H-3 (John A. Burns Freeway) project in Hawaii was first conceived in the 1960s, opened in 1997, and resulted in a new 16-mile-long highway that traverses the Koolau mountain range on the island of Oahu (Figure 5.50). The highway connects Pearl Harbor Naval Base on the southwest (leeward) side of Oahu to the Kaneohe Marine Corps Naval Air Station on the east (windward) side. Construction included two 1-mile-long highway tunnels, two 300-foot-long cut-and-cover tunnels, four 1-mile long viaducts, 3 miles of four-lane bridges, 7 miles of construction access roads, and three major interchanges. The cut-and-cover tunnel is located at Hospital Rock. The Tetsuo Harano Tunnels were mined out of volcanic basalt. Large multi-story buildings were constructed and "sculpted"

Figure 5.50 | Interstate H-3 Project in Hawaii

into the hillside at each portal to house the extensive ventilation fans and equipment for safe operation of the tunnels. Great care was given to blending these facilities into the native vegetation and environment (Figures 5.51 and 5.52). The scale of these large tunnels is evident in Figure 5.53.

Although most highway tunnels have twin tubes with two lanes in each direction, the Fort McHenry Tunnel has four tubes with two lanes each. It was opened in 1985 and closed a gap on Interstate 95 between Maine and Florida under the Baltimore Harbor in Maryland (Figure 5.54). The tunnel connects the Locust Point and Canton areas of Baltimore, crossing under the Patapsco River, just south of historic Fort McHenry. The immersed-tube method of tunnel construction was used by floating and sinking twin binocular steel tubes into place after dredging a trench at the bottom of the harbor. A dredge-disposal site for materials removed from the tunnel trench was created at nearby Canton/Seagirt. The resulting 136 usable acres were developed later by the Maryland Transportation Authority (MdTA). The result was the Seagirt Marine Terminal, which opened for business in 1990. Each two-lane roadway is 26 feet wide and 12.5 feet high. The tunnel is 7,200 feet long, and the lowest point is 107 feet below water level (MdTA 2015).

Interstate 395 is a spur interstate called the Center Leg Freeway that branches off of Interstate 95 and 495 (the Beltway) in the heart of Washington, D.C. The Center Leg Freeway consists of a series of tunnels that pass underneath the U.S. Capitol Building grounds and underneath the U.S. Department of Labor headquarters before reemerging at a T intersection with U.S. Highway 50, known locally as New York Avenue. Built for four lanes and two shoulders in each direction, the twin-tube, cut-and-cover Mall Tunnel measures 3,400 feet long and 132 feet wide. The tunnel has reinforced concrete walls and closely spaced transverse steel girders across the ceiling with a concrete roof slab. A concrete gravity slab helps the structure resist hydrostatic pressure as it is below sea level. Planning for the Center Leg Freeway began as early as the 1940s and construction was not completed until 1973.

The Papago Freeway Tunnel runs along Interstate 10 from North Third Avenue to North Third Street in downtown Phoenix, Arizona. Interstate 10 connects Santa Monica, California, with Jacksonville, Florida. The tunnel has twin tubes 2,887 feet long and has five lanes of traffic in each direction. It was opened in 1990. This tunnel highlights the popular debate about when a highway tunnel is really a highway tunnel. Technically, this "tunnel" passes underneath 19 side-by-side bridges. So maybe it should be more correctly called an extremely long bridge underpass, but calling it a tunnel is a lot simpler. The bridges are part of a large urban park called the Margaret T. Hance Park in honor of Phoenix's first female mayor. Running the freeway belowground and creating a park above was an ideal way of completing this part of the Interstate without bifurcating the neighborhood with a roadway or bridge structure.

Figure 5.51 | **Tetsuo Harano Tunnel, North Halawa Ventilation Buildings**

Figure 5.52 | **Architectural Features of the Tetsuo Harano Tunnel**

The second Mount Baker Ridge Tunnel opened in 1989 and consisted of a 63-foot-diameter soft ground tunnel and adjoining covered structures for the Seattle portion of Interstate 90. The $139 million project was a critical component of the $1.46 billion program to complete a 7-mile stretch of I-90 from Seattle to Bellevue, Washington. The major components for the project were the 1,476-foot-long tunnel through Mount Baker Ridge and the construction of 1,900 feet of cut-and-cover tunnel in Rainier Valley that reconnects the community that had been divided by the freeway. Because of the difficulty of constructing such a large tunnel by conventional methods, a "stacked-drift" method was developed. An articulated or semi-flexible tunnel lining consisting of 24 concrete-filled drifts was first constructed, forming a compression ring, followed by removal of the soil core. Since less of the tunnel roof or crown is exposed at one time and each drift is immediately backfilled with concrete, the method resulted in minimal distortion of the liner and little disturbance to the ground above. The tunnel is equipped with ventilation, fire suppression, emergency telephones, fire detection and alarm, closed-circuit TV monitoring, and an AM/FM rebroadcast system. Equipment for these systems, along with the electrical distribution, is housed in underground buildings alongside the cut-and-cover tunnel. The portal structure shown in Figure 5.55 boasts being the "Portal to the Pacific." The American Society of Civil Engineers (ASCE) has recognized the Mount Baker Ridge Tunnel as a historic landmark (Figure 5.56).

Figure 5.53 | **Cross Section of the Tetsuo Harano Tunnel**

Figure 5.54 | **Fort McHenry Tunnel in Baltimore**

Figure 5.55 | **Portal to the Pacific at the Second Mount Baker Ridge Tunnel**

Figure 5.56 | **ASCE Recognition Plaque for the Mount Baker Ridge Tunnel**

During the end of the twentieth century, no other highway tunnel project received more notoriety than the "Big Dig" in Boston, Massachusetts. That was the unofficial name for the Central Artery/Tunnel Project. In the 1950s, the Central Artery, also called the John F. Fitzgerald Expressway, was constructed consisting most noticeably of a double-deck six-lane viaduct that became a significant disruption to passage between Boston's neighborhoods including its North End. Although it was originally intended to build other highway projects to enhance and complement the Central Artery, these projects never came to fruition, leading to major traffic congestion problems in addition to its unsightly and disruptive nature. The Central Artery carried Interstate 93, U.S. Route 1, and State Route 3 through Boston. Merging of north–south traffic and east–west traffic through the heart of downtown created excessive congestion. In the early 1980s, serious planning efforts began to untangle this mess of traffic, resulting in the design and construction of new bridges and tunnels and the demolition of the viaduct that so many people hated and called the "Green Monster," a reference to Fenway Park in Boston. The two main components of the new project consisted of depressing the viaduct underground and extending U.S. 1 from South Boston to Logan Airport.

The Central Artery/Tunnel Project was constructed between 1991 and 2006 and cost more than $14 billion. It was the most expensive highway project in the history of the United States. It consisted of new tunnels that added traffic capacity under Boston Harbor, extended Interstate 90, and reconstructed Interstate 93 through downtown Boston. Bridges were also constructed to increase capacity across the Charles River. The Boston Harbor crossing is 1.6 miles long, including twelve football-field length steel immersed-tube sections. This tunnel is named after Ted Williams, a famous Boston Red Sox baseball player (Figure 5.57). As a result of the extension of Interstate 90, the

CHAPTER FIVE | HIGHWAY TUNNELS 269

Figure 5.57 | **Ted Williams Tunnel of the Central Artery Project in Boston**

Massachusetts Turnpike, or Mass Pike, now runs 138 miles from the New York state border to Route 1A in East Boston, and on a national scale between Seattle, Washington, and Logan Airport. This new highway connection now runs from its previous terminus at Interstate 93 near South Station to underneath railroad tracks, the Fort Point Channel, and South Boston before connecting to the Ted Williams Tunnel. A new interchange in South Boston also provides direct access to the center of a vital new development area for the Boston seaport, which features the newly opened Massachusetts Convention Center. This part of the project required tunnel jacking, the construction of a casting basin for immersed-tube tunneling, and cut-and-cover tunnel construction (Figures 5.58 and 5.59). The tunnel jacking was performed on the largest tunnel in North America at that time, a significant achievement. Finally, the Interstate 93 viaduct itself was replaced by the 1.5-mile-long Thomas P. O'Neill Jr. Tunnel. Before this heavy construction began, utilities had to be relocated and mitigation measures put in place (Figure 5.60). Then slurry wall construction began in the mid-1990s, which required underpinning of the existing elevated Central Artery before excavation. Once I-93 North opened under the footprint of the elevated Central Artery, Big Dig crews began demolishing the aging elevated highway. What used to be a dark and dangerous overhead structure that blocked the beauty of Boston Harbor both physically and visually from downtown is now an inviting promenade used extensively by pedestrians, tourists, and a multitude of recreationists.

TWENTY-FIRST-CENTURY TUNNELS

New tunneling methods and environmental systems are being perfected on current and future highway tunnel projects across America. Some of the more routine highway tunnel projects completed after the turn of the century include the Wolf Creek

Figure 5.58 | **Casting Basin for the Central Artery Project Jacked Tunnels in Boston**

Figure 5.59 | **Cross Section of the Central Artery Project Jacked Tunnels**

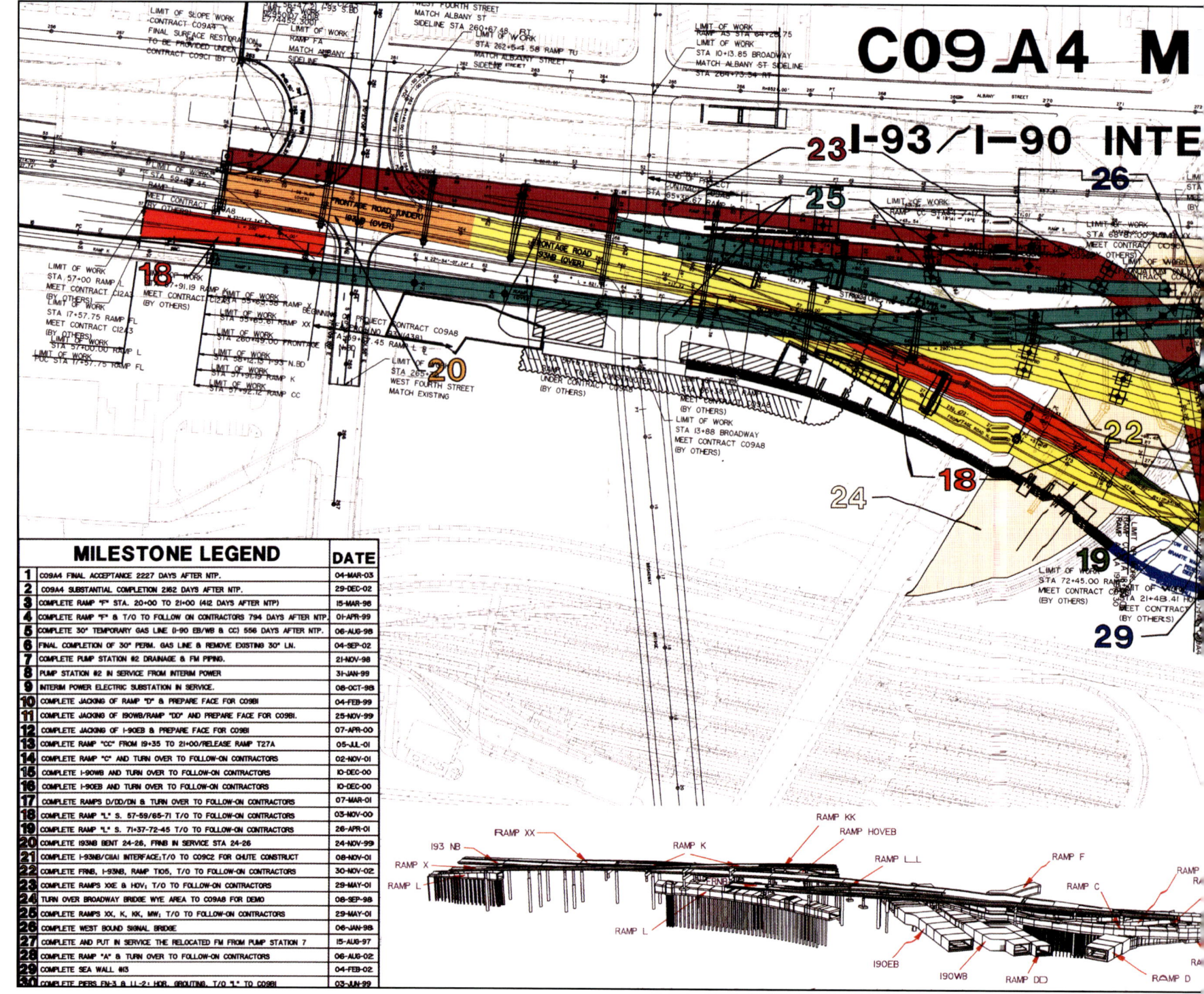

CHAPTER FIVE | HIGHWAY TUNNELS

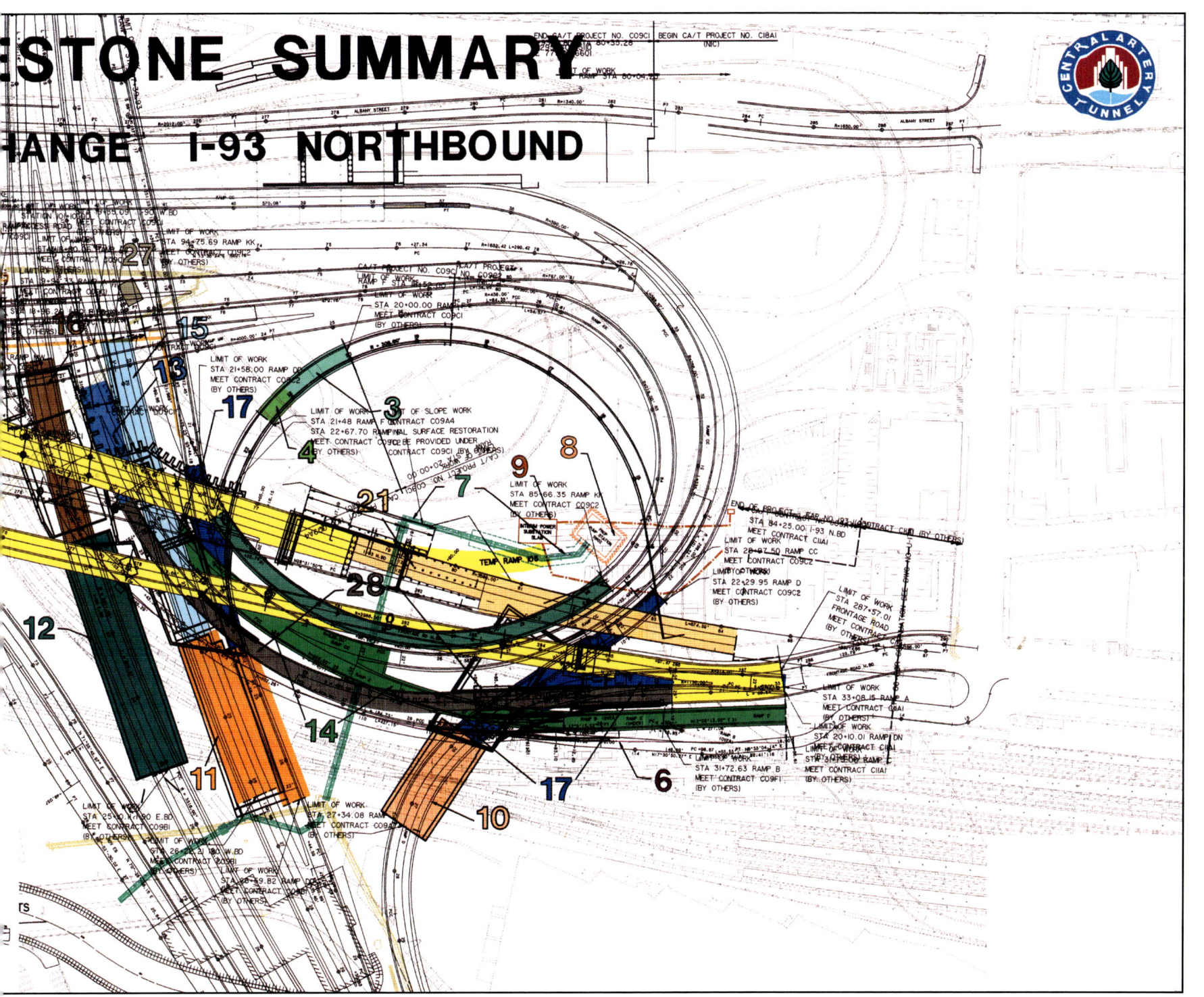

Figure 5.60 | **Plan and Schedule Milestones for Portions of the Central Artery Project**

Figure 5.61 | **Excavation of Wolf Creek Pass Tunnel in Colorado**

Figure 5.62 | **Route 1 Devil's Slide Tunnel near San Francisco**

Pass Tunnel in Colorado and the New Jersey Route 29 Tunnel. U.S. Route 160 runs between Tuba City, Arizona, and Poplar Bluff, Missouri. It passes through Colorado and crosses the Continental Divide at Wolf Creek Pass. The highway was modified for safer driving conditions and expanded from two lanes to four lanes in 2006. One of these modifications included the construction of the 916-foot-long Wolf Creek Pass Tunnel (Figure 5.61). The Route 29 project alleviated traffic congestion by connecting the Interstate loop around Trenton, New Jersey, with a four-lane highway. The highway runs along the Delaware River in a 2,400-ft tunnel section constructed by cut-and-cover. The Route 29 Tunnel opened in 2000 and allows for the four-lane roadway to be overlaid with a park, providing bicycle paths and walkways for waterfront beautification.

The Devil's Slide (Tom Lantos) Tunnel was constructed to bypass an unstable section of U.S. Highway 1 located just south of Pacifica, California, that was very prone to rockslides and ground movement (Devil's Slide). The ribbon-cutting ceremony for the project was held on March 25, 2013. The bypass was accomplished by moving the two-lane highway inland from the Pacific Ocean and the unstable area. It is about 1 mile long and consists of twin-mined 4,200-foot-long tunnels through the San Pedro Mountain, a 1,500-foot-long bridge approach on the northern end and a 1,000-foot-long at-grade approach on the southern end. The tunnels are 30 feet high and 22 feet wide (Figure 5.62).

Only one lane of traffic with a breakdown shoulder is provided in the tunnels. One very unique feature of these highway tunnels is that they are accessible to bicycle riders.

The southern approach to the Golden Gate Bridge in San Francisco, California, is known as Doyle Drive, located just east of where U.S. Highway 101 and California State Route 1 intersect. After the Loma Prieta earthquake in 1989, the viaduct that carries Doyle Drive to the Golden Gate Bridge was judged to be too old and seismically vulnerable, and plans were made for replacement of the structure. As part of this project, Doyle Drive was renamed the Presidio Parkway. It is located in the Presidio, a park and former military base on the northern tip of San Francisco. The project consists of new roads, viaducts, and two cut-and-cover tunnels (Figure 5.63). The Battery Tunnel is approximately 850 feet long and the Main Post Tunnel is about 1,000 feet long. Both tunnels will have extensive landscaping above to blend in with the rest of the park and provide recreational use.

Economic growth in downtown Miami, Florida, was being stymied during the beginning of the twenty-first century in part by commercial traffic on busy city streets that was heading to the Port of Miami in Biscayne Bay. The Port of Miami is one of the most significant economic generators in south Florida, catering to numerous cruise lines and vast amounts of containerized cargo. The Port of Miami Tunnel project was designed to relieve city streets of car and truck traffic serving the cruise and cargo ships that use this port. The tunnel itself on State Road 887, now called the PortMiami Tunnel, connects the MacArthur Causeway on Watson Island with the Port of Miami on Dodge Island (Figure 5.64). These twin two-lane tunnels are 4,200 feet long and 43 feet wide, with 15-foot-high clearance, and are for truck use only.

Figure 5.63 | **Doyle Drive Tunnel near San Francisco**

Figure 5.64 | **PortMiami Tunnel**

The replacement for the Alaskan Way Viaduct will be a tunnel constructed with what was once the world's largest tunnel boring machine (TBM). As a single-tube, double-deck highway tunnel, it will replace the earthquake-damaged Alaskan Way Viaduct across the waterfront of Seattle (Figure 5.65). Carrying approximately 90,000 vehicles per day, the existing viaduct was designed for a 50-year design life. Built in the 1950s, the viaduct does not meet current seismic or roadway design standards. The Washington State Department of Transportation, City of Seattle, King County, Port of Seattle, and the Federal Highway Administration have partnered to develop and construct a replacement that will meet current standards and improve long-term mobility. The viaduct is located in an important and sensitive location—an area of substantial and continuous economic and tourist activity, as well as the site of the Washington State Ferries water access. In July 2013, TBM "Bertha" began boring for the 57-foot-diameter, 9,270-foot-long stacked roadway tunnel (Figure 5.66). When launched, Bertha was the world's largest soft-ground TBM. Her size was recently eclipsed by less than 5 inches by the Tuen Mun-Chek Lap Kok Tunnel TBM in Hong Kong. The State Route 99 (SR 99) Tunnel is being lined with precast concrete segments, as shown in Figure 5.67.

Another highway project under construction that includes a tunnel is the Ohio River Bridges Project connecting Ohio and Kentucky. There are two sections of the project. The Downtown Crossing is between Louisville, Kentucky, and Jeffersonville, Indiana, and includes building a new I-65 bridge with six northbound lanes, reconfiguring Spaghetti Junction (I-64, I-65, and I-71), reconfiguring Indiana roadway and bridge approaches, and rehabilitating the Kennedy Bridge (I-65) with six southbound lanes. The East End Crossing is from the intersection of SR 841/I-265 and I-71 in Kentucky to the SR 265/SR 62 interchange in Indiana. Work includes building a new East End Bridge

8 miles upstream from downtown Louisville, extending the Snyder Freeway (SR 841/I-265) in Kentucky to the new bridge, including a 1,700-foot tunnel under the Drumanard Estate in Prospect and constructing a new 4-mile highway in Indiana that will extend the Lee Hamilton Highway (SR 62/SR 265) to the new bridge. The twin tunnels will each carry two lanes of traffic.

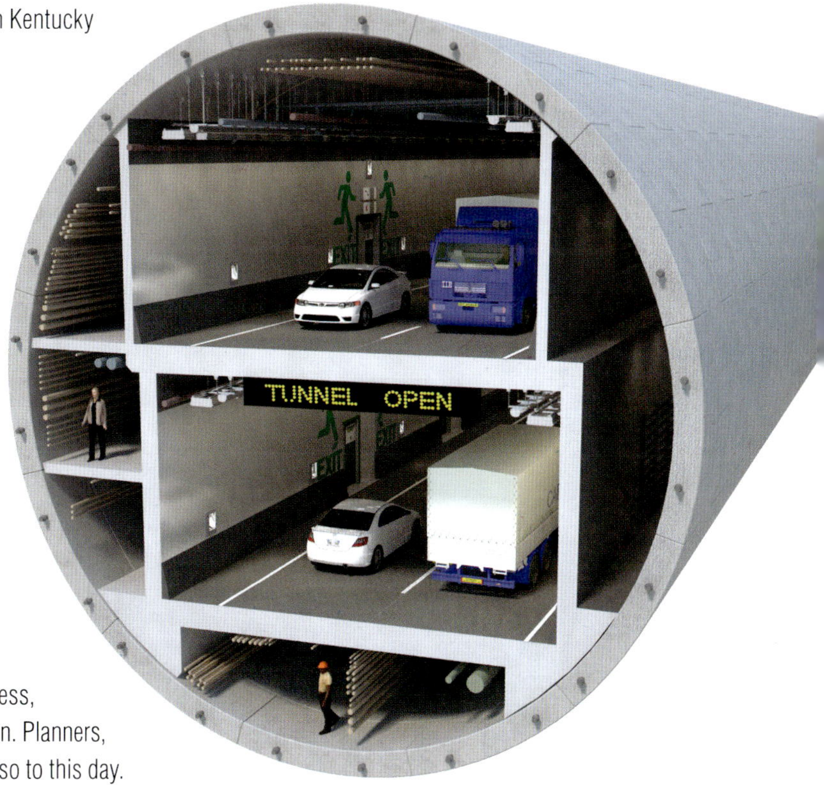

FUTURE HIGHWAY TUNNEL PROJECTS

Across the United States from New York to California, tunnels continue to be a viable construction method to take vehicles across waterways, through mountains, and past environmentally sensitive areas. Future highway tunnel projects are currently in the planning stages to complete Interstate 710 west of Los Angeles, ease traffic congestion on Interstate 405 through Sepulveda Pass north of Los Angeles, provide more efficient freight truck traffic across the harbor in New York, and double the traffic capacity on the Lyndon B. Johnson Freeway in Dallas.

BENEFITS, INNOVATIONS, AND THE FUTURE

Highway tunnels are beneficial in a number of ways, including provision of more efficient access, environmental improvements, economic development and prosperity, and vulnerability reduction. Planners, designers, and constructors have innovated over the past 100-plus years and continue to do so to this day. Longer and larger tunnels are being conceived. More and more contractual delivery methods are being utilized in the industry, including single teams that both design and build tunnels, single teams acting as the contractors as well as the managers of construction, and public–private partnerships that plan, finance, build, and operate tunnels for a suitable period of time to recoup their investment and then turn the facility over to a public agency entity. Other innovations include energy savings and significant improvements to the environment. Future developments are likely to be in the areas of energy conservation, economical construction, and homeland security issues.

Figure 5.65 | **Cross Section Through the SR 99 Tunnel in Seattle**

Figure 5.66 | **Fully Assembled TBM Used for the SR 99 Tunnel**

Figure 5.67 | **Precast Segmental Concrete Lining for the SR 99 Tunnel**

REFERENCES

Abramson, L.W., and Slakey, D.M. 1990. Highway tunnel linings. In *Proceedings of the International Symposium on Unique Underground Structures*, Vol. 1. Denver, CO: Colorado School of Mines Press.

ADOT (Alaska Department of Transportation & Public Facilities). 2015. Anton Anderson Memorial Tunnel—Tunnel history. www.dot.state.ak.us.

ASCE (American Society of Civil Engineers). 2007. Lacey V. Murrow Memorial Bridge and Mount Baker Ridge Tunnels. ASCE History and Heritage International Landmark Nomination by the American Society of Civil Engineers Seattle Section, August 23.

Caltrans (California Department of Transportation). 2015. Yerba Buena Island transition structure. The San Francisco–Oakland Bay Bridge seismic safety projects. www.dot.ca.gov.

CDOT (Colorado Department of Transportation). 2015. Details on the Eisenhower Memorial Bore. www.codot.gov.

Gonzales-Hardy, R. 2010. *San Francisco's Broadway Tunnel*. Lulu.com.

Lange, R.S. (March 1993). *National Historic Landmark Nomination: Holland Tunnel*. Washington, DC: National Park Service.

MdTA (Maryland Department of Transportation). 2015. Fort McHenry tunnel history. www.mdta.maryland.gov.

Nycroads.com. 2015. Queens–Midtown Tunnel—Historic Overview. www.nycroads.com.

Pennsylvania Turnpike. 2015. 75 Years of Turnpike History—PA Turnpike. www.paturnpike.com.

Richardson, E. 2008. Third Street Tunnel: A Primer. www.blogdowntown.com. Southern California Public Radio. September 5.

Richardson, H.W., and Mayo, R.S. 1975. *Practical Tunnel Driving*. New York: McGraw-Hill.

Row, K.S., LaDow, E., and Moler, S. 2004. Glenwood Canyon 12 years later. *Public Roads* 67(5).

Russell, H.A. 2002. Rehabilitation of the Sumner/Callahan Tunnels. *Concrete Repair Bulletin* (May/June):8–11.

Solon, M., and McCosker, M. 2014. *Building the Caldecott Tunnel*. Images of America Series. Charleston, SC: Arcadia Publishing.

TBTA (Triborough Bridge and Tunnel Authority). 2015. Queens–Midtown Tunnel Turns 70. November 12, 2010. www.mta.info.

Tilton, E.G. 1915. Method of constructing rock tunnel of 50-ft. clear width, Stockton St., San Francisco. *Engineering and Contracting* XLIII(5):93.

VDOT (Virginia Department of Transportation). 2015. Project History—Midtown Tunnel. www.virginiadot.org.

Wallace, K. 1949. The City's Tunnels—When S.F. Can't Go Over, It Goes Under Its Hills. *San Francisco Chronicle* (March 27): 1, 10.

Weingroff, R.F. 2011. Highway history: The Lincoln Highway. Washington, DC: Federal Highway Administration. www.fhwa.dot.gov.

6 WATER TUNNELS

Water is essential for our sustenance—when we turn on a faucet in our home, our expectation is that an ample volume of clean drinking water will naturally flow from the spigot.

With our demand for limitless amounts of water as a society, we generally have not considered how water is supplied, and how that supply keeps up with demand. More than 313 million people in the United States use public supply withdrawals in excess of 42 billion gallons per day, as estimated by the U.S. Geological Survey in 2010 (Maupin et al. 2014). To meet society's needs and demands, a large water supply infrastructure network has been constructed, of which tunnels are a strategic component for transmission and distribution. Tunnels fall under a vast infrastructure category that includes dams and reservoirs, groundwater wells and fields, pumping stations, water treatment plants, hydropower, and other conveyance and distribution system components.

The U.S. water distribution system consists of more than 1.8 million miles, a distance equivalent to circling the earth more than 70 times at the equator, and delivers water to industry, businesses, and residents, and provides irrigation water for agricultural and horticultural practices. To make it possible to supply water to towns and cities, tunnels form a critical component connecting to this vast network of underground aqueducts and pipelines. With more than 800 miles of water tunnels in the United States (including some tunnels associated with reservoirs and hydropower), these tunnels have enabled physical barriers such as mountains, rivers, major highways, and urban development to be crossed without disruption while supplying clean water and reducing the potential risk of diseases.

The evolution of tunneling, and the innovative practices and techniques that have been developed, are strongly centered in the history of our nation's supply and demand for water.

Within the United States, roughly two-thirds of our water is supplied from surface water sources while one-third is from groundwater. Initially, the water sources for the largest cities in the United States came from nearby lakes and rivers. However, as population centers grew and these water sources became contaminated, it became necessary to seek water sources outside the city limits and to tap into more distant watersheds. Throughout the last 200 years, four metropolitan areas are significant in terms of milestone development in water infrastructure by means of the level of sophistication and innovation used at the time: New York City (1842) and Boston (1848) in the east, Chicago (1842) in the Great Lakes region, and Los Angeles (1904) and San Francisco (1914) in the west. These four municipalities all sought distant, more protected water sources to supply water to their growing cities. Development of these water sources, with sophisticated water transmission and distribution systems that required long aqueducts and tunnels, was designed and constructed to bring clean water supplies to the urban areas.

For as long as distribution systems have been constructed, tunnels have played a primary, if not the most important, part of the distribution system. A few of the main tunnels that support major metropolitan areas, including tunnels up to 30 feet in diameter, are highlighted here. The Los Angeles Aqueduct required 142 tunnels totaling 43 miles. The longest intermountain or diversion tunnel is the 13.1-mile-long Alva B. Adams Tunnel, which brings water from the western slopes of the Colorado River basin to the eastern Front Range in Colorado. The longest and most extensive water distribution tunnel in the United States and the world is the Delaware Aqueduct Tunnel in New York (84 miles).

Freshwater supply in larger cities dwindled during the nineteenth century and created significant challenges as America's population grew rapidly. The imbalance of supply and demand increased; therefore, the larger cities began to develop water and sewage infrastructure, but since many of the nearby lakes and rivers were the sources for fresh water, these sources became contaminated through sewage and industrial waste disposal. The unsanitary conditions of the water sources caused an increase in disease as epidemics of dysentery, yellow fever, and cholera ravaged cities such as Chicago and New York City over successive years. The polluted water sources, overcrowded

---

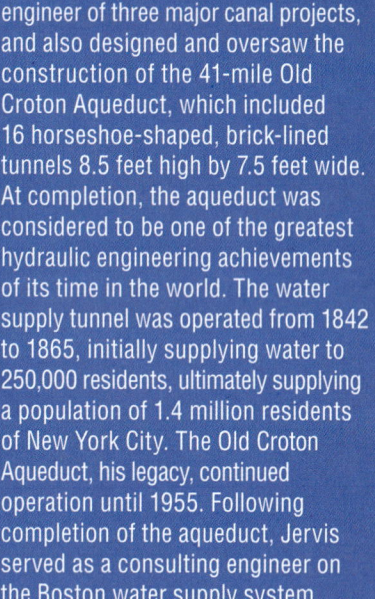

John B. Jervis (1795–1885)

Jervis was a prominent civil engineer who designed and supervised the construction of five of America's earliest railroads, was chief engineer of three major canal projects, and also designed and oversaw the construction of the 41-mile Old Croton Aqueduct, which included 16 horseshoe-shaped, brick-lined tunnels 8.5 feet high by 7.5 feet wide. At completion, the aqueduct was considered to be one of the greatest hydraulic engineering achievements of its time in the world. The water supply tunnel was operated from 1842 to 1865, initially supplying water to 250,000 residents, ultimately supplying a population of 1.4 million residents of New York City. The Old Croton Aqueduct, his legacy, continued operation until 1955. Following completion of the aqueduct, Jervis served as a consulting engineer on the Boston water supply system.

Previous Page | **Lake Mead Intake Structure Being Barged to Excavation Location** cities, lack of sewers, general public ignorance of basic sanitary conditions, and the existence of polluting industries contributed to an unprecedented mortality rate in the nineteenth century in America, together with the increasing demand for clean water.

PIONEERS OF MAJOR WATER SYSTEMS

Several engineering visionaries pioneered the water supply distribution systems to bring water to our largest cities, which included major and strategic tunnels to accomplish this objective. Many of these larger distribution systems are attributable to several key civil and water resources engineers who solved the early challenges, including lack of capacity for the growing population centers and polluted surface waters.

Four prominent engineers of the modern water tunnel system brought innovations and solutions to the most challenging issues of their time. Their accomplishments span two centuries—John Bloomfield Jervis and Ellis Sylvester Chesbrough in the nineteenth century, and Michael O'Shaughnessy and William Mulholland in the early twentieth century. Each of their contributions significantly affected how our current systems are maintained and operated today. These men are pioneers in civil engineering.

In the late 1840s, Jervis selected Chesbrough as a water resources engineer to work on the construction of Boston's new water system. As its first chief engineer, Chesbrough's accomplishments included centralized waterworks, sewers, streets, and harbor facilities. As a result of his Boston accomplishments, Chesbrough moved to Chicago to solve its environmental problems following a succession of cholera and dysentery epidemics. Initially in charge of solving the city's sewage and drainage problems, he was also responsible for the water supply system. He observed that the sewage was being discharged into Lake Michigan and that this could become a danger to the shoreline and the water intake located a few hundred feet offshore. Because of the growing demand of the Chicago metropolitan community on Lake Michigan, both as a source of fresh water and for wastewater disposal, Chesbrough proposed the construction of a water intake tunnel 2 miles out into the lake. The new water intake tunnel was completed in 1867, and being the longest of its day was described as the "eighth wonder of the world." Chesbrough's legacy of transforming a single water source into a dual system was insurmountable at the time and proved to generate more national attention on the importance of water infrastructure for large metropolitan centers.

Not only does the demand for water influence infrastructure, natural disasters have a history of driving innovation in design and construction. After the 1906 earthquake, fire, and the failure of the water system to curtail widespread destruction, San Francisco refocused its attention on building a secure, reliable water source. Michael O'Shaughnessy, best known for engineering one of the largest and most controversial projects in the United States, was hired by San Francisco Mayor James Rolph Jr. as the city engineer in 1912 (Figure 6.1). He led the design and construction of the massive Hetch Hetchy Regional Water System, which involved building several dams and a hydropower plant

Figure 6.1 | **O'Shaughnessy's Legacy Included Dams, Tunnels, and Reservoirs**

a 68-mile-long railroad, and a 156-mile-long aqueduct that included 85 miles of tunnels. O'Shaughnessy, nicknamed "The Chief," was a man of action. His design for the water system was driven by hydraulics, economics, durability, and capacity for future expansion. Most significant to the project was his decision to promote a gravity flow system and choosing tunnels rather than pipelines for the longevity, capacity, and ease of maintenance offered by the tunnels, including the Mountain, Foothill, Coast Range, and Irvington Tunnels. As recognition of his significant contributions to Hetch Hetchy, one of the two main dams in the system is named after him, O'Shaughnessy Dam. In addition, he supervised the construction of several other tunnels, including the Stockton Street and Twin Peaks Tunnels in the city of San Francisco (Hanson 2005).

William Mulholland began his career in the Los Angeles area as a ditch digger with a private water company, the Los Angeles City Water Company, which was taken over by the city. He ultimately became the chief engineer of the Los Angeles Department of Water and Power. As Los Angeles boomed, Mulholland realized that the Los Angeles River could not continue to supply fresh water for the city to meet further growth and there would be no Los Angeles as we know it. The concept of tunneling was a major driver in Los Angeles urban planning at the time. In 1904, Mulholland and former mayor Fred Eaton came up with the solution to purchase land in Owens Valley and build a 233-mile-long aqueduct, including 43 miles of tunnel, to bring water from the snowmelt of the eastern Sierra Nevada. One of the largest public works projects in its day, it was constructed from 1908 to 1913 and was completed within budget. Mulholland organized the project into separate divisions, and he instituted a bonus system where each man would get 40 cents added to his day's pay for every foot that the crew exceeded the daily average. The project also used its own workforce instead of contracting out the work. Many tunnel production records were set; for example, Tunnel 17, at 1,061 feet of tunnel achieved in a month, was a world record for tunneling in sandstone, and Elizabeth Tunnel achieved an American record of 604 feet in a month and averaged overall more than 4 miles per month. Mulholland estimated that the entire project was built for $22 per foot. The completion of the project provided a reliable water supply, a key factor that enabled Los Angeles to become the city it is today, one of the largest urban metropolitan centers in the nation.

WATER TUNNEL SYSTEMS OF THE NORTHEAST

Early settlement of the Northeast region was along coastal areas and major rivers where sources of clean water were abundant at local springs or waterways. As the population grew, additional water sources were found and water access became centralized with construction of some rudimentary piping distribution systems.

Early Water Systems of Boston and New York City

With demand growing, reservoirs were constructed for water storage such as Jamaica Pond in Boston. New York City had several reservoirs built near springs at various locations. As population growth continued in the mid-1800s, the water systems were expanded to new reservoirs located longer distances away from the population centers and connected to the city areas by aqueducts that were generally made of masonry, portions of which were shallow buried conduits and some were on elevated structures and bridges.

The first such construction involved the 41-mile-long Old Croton Aqueduct for New York City completed in 1842 (Figure 6.2) and the Cochituate Aqueduct in Boston completed in 1848. Cost and risk considerations were important factors in decisions regarding whether to use tunnels for water system conveyance versus more traditional construction methods, such as near-surface pipelines, aqueducts, or bridge-supported aqueducts.

William Mulholland (1855–1935)

As water superintendent for the City of Los Angeles, Mulholland designed and oversaw construction of the 233-mile-long Los Angeles Aqueduct, carrying water from the Owens Valley to the San Fernando Valley. Opened on November 5, 1913, the aqueduct includes 164 tunnels and 175 miles of pipe and channels. Upon its opening, Mulholland opined, "There it is! Take it!"

CHAPTER SIX | WATER TUNNELS 283

The Old Croton Aqueduct included a crossing of the Harlem River that was made by a high bridge. Although a tunnel was considered to be more cost effective, tunnel construction was believed to be more risky at the time.

With these examples, we can see from history how natural disasters shaped innovation in infrastructure. Water demand increased because of population growth as well as a greater demand for advanced firefighting ability in the congested city environments, most notably the Great Fires in Chicago, Boston, and other large cities between 1865 and 1875. Identifying a need to manage water more effectively, public water boards were created in the late-1800s to construct and operate larger and more reliable water supply systems.

In metropolitan Boston, the Chestnut Hill Reservoir and the Sudbury Aqueduct were constructed and completed in 1878, and the Wachusett Aqueduct was completed in 1896.

The Sudbury Aqueduct included two tunnel sections along its route with the longest being just under a mile in Newton, Massachusetts. At the time, tunnels were being used more for railroad construction, such as the Hoosac Tunnel in western Massachusetts between 1855 and 1876. Thus, a tunnel approach for raw water transmission to supply water to major cities was a natural progression.

During this period, the demand for water storage resulted in construction of more reservoirs, which in many cases involved construction of dams. However, this was not deemed to be a long-term solution due to several recent failures. Between 1874 and 1890, several major dams failed, including the famous South Fork Dam failure during the Johnstown Flood in Pennsylvania in 1889, resulting in numerous deaths and substantial damage. Improvements in design and construction of the dams and their siting in more remote locations away from heavily populated areas necessitated additional and longer conveyance systems. In the late 1800s and early 1900s, increased immigration and substantial improvements in plumbing further

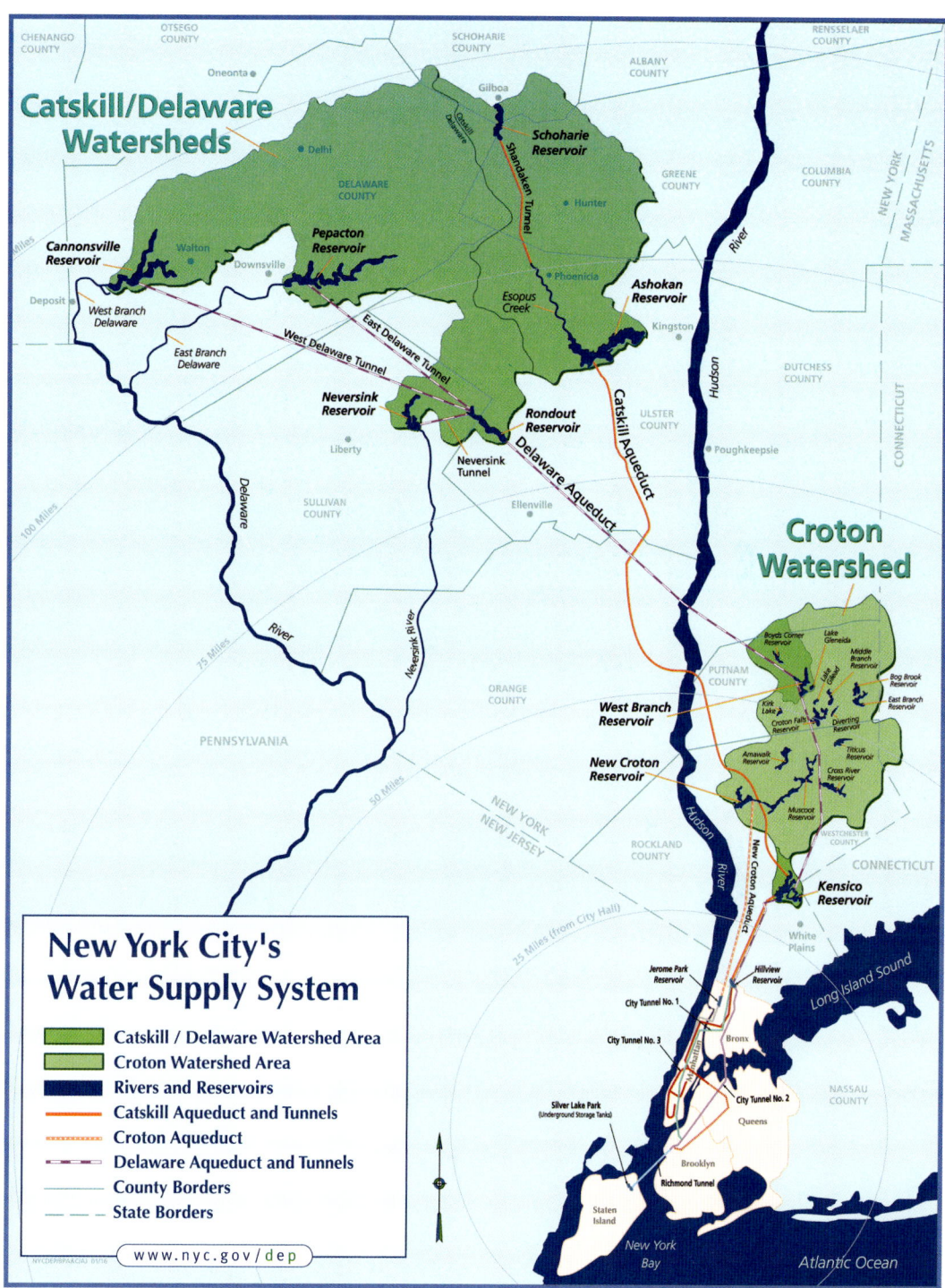

Figure 6.2 | **New York City Water Supply System**

Figure 6.3 | Mystic River Tunnel Construction in 1900

increased water demand and stressed the existing water systems. More frequent droughts and dry weather also diminished water supplies. In Boston, the Wachusett Reservoir and the Wachusett Aqueduct Tunnel were constructed between 1897 and 1903, which pushed the metropolitan Boston water supply system further westward (approximately 36 miles) for the third time in a span of only 55 years. In 1900, construction also began on the Mystic River Tunnel using a 9-foot-diameter, 4.5-foot-long hydraulic shield to excavate the soft-ground tunnel beneath the river (Figure 6.3).

In New York City, the Public Water Board in 1884 initiated planning for the New Croton Aqueduct, which was placed into operation in 1890. The New Croton Aqueduct was a 33-mile-long horseshoe-shaped masonry tunnel with three times the capacity of the Old Croton Aqueduct. Upon its completion, it was the largest and longest water conduit in the world. It included 40 vertical shafts for construction access and was constructed at depths to 391 feet. It crossed beneath the Harlem River by tunnel near the Old Croton Aqueduct High Bridge. After several failed attempts, the tunnel was successfully completed at a greater depth than originally planned.

The Catskills—Public Health and Environment Drives Treatment and Distribution

In the early 1880s, it became clear to public health officials that several major diseases were attributable to poor water treatment and distribution. Cholera and typhoid were rampant as waterborne illnesses. These epidemics were substantially reduced by development of a clean and reliable water source with distribution systems that could not be polluted. Closed aqueducts and tunnels were preferred as transmission networks. In 1886, *Scientific American* recommended use of the Catskill watershed for water supply due to its remoteness from the general population, which provided a clean supply of water.

In 1907, New York City began work to expand its water system to the Catskills. The Catskill Aqueduct was 92 miles long and included shallow, concrete-lined "grade tunnels" in 24 sections totaling 14 miles, plus concrete-lined deeper "pressure tunnels" in seven tunnels totaling 17 miles. The Catskill Project was as significant to the water supply as the Panama Canal construction was to transportation at the time. The shallow tunnels were horseshoe-shaped cross sections and the pressure tunnels were circular (14 to 16 feet in diameter) because of the hydraulic pressures at extreme depths. Generally the pressure tunnels were constructed at depths between 300 and 700 feet; however, the crossing of the Hudson River was up to 1,100 feet deep. Nevertheless, the most difficult tunnel section was the Rondout Pressure Tunnel under Rondout Creek and its valley. The Rondout Tunnel was 23,608 feet long and 12 different rock types were encountered during excavation, which presented many challenges in excavation and initial support for the tunnel.

Transmitting the water from the reservoir to the city through an aqueduct was important; however, as the city grew, distributing the water within the city was also essential. In 1911, work was started on City Water Tunnel No. 1, which was put into operation in 1917 for conveyance of Catskill water into the New York City urban area (Figure 6.4). At 18 miles in length, City Water Tunnel No. 1 was the longest pressurized water tunnel in the world. It was constructed at depths of 250 to 757 feet and had 24 vertical shafts. The use of high explosives in an urban environment presented unique challenges; however, the construction went largely unnoticed by the public. City Water Tunnel No. 1 was constructed by workers from Local 147, proudly known as the "sandhogs," who risked the onset of caisson disease or "the bends" while working under pressurized environments. These risks were mitigated by innovation in construction techniques that had been developed.

Figure 6.4 | **Assembling Steel Lining Rings, 1914**

These tunnels were constructed in part during World War I, and it was fortunate that both Boston and New York City had the foresight to expand their watersheds and water conveyance/transmission systems. Through the 1920s and 1930s, extended periods of drought were experienced with minimal recharge of the water supply possible. This time was perhaps the greatest growth period for the water supply systems of the Boston and New York City metropolitan areas. Demand was strong and there was an increased need for clean water for improved health and because of continuing increases in population growth. The Great Depression, starting in 1929, became the catalyst for obtaining substantial government funding of public works projects.

At the time of its completion in 1939, the 1.26-million-acre-foot Quabbin Reservoir, located 75 miles west of Boston, was considered to be the world's largest human-made reservoir solely utilized for water supply (Figure 6.5). (Lake Mead in Arizona and Nevada was the largest human-made water supply project at 31 million acre-feet but was also utilized for power generation.) During this period, Massachusetts began construction of the Wachusett-Coldbrook Tunnel as the eastern portion of the Quabbin Tunnel, completed in 1933; and construction of the Quabbin Reservoir was performed between 1936 and 1939. While the Quabbin Tunnel was under construction, the City of Springfield, Massachusetts, constructed its Cobble Mountain Reservoir with a tunnel discharge that is one of America's first pressurized deep tunnels in rock.

For the New York City water supply system, major expansion was also underway within the city and as far away as the Delaware River. This system expansion included the completion of the Shandaken Tunnel (1928), City Water Tunnel No. 2 (1935), and the Delaware Aqueduct Tunnel (1944). With completion of City Water Tunnel No. 2 (Figure 6.6), New York City again activated the longest water tunnel in the world at 20 miles long, up to 766 feet deep, and with a diameter of 17 feet. The project was delayed, however, in the early 1930s because of the Great Depression.

Figure 6.5 | **Excavation of Quabbin Tunnel, 1930**

Figure 6.6 | Innovative Cross-Section Survey Tool, 1932

Figure 6.7 | **Typical Drill Jumbo for Drill-and-Blast Excavation, 1949**

Figure 6.8 | **Concrete Delivery for Lining, 1956**

The City Water Tunnel No. 2 record for the longest water tunnel in the world lasted only several years. With a length of 84 miles, the Delaware Aqueduct Tunnel was an unprecedented worldwide undertaking that far exceeded the length of any tunnels previously constructed (1944) and continues to be the longest water tunnel in the world. Planning for the use of the Delaware River and watershed began in 1920, when the chief engineer, J. Waldo Smith of the New York Board of Water Supply, forecast that the demand for water in New York City by 1932 would be greater than the Catskill watershed supply. Use of this supply, however, would affect the water supply to Pennsylvania and New Jersey, which led to major litigation and agreements at the state and national levels, culminating in a historic 1931 U.S. Supreme Court decision.

During this time, innovations in engineering and construction created many efficiencies in tunneling. Mechanized technologies provided increasing production times for drill-and-blast excavation and concrete lining of the Delaware Aqueduct Tunnel, including the use of rail-mounted drilling machines or "jumbos," track-mounted mucking machines, and pneumatically powered drills (Figures 6.7 and 6.8).

In addition to the application of advancing innovations in tunneling methods for developing the water conveyance network, specialized distribution control valves were required both to divert water flows where it was needed and to operate the distribution system under extreme hydraulic pressures. These required installation of massive, precisely engineered and fabricated mechanical valves, as shown in Figure 6.9.

With construction starting in 1936, the Delaware Aqueduct Tunnel was comprised of three tunnel sections: Rondout–West Branch Tunnel (48 miles), shown in Figure 6.10; West Branch–Kensico Tunnel (23 miles); and Kensico–Hill View Tunnel (13 miles). As was the case with City Water Tunnel No. 2, the Delaware Aqueduct Tunnel experienced delays due to funding and the Great Depression.

Figure 6.9 | Installation of 96-Inch-Diameter Steel Butterfly Valve, 1985

Figure 6.10 | Delaware Aqueduct, Rondout–West Branch Tunnel, 1939

Figure 6.11 | **New York City Water Supply Tunnels System**

Explosive Growth in Water Tunnels After World War II

World War II delayed further expansion of the respective water supply systems for Boston and New York City until after the war. Several societal changes that occurred after World War II had significant impacts on the water supply systems. The baby boom and the population spread to the suburbs in the 1950s and 1960s, resulting in the need to upgrade the systems in the areas surrounding major cities. The demand for redundant water transmission pipelines for reliability as well as improvements in operating pressures was necessary for sophisticated water infrastructure to continue moving forward.

Perhaps the most significant project that still has an effect today is City Water Tunnel No. 3. New York City started construction on this tunnel in 1970 and it has been such a major undertaking that the tunnel will not be fully completed until 2020 or later—a 50-year program (that at the time was) considered to be one of the largest capital construction projects in New York City history at more than $6 billion (Figure 6.11). The tunnel was first proposed in 1954 and was to be constructed in four stages. The first stage, a 12.5-mile-long, 20-to-24-foot-diameter tunnel was completed in 1998 after a three-year suspension of the project in the mid-1970s due to economic and contractual issues as well as other controversies. The second stage, comprised of a 10-mile-long tunnel between Brooklyn and Queens and a 5-mile-long tunnel within Manhattan, started construction in 1993. These two stages were followed by the third stage, 16-mile-long Kensico–City Tunnel, and a 14-mile-long fourth-stage tunnel from Hillview Reservoir through the Bronx and Queens to join with the second-stage tunnel.

One innovative construction technique that was used in the second-stage tunnel was the introduction of tunnel boring machines (TBMs). A TBM was used for excavation of the second-stage tunnel, which greatly improved excavation rates; it used only 25% of the labor compared to the conventional drill-and-blast excavation method. In 1962, a TBM known as the Alkirk hard-rock tunneler was used for initial excavation of the 5-mile-long Richmond Tunnel being constructed between Brooklyn and Staten Island. The TBM, however, had numerous mechanical breakdowns and was considered very inefficient. Consequently, the Richmond Tunnel was completed in 1970 by conventional drill-and-blast excavation. Figure 6.12 shows the disc-cutter pathways embedded in the tunnel excavation face in New York City Tunnel No. 3.

Boston started a major program for redundant distribution tunnels within the innermost metropolitan areas. Tunnel construction for water tunnels commenced in Boston in the late 1940s and has continued through present day. City Tunnel was the first water tunnel started in 1947 and completed in 1951, followed by the City Tunnel Extension completed in 1956. Soon after, the Malden Tunnel was also completed in 1958. The Cosgrove Tunnel was completed in 1965, and by the 1970s, similar to New York City, the use of TBM technology had been introduced to water tunnel construction. A TBM named "The Mole" was used to excavate the Dorchester Tunnel, which was completed in 1974 (Figures 6.13 and 6.14).

Figure 6.12 | **Disc-Cutter Tracings, 1980s**

THE HISTORY OF TUNNELING IN THE UNITED STATES

Figure 6.13 | **"The Mole" Excavated the Dorchester Tunnel, 1970**

Hultman Aqueduct: Lessons Learned

As a result of this extensive and ongoing tunneling program, the Boston community was provided with a robust water supply network with a redundant network of tunnels to deliver a reliable supply of water to 40 surrounding communities and approximately 2.5 million people. However, the system had a weak link in its network that needed to be corrected. For a 17.6-mile-long section, approximately 85% of the water to metropolitan Boston was being conveyed by a single near-surface 12-foot-diameter pipeline, the Hultman Aqueduct, which was leaking in several locations. It was estimated that the economic damage to metropolitan Boston would be enormous if this aqueduct experienced a major break while in service. The Hultman Aqueduct was completed in 1939, but construction of a second aqueduct to provide system redundancy had been halted with the onset of World War II.

In 1989, the Massachusetts Water Resources Authority began planning for an additional water conveyance system to enable the Hultman Aqueduct to be taken

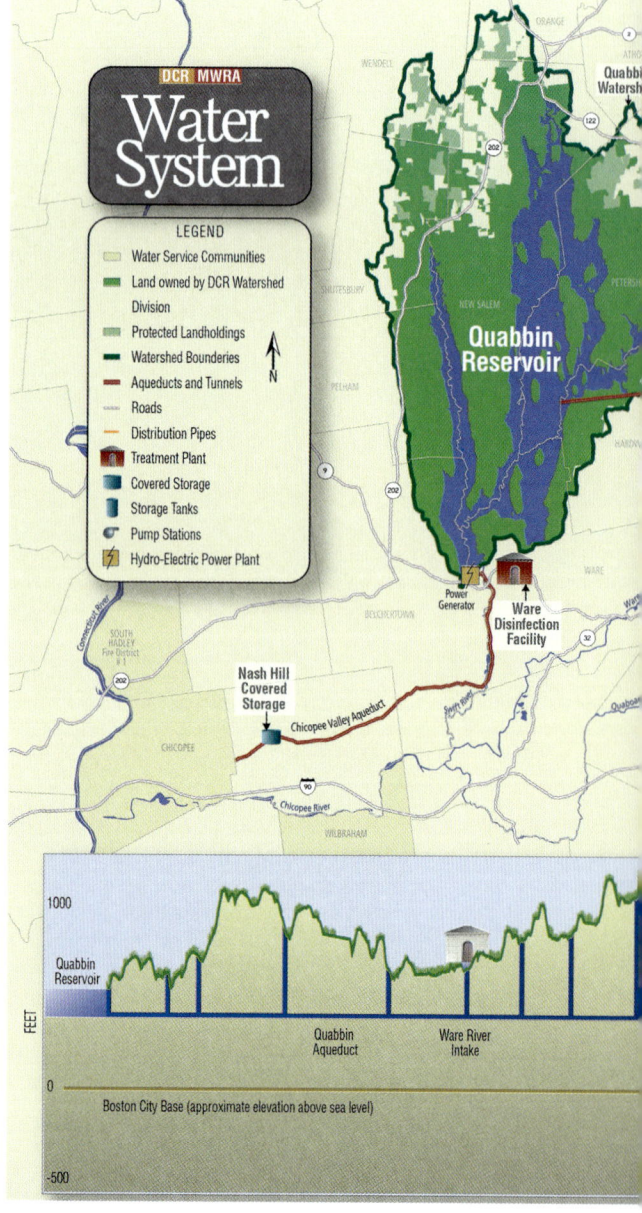

out of service, repaired, and then put back in service. The original plan was for construction of a short tunnel segment as well as rehabilitation of the Sudbury Aqueduct. The Sudbury Aqueduct had been taken out of service many years before and was available only for emergency service if needed. Engineers reviewed this plan and proposed an alternative that resulted in the construction of the $600 million MetroWest Water Supply Tunnel. This included interconnections to several municipal water supply systems and the Hultman Aqueduct along its route, as well as two covered storage facilities (Figure 6.15).

The project was completed on schedule and within budget, only slightly below the engineer's estimate. Also noteworthy, the $600 million project cost was equivalent to 10 days' value of the economic impact in the event of a failure to the non-redundant Hultman Aqueduct.

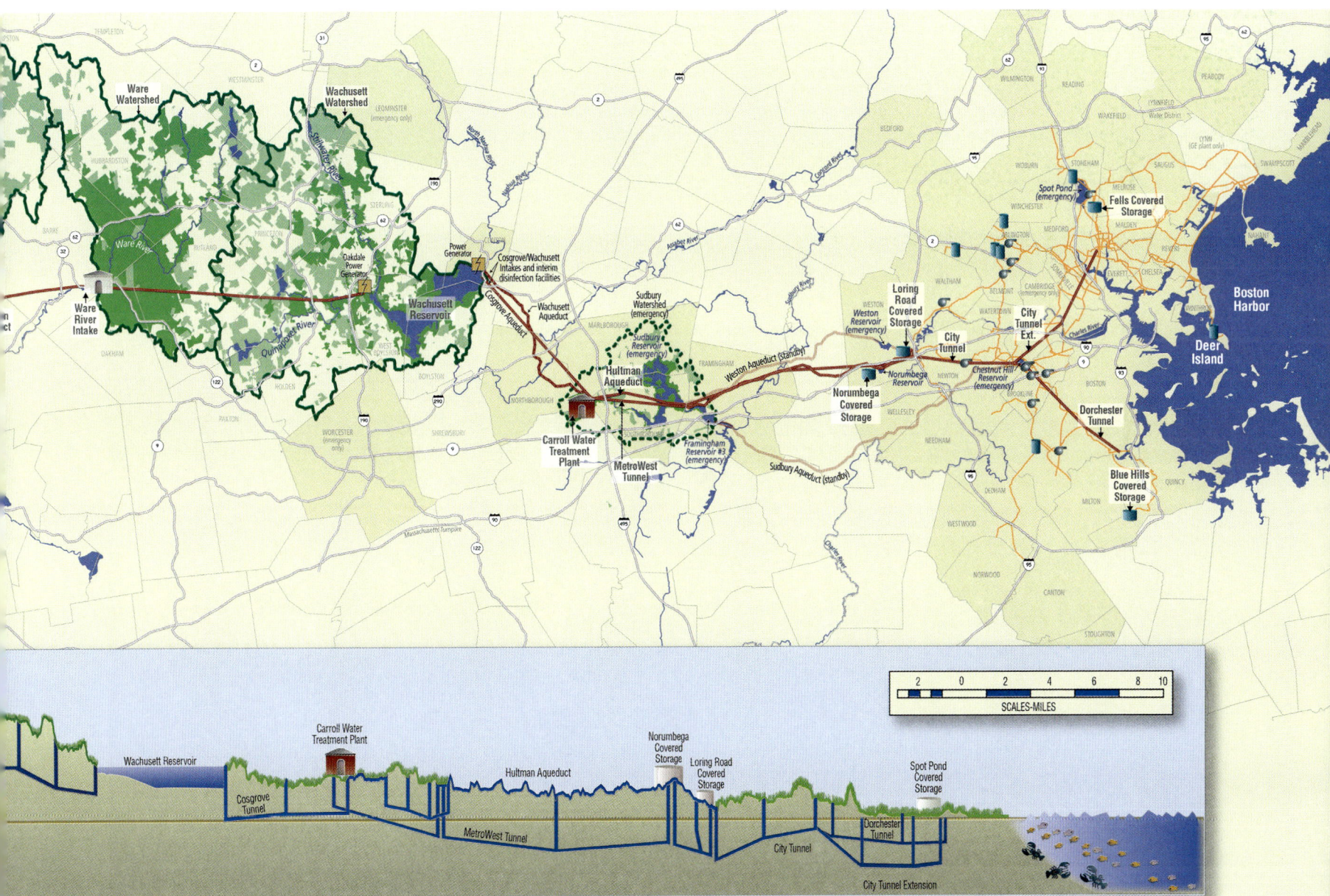

Figure 6.14 | **Boston Water Supply Tunnels**

With the aging of our vital water infrastructure today, water departments and agencies throughout the United States are placing significant emphasis on inspection and rehabilitation of these important lifelines. In New York City, the Department of Environmental Protection has been performing inspections, evaluations, and rehabilitation of the New Croton Aqueduct Tunnel and shafts since 1993, with the most recent phase completed in 2013. More than 120 years old, the 33-mile-long tunnel and 33 shafts are constructed of masonry brick-and-mortar courses, 12 to 24 inches thick, with mortar rubble material filling the annular space between the rock surface and brick lining (Figure 6.16). The rehabilitation effort to revitalize the vital links to New York City included tunnel and shaft liner contact grouting, consolidation grouting, joint and crack repairs, and limited brick-and-mortar replacement (Figure 6.17).

WATER TUNNEL SYSTEMS OF THE GREAT LAKES

Chicago—First Intake Tunnel in the United States

With the rapid growth of the city of Chicago and the pollution of its primary water source, the Chicago River, in 1842, Chicago began obtaining the first water from Lake Michigan through the construction of the first of two cribs (water protection structures) 150 feet offshore. This Lake Michigan water source proved to be inadequate and a second crib 600 feet offshore was constructed and put into operation in 1854.

Figure 6.15 | **MetroWest Tunnel Formwork for Cast-in-Place Concrete Lining**

As Chicago became more industrialized and commercialized, demand for a sophisticated wastewater system began. Residential waste, which was being dumped into the river and had made its way into Lake Michigan, was contaminating the water supply. Cholera, dysentery, and typhoid epidemics occurred over successive years, and as a result, 5% of the population died in 1854. Chicago's Chief Engineer Ellis S. Chesbrough, who designed the city's sewer system, studied the water supply systems in Europe and developed recommendations to improve Chicago's system. He proposed constructing a 5-foot-diameter brick-lined tunnel beneath Lake Michigan to a distance of 2 miles offshore. Thus was born Chicago's tunneling industry and the first water supply tunnel and raw water intake in America. Tunnels would become a major part of the Chicago area infrastructure system, ultimately leading to more than 76 miles of water tunnels.

The first water intake tunnel, the Chicago Lake Tunnel, extended 10,587 feet under Lake Michigan to Two-Mile Crib (Chicago DWS 1956; ASCE 2003), as shown in Figure 6.18. Tunnel construction began in March 1864, excavating from the shaft at the famous Chicago Water Tower (the only building to later survive the Great Chicago Fire of 1871) toward the intake crib shaft. The excavation of the 5-foot-diameter brick-lined tunnel was completed in November 1865. A unique component to the construction involved the tunnel being excavated by hand in blue clay, encountering occasional sections with gravel deposits, gas pockets, and boulders, at a depth of 60 feet below the lake bed. Construction was achieved without any serious injuries (Pikarsky 1971). Given that this work was performed before the Occupational Safety and Health Administration era, work progressed 24 hours a day, six days per week, averaging about 12 feet each day at a considerable risk, considering the working practices that were typical of the day. Two shifts excavated the tunnel, removing the excavated material or "muck" on small mule-drawn railcars. One shift of masons placed the brick tunnel lining, with two layers, using two semicircular arches to form both the top and bottom of the tunnel. The lower semicircular arch was built first, about 6 feet ahead of the crown arch (Figure 6.19). The excavation was carried out by two miners working side by side. As the excavation proceeded, the masons built the brick lining 10 to 20 feet behind the heading.

At the lake shaft, a crib was constructed of timbers, bolted together with iron rods in a pentagonal shape nearly 100 feet across at a location where the lake was 30 feet deep. The perimeter was built as three separate walls that were bolted and sealed or caulked and tarred. Generally, the annulus between the outer crib walls was filled with stone or concrete.

After completing nearly 50% of the tunnel from the shore, a second heading was begun from Two-Mile Crib. When the two tunnels met, the tunnels were only 7 inches out of alignment, a remarkable achievement given the technology available (Christenson 1973). The final brick of stone, a marble stone, was placed by the mayor of Chicago, who was accompanied by the Board of Public Works and other dignitaries who had traveled by muck car to the connection point. Shortly after the end of the Civil War, the first water supply tunnel in the United States was put into service in March 1867, bringing 50 million gallons per day of unpolluted water to a new pumping station and water tower at Chicago and Michigan Avenues. The first intake tunnel in the United States brought international acclaim to Ellis Chesbrough and, as an engineered system, was described as the "eighth wonder of the world" (ASCE 2003).

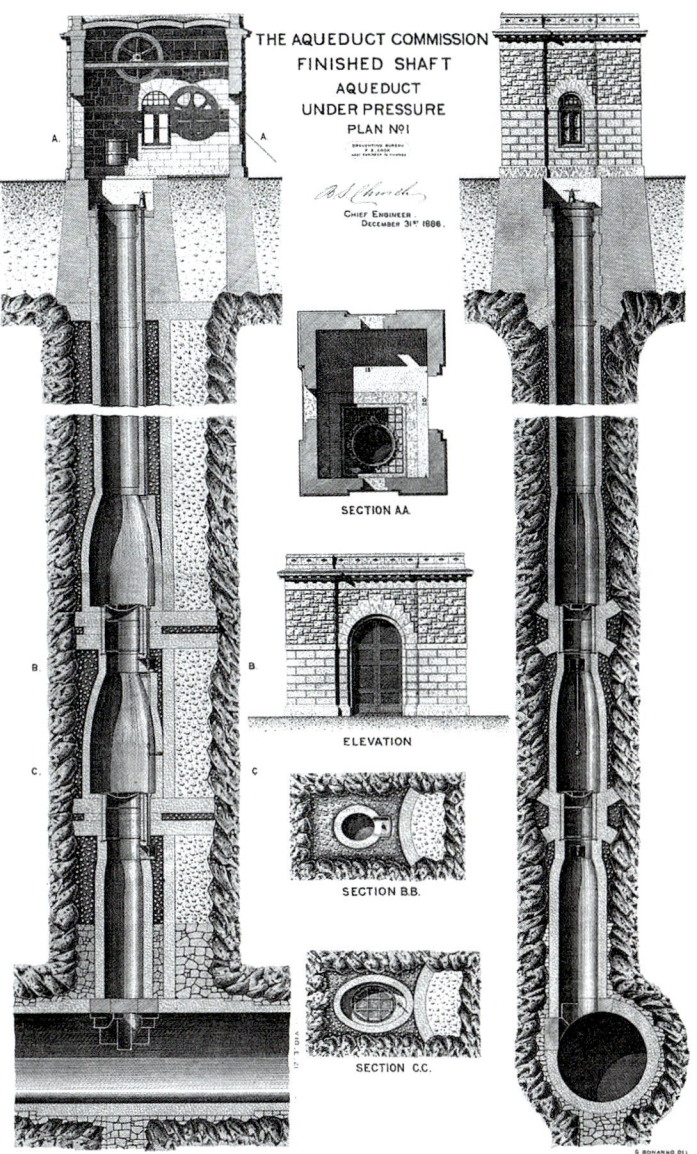

Figure 6.16 | **New Croton Aqueduct Tunnel and Shaft Details**

Figure 6.17 | **Contact Grouting of Brick-Lined Tunnel**

Figure 6.18 | **Construction of Chicago Lake Tunnel Using Two Headings**

Figure 6.19 | **Sketch of Chicago Lake Brick-Lined Tunnel**

Within a decade of completing the water supply project, demand exceeded the capacity of the first tunnel, resulting in a second tunnel, the Cross-Town Tunnel, being completed in 1874. This tunnel paralleled the first tunnel from Two-Mile Crib to shore before trending southwest to the 22nd Street Pumping Station. The Cross-Town Tunnel was also excavated in the blue clay and was 7 feet in diameter, brick lined, and 6 miles long (Chicago Bureau of Public Efficiency 1917). The land-based tunnel segment was later abandoned because the tunnel was an obstruction to planned deep building foundations. In 1887, a third tunnel, the North Shore Extension Tunnel, was constructed and ultimately connected to the Two-Mile Crib, north of the Chicago Lake Tunnel.

A new system of water intake tunnels and cribs was constructed to the south, called the Polk Street Tunnel, which extends farther out beneath the lake to the Four-Mile Crib (Chicago Bureau of Public Efficiency 1917; Figure 6.20). The east–west tunnel beneath the lake bifurcated once on shore; a south segment extended south to the 14th Street Station while a north segment extended the Harrison Street Station. The land portion of this tunnel, like the Cross-Town Tunnel, was cut diagonally beneath private property, which interfered with the deep building foundations. Ultimately, a new Polk Street Tunnel was constructed beneath city streets in 1907.

By 1900, 30 miles of brick-lined tunnels, 5 to 10 feet in diameter, had been constructed in the soils beneath the city and Lake Michigan. Chicago had become a model from which other midwestern cities would follow. Growing cities like Cleveland, Toledo, and Milwaukee looked to the Great Lakes municipalities for intake water supply tunnels.

Beginning in 1900, brick was no longer used for lining the tunnels, and all future tunnels were constructed with a concrete lining (Chicago DWS 1956). The first concrete-lined tunnel in Chicago was the Blue Island Avenue Tunnel, a 26,550-foot-long tunnel that was constructed in 1909 to replace the Cross-Town Tunnel. The alignment of the Blue Island Avenue Tunnel beneath city streets was an 8-foot finished diameter horseshoe-shaped tunnel excavated primarily in the stiff blue clay, with little water encountered, at 34 to 62 feet deep (Figure 6.21). The average monthly excavation rate was 278 feet per heading, with a maximum daily rate of 21 feet (ENR 1908). Steel forms were used and concrete was shoveled into the crown and rodded into place. A 10-inch-thick lining was placed in this fashion and was covered by a thin layer of cement plaster (0.5 inch). The lining was not reinforced, except for a shallow cover area beneath the Chicago River 400 feet long and at shaft connections. In these areas, 1-inch steel bars were incorporated into the lining. The Blue Island

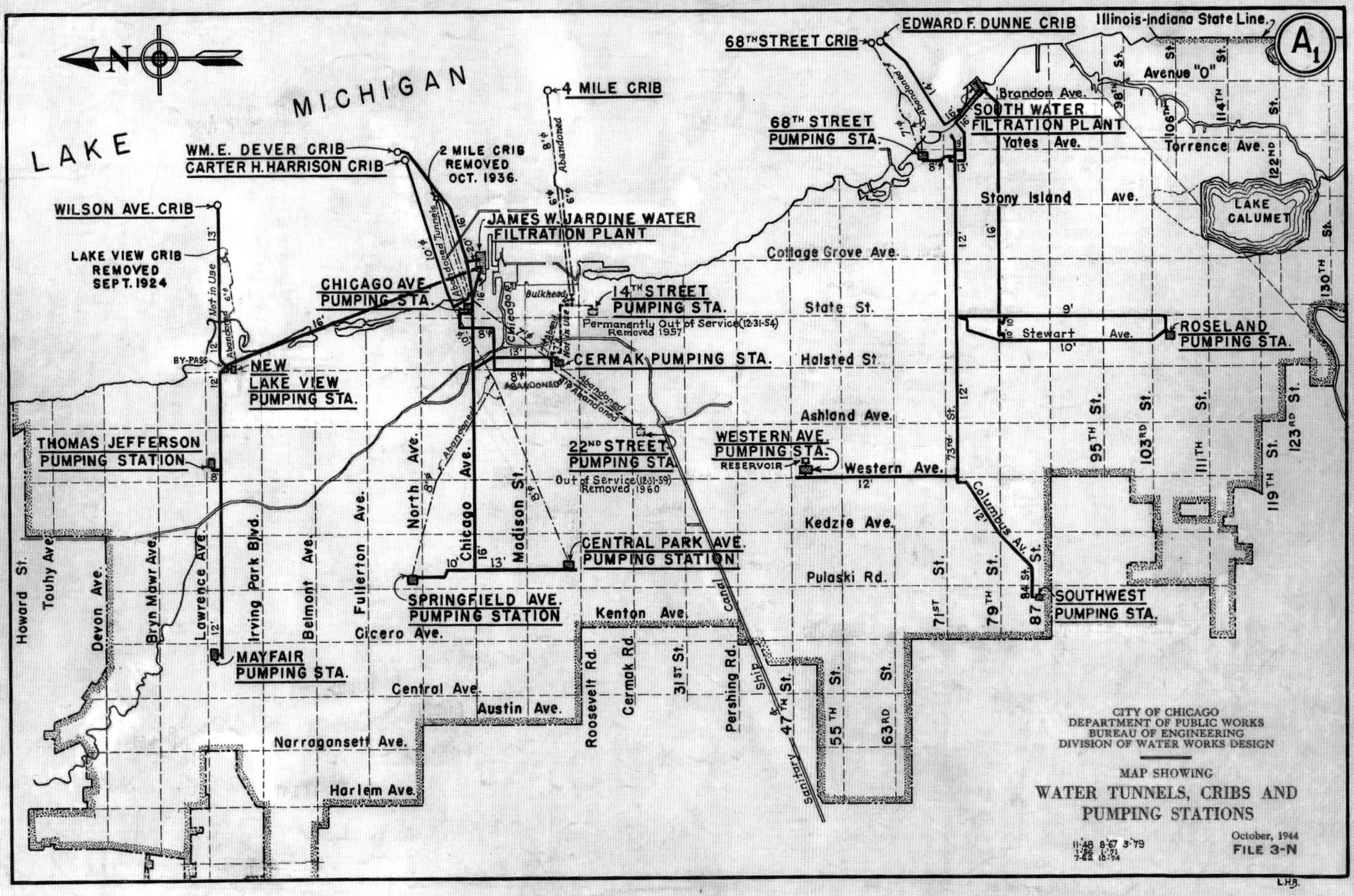

Figure 6.20 | **Chicago Land and Lake Tunnels**

Avenue Tunnel draws its supply of water from Two-Mile Crib. When the tunnel was first dewatered in 1932, after about a quarter of a century in operation, the concrete lining was found to be in structurally sound condition (ENR 1933).

As the city of Chicago continued to grow, additional tunnels were built that connected to new pumping stations. However, difficulties were experienced with tunnel construction due to the poor soil conditions encountered. In an effort to mitigate the problems with the soft-ground tunnels, city engineers resolved that, wherever feasible, future tunnels would be constructed in rock. The rock would create a stronger foundation, ensuring greater longevity of the tunnel structure. This was an important strategic decision by the city that would have a dramatic change on the type of tunneling performed thereafter.

The first of the tunnels to be constructed in bedrock was the Southwest Land and Lake Tunnel located on the south side of the city and connected to the Edward

Figure 6.21 | **Blue Island Tunnel West Heading Support of Excavation, 1907**

Figure 6.22 | **Southwest Land and Lake Tunnel, Finished Concrete Lining, 1911**

Dunne Crib (Chicago Bureau of Public Efficiency 1917; Pikarsky 1971). Tunnel construction began in 1906 and was completed in 1911. The tunnel depth varied from 102 feet beneath the city to 160 feet beneath Lake Michigan. Multiple tunnel headings were used to excavate the Southwest Lake and Land Tunnel with as many as six headings being excavated at one time. The horseshoe-shaped tunnel section for the land segment was between 9 and 12 feet wide while the underwater segment section was 14 feet. Conventional drill-and-blast techniques with a top-heading-and-bench excavation method were used to advance tunnel excavation. (The "top heading and bench" excavation method involves drilling, blasting, and excavating the top portion of the tunnel first followed by the bottom or bench.) Blasting techniques left a lot to be desired with large volumes of over-excavated ground or "overbreak" occurring, making the tunnel slightly greater in dimension than was required. Conversely, sometimes insufficient volume was excavated by the blast, and considerable trimming was also required to remove rock "tights" to achieve the desired cross section. For the 14-foot-wide tunnel section beneath the lake, significant water inflows were encountered, which reduced the tunnel advance rate as low as 4 feet per month. It is noteworthy, however, that a maximum excavation rate of 416 feet per month was achieved for a single heading, which helped to compensate for lost time (Koncza et al. 1980; Pikarsky 1971).

The placement of the concrete lining for the Southwest Land and Lake Tunnel varied according to the size of the tunnel section: lining thicknesses of 12, 9, and 7 inches were constructed for the 14-, 12-, and 9-foot wide tunnels, respectively (Figure 6.22). Several lessons were learned about placing a concrete lining in tunnels in Chicago, as the monthly production rate for the lake segment topped out at 1,114 lineal feet for the 14-foot-wide tunnel, whereas almost double that rate was achieved in the two smaller cross-sectional tunnels beneath the city. Challenges due to water bleeding and seeping through the concrete lining occurred, sabotaging attempts at applying the cement plaster covering and in some cases making the lining work almost futile.

Since the time of the first rock tunnel for water supply, several deep rock tunnels to the intake cribs and beneath the city have been completed. The Chicago Avenue Tunnel, completed in 1935, is the latest lake tunnel. This tunnel is horseshoe shaped, 16 feet in diameter, and is 170 to 190 feet below lake level for the section underneath the lake. The tunnel was excavated through limestone using drill-and-blast methods and is fully concrete lined.

In the 1950s, two additional water tunnels were added, the 4.8-mile-long 79th Street Tunnel and the approximately 5.3-mile-long Wilson Avenue to Central Water Filtration Plant Tunnel. The 79th Street Tunnel was one of the most difficult tunnels to construct in Chicago because of the rock conditions. Both were 16-foot-wide horseshoe-shaped tunnels excavated using drill-and-blast methods. In the east heading of the 79th Street Tunnel, which is the shallower of the two headings at 109 to 155 feet below the ground surface, the rock conditions varied from soft limestone with clay seams to blocky rock containing clay pockets and seams. In places there was zero standup time and significant overbreak, and steel supports were required for more than 7,000 feet (Pikarsky 1971). While mining progress was slow on the 79th Street Tunnel, the Wilson Avenue to Central Water Filtration Plant Tunnel had the best mining progress recorded—3,165 feet were excavated on two headings in October 1957.

In 1995, the 79th Street Tunnel was extended 3.6 miles to the west to increase capacity and reduce operating costs for the pumping stations (Pestonatto et al. 2003). The tunnel was excavated in limestone bedrock using an 18.25-foot-diameter Robbins hard-rock TBM. The contractor had set several excavation records at the time, achieving a best-month record of 4,963 lineal feet. This tunnel was concrete lined (Figure 6.23), a gate shaft was constructed, and a live tunnel connection was made to the existing water tunnel system.

Today in the Chicago area there are approximately 76 miles of water tunnels in service, or available for service, and of these, 61 miles were constructed in rock. They vary in size from 9 to 20 feet; the majority of the tunnels are in the 12-to-16-foot range.

The intake cribs, located where the shafts are constructed in the lake at the tunnel ends, are massive structures built on top of the lake bed to withstand the huge storm waves on Lake Michigan and the external water pressure due to the lake level. The initial cribs were constructed of wood timbers bolted together in a multiple-wall configuration. Later, the cribs would be constructed of wood timbers and a steel structure with the space between the inner and outer walls filled with stone or concrete to hold the crib in place. The diameter of the cribs reached 112 feet with an inside diameter of 60 feet or so. Within the crib, one or two vertical shafts up to 16 feet in diameter (e.g., Dunne Crib) would be excavated to the deep rock tunnels up to 190 feet below lake level. Currently, the City of Chicago is serviced by four intake cribs (Four-Mile, 68th Street, Edward F. Dunne, and William E. Dever), located 2 to 3 miles offshore.

Figure 6.23 | **Placement of Reinforced Concrete Lining for 79th Street Tunnel Extension**

With the invasion of zebra mussels into Lake Michigan in 1989, concerns developed that the three intake tunnels could become clogged. In 1997, the Northeast Lake Tunnel (formerly known as the Northwest Land and Lake Tunnel, constructed in the late 1890s), a 14,200-foot-long, 10-foot-diameter, hand-mined soft-ground tunnel was dewatered for inspection and installation of a zebra mussel abatement system. In July 1997, a breach occurred in a century-old brick-lined water supply tunnel beneath Lake Michigan. Since the Northeast Lake Tunnel is also hydraulically connected to a lower Chicago Avenue Tunnel, a concrete-lined water tunnel in rock, two water intake tunnels were now out of service. With the sensitivity of collapsed tunnels following the Great Chicago Flood of April 1992, the city moved quickly to abandon the upper Northeast Lake Tunnel to protect the city's water supply system and to eliminate any possibility of land-side leakage that might cause subsidence or road stability problems on Lake Shore Drive, a major transportation artery in Chicago (ENR 1998). The Northeast Lake Tunnel was sealed off and the damaged tunnel was abandoned. Two 90-foot deep temporary shafts were excavated at the shore to penetrate the brick-lined tunnel and install a concrete bulkhead to seal off the lake and to backfill the vertical shaft connection to the lower Chicago Avenue (rock) Tunnel. In addition, the Northeast Lake Tunnel was sealed at the Harrison Crib intake end with concrete. Figure 6.24 shows the concrete mixer trucks being barged to the Harrison Crib to seal the damaged brick-lined tunnel.

Figure 6.24 | **Concrete Mixer Trucks Being Barged to Harrison Crib**

Cleveland—Inaugural Concrete Segmental Lining in the United States

The demand for tunnels into Lake Erie began around 1876 when, similar to Chicago, there was a need for separate water and wastewater systems due to increased sewage flow into the water supply.

Increased population growth in Cleveland was driving the demand for clean water. The City of Cleveland looked to draw water from Lake Erie, the fourth largest lake of the Great Lakes. Similar to Chicago, a water intake tunnel built a considerable distance out into the lake was required because of the increasing discharge of sewage flows along the shore that had created a health hazard.

Figure 6.25 | Excavation near Crib No. 2 of Cleveland Lake Tunnel

The first water tunnel in Cleveland, Intake No. 2, was constructed to Intake Crib No. 4 approximately 6,600 feet offshore, which connected to the Division Street Pumping Station. As water supply needs increased, a second tunnel, Intake No. 3, 7 feet in diameter, was excavated beneath the lake to a distance of 9,117 feet. Both tunnels were lined with brick. In 1896, Cleveland's Water Works began excavation of a new 9-foot-diameter tunnel, Intake Tunnel No. 4, approximately 26,000 feet out to Five-Mile Crib (Kirtland Crib) to bring fresh water into a newly built pumping station at Kirtland Street. The construction of the new intake tunnel would take eight years with completion in 1904 (Figure 6.25).

The excavation of tunnels in Cleveland was not without hazards and challenges. In addition to working with compressed air to maintain the tunnel heading, and the perils of decompression sickness in the early years, pockets of methane gas were encountered. Five disasters occurred during construction of the initial four water intake tunnels. During the construction of Intake Tunnel No. 4 to Five-Mile Crib, four gas explosions occurred in 1898, 1901, and 1902 (Bellamy 1995). These devastating events took the lives of 28 tunnel workers. Additionally, in August 1901, a fire was ignited by cinders from a boiler stack in a temporary crib built to house the workers; five men burned to death, three drowned trying to escape the crib, and one rescuer lost his life trying to reach workmen in the tunnel. The advancement of water infrastructure was not without huge risks.

Figure 6.26 | **Excavation of Cleveland's West Side Tunnel**

Another issue came up soon after these tunnels were completed. Increasing pollution from sewage along the shores of Lake Erie made it necessary to go even farther out into the lake. In 1914, a new tunnel, the fourth, was needed to connect a new shaft at Crib No. 5, 5 miles out into the lake. The West Side Tunnel linked up to the Division Avenue Pumping Station with a separate connection to Crib No. 4 (Figure 6.26). The shaft at Crib No. 5 was excavated to a depth of 40 to 50 feet beneath the lake bed. The 10-foot-diameter tunnel was excavated through the sand, gravel, and clay material using a hydraulic shield. The tunnel was lined with interlocking concrete segments (Bellamy 1995) invented by Walter Parmley, an engineer and contractor, who specialized in large-sewer and soft-ground tunnel construction (Figure 6.27). The concrete segments consisted of 1-foot-wide reinforced concrete blocks or segments, with six pieces composing a ring. The completed Cleveland Intake No. 5 Tunnel is shown in Figure 6.28. This is believed to be the first use of a precast concrete segmental lining in the United States.

The Parmley segment lining configuration was such that the transverse joints were offset while the longitudinal joints were continuous. The segments were cast with the bottom segment containing a single reinforcing bar while the upper blocks contained two reinforcing bars. Each segment weighed approximately 1,075 pounds and for larger tunnels were installed using an erector. The Parmley segments were also used in sewer tunnels (Parmley 1927).

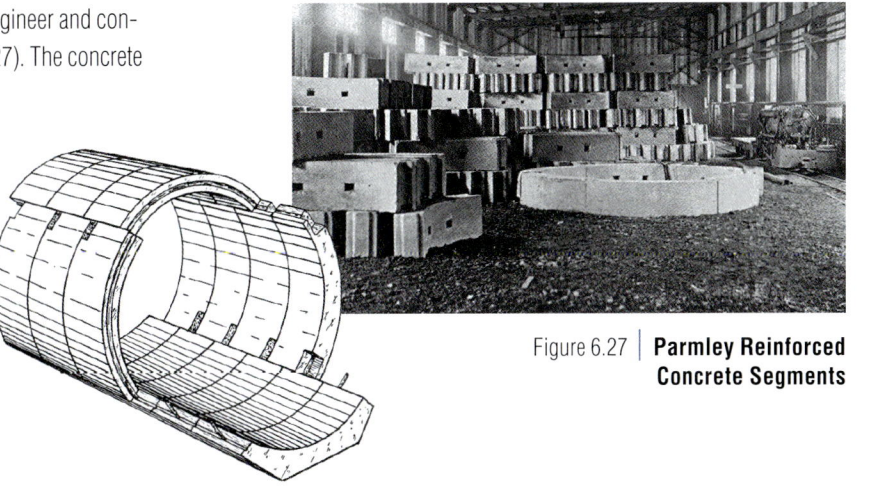

Figure 6.27 | **Parmley Reinforced Concrete Segments**

Figure 6.28 | **Completed Cleveland Intake Tunnel No. 5**

Figure 6.29 | **Morgan's Breathing Apparatus**

The final disaster took place 12 years later, in July 1916, during construction of Intake Tunnel No. 5, the West Side Tunnel, when workmen digging the 10-foot-diameter tunnel hit a pocket of natural gas. A spark triggered an explosion, killing 11 miners and 10 would-be rescuers who were overcome by gas when they entered the pressurized tunnel. The hero was Clevelander Garrett Morgan, inventor of the breathing device shown in Figure 6.29, who entered the gas-filled tunnel to save lives (Bellamy 1995).

WATER TUNNEL SYSTEMS OF CALIFORNIA

The late 1800s and early 1900s saw rapid urban growth of San Francisco and Los Angeles. These cities obtained their fresh water locally, through private water companies who obtained the water primarily from nearby streams and rivers. However, the increasing freshwater demands of these communities saw the development of San Francisco's "water wars" and the need to look beyond the city limits for water sources. As a consequence, these communities began to look for alternative freshwater supplies to the east and north, looking to the Sierra Nevada and Owens Valley, respectively.

San Francisco's Hetch Hetchy Regional Water System— A Rare Surface Water Source

San Francisco's water supply originates from the pristine Sierra Nevada snowmelt, which is captured behind O'Shaughnessy Dam in Hetch Hetchy Reservoir located in Yosemite National Park. The concrete arch gravity dam stores up to 360,360 acre-feet of runoff from the Tuolumne River basin. The raw water then flows by gravity through a series of tunnels and pipelines 160 miles westward through the farmlands of the San Joaquin Valley, into rural Alameda County, across southern San Francisco Bay, through several cities, and into the City of San Francisco (Figure 6.30). The Hetch Hetchy system supplies approximately 85% of San Francisco's current daily water demand with the remainder supplied from local reservoirs. Moreover, the Hetch Hetchy system is one of the few surface water sources in the United States that is legally delivering unfiltered water to customers, with only disinfection required for potability.

San Francisco's interest in the Tuolumne River as a water supply for the city dates from May 1882. J.P. Dart, an engineer for San Francisco and the Tuolumne County Water Company, first proposed bringing Tuolumne River water from Jackson to San Francisco. This idea never caught on at the time, probably because of the long 160-mile distance and the great investment in infrastructure that would be required. However, the idea never went away. The U.S. Geological Survey published their *Annual Report 1899–1900*, which included a study recommending the Hetch Hetchy as a water source for San Francisco. In 1901, Mayor James D. Phelan filed the first application with the U.S. Interior Secretary for water and reservoir rights at Hetch Hetchy Valley and Lake Eleanor. Phelan originally pursued the application for water rights as a private citizen for fear it would have drawn the attention of local and national interests opposed to the application. It was eventually rejected in 1903 (Hanson 2005).

The ever-growing demand for fresh water became increasingly apparent following the magnitude 7.8 earthquake of April 18, 1906, along the San Andreas Fault that caused widespread devastation in San Francisco. The most dramatic impact was the failure of the city water system—the rupturing of water pipes and failures of water storage facilities throughout the city. The city lost the ability to fight the many fires that resulted from broken gas mains following the earthquake (Figure 6.31). San Francisco burned uncontrollably for three days until more than 80% of the city had burned to the ground. An estimated 3,000 people lost their lives and countless more sustained injuries.

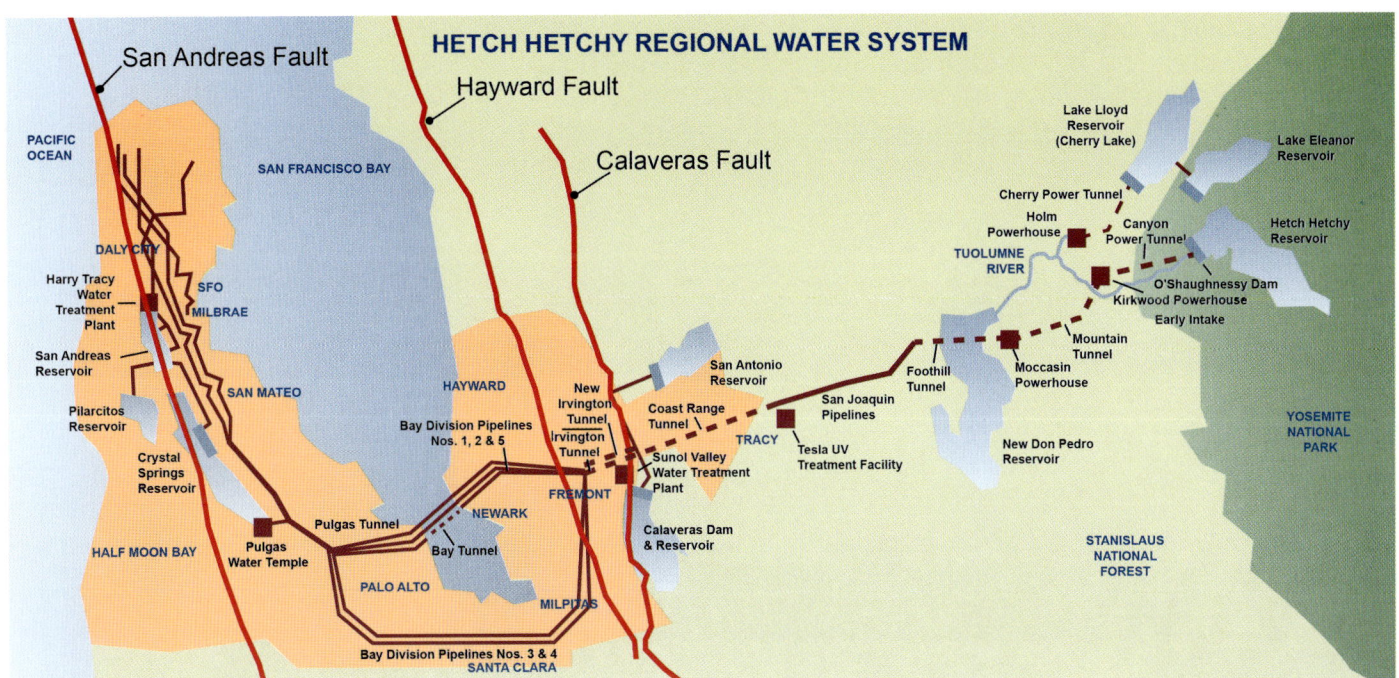

Figure 6.30 | San Francisco's Hetch Hetchy Regional Water System

In the aftermath of the great earthquake and fire, there was a renewed sense of urgency on the part of civic leaders to find a reliable water source for the city. In April 1908, Marsden Manson, city engineer, filed Hetch Hetchy maps for a new application with U.S. Interior Secretary James Garfield. In May 1908, Secretary Garfield granted the city limited permission for reservoirs, dams, aqueducts, and rights-of-way at Hetch Hetchy and Lake Eleanor. The city acted quickly, approving a $600,000 bond issue to purchase the land around Hetch Hetchy Valley and Lake Eleanor. In addition, San Francisco voters approved a $45 million bond measure that passed by a 20:1 margin in 1910 for construction of a Hetch Hetchy water system. In 1912, John Ripley Freeman, a well-known hydraulics engineer hired by the city, published a visionary report that became the master plan which compared the options and recommended the Hetch Hetchy water system for San Francisco that was later designed and built by Michael O'Shaughnessy (Hanson 2005; Hennessey 2012).

Political Opposition to Hetch Hetchy

The movement of the city to secure the Hetch Hetchy Project rights did not go unnoticed. Opposition soon came from four sources: the Spring Valley Water Company with the controlling interest in all San Francisco Bay Area water sources; the Turlock and Modesto Irrigation Districts, who saw the Hetch Hetchy Project as an infringement of their water rights; promoters of water rights, including Sierra Ditch and Water Company; and the National Park Service and environmentalists, led by naturalist John Muir, who argued for the preservation of the pristine wilderness and environment of the Hetch Hetchy Valley.

In 1910, Richard Ballinger took office as the new U.S. Interior Secretary and he issued an order to San Francisco to show cause as to why the prior permit issued by Secretary Garfield for Hetch Hetchy should not be revoked. It soon became apparent that any permit from the Interior Secretary would be subject to the political whims of succeeding national administrations. The remedy would be a grant of the water rights from the U.S. Congress, forever guaranteeing the rights (Hanson 2005).

Garrett Augustus Morgan (1877–1963)

Morgan was an African-American inventor and community leader in Cleveland, Ohio. Among his many inventions, which include the first traffic signal and hair care products, he is recognized as the inventor of a "rescue hood" breathing apparatus, which was used in a heroic rescue of workers trapped within a water intake tunnel beneath Lake Erie. He and his brother Frank and two others entered the gas-filled tunnel at the request of the Cleveland police. Sustained by the breathing apparatus that drew air from beneath the smoke layer in the tunnel via a series of tubes attached to the hood, they rescued two men and recovered four bodies. However, officials from the U.S. Bureau of Mines stopped further rescue and recovery efforts.

Figure 6.31 | **The 1906 San Francisco Fire Burned Uncontrollably for Three Days**

The battle in Congress began on April 7, 1913, when Representative John Edward Raker (Figure 6.32) introduced HR112 on the floor of the House of Representatives. The bill passed the House with the provision to preserve the prior Tuolumne River water rights owned by the Modesto and Turlock Irrigation Districts. Once it moved on to the Senate, the bill took on a national debate.

Scores of letters from all over the nation, both pro and con, were written into the Congressional Record, coming in from entities such as the Spring Valley Water Company, John Muir and environmentalists, and others. The *New York Times* editorialized against the project. William Randolph Hearst and the *San Francisco Examiner* had significant involvement and published statements from the Vice President, the Secretary of State, and the Secretaries of Interior and Agriculture all in support of the Hetch Hetchy Project. Soon after, President Woodrow Wilson signed the Raker Act into law, and the City of San Francisco approved the Hetch Hetchy Project in the spring of 1914 (Simpson 2005).

The Hetch Hetchy Regional Water System

Construction of the Hetch Hetchy system, which began in 1914, was preceded by the building of three key infrastructure facilities, including a 68-mile-long railroad supply line; a sawmill that produced lumber for constructing camps, formwork, and timber supports for the tunnels; and the Early Intake Powerhouse to supply electric power to run the drills and lights in the tunnels, power other equipment, and provide illumination for the camp and construction sites. Once the infrastructure was in place, the focus became construction of Mountain Tunnel and O'Shaughnessy Dam followed by construction of Foothill, Coast Range, and Irvington Tunnels—completed by 1934. More than 60 miles of the Hetch Hetchy water system was comprised of tunnels. Later in 1965, the 10.4-mile-long Canyon Power Tunnel was completed to directly convey the water to the Mountain Tunnel from O'Shaughnessy Dam (Hanson 2005; Simpson 2005).

Figure 6.32 | **John E. Raker**

Mountain Tunnel. The 19-mile-long Mountain Tunnel was constructed between 1917 and 1925 from the Early Intake Reservoir, downstream of O'Shaughnessy Dam, to Priest Reservoir above the town of Moccasin. The tunnel was constructed at an average depth of 1,000 feet below the ground surface. About 38% of the tunnel is unlined with an internal diameter of 13.5 feet (Figure 6.33). The remainder of the tunnel is lined with concrete to a 10-foot finished diameter.

Contracts were put out to bid, but the bids were considered too extravagant by O'Shaughnessy and subsequently rejected. Instead, O'Shaughnessy decided to construct the tunnel with City of San Francisco crews. Only the best laborers were picked for this challenging assignment.

Work on Mountain Tunnel started in the summer of 1917 from the Early Intake. What was rather innovative at the time was the tunnel excavation sequencing that was achieved from twelve excavation headings along the tunnel alignment. Four excavation headings were started from portals, four were from adit passageways into the main tunnel, and four were from the bottom of deep vertical shafts dug at two intermediate locations along the alignment. The Second Garrotte Shaft was 786 feet deep, and the Big Creek Shaft was 646 feet deep (Figure 6.34). It was a great challenge for the engineers to constantly check their surveys and match up each of the twelve headings, both horizontally and vertically. It was a masterful and satisfying accomplishment when the mining crews working toward each other "holed through" at the same line and grade along the long tunnel alignment.

Construction equipment used by city crews included using electric drills, which greatly facilitated the drill-and-blast excavation of the tunnel through solid granite. Tunnel muck was removed by small electric locomotive and muck cars (Figure 6.35). The tunnel had a horseshoe-shaped cross section with a flat bottom to facilitate construction access by equipment and workers. A reliable source of hydroelectric power was available from the Early Intake Powerhouse in May 1918. A much larger second power source, Moccasin Powerhouse, was added in 1925 at the downstream end of Mountain Tunnel at Priest Dam and Reservoir. Mountain Tunnel was completed in 1925 for $25 million (Hanson 2005; Hennessey 2012).

Foothill Tunnel. The 15.9-mile Foothill Tunnel was constructed from 1926 to 1929 starting in the town of Moccasin to Oakdale Portal, where connections were made to the San Joaquin pipelines. The tunnel was divided into two reaches by Don Pedro Reservoir where the Red Mountain Bar Siphon continues beneath the reservoir.

Approximately 50% of the inverted horseshoe-shaped tunnel is unlined and mostly free-standing. The unlined sections varied from 15 to 18 feet wide by 14 to 16.5 feet high

Figure 6.33 | **Electric-Powered Mucking Machine Loading a Muck Car**

Figure 6.34 - Mountain Tunnel Construction at Bottom of Big Creek Shaft

(Figure 6.36). The concrete-lined sections varied from 10 to 12.3 feet wide by 10 to 13 feet high. Similar to Mountain Tunnel, Foothill Tunnel construction was from ten headings—two from the portals, four from two shafts at Hetch Hetchy Junction and Rock River, and four from adit passageways into the main tunnel. Excavation was by drill-and-blast method, facilitated by a narrow gauge muck train to haul rock and debris from the Brown Adit to the east bank of the Tuolumne River. Challenges during construction included the crossing of Tuolumne River—solved by the installation of 770 feet of inverted siphon pipe 9.5 feet in diameter (Hanson 2005; Jacobs Associates and Korbin 2008).

This tunnel featured some noteworthy accomplishments. City crews achieved a U.S. record of 803 feet in one month of tunnel excavation in September 1926 in the tunnel heading east of Hetch Hetchy Junction Shaft. In addition, a pioneering cableway approach was used to move labor and supplies from the main Hetch

Figure 6.35 | **Electric Locomotives and Muck Trains Hauling the Excavated Rock**

Figure 6.36 | **Priest Tunnel Excavation by Top Heading and Bench Method**

Hetchy Railroad across the Tuolumne River to a narrow gauge railroad that served the tunnel (Figure 6.37). This cableway innovation was later used during construction of the Hoover Dam and the Panama Canal. City crews completed the nearly 16-mile-long tunnel in 1929 at a cost of $8 million (Hanson 2005).

Irvington Tunnel. The 3.4 mile-long Irvington Tunnel was constructed from 1927 to 1932 from the Alameda Siphons to Irvington Portal near Mission San Jose in Fremont, California. Construction was from two headings—one from each portal. Excavation was primarily by drill-and-blast method, supplemented by manual spading. The tunnel was excavated through mostly Briones sandstone, shales, softer rock/soil composite materials, and clayey gouge material that filled the seven secondary faults crossing the alignment. The initial support system consisted of 18-inch-square, heavy timber sets with wood lagging or boards for ground support around a horseshoe ring and across the invert (Figure 6.38). The tunnel has a cast-in-place concrete lining with a finished horseshoe diameter of 10.5 feet. The concrete was unreinforced but the lining was as much as 3 feet thick in some locations to arrest swelling or squeezing ground.

Heavy groundwater inflows were challenging and treacherous during construction. Inflows were as high as 2,000 gallons per minute, requiring the construction of concrete bulkheads at several locations to stabilize the excavation and prevent the loss of the opening due to running ground conditions. Methane and hydrogen sulfide gases were also encountered during excavation, but no explosive calamities and injuries occurred during the work (Hanson 2005).

Coast Range Tunnel. This 25-mile-long tunnel was constructed between 1927 and 1934 from Tesla Portal to Alameda Siphons near Alameda Creek. At the time of its construction, the 18-foot-diameter tunnel was the longest continuous tunnel in the world and was constructed at an average depth of 2,500 feet under the Diablo Range mountains. Multiple headings were again required, two from the portals and five from vertical shafts. Tunnel excavation was facilitated by electric-powered drilling equipment using the drill-and-blast method and locomotives for tunnel muck removal. Because of the ground conditions, the initial support system consisted of 18-inch square, heavy timber sets around a horseshoe ring and across the invert (Figure 6.39). The tunnel was lined with unreinforced concrete.

Challenges during construction included flammable methane gas, groundwater inflows, quicksand, and swelling ground. Methane gas proved to be deadly to the tunnel workers. On July 17, 1931, in the tunnel section east of the Mitchell Shaft, an explosion occurred, killing 12 miners. Squeezing or swelling ground was also significant. Some sections of the tunnel bore were reduced by 3 feet all around in one day, and several days later, the tunnel had nearly closed entirely.

Figure 6.37 | Cableways for Transporting Labor and Supplies over Obstacles

Figure 6.38 | **Irvington Tunnel Construction, Placing Invert Concrete, 1931**

Figure 6.39 | **Concrete Was Typically Batched in Drum Mixers Inside the Tunnel**

Figure 6.40 | **O'Shaughnessy and Mayor Rolph (top center) Celebrate the Coast Range Tunnel Hole-Through, January 5, 1934**

A number of major innovations were introduced for these challenging ground conditions. The design included flexible joints in some of the concrete-lined sections to accommodate earthquake movements. To solve the problem of swelling ground, the tunnel engineers over-excavated the tunnel diameter to create extra annular space around the tunnel to accommodate sprayed rings of gunite, a form of spray-applied concrete, on the tunnel walls and ground swelling. City crews completed the tunnel in 1934 for $28 million. This tunnel segment was the last component to be completed for the 160-mile-long Hetch Hetchy system (Figure 6.40; Hanson 2005).

Other Main System Facilities. Key elements of the Hetch Hetchy system also include O'Shaughnessy Dam construction, a 227-foot-high cyclopean concrete gravity arch dam—reputed to be the largest human-made structure at the time on the West Coast—which began in 1919 and was completed in 1923 (Figure 6.41). Hetch Hetchy Reservoir had an initial storage capacity of 206,000 acre-feet, but the dam was subsequently raised by 85.5 feet in 1938 (as originally planned), which increased the storage capacity to 360,360 acre-feet (Hennessey 2012).

Pipelines filled in the gaps between tunnel segments to complete the 160-mile Hetch Hetchy conveyance system with the 47.5-mile stretch across the farmlands and cattle ranches of San Joaquin Valley, between the Foothill and Coast Range Tunnels, and west of Irvington Tunnel, where the tunnel connects to the Bay Division pipelines to transport the water supply west across southern San Francisco Bay and, more recently, south around the Bay.

Canyon Power Tunnel. This 10.4-mile-long tunnel connects to O'Shaughnessy Dam and conveys water through a 9-foot-diameter steel pipe encased in concrete to the Kirkwood Powerhouse, located above Mountain Tunnel. The Canyon Power Tunnel was not constructed in the O'Shaughnessy era but was completed in 1965 at a cost of $11 million. Before the tunnel was built, water flowed from O'Shaughnessy Dam down the Tuolumne River and entered a downstream diversion structure, Early Intake. Canyon Power Tunnel has a horseshoe-shaped cross section, 14 feet wide by 14.5 feet high, and was excavated by the drill-and-blast methods. The vertical alignment is up to 2,000 feet below the ground surface. More than 90% of this tunnel was mined through granite but was not lined with concrete, leaving an irregular exposed rock surface along the tunnel wall and invert (Hanson 2005; Simpson 2005).

Recent Water System Improvement Program

On October 17, 1989, the Loma Prieta earthquake struck the San Francisco Bay Area. Just as the 1906 San Francisco earthquake and fire served as an impetus for the development of the original Hetch Hetchy system, the magnitude 6.9 Loma Prieta earthquake exposed the vulnerabilities of the existing water system to the damage and disruptive effects of earthquakes. Water mains ruptured in the Marina District of San Francisco, hindering the ability to fight the fires that occurred after the earthquake. This event provided the incentive to increase the reliability of the system through a new improvement program.

Figure 6.41 | O'Shaughnessy Dam Under Construction at Night

THE HISTORY OF TUNNELING IN THE UNITED STATES

Figure 6.42 | **Steel Liner Insertion into New Crystal Springs Bypass Tunnel**

San Francisco planners began a vulnerability assessment of the water system assets to earthquakes, landslides, fires, floods, and power outages. The outcome of this assessment led the implementation of a new Water System Improvement Program to enhance seismic, delivery, water quality, and water supply reliability. With a primary focus on seismic and delivery reliability, among the improvement projects were three new tunnels constructed and added to the water system between 2009 and 2015: the New Crystal Springs Bypass Tunnel, the Bay Tunnel, and the New Irvington Tunnel (SFPUC 2015).

New Crystal Springs Bypass Tunnel. This 4,200-foot-long tunnel provides a seismic-resistant connection between the existing Crystal Springs Bypass Pipeline, the Crystal Springs Pipeline No. 2, and the Sunset Reservoir Supply Line near Crystal Springs Reservoir on the San Francisco Peninsula. It has an 8-foot-diameter welded steel pipe liner finished with cement mortar for corrosion protection (Figure 6.42). Vertical shafts are located at both ends of the tunnel for access by the 12-foot-diameter TBM used for construction (SFPUC 2015).

Bay Tunnel. The 5-mile-long Bay Tunnel is the first tunnel excavated under San Francisco Bay from Newark to Palo Alto. Tunnel construction was completed by an earth pressure balance TBM about 75 feet below the bay's water level. The tunnel is lined with precast concrete segments that serve as the initial support system (Figure 6.43). A 9-foot-diameter welded steel pipe was inserted into the tunnel as the final liner with an interior cement mortar finish to protect the liner from corrosion. The annular space between the concrete segments and the steel pipe was backfilled with pump-injected cellular concrete. The tunnel connects the pipe manifold to the Bay Division Pipelines Nos. 1 and 2 and a new Pipeline No. 5 to the east, upstream. The entire 5 miles of tunnel was completed without intermediate shafts starting from the Newark Portal (SFPUC 2015).

New Irvington Tunnel. The 3.5-mile-long New Irvington Tunnel is parallel to the old Irvington Tunnel and constructed within the same underground right-of-way. Similar to the old tunnel, the new tunnel is located about 1 mile east of the Hayward Fault and about one-half mile west of the Calaveras Fault. Also similar to the old tunnel, the New Irvington Tunnel was constructed by conventional excavations methods, including with roadheaders (Figure 6.44), and drill-and-blast for the Briones sandstone and hard rock that was initially supported by wood lagging. Unlike the old tunnel, this 12-foot by 14-foot horseshoe-shaped tunnel used steel beam sets instead of heavy timbers for the initial support system, and had an 8.5-foot-diameter steel pipe that was finished with a smooth cement mortar lining.

Figure 6.43 | **Bay Tunnel Supply Train Transporting Precast Concrete Liner Sections to the TBM**

As would be expected given the close proximity to the old tunnel, the New Irvington Tunnel encountered similar ground condition challenges, gassy ground conditions due to the presence of flammable methane and noxious hydrogen sulfide gases, groundwater inflows, and faults zones with squeezing ground. Pre-excavation grouting was required throughout the tunnel and required the injection of 7.8 million pounds of cement. Squeezing ground required re-mining and further supporting of some sections of tunnel before the final steel pipe liner could be installed.

Figure 6.44 | New Irvington Tunnel Construction by Roadheader

CHAPTER SIX | WATER TUNNELS

Southern California's Water Supply Tunnels

In Southern California there are three major imported water conveyance systems: the Los Angeles Aqueduct owned and operated by the Los Angeles Department of Water and Power bringing in water from the Owens Valley, the Colorado River Aqueduct owned and operated by the Metropolitan Water District of Southern California bringing in water from the Colorado River, and the California Aqueduct owned and operated by the Department of Water Resources and bringing in water from Northern California (Figure 6.45). All three water systems cross the San Andreas Fault and other significant faults as they bring water to Southern California, creating challenging construction conditions at the fault crossings. Each of these systems is highly dependent on tunnels where each tunnel has a unique story to share. These are stories of technological innovation, construction challenges, and heroic efforts that helped to shape Southern California's underground water infrastructure and the growth of its metropolitan areas.

Most of the tunnels in California have been built for water supply to accommodate the needs of more than 35 million people. These tunnels are a vital addition to California's water infrastructure in that they are needed to help the expansion of agriculture and in meeting the demands of a growing population. California's major water problem is that more than 70% of precipitation falls in the northern half of the state, but almost 70% of the state's population lives in the southern half, an area that is mostly semiarid desert.

Tunnels have influenced the flow of water in California for more than 100 years. Their construction history gives a glimpse into the engineering dedication required to build these human-made marvels.

Los Angeles Aqueduct

When completed in 1913, the Los Angeles (LA) Aqueduct was considered a great engineering achievement, second only to the Panama Canal.

Figure 6.45 | California Water and Aqueduct Systems

Figure 6.46 | **Elizabeth Tunnel Muck Removal During Early Excavation**

The LA Aqueduct was designed and built by the Los Angeles Water Department and it was originally named the Bureau of Los Angeles Aqueduct. Under the supervision of Chief Engineer William Mulholland, the aqueduct was built between 1905 and 1913 at a cost of $23 million. The aqueduct tapped into the waters of the Owens River and delivered water 233 miles south to Los Angeles (State of California 1974a). With the rapidly decreasing supply of water from the Los Angeles River and groundwater sources around 1900, the LA Aqueduct was the saving grace to enable the city to grow. The most difficult part of the construction was tunneling. There were 142 tunnels initially totaling 43 miles in length. Two of the more important tunnels of the LA Aqueduct are the Elizabeth and Mono Craters Tunnels.

A century later, it continues to be a marvel in modern engineering (Water and Power Associates 2015). The aqueduct is now 338 miles long, including 164 tunnels totaling 74 miles (Proctor 1998).

Elizabeth Tunnel. The LA Aqueduct draws water from the Owens Valley, which is conveyed through a series of canals, pipelines, tunnels, and siphons, and travels under the San Gabriel Mountains by way of the Elizabeth Tunnel to deliver water to Los Angeles. The Elizabeth Tunnel was the longest tunnel on the project with a length of more than 5 miles. The workers had to drive this pressure tunnel through relentless and unpredictable granitic rock and through the San Andreas Fault (Nadeau 1950).

Excavation of the 10-foot by 12-foot tunnel began by hand drilling at the south portal on October 5, 1907, and at the north portal on November 1, 1907 (Figure 6.46). Inspired by friendly competition, the workers at both portals competed at conquering the impenetrable and difficult ground conditions. The construction slowly continued by hand until adequate machinery could be installed (Figure 6.47). Consequently, new equipment was placed at each portal where it facilitated a drastic increase in progress. Such equipment included air compressors to power the pneumatic drills and the first Caterpillar tractor ever built.

With production incentives, the Elizabeth Tunnel set the world record for hard-rock tunneling (Figure 6.48) with 604 feet excavated in single month (Standiford 2015). With tunneling underway up and down the line, the continuous assault from the workers' battering of the granite and the temptation of early completion bonuses in the first 11 months of construction of the LA Aqueduct, 22 miles of tunnel were driven. Ultimately, the miners beat the deadline set by the Board of Engineers by completing all tunnel excavations 20 months ahead of schedule (Jacques 1939; Standiford 2015).

Just 1,117 feet from the north portal, the Elizabeth Tunnel penetrated the San Andreas Fault. At that location, the workers broke into a large fissure filled with sand and water. The fissure emptied its contents of sand and water into the tunnel, causing massive flooding and cave-ins. The workers ran for their lives toward the portal entrance. When excavation resumed, the miners were unable to excavate through the saturated muck. Consequently, tunnel construction was stopped for a month and a half while an auxiliary 3,000-foot-deep shaft was constructed beyond the fault. From the new shaft, a new heading was launched to approach the dangerous ground more guardedly from the south. The problems with flooding and cave-ins persisted as the workers mined through the San Andreas Fault gouge zone, forcing them to run for their lives as the tunnel again filled with water. The danger of flooding from the rising groundwater and the cave-ins from loose sand were finally overcome by driving overlapping steel rails in advance of the heading and closely following with careful excavation and timbering for ground support.

The concrete lining of the tunnel was started in May 1911. The lining began at the center of the tunnel and proceeded on two headings to place the lining toward both the north and south portals at the same time. Rock crusher and mixing plants erected at both portals speeded up the process of lining the tunnel. In some sections of the tunnel, steel I-beams were concreted into position to replace timber lagging and to provide better support. The tunnel was completed in 1912.

Mono Craters Tunnel. The newer Mono Craters Tunnel experienced many of the typical challenges involved in tunneling during the 1940s. This tunnel was constructed in the second phase of the LA Aqueduct, taking the water that feeds Mono Lake and delivering it to the portion of the LA Aqueduct completed during the first phase of the project in the early 1900s.

Figure 6.47 | **Elizabeth Tunnel Drill-and-Blast Crew**

Figure 6.48 | **Elizabeth Tunnel Portal Crew After Setting Hard-Rock Footage Record in Early 1900s**

Figure 6.49 | Mono Craters Tunnel Survey of Alignment

The Mono Craters Tunnel, 59,812 feet long, was constructed through the mesa separating the Mono Basin and the Owens River watershed (Figure 6.49). The landscape had been shaped by ground moraine deposited by mountain glaciers and volcanic activity that resulted in a large amount of volcanic ash of varying thicknesses and volcanic cones. These geologic conditions created a hazardous, dangerous, and life-threatening work environment for the tunnelers.

Various shafts were needed to construct the Mono Craters Tunnel and it was within one of these shafts where the first signs of trouble occurred. A problem developed in Shaft No. 1 (896 feet deep) when the shaft came into contact with groundwater. Beneath this depth, there was a shift from a stable material to a highly fragmented material. The high groundwater pressure surrounding the fragmented soil caused significant amounts of soil to boil up into the shaft at alarming rates. Despite the large amounts of water being pumped out of the shaft, this problem was exacerbated the deeper they excavated. This unpredictable danger lurked for the entire drilling depth of the shaft (Dunn 1940). After Shafts No. 1 and No. 2 were completed, the next problem occurred during the boring of the west portal through the volcanic cones. As the workers progressed, they ran into water and then, after some distance, ran into low levels of carbon dioxide gas. Assuming it was of no importance, the workers continued their progress further not knowing that the water and carbon dioxide levels were increasing. It was this complication that led to the construction of a third and final shaft to provide ventilation in an otherwise inhospitable environment for the miners.

To build this necessary addition to the aqueduct, a formidable barrier of young volcanic cones had to be tackled. The large amount of volcanic ash of varying depth made it nearly impossible to determine the underlying rock conditions before construction (Dunn 1940).

Large quantities of warm water (as high as 97°F) and carbon dioxide caused run-ins that filled the tunnel for hundreds of feet. The volume of water was so great that in some sections of the tunnel, the water broke through the lining and flooded the tunnels over a significant length (Dunn 1940).

Figure 6.50 | **Mono Craters Tunnel Initial Support System**

The construction of the Mono Craters Tunnel (Figure 6.50) showcased the difficulties and dangers involved in tunneling, highlighted the resolve of the workers involved in the process, and used an impressive collection of tools to combat the varying thicknesses of the volcanic cones (Gruen 1998; Figure 6.51).

Colorado River Aqueduct

A period of faster-than-expected population growth following the completion of the LA Aqueduct in 1913, as well as a lingering drought, spurred Los Angeles city officials to search for another supply of water by the early 1920s. William Mulholland and other engineers from the Los Angeles Department of Water and Power started looking at the feasibility of tapping the Colorado River for water. Realizing that one city could not afford to build an aqueduct of the magnitude required to move water from the Colorado River, the California Legislature established the Metropolitan Water District Act in 1928 with the primary purpose to construct and operate the 242-mile Colorado River Aqueduct (MWD 1978).

Before construction work began, seven years of grueling survey work was performed to determine the most economical route for the aqueduct that had the lowest combination of construction costs, operating expenses, and construction safety issues. Both Mulholland and the consultants hired by the Metropolitan Water District of Southern California (Metropolitan) recommended the Parker

Figure 6.51 | **Installation of Final Tunnel Support and Liner at the Mono Craters Tunnel**

Route, arguing that it required less tunneling than other routes while managing to avoid most of the major active faults that crossed the region (MWD 2005).

Although the 242-mile-long Parker Route did require less tunneling than other routes, it still required 29 tunnels—a total of 92 miles—to extend through the mountains by the time it was completed in 1941. Engineers designed the tunnels to carry the full capacity flow of 1,605 ft^3/s (cubic feet per second).

The Colorado River Aqueduct is key to the vitality of Southern California while population growth continues to exceed the water capabilities of the LA Aqueduct. By transporting the water through 98 canals, 53 conduits, 146 siphons, 29 tunnels, five pumping plants, and four reservoirs, the Colorado River Aqueduct provides the crucial link between the Colorado River and Southern California communities.

San Jacinto Tunnel. One unique tunnel on the Colorado River Aqueduct is the San Jacinto Tunnel (Thompson 1966). The San Jacinto Tunnel required the longest distance between access points of all aqueduct tunnels at 8.23 miles. Workers would have to tunnel through granitic rock and less fractured sections of older metamorphic rock to connect the two portals (Weymouth 1939).

Construction of the Colorado River Aqueduct began in May 1933 with the sinking of the Potrero Shaft. At the onset, workers began encountering massive internal water flows where the shaft crossed a fault zone, only 160 feet away from the shaft (Figure 6.52). While early investigations hinted toward the presence of faults within the tunnel alignment, the geologic investigations conducted before tunneling began did not encounter the large amounts of water hidden within the rock (Weymouth 1937). Although inconvenient, the water flows did not stop construction until workers encountered a fault in July 1934 at the west portal. Water and debris began rushing into the tunnel and rose to a height in the Potrero Shaft that forced workers to scramble up ladders to safety as the tunnel filled rapidly behind them with a deadly slurry of water and rock (Burkholder 1939).

Figure 6.52 | **San Jacinto Tunnel Placing Sandbags for Bulkhead Construction, 1935**

Figure 6.53 | **San Jacinto Tunnel Mucking Machine**

Two months later, when water was still being pumped out of the tunnel, a discharge pipe broke and water flooded the Potrero Shaft once again. Work on the tunnel did not resume until November 1934 when the water was finally pumped out (Figure 6.53). However, this was not the end of the water problems, as another fault was encountered while tunneling from the west portal. In a sudden onslaught, more than 15,000 gallons of water per minute began flowing into the tunnel. It flooded the Potrero Shaft, halting any progression as the water infiltrated and filled the tunnel once again (MWD 2005).

In February 1935, Metropolitan fired the contractor and hired its own workers to finish the tunnel because after 18 months of construction, only 2 miles of the San Jacinto Tunnel had been driven. Larger, electrically powered pumps were brought in by Metropolitan to control the water that continued to flow into the tunnel. The original equipment used by the contractor consisted of deep-well turbine pumps suspended in the shafts or in the tunnel (Figure 6.54). Metropolitan built sealed, water-tight chambers at the bottom of each shaft to house four-stage horizontal centrifugal pumps. A new horizontal adit, the Lawrence Adit, was built to increase construction speed along the 6-mile-long central section (Weymouth 1937).

After the difficulties from three floods inside the tunnel, a comprehensive investigation was put into action to map out any faults lying within the tunnel alignment. The results of the study concluded that more than 21 faults lay before the workers, with the buried main trace of the San Jacinto Fault lying just inside of the west portal. To combat future flooding, two smaller parallel tunnels (commonly referred to as pilot tunnels) were excavated ahead of the mile-long adit to identify potential water flows in an attempt to head off any surprise water assaults. After a short labor dispute between Metropolitan and the tunnel workers that caused the work to shut down in the fall of 1937, the tunnel was finally completed in November 1938 (MWD 2005).

Tunnel construction in the 1930s followed what had become standard practice from techniques developed concurrently at the Hoover (Boulder) Dam project. Initial construction on aqueduct tunnels used the top-heading-and-bench excavation method. As construction progressed, excavation equipment and techniques were improved to increase the pace of tunneling—a credit to the ingenuity of the team in charge of driving the San Jacinto Tunnel. Workers installed railroad tracks to carry large drill jumbos outfitted with fast automatic-feed drills that allowed full-face excavation of the tunnel (Figure 6.55). Explosives were packed into the holes created by the drills (Figure 6.56), and a staggered formation that controlled detonations was used so that the center then the top, sides, and bottom of the tunnel exploded sequentially. Large fans cleared the air in the tunnel and within 30 minutes of the blast, workers returned to the tunnel face to clear the excavated ground, called "muck" or rubble, in the vicinity of the blast. Larger and faster muck haulage cars, along with the development of the "California switch" to enable muck cars to pass each other within the tunnel on a moveable double-rail track, allowed tunnel construction to proceed at a faster pace (Arnold 1966; Figures 6.57 and 6.58).

Figure 6.54 | **San Jacinto Tunnel Discharge Water from Potrero East**

Figure 6.55 | **Drill Jumbo or Carriage**

Figure 6.56 | **Typical Drilling Machine**

A number of tunneling speed records were broken during construction of the aqueduct tunnels, despite the numerous complications that tunnelers faced in their endeavors. Initial planning for tunnel construction estimated that an average of 4 to 5 feet could be excavated per eight-hour shift (Weymouth 1937). However, the overall average was nearly 6 feet per shift, with several tunnel crews excavating higher than the average. The San Jacinto Tunnel stands as a pinnacle of advancement in tunneling techniques. The technical improvements developed during the driving of this tunnel shaped the future of tunnel construction techniques since the 1930s (MWD 1978).

State Water Project (California Aqueduct)

The California (CA) Aqueduct of the California State Water Project delivers water from Northern California to Southern California. Southern California is dependent on transported water from Northern California and the CA Aqueduct facilities to meet the needs of those living in the southern part of the state. The CA Aqueduct, a vital infrastructure link that ensures stable and flexible water delivery from north to south, was planned, designed, and constructed by the California Department of Water Resources to supply supplemental water to approximately 25 million people and 750,000 acres of irrigated farmland (State of California 1974a). The aqueduct was completed in 1972 with a total length of 444 miles, of which 24 miles were constructed as a series of tunnels. The tunnels connected to the CA Aqueduct presented in this section are the Tehachapi, San Fernando (Sylmar), San Bernardino, and Arrowhead, which are all owned and operated by Metropolitan.

Tehachapi Tunnels

The Tehachapi Tunnels are a part of the Tehachapi Division of the California State Water Project. The tunnels travel through the Tehachapi Mountains, weaving through the massive mountain range to deliver water from Northern California to Southern California (State of California 1974a). Three tunnels comprise the Tehachapi

Figure 6.57 | California Switch Allows Locomotive to Pass One Another, 1934

Figure 6.58 | San Jacinto Tunnel Operations at Bottom of Potrero Shaft

Tunnels, designated No. 1 (1.5 miles long), No. 2 (0.5 mile long), and No. 3 (1.1 miles long). The 23.5-foot finished-diameter tunnels were constructed between 1966 and 1969.

These tunnels passed through a formidable mountain range while simultaneously navigating a vast maze of six major faults. The construction workers on the project progressed through the mountain range at an average rate of nearly 26 feet per day for Tunnels No. 1 and No. 2. It was during the excavation of Tunnel No. 3 that the workers came face to face with their most difficult challenge. While tunneling, the workers broke into a fault block that happened to be directly in between the northern and southern branches of the Garlock Fault (the second largest fault in California). They had to mine through faulted and crushed rock while large amounts of groundwater flowed into the tunnel from the surrounding strata. This water made tunneling extremely difficult and slowed the progression to 12.9 feet per day. The tunnels were later lined with unreinforced concrete (Figure 6.59).

In addition to the interior problems of the mountain, the exterior proved to be just as troublesome. During construction, a 1,300-cubic-yard landslide occurred immediately west of Tunnel No. 3's portal. As the years progressed, different mitigation techniques were applied to maintain the stability of the landslide area, but these seemed to exacerbate the problem. A concrete slurry mix was applied to the landslide, but instead of solidifying the slope, it lubricated the rock instead. The rock proceeded to slide into the tunnel and distorted the steel ribs by as much as 5 inches (State of California 1974b).

In 1969, the Tehachapi Tunnels were completed, having successfully overcome some of the most challenging tunneling conditions ever attempted to date.

San Fernando (Sylmar) Tunnel

The West Branch of the CA Aqueduct delivers water to Castaic Lake, the terminus point of the State Water Project, then travels south in Metropolitan's Foothill Feeder. The San Fernando Tunnel, also called the Sylmar Tunnel, is a component of the Foothill Feeder. The Foothill Feeder delivers water to Metropolitan's water treatment plants and surrounding underground water basins.

The San Fernando Tunnel is 5.5 miles long, 18 feet in diameter, and crosses the Santa Susana thrust fault. Coincidently, this tunnel was under construction at the time of the 1971 Sylmar/San Fernando earthquake but showed no signs of internal damage. However, a post-earthquake survey showed that the south portal, near the San Fernando Fault segment, was nearly 7.5 feet higher than before the quake.

Figure 6.59 | **Tehachapi Tunnel Liner Installation**

The construction workers showed exceptional forward progress as they reached a tunneling rate of 277 feet per day during construction using a piece of machinery called the "digger shield," which was designed specifically for the San Fernando Tunnel. The digger shield was a 225-ton tube, 140 feet long and 21 feet in diameter. This machine facilitated rapid progress because it could be used as a temporary support system and a tunneling device. Once the machine propelled its way through 4 feet of rock, it would install unreinforced cast-in-place concrete supports that the workers would then attach together to complete the support system. So many factors promoted the accelerated excavation of the tunnel that safety was not a very high priority for the construction company. Metropolitan's promised bonuses to the construction company for early completion was a great motivator for the workers and contractor (State of California 1974a; Zavattero 1978).

The workers encountered various complications, such as unstable ground from a recent earthquake and excessive groundwater inflow, but the most tragic problem occurred during the excavation phase of the tunnel due to the presence of hazardous gases. Hazardous gases within the mountain were expected after the recent Sylmar/San Fernando earthquake (on the Santa Susana thrust fault). The gases seeped through cracks produced in bedrock by the earthquake and slowly infiltrated into the working area of the tunnel. On June 24, 1971, 175 feet below the sleeping town of Sylmar, California, an explosion occurred during tunnel excavation. Unfortunately, alerts about unsafe conditions had been ignored on several occasions, and as a result, the explosion killed 17 of the 18 men that were working in the tunnel during that shift. This tragic event did not heed the small explosion with a flash fire the night before that injured four men. Despite the seven safety orders brought up by the site safety engineer, construction continued on the tunnel. The tunnel was 77% complete when the explosion occurred. In its wake, it left a tragic tale that marred the once exceptional tunneling progress. The explosion resulted in a two-year delay and a $9,215,796 increase in cost (State of California 1974b; Zavattero 1978; Proctor 2002).

The tragedy of that June night would not be forgotten, and it revolutionized safety systems and tunneling procedures. The tragedy of the San Fernando Tunnel explosion generated the toughest mining and tunneling regulations to date. The Tom Carrell Memorial Tunnel and Mine Safety Act of 1972 was passed soon afterward, and a bill that increased the punishment of gross negligence from a misdemeanor to a felony was also passed (Zavattero 1978). Sacrificing the safety of the project for increased productivity and profits will haunt the tragic memory of the San Fernando Tunnel forever (Nadeau 1950).

San Bernardino Tunnel

The San Bernardino Tunnel is part of the East Branch of the CA Aqueduct and was constructed in the late 1960s. It is 4 miles long and has an excavated diameter of 16 feet (Figure 6.60). As a part of the California State Water Project, this tunnel is part of the transmission system that takes water from Silverwood Lake and delivers it into the Devil Canyon Power Plant (State of California 1974a).

Construction workers on this project had to blast more than 160,000 cubic yards of the massive hard-rock mountain range (State of California 1974b). To support the newly excavated tunnel face, the workers used steel ribs with spilings and invert struts as an initial means of support to hold the weight of the rock in place (Figure 6.61). This support system was later replaced with more permanent unreinforced cast-in-place concrete (State of California 1967). The average rates of advance from the north and south portals were 14.1 and 34.1 feet per day, respectively. In the last 3,400 feet of tunnel excavated from the south portal, the contractor averaged 39.9 feet per day. The overall rate of advancement for the two headings was 22.6 feet per day (MWD 1995).

The San Bernardino Mountains are composed of igneous and metamorphic rocks of the late Mesozoic Age with alluvium deposits of varying ages. To penetrate this solid barrier, workers would have to traverse a labyrinth of unknown faults (State of California 1974b). However, the rock itself would not be the greatest danger the workers would face during this project. The real danger plaguing the construction of this tunnel was groundwater. The San Bernardino Mountains have an abundance of groundwater stored in the mountain's fractured and jointed rock mass (USNCTT 1984).

To determine the extent of groundwater problems within the region, engineers in charge of the project performed early alignment studies (USNCTT 1984). With 27 undeveloped springs and 54 developed wells and springs, the workers knew that groundwater would be a concern; however, the tests did not prepare them for the massive and destructive onslaught of groundwater flows encountered during tunnel excavation (State of California 1974b; USBR 2015).

Figure 6.60 | **Location of the San Bernardino Tunnel**

Figure 6.61 | **San Bernardino Tunnel, Installation of Support**

Originally expected to cost $24,902,072, the project would come in at $3.5 million more than the contractor's bid (State of California 1974a, 1974b). The reason for the high cost was primarily due to the complications presented by the groundwater. Despite the mitigations to prevent groundwater impacts, the pilot holes installed ahead of tunnel progression for early warning of large water inflows and ground conditions were unable to properly account for the massive amounts of water as it seeped in through multiple fractures. The uncontrollable flooding interrupted the tunnel construction multiple times and completely halted the drive from the north portal. This resulted in the significant increase in cost. Once the north portal tunnel flooded, the workers shifted gears to drive from the south portal through the mountain by taking advantage of gravity to drain the tunnel as they mined forward. The variable inflows of groundwater at high pressures were such an influential and destructive complication to the project that it changed the design and construction of tunnels from that point onward (USNCTT 1984; USBR 2015).

To manage the groundwater inflows, which could reach a rate of up to 2,000 gallons per minute, a combination of cut-off and consolidation grouting was used to stop the groundwater within the rock (State of California 1974b). During tunnel advance from the north and south portals, the contractor performed cut-off grouting ahead of the face (pre-excavation grouting) when moderate to high water flows were encountered. In several locations, grouting was conducted both to reduce water inflows and to consolidate the rock (MWD 1995). A total of 34 grouting stations in the north and south portal tunnels were needed to combat the barrage of groundwater; and at times even this was not enough. In one section of tunnel, workers lost nearly two months of progress because of the tremendous volume of groundwater inflow into the tunnel such that they were forced to construct an 8-foot-thick concrete bulkhead with various pipes to divert the water around the main work site (USNCTT 1984).

In January and February 1969, a phenomenal 48 inches of rain fell within 35 days in the San Bernardino Mountains. When the first storm hit, the north portal heading of the San Bernardino Tunnel was being excavated. The protective levee around the portal was overtopped, and surface water started pouring into the tunnel. An alert equipment operator drove his skiploader into the fast-flooding tunnel and rescued the terrified crew who would have otherwise drowned (Proctor 1998).

Prior to its use in this tunnel, pre-excavation grouting was not commonly used as a tunneling technique for control of groundwater in advance of tunnel excavation. As a result of the San Bernardino Tunnel project, pre-excavation grouting has now become a popular mitigation technique used for controlling groundwater seepage in the construction of new tunnels.

Arrowhead Tunnels

The Arrowhead Tunnels, components of the Inland Feeder Project, consist of two tunnels, the Arrowhead West Tunnel (4.1 miles long) and the Arrowhead East Tunnel (5.8 miles long). These tunnels represent the final portion of a 44-mile-long water conveyance facility that brings up to 1,000 ft³/s of water into Southern California. The tunnels were excavated using 19-foot-diameter tunnel boring machines, or TBMs (Figure 6.62). Both tunnels have a finished inside diameter of 12 feet and extend from the Devil Canyon Power Plant to a portal in City Creek Canyon, north of the city of Highland. The tunnel alignments extend east-southeast beneath the foothills of the San Bernardino Mountains, paralleling the mountain face and passing beneath and across several canyons. Construction began in 1997 and the project was completed in 2009.

Figure 6.62 | **Arrowhead Tunnel Shielded TBMs for a One-Pass System**

As the owner of the project, Metropolitan initiated construction of the Arrowhead Tunnels in spring 1997, starting with the Arrowhead East Tunnel. After mining about 8,000 feet from the City Creek Portal site, construction was halted because of elevated groundwater inflows (Figure 6.63). The U.S. Forest Service's special use permit restricted the volume of groundwater inflows during tunnel construction as a measure to protect the surrounding groundwater resources.

Metropolitan therefore redesigned the tunnels. The redesign specified construction measures intended to limit groundwater inflows. The redesign required two new TBMs; pre-excavation probing and grouting; a bolted, gasketed, precast concrete segmented initial lining; and an impervious welded steel or steel-reinforced concrete pipe final lining. The bolted and gasketed initial lining was designed to limit inflows in the advancing tunnels, significantly reducing sustained groundwater discharges from the portals (Figure 6.64).

The Arrowhead Tunnels lie near the base of the San Bernardino Mountains and the tunnel alignments are parallel to and less than 0.6 miles from the San Andreas Fault. Although the San Andreas does not intersect the tunnel alignment, there are numerous significant splay faults that cross the tunnel alignment. With the tunnels at considerable depth below the groundwater table, high hydrostatic pressures, as high as 300 psi, were encountered during excavation. Because of the range and severity of ground and groundwater conditions actually encountered, pre-excavation grouting was essential to successfully complete the mining of the tunnel (Fulcher et al. 2007).

Among the risks of working in the dry canyons of Southern California were fire and flood (and landslides). Seasonal dry periods elevated the risk of fire to the point where frequent and unexpected forest fires delayed the job. In 2003, for example, a massive flood and landslide in a nearby canyon inundated the Arrowhead West Tunnel Portal, resulting in a three-month delay to the completion of this tunnel alone.

WATER SUPPLY TUNNELS IN SOUTHERN NEVADA

The earliest inhabitants of the Las Vegas Valley in Southern Nevada relied on a few small natural springs for all of their water supply. This supply was adequate for the small native population and a few settlers to survive with their limited crops and animals. After the town of Las Vegas was founded by the railroad in 1905, the water supply began to be augmented by drilled wells, some of which were artesian and some were pumped. The initial tiny population

Figure 6.63 | **Water Inflow in the Arrowhead East Tunnel**

Figure 6.64 | **Arrowhead Tunnel After Construction of the Segmental Lining**

of Las Vegas grew steadily during the early part of the twentieth century and continued to meet its water needs with groundwater. As more water was pumped from wells, the spring flow declined and the water table fell. By 1940, with a valley population of about 8,000 people, water use had increased to about 20,000 acre-feet per year or an average of approximately about 18 MGD (million gallons per day). The groundwater aquifer was capable of sustaining this amount of water use, but the community was still growing and straining the capacity of the water system. It was generally understood that additional water supplies and facilities would eventually be required if growth were to continue.

To support a war-time magnesium processing facility in the Las Vegas valley, the United States government installed pumps on a structure built over Lake Mead behind Hoover Dam, along with 16 miles of 40-inch-diameter pipeline and a booster pumping station, to convey water from the lake to the industrial complex and its supporting community. With a system capacity of over 30 MGD, this was the first delivery of Colorado River water to the Las Vegas valley. The success of this water delivery system prompted support for construction of other Lake Mead water conveyance facilities to meet the future water needs of Southern Nevada.

Between 1968 and 1971, the State of Nevada and the federal government together undertook design and construction of facilities that would treat and deliver up to 200 MGD of water to the communities of Boulder City and the Las Vegas valley. With this assurance of a reliable water supply for the long term, development in the region and the population of southern Nevada continued to grow at a rapid rate. To meet the steady increase in water demand, the water system was expanded in phases during subsequent decades until the regional water supply system, now owned and operated by the Southern Nevada Water Authority (SNWA), reached a capacity of 900 MGD.

Eventually, 2 million residents of the Las Vegas Valley came to rely on the Colorado River for 90% of their drinking water supply, drawn from Lake Mead through two intakes, Lake Mead Intakes No. 1 and No. 2, and transmitted to two water treatment facilities for treatment and delivery to the community. As early as 2003, declining lake levels resulting from an ongoing, severe drought in the Colorado River basin raised concerns about the future operability of one of the existing intakes. By May 2005, the SNWA approved a project for design and construction of a third intake in Lake Mead with the primary objective of protecting Southern Nevada's water supply from significant loss of system capacity resulting from a continued decline in lake levels. The third intake was commissioned in September 2015 and, combined with a new Low Lake Level Pumping Station currently in construction, is deep enough to function at the lowest expected lake levels and draw high-quality water.

Lake Mead Intake No. 1

Lake Mead, impounded by Hoover Dam when full, is the largest reservoir in the United States in terms of maximum volume and is a critical part of the storage system on the Colorado River that provides physical and economic sustenance to nearly 30 million people. However, Lake Mead water levels rise and fall in response to cycles of drought, flood, and water demand. The impact of the current drought cycle is demonstrated in the photo of the water intake facility built in the 1940s on the south end of Saddle Island and still in operation today (Figure 6.65).

Figure 6.65 | **Lake Mead Intake Facility from 1940s**

In the 1960s, the U.S. Bureau of Reclamation undertook design of the next Lake Mead intake, consisting of a submerged tunnel system that would convey water to a deep well pumping station and water treatment facility.

This intake system, referred to as Intake No. 1 and Intake Pumping Station No. 1 has several component parts. The underground portion of the pumping station includes a rectangular 10-foot by 22-foot access shaft and an array of 20 vertical pump well shafts that extend down 220 feet to an underground forebay 22 feet wide by 12 feet high and 150 feet long. The intake is a 1,400-foot-long, 13-foot by 13-foot tunnel excavated from the pumping station forebay chamber to an intake shaft east of Saddle Island. The intake shaft was drilled into the intake tunnel from a barge anchored in Lake Mead. An inclined adit was excavated between the forebay and a portal located southwest of the pumping station to facilitate construction access to the underground work, but was plugged after construction was complete. All excavation, except for the drilled well shafts and intake shaft, was accomplished by conventional drill-and-blast methods. As part of the third intake connection work, the Intake No. 1 tunnel was dewatered and inspected in January 2014. After being in service for 43 years, the tunnel and shotcrete lining were found to be in good condition (Figure 6.66).

Figure 6.66 | **Intake Tunnel No. 1, First Inspection Since Going into Service in 1970**

Lake Mead Intake No. 2

In the early 1990s, SNWA was formed to manage the regional water supply, treatment, and distribution to the Las Vegas metropolitan area. SNWA undertook construction of Intake No. 2 and Intake Pumping Station No. 2 as part of a major system expansion. Its design followed the model of Intake No. 1 and was constructed about 400 feet south of Intake No. 1; however, Intake No. 2 was designed to take water from 50 feet deeper in Lake Mead (Figure 6.67). Construction work on Intake No. 2 started in 1998 and was completed in 2000.

Like Intake No. 1, the access shaft, forebay, and tunnels for Intake No. 2 were constructed by conventional drill-and-blast excavation methods and supported by rock bolts and fiber-reinforced shotcrete (Figure 6.68). For the access shaft, the excavation was lined with cast-in-place concrete. In 2012, again as part of the third intake connection work, the Intake No. 2 was dewatered, inspected, and found to be in excellent condition.

Intake No. 2 was constructed in Lake Mead by the installation of secant piles to bedrock and drilling holes through a lake bottom template using a barge-mounted drill. After excavating a 13-foot-diameter, 60-foot-deep excavation,

Figure 6.67 | **Profile of Lake Mead Intake Tunnels No. 1 and No. 2**

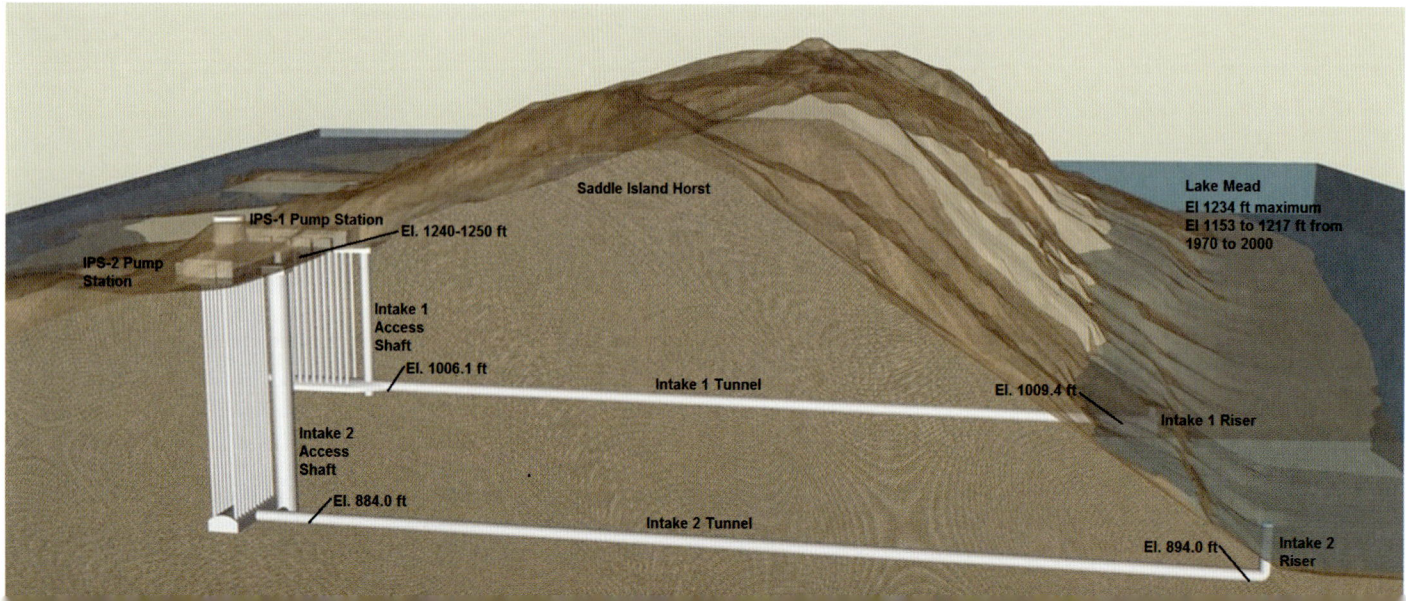

a 12-foot-diameter steel riser casing with bulkhead plates was placed and the annulus grouted with tremie concrete. When the intake tunnel reached the intake riser, excavation was turned upward to complete the connection with the intake riser (Figure 6.69).

Lake Mead Intake Tunnel No. 3

The persistent drought on the Colorado River basin that begin in the year 2000 caused a precipitous decline in Lake Mead water levels that prompted concern about the potential inability to continue operation of the more shallow Intake No. 1. This concern led to the decision to construct a third intake much deeper than the two previous ones. The location for Intake No. 3 was selected based on studies of the lake-bottom depth contours and water quality measurements and modeling under potential future conditions (Figure 6.70). Three alignments for an intake tunnel to that location were considered and the final alignment selected was more curved, longer, and the more northern alignment through the Tertiary age sedimentary rock. This alignment was selected to avoid less desirable rock quality associated with the more direct, shorter tunnel alignments.

Three construction contracts were awarded for Lake Mead Intake Tunnel No. 3. An approximately $500 million design–build contract was for construction of a 30-foot-diameter, approximately 600-foot-deep access shaft; 20-foot-diameter, 14,577-foot-long deep-level tunnel; and a lake-bottom intake structure drawing water from the deeper reaches of Lake Mead. The deep tunnel under the lake was excavated by a specially designed tunnel boring machine, or TBM. Additionally, two conventional design–bid–build contracts were awarded for a 22-foot-diameter, 380-foot-deep gate shaft (Figure 6.71) and 403-foot-long connecting tunnel to Intake Pump Station 2, as well as a 2,720-foot-long, 16-foot by 16-foot modified horseshoe tunnel to connect to both the gate shaft and Intake Tunnel No. 3. All the underground excavations for the design–bid–build contracts were

Figure 6.68 | **Installation of Grouted Mechanical Rock Bolts and Steel Straps, 1999**

Figure 6.69 | **Intake No. 2 Riser During Installation**

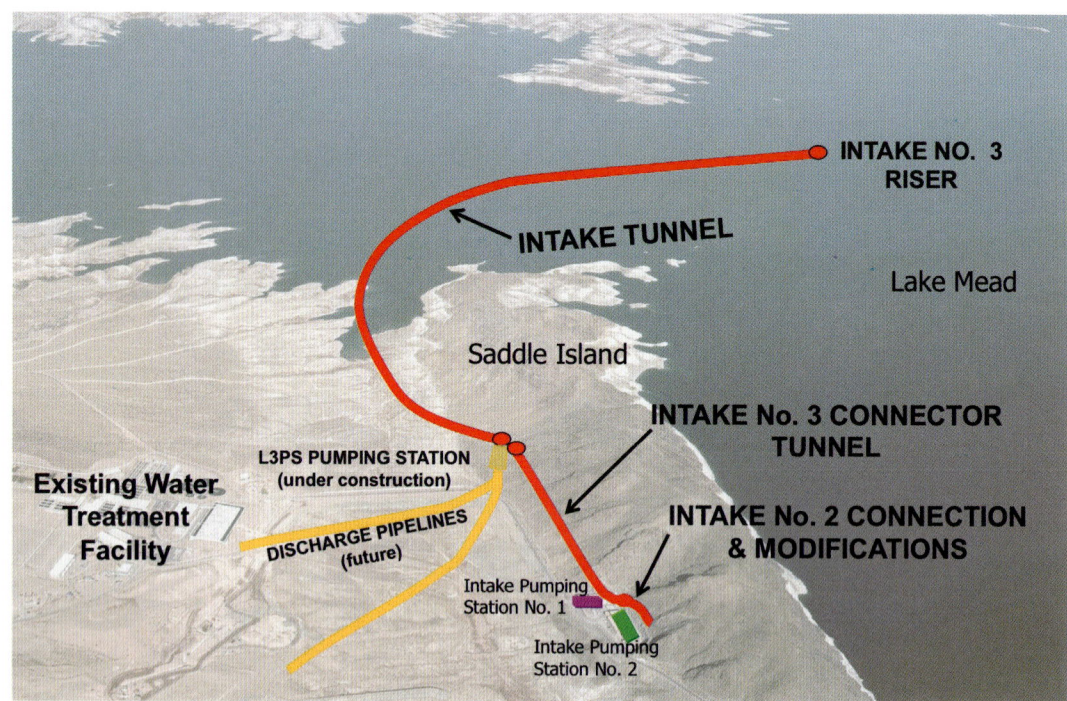

Figure 6.70 | **Aerial View of Lake Mead Intakes and Tunnels**

Figure 6.71 | **Stainless-Steel Isolation Gate Structure Between Intake Tunnels No. 2 and No. 3**

completed by conventional drill-and-blast methods and were supported with permanent rock bolts and fiber-reinforced shotcrete methods that avoided the much higher cost of a concrete lining.

Herrenknecht was selected to design and fabricate the 23.5-foot-diameter mixshield TBM for excavation of Intake Tunnel No. 3 based on their knowledge gained on a variety of worldwide projects with high groundwater pressures and variable ground conditions. After getting through the difficulties of completing the starter tunnel, the TBM was launched in December 2011 and the tunnel excavation was completed in December 2014. The dual-mode TBM, which was designed to operate under hydrostatic pressures of up to 250 psi, was operated at pressures as high as 200 psi (closed mode) for long periods of time. Approximately 50% of the tunnel was bored in closed mode, 4,300 feet of which was excavated at chamber pressures of 175 psi or more, a world record. The tunnel was supported by precast bolted and gasketed concrete segments installed behind the TBM as the machine advanced forward.

Underwater excavation of the lake bottom to 70 feet deep at the intake location in nearly 300 feet of water was required to provide the space and foundation for the intake structure on the lake bed (Figure 6.72). Excavation methods included clamshell for taking away the soil and blasting for removing the basalt rock. A 97-foot-tall, 28-foot-wide, and about 63-foot-long composite reinforced concrete structure with a 14-foot inside-diameter stainless-steel riser with temporary bulkhead was constructed at a shoreline barge, floated to the intake location where it was lowered to the lake bed, and encased in more than 11,300 cubic yards of concrete (see chapter opener photo and Figure 6.73). The TBM docked with the intake structure by excavating through one of the concrete walls, thereby completing one of the most challenging tunnel projects in recent times (Figure 6.74).

Figure 6.72 | **Intake Structure**

Figure 6.73 | Barge Operation for Installation and Placement of Concrete to Encase the Intake Structure and (inset) Concrete Trucks Barged to Intake Structure

CHAPTER SIX | WATER TUNNELS

IRRIGATION TUNNELS

In the western United States, the increasing demand for agricultural products for a growing population suffering from water shortages continually challenges irrigators. The U.S. Reclamation Service—established in 1902 and renamed the Bureau of Reclamation in 1923—is responsible for much of the development of water conveyance, storage, and power structures in the 17 western states. This includes the western slope of the Continental Divide, in the Colorado River basin, and in Utah. In addition, the Bureau of Reclamation was responsible for development of the irrigation systems. In the 1900s, five irrigation tunnels were constructed. Three major irrigation tunnels changed the West: the Gunnison, Strawberry, and Alva B. Adams Tunnels.

Gunnison Tunnel

The earliest of the irrigation tunnels built by the U.S. Reclamation Service was the Gunnison Tunnel, constructed between 1904 and 1909 as part of the Uncompahgre Project in western Colorado. The Gunnison Tunnel diverts water from the Gunnison River and conveys it via the tunnel to the semiarid Uncompahgre River basin to the west. Tunnel construction was initiated by the State of Colorado, but after running out of funds, the project was taken over by the new U.S. Reclamation Service. Construction of the tunnel resumed in 1905 and two significant advances in tunneling technology were used to excavate the tunnel—jackhammers powered by compressed air replaced hand-turned drill bits and single- or double-jackhammering, and dynamite replaced black powder for blasting. These advances made the work safer, easier, and reduced the number of injuries during the project. The tunnel had an excavated cross section of 11 feet wide by 12 feet high with an arch roof (Figure 6.75). Many who mined the tunnel were Appalachian coal miners who had traveled west for work. During tunnel excavation, two drilling records were set for the number of feet driven during a month.

Figure 6.74 | **TBM Hole-Through to Intake Structure**

The record for driving through granite was established at 449 feet, and the record for driving through shale was 824 feet, both accomplished in a one-month period. However excavation of the tunnel was not without its challenges: seeping groundwater, poisonous gases, excessive temperatures, and the presence of clay, sand, shale, and a fractured fault zone led to the suspension of tunneling for six months so a 400-foot-deep ventilation shaft could be driven down to the tunnel level. After excavation was complete, the tunnel was lined with concrete.

The diverted water for the Gunnison Tunnel had a capacity of 1,000 ft^3/s and was used to initially irrigate about 146,000 acres of farmland around Montrose, Olathe, and Delta, Colorado. The water supply provided by the Gunnison Tunnel has allowed for agricultural development and population growth of the Uncompahgre Valley.

When it was completed in 1909, the 5.8-mile-long Gunnison Tunnel was the longest irrigation tunnel in the world and was constructed at a cost of $2,905,307. Figure 6.76 shows the support system consisting of steel ribs and lagging, and the crews scaling loose rock. The Gunnison Tunnel was designated by the American Society of Civil Engineers as a National Historic Civil Engineering Landmark in 1972.

Strawberry and Alva B. Adams Tunnels

Two additional historically significant irrigation tunnels in the western United States include the Strawberry Tunnel in Utah and the Alva B. Adams Tunnel in Colorado. These tunnels are worth mentioning because of their contributions toward solving major water supply needs for communities in remote locations. Both tunnels diverted water from one basin to irrigate lands in valleys or basins where water was less abundant.

Strawberry Tunnel, the Bureau of Reclamation's second oldest tunnel at 19,091 feet long, was constructed from Strawberry Reservoir beneath the Wasatch

Figure 6.75 | **Gunnison Tunnel Portal, Congressional Party Site Visit**

Figure 6.76 | **Gunnison Tunnel**

Figure 6.77 | **Water Inflow Encountered During Excavation of the Strawberry Tunnel**

Figure 6.78 | Miners at Strawberry Tunnel Portal, Circa 1907

Mountains west to the Diamond Fork of the Spanish Fork River, through the Wasatch Divide as part of the Strawberry Valley Project. The tunnel construction was started in 1906 and completed in 1912. Strawberry Dam and Indian Creek Dike, which impounded Strawberry Reservoir (capacity of 283,000 acre-feet), were constructed between 1908 and 1913. During the excavation of the tunnel, which was excavated on two headings from each portal, significant water inflow was encountered that slowed construction (Figure 6.77). The water-bearing sandstone became a major problem for work crews trying to advance the tunnel. On several occasions in the tunnel heading driven westward and downgradient from the east portal, the pumps that moved the water ceased operation, causing the tunnel to flood temporarily. Eventually, crews overcame the water inflow, and when the headings met, the tunnel centerlines were within inches. Strawberry Tunnel is a 7-foot-wide and 9-foot-high concrete-lined tunnel that has a capacity of 600 ft^3/s (Figure 6.78). The diverted water irrigated about 45,000 acres of land located southeast of Utah Lake annually.

The Alva B. Adams Tunnel is the principal component of the largest transmountain water project (the Colorado–Big Thompson Project) that transfers water from the western slope in the Colorado River basin to users on the Front Range in (eastern) Colorado. The tunnel is 13.1 miles long and is concrete lined with an internal diameter of 9.75 feet. Within the tunnel excavation and other associated underground structures for the project (for example, the gate shaft), heavy steel supports were required in localized areas (Figure 6.79).

The tunnel drops 109 feet in elevation along its length and runs in a straight line under the Continental Divide from west to east, passing under Rocky Mountain National Park. At its deepest point, the tunnel is about 3,800 feet below the surface of the mountain peaks. Construction began in June 1940, but was suspended for a period during World War II and was completed in March 1947. The tunnel was named for its chief advocate, Colorado's U.S. Senator Alva B. Adams.

Tunnels are an essential component in successfully transmitting water for hydroelectric power from remote locations to large metropolitan areas. This is the case in Colorado, from the "High Country" down to the Front Range. Waters from the western slope of the Colorado River basin are collected from Windy Gap Reservoir, Willow Creek Reservoir, and other drainages from the North Fork of the Colorado River and pumped to Lake Granby, just 52 miles from Denver. From Lake Granby, the water is pumped into Shadow Mountain Reservoir and then flows by gravity to Grand Lake, Colorado's largest natural body of water, from where it flows through the Alva B. Adams Tunnel. Lake Granby is the second largest reservoir in the state of Colorado with a nominal capacity of 470,000 acre-feet. Once the water reaches the Front Range on the eastern side of the Rocky Mountains, it is used to generate electricity as it falls almost half a mile through five power plants on its way down the Front Range where it serves 4.5 million people with reliable power.

To transport power from the Front Range reservoirs and power plants back to the pumps on the western slope, a 5-inch-diameter, watertight, nitrogen-filled conduit carrying a 69,000-volt electric power transmission line is mounted on the roof of the Alva B. Adams Tunnel. This transmission line is used to power the pumps on the western side of the tunnel and surplus power is sold. West slope water is stored on the Front Range in Mary's Lake, Lake Estes, Carter Lake (all reservoirs), Flatiron Reservoir, Horsetooth Reservoir, and Boulder Reservoir for power generation or release after the peak snowmelt months of May to July have passed.

The Alva B. Adams Tunnel is capable of transporting 550 ft^3/s of water. An average annual total of 213,000 acre-feet of water is diverted through the tunnel from the west to the east slope per year. The tunnel can flow as much as 1,100 acre-feet in one day. The water transferred through the tunnel is considered part of the Colorado River water allocation of 3,880,000 acre-feet per year as agreed to in the Colorado River Compact. This 1922 agreement among seven states in the basin of the Colorado River in the American Southwest governs the allocation of the water rights to the river's water among the parties of the interstate compact.

CHAPTER SIX | WATER TUNNELS | 343

Figure 6.79 | **Heavy Support Being Installed in the Alva B. Adams Tunnel Gate Chamber, 1939**

TUNNELS AND UNDERGROUND SPACE ASSOCIATED WITH DAMS AND HYDROPOWER

Tunnels, shafts, and underground spaces are often associated with dams and hydropower projects. Some serve multi-functional purposes, such as power generation and water supply (Hoover Dam); power generation and irrigation (Strawberry Valley Project); irrigation and water supply (Colorado–Big Thompson Project); and flood control, power generation, and navigation (Oahe Dam) as part of our nation's infrastructure. In addition, many of the reservoirs created are used for recreation.

Underground spaces for permanent facilities may include headrace and power tunnels, penstocks, outlet tunnels, and tailrace tunnels; intake, surge and gate shafts; and underground chambers for power stations and transformers. In addition to the water-related underground structures, there are often vast networks of access, drainage, and grouting tunnels or galleries. Tunnels and underground spaces are more often associated with pumped-storage projects and complex systems of dams, reservoirs, power stations, tunnels, and pipelines bringing water from mountainous areas to population centers.

Snoqualmie Falls Underground Powerhouse

The falls on the Snoqualmie River in the foothills of the Cascade Range, about 24 miles east of Seattle, is the location of a hydropower plant that features the world's first underground powerhouse station. The power station began as one man's vision. Charles Baker believed he could harness the energy of one of the Pacific Northwest's most powerful waterfalls. At 270 feet high, Snoqualmie Falls is nearly twice the height of Niagara Falls. The substantial precipitation that accumulates in the basin (90 inches per year), the solid bedrock, and the great height of the falls made this an ideal site for hydropower development. Baker also realized that the clouds of water mist created by the falls, which coated the landscape year-round and would freeze to ice in cold weather, would not be a suitable site for a surface powerhouse. Consequently, he concluded that the power station should be completely underground.

Figure 6.80 | **Snoqualmie Falls Longitudinal Profile Showing the Cavern and Shaft**

Construction began in 1897 with the excavation of a 270-foot-deep shaft on the south bank of the river above the falls using conventional drill-and-blast methods. Crews sank the 10-foot by 27-foot-wide vertical shaft in solid andesite using a steam-powered air compressor that fed 10 pneumatic drills. When completed, the shaft was used to convey flows through the penstock as well as for elevator access to the powerhouse cavern below. At the same time that the shaft was being excavated, a 12-foot-wide by 21-foot-high by 650-foot-long tunnel was driven upstream from the base of the falls to connect to the shaft (Figure 6.80). Before beginning the horizontal tunnel, crews had to build an air line from the compressor downstream and down the cliff while working in the wind-whipped spray and while being suspended in the air. Where the two excavations met, a large underground cavern, 30 feet high, 40 feet wide, and 200 feet long, was excavated using drill-and-blast methods (Figure 6.81). The powerhouse cavern was dry and unlined except for the concrete slab placed in the invert, and the sidewalls were covered with a white wash.

Figure 6.81 | **Snoqualmie Falls Underground Powerhouse Cavern**

Four generating units, each rated at 1.5 MW (megawatts) capacity, were installed and the project was completed in 1898. In 1905, a fifth unit, a 5-MW Francis turbine was installed, which increased the capacity to 11 MW (Figure 6.82). Later in 1910, a second intake tunnel with a penstock and a single-unit surface powerhouse were constructed on the north side of the river. In 1956–1957, the surface power station was expanded to include a second penstock and unit. The total capacity of Plant No. 2 was increased to 33.4 MW, bringing the combined capacity to 44.4 MW.

Figure 6.82 | **Snoqualmie Falls Underground Cavern with Four Generating Units**

Figure 6.83 | **Ward Tunnel, Crews Eating a Hot Meal on the Grub Train**

In 2013, several improvements and upgrades were made, which increased the total capacity of the plant to 54.4 MW.

Ward Tunnel, Big Creek Hydroelectric Project

The Big Creek hydroelectric system is comprised of a vast network of dams (27), lakes/reservoirs (6), connecting tunnels (8), and penstocks and powerhouses (9) that were constructed in four phases from 1911 to 1995. A number of tunnels divert water from surface reservoirs or lakes to power stations downstream. The Ward (Florence) Tunnel, a 13.5-mile-long rock tunnel, is a major component of the Big Creek system on the upper San Joaquin River in the Sierra Nevada of central California. The construction of the Big Creek system was begun by Pacific Light and Power Company in 1911 and all rights and properties were conveyed to Southern California Edison Company in 1917, who completed the construction and continues to operate the project today.

The excavation of the Ward Tunnel was begun at both ends, and because it was so deep underground, work was able to proceed year-round. This was especially important because of the short construction season on the surface at this high-altitude location. As the tunnel was advanced, a locomotive train was used to bring men and supplies in and out of the tunnel. Ultimately, as the distance to the portals became too long to return to the surface for mid-shift meals, a flatcar with tables and benches brought the hot lunch meal to the miners (Figure 6.83).

Figure 6.84 | **Two Portals Meet in the Middle at Hole-Through of Ward Tunnel, 1924**

The purpose of the Ward Tunnel is to convey water from the highest lakes in the system—Lake Edison as well as Florence Lake, both above elevation 7,350 feet—to Huntington Lake. The rock tunnel was excavated using drill-and-blast methods through massive granite from both ends (portals), and the excavations met at the middle of the tunnel (Figure 6.84). Tunnel construction started in November 1920 and was completed in April 1925. At the time of its completion, Ward Tunnel was the longest tunnel in the world. In 1954, Portal Powerhouse, which was not housed in a building, was constructed at the outlet end of Ward Tunnel, just upstream of Huntington Lake.

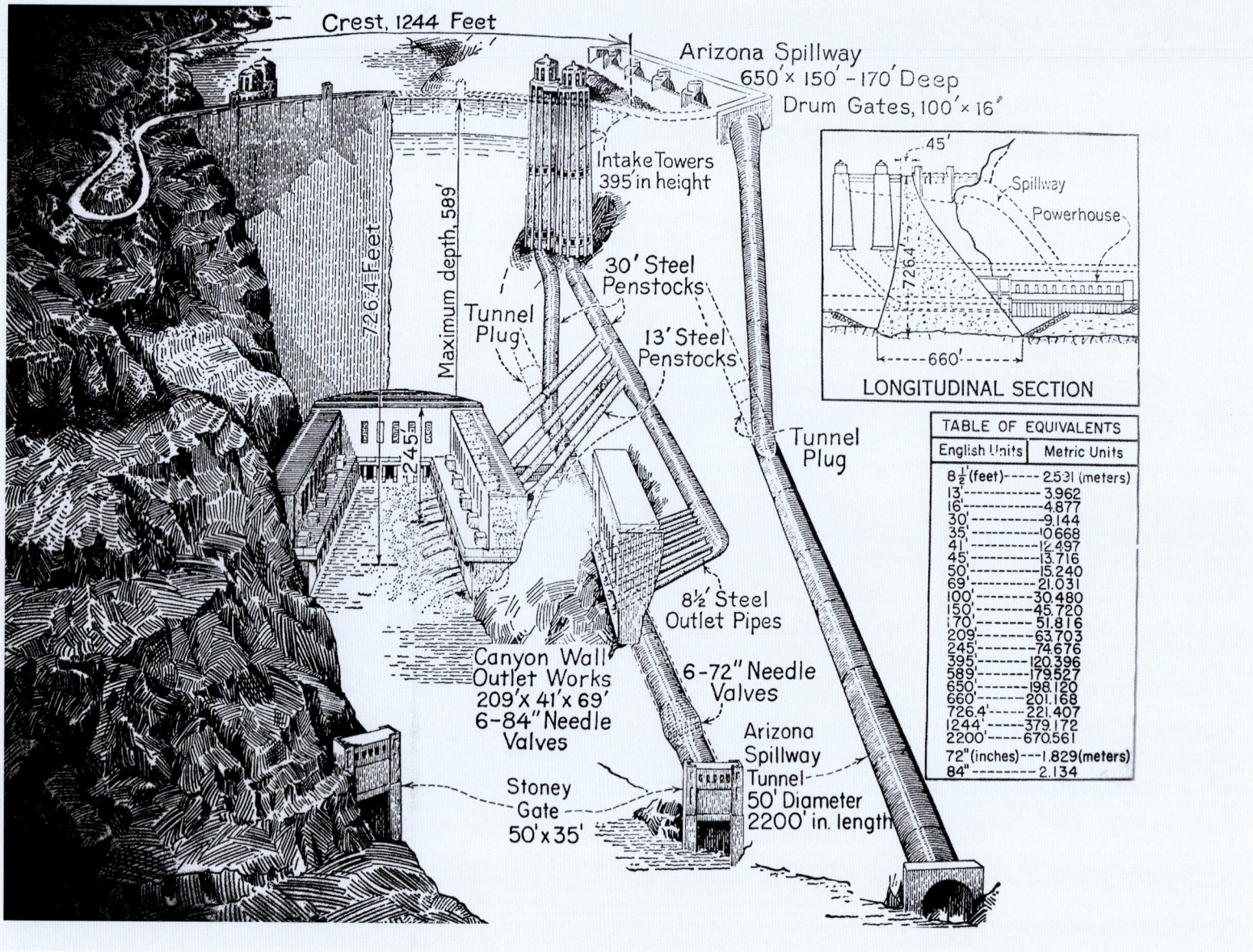

Figure 6.85 | **Boulder Dam Tunnel Schematic for Left Abutment**

Boulder Canyon Project (Hoover Dam) Diversion Tunnels

Boulder Dam (renamed Hoover Dam in 1947) is a concrete arch dam near Las Vegas that impounds Lake Mead. It was the tallest dam in the world when it was completed in 1935. At 726 feet high, it remains the tallest concrete dam in the United States and has the largest reservoir volume at full capacity.

One of the key elements in the formation of Hoover Dam was the diversion of the Colorado River to allow construction of the dam and powerhouse in dry conditions. Tunnels were required for river diversion, spillway discharge, and power generation (Figure 6.85).

Because of the wide range in seasonal flows that could be realized during construction, four 50-foot-diameter diversion tunnels were constructed, two on each side of the river, to handle river flows up to 200,000 ft^3/s.

The circular tunnels were excavated in solid rock to 56 feet in diameter and then were lined with a 3-foot-thick reinforced concrete lining. These were the largest diameter tunnels in the United States at the time, only recently surpassed by Seattle's larger diameter SR 99 transportation tunnel, a TBM tunnel, in 2014.

Because of the large diameter, the diversion tunnels were excavated by removing a pilot drift in the center of the crown, then the two sides, followed by enlargement to the full dimension by removal of the lower benches to the invert level. Drilling of the blastholes was accomplished with pneumatically powered, hand-operated jackhammers or "jack leg" drills. As a means to increase production, a full-face excavation was made possible by creating the first drill jumbo (Figure 6.86), which enabled four levels of drilling to be undertaken simultaneously using 24 to 30 drills. Because the tunnels had to be completed prior to spring floods, excavation of the four tunnels, approximately 16,000 feet (3 miles) in total length, was completed in less than eight months. The combined length of the diversion tunnels was nearly twice as long as that being completed by Seattle's "Bertha" for the SR 99 Project, a transportation tunnel.

Figure 6.86 | **Drill Jumbo for Diversion Tunnel Excavation at Hoover Dam**

The concrete lining of the diversion tunnels was accomplished in two stages. First, the tunnel invert was lined with concrete, and then the remainder of the tunnel, including the sidewalls and the crown, was lined with a traveling form—the largest ever used. The steel-frame carriage shown in Figure 6.87 supports the 110-degree top-arch form. An asphaltic coating was applied to the concrete immediately upon removal of the form to prevent rapid loss of the concrete's moisture content and concrete cracking (Figure 6.88).

Figure 6.87 | **Hoover Dam Diversion Tunnel Lining Arch Form and Carrier**

Figure 6.88 | **Finished Concrete Lining of Hoover Dam Spillway Tunnel**

Figure 6.89 | Oahe Dam, Robbins First Rock Tunneling Machine, the "Mittry Mole"

The four diversion tunnels also fulfilled multiple purposes. In addition to diverting the river during construction, the tunnels were subsequently converted to serve operational purposes. Each of the tunnels was plugged with concrete and two were modified (one on each side of the river) to encase a 30-foot-diameter steel penstock in each tunnel that would provide flows to the powerhouses to generate hydroelectric power. The remaining two diversion tunnels were used to connect the spillways to the river downstream to bypass flows during flood events.

Oahe Dam Project

Construction of Oahe Dam, specifically the tunnels, saw the birth of modern mechanized tunneling. Located in central South Dakota on the Missouri River, Oahe Dam was a roller-fill embankment dam, 242 feet high and 9,300 feet long. The multipurpose reservoir is used for flood control, irrigation, navigation, power generation, recreation, and wildlife preservation. Flood regulation and control is provided by six circular tunnels, each 19.75 feet in diameter and 3,408 to 3,571 feet long on the right abutment. Six intake tunnels, 24 feet in diameter, are located on the left abutment.

In 1952, James S. Robbins was approached by Mittry Construction Company to build a large jumbo-type structure with a coal cutter to cut out a smooth, circular tunnel profile prior to excavation of the rock within. After reviewing the project, Robbins recommended a more practical approach: using a full-face rotary boring machine. With the development of the Robbins 25.75-foot-diameter full-face tunneling machine, designated Model 910-101 and dubbed the "Mittry Mole," the era of modern mechanical tunnel excavation began. This first tunneling machine had counter-rotating inner and outer heads equipped with fixed tungsten carbide drag picks and parallel rows of free-rolling steel discs, the inception of the first disc cutters (Figure 6.89). The steel discs were set slightly behind the fixed cutters, based on the idea that the discs would fracture the ridges of rock left between the grooves cut by the picks.

The Mittry Mole was delivered in 1953 and began excavation of the power tunnel in the soft Pierre shale. The excavation of the rock tunnel achieved better-than-anticipated advance rates. Consequently, three additional tunneling machines were procured and used to complete the excavation of the six diversion/outlet tunnels and six power tunnels at the Oahe Dam site. Figure 6.90 shows the third TBM used during excavation, the Robbins Model 351-107 at 29.5 feet in diameter with a single rotating head. Advance rates for the TBMs were as high as 61 feet for an eight-hour shift and 161 feet in a day, both new world records. During excavation of four of the tunnels, 362 fixed drag pick cutters were consumed; however, only six disc cutters were replaced, and then only because of seal and bearing failure and not because of actual cutter wear (Foley 2009).

The success of the first two full-face tunneling machines at Oahe Dam lead to the use of smaller-diameter TBMs for sewer tunnels in Pittsburgh, Chicago, and Toronto in the 1950s and ultimately a major breakthrough for driving tunnels in harder rock.

Figure 6.90 | **Third Oahe Dam Robbins TBM**

Helms Pumped Storage Plant

Helms Pumped Storage Plant, owned by Pacific Gas and Electric Company, is located about 50 miles east of Fresno, California, in the Sierra Nevada. After being planned in the early 1970s, construction on the plant began in June 1977, with commercial operations beginning on June 30, 1984.

The pumped storage operates by moving water between an upper reservoir (Courtright) and a lower reservoir (Wishon), as shown in Figure 6.91. Connecting the reservoirs, in order from upper to lower, is first a 10,511-foot-long headrace tunnel, which turns into a 2,248-foot-long inclined shaft, which drops in elevation and trifurcates into three individual penstocks that feed the pump generators in the underground power station. After the water is used to generate electricity, it is discharged into the lower reservoir via a 3,797-foot-long tailrace tunnel. The difference in elevation between the reservoirs affords an effective variable hydraulic head (drop of the water) of 1,470 to 1,744 feet.

During bidding, contractors priced both a vertical and inclined shaft. For economic reasons, the inclined shaft was selected for construction. The excavation of the inclined shaft, 55 degrees above the horizontal, was unprecedented at the time, requiring unusual excavation and concreting techniques (Stassburger et al. 1984). Excavation techniques included the drilling of a pilot bore and use of a raise bore machine, both of which meandered off alignment in the lower portion of the shaft. The shaft was then excavated and slashed to the full width. During the lining operation, movement of the 60-foot-long telescoping concrete forms was made difficult because of the meandering alignment of the inclined shaft. Additional challenges occurred when casting the concrete liner in 25-foot lifts due to the need for very long concrete delivery pipelines and the significant drop (head) in pumping and placing the concrete. Consequently, the lower lining had to be placed by pumping the concrete up from the bottom of the shaft (penstock tunnel).

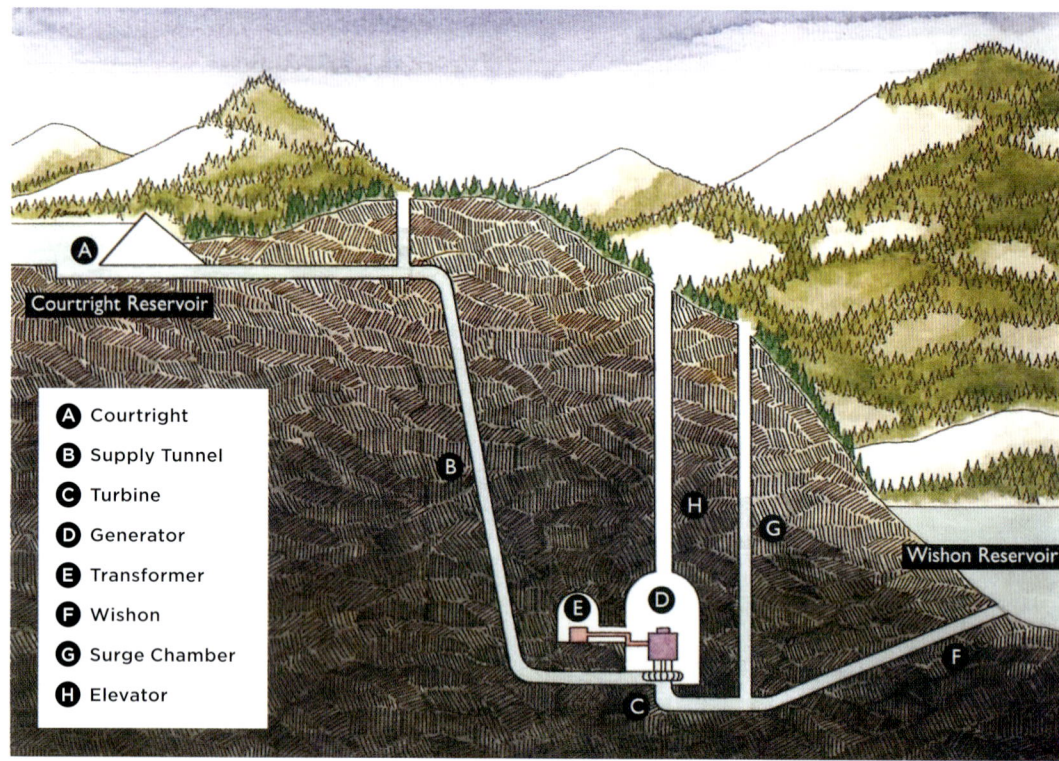

Figure 6.91 | **Helms Pumped Storage Plant Longitudinal Profile**

Figure 6.92 | **Helms Underground Powerhouse Chamber, Unlined with Suspended Ceiling**

An underground chamber was also excavated for the large power station that was 326 feet long (about the length of a football field), 142 feet high (about as high as a 10-story building), and 83 feet wide. This chamber housed three 404-MW Francis pump-turbine generators. The underground chamber was unlined but has a suspended ceiling to catch any drips or seepage (Figure 6.92). The power station has an installed capacity of 1,212 MW of electricity, enough to power 900,000 homes in the cities of Fresno and Oakland.

Bath County Pumped Storage Project

The Bath County Pumped Storage Project, the largest pumped storage facility in the world, is located in the northern corner of Bath County, Virginia, near the border with West Virginia. The station has a generation capacity of 3,003 MW—described as the "largest battery in the world." It is roughly equal to the size of three nuclear reactors and provides electricity to more than 1 million households. Construction on the power station began in March 1977 and was completed in December 1985 at a cost of $1.6 billion (Figure 6.93).

The station consists of upper and lower reservoirs separated by about 1,260 feet in elevation, the powerhouse, and power tunnels and penstocks. The major portion of the 20-story powerhouse is submerged beneath the lower reservoir. Excavation of the powerhouse required removal of about 497,000 cubic yards of rock, with rock cuts in the excavation ranging from 70 to 340 feet. The powerhouse is equipped with six 462-MW vertical reversible Francis unit turbines.

The tunnel and penstock system is comprised of three power tunnels with a diameter of 28.5 feet, three 900-foot-high shafts, and three lower power tunnels that bifurcate into two penstocks (for a total of six) before reaching the turbines (Figures 6.94 and 6.95). The tunnel and penstock system includes 4.5 miles of tunnels. Each of the three power conduits or tunnels is made up of the upper low-pressure tunnel, the vertical power shaft, the lower high-pressure tunnel, and two steel-lined penstocks, which are the largest pressurized penstocks. The penstocks are 18 feet in diameter, tapering to 9 feet in diameter at the spherical valves. The maximum static head, or pressure, from the top of the water delivery system to the turbines is 1,320 feet.

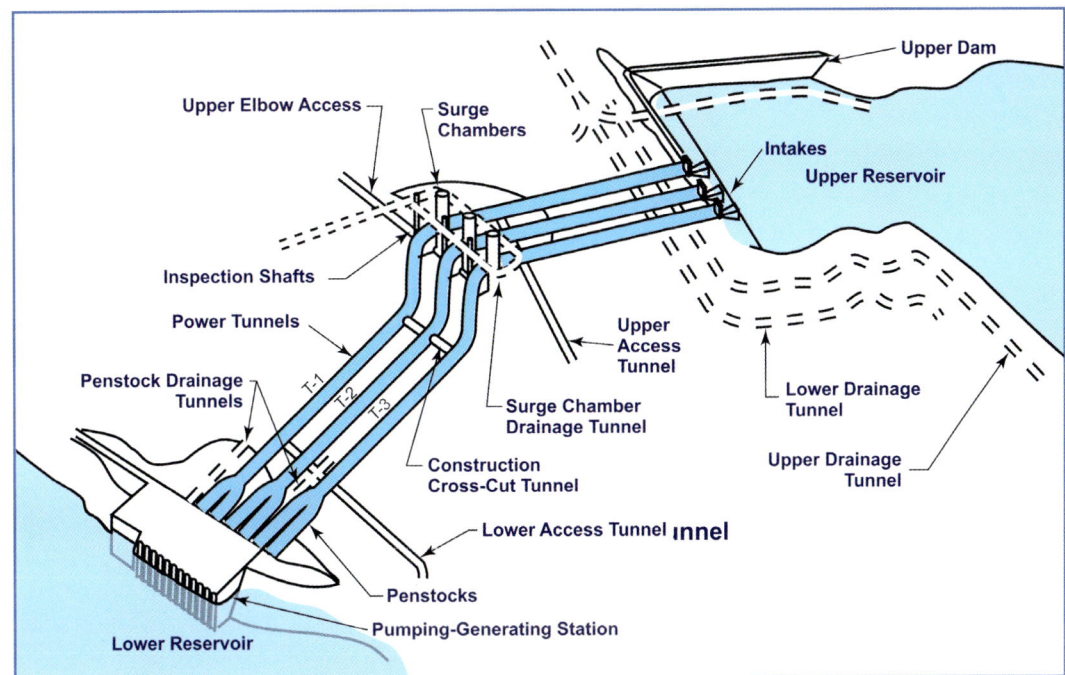

Figure 6.93 | **Bath County Pumped Storage Project Profile**

THE HISTORY OF TUNNELING IN THE UNITED STATES

Figure 6.94 | **Backslope, Five of Six Penstock Tunnel Excavations**

Figure 6.95 | **Concrete Placement for Penstock Tunnels**

SUMMARY

The advancement and evolution of tunneling for water supply in the United States had developed in response to population growth, expansion of commercial centers, scarcity of water, health problems from pollution and disease, and support of agricultural growth. The pioneers and planners such as Jervis, Chesbrough, Freeman, O'Shaughnessy, and Mulholland, as well as the Bureau of Reclamation, developed the visions and laid the plans in the 1800s and early 1900s using tunnels to convey raw water from the lakes and the mountains to the metropolitan areas. As part of the complex water systems, many of the projects included dams, reservoirs, power stations for power generation, and tunnels and underground chambers, thus delivering multi-purpose solutions to meet the public and industrial needs.

From the early nineteenth century to the present day in New York, Boston, Chicago, San Francisco, and Los Angeles, tunneling has experienced extreme highs and some disappointing lows. Each tunneling example in this chapter showcases a diversity of circumstances and geographic challenges from which great engineering technologies and innovations have originated. However, there are common themes. From the Northeast, the Midwest, and the West, high demands were addressed through great risks that were taken to meet future needs that were sustainable, economical, and environmentally prudent.

Growing populations, climate change, disease, and politics will forever influence the drive for access to water. Meeting the needs and demands of existing and growing populations will always be the number-one priority for engineers, community planners, politicians, policy makers, and water boards. Every time a decision is made about finding, accessing, and delivering water, the solution will reside in tunneling design, technology, equipment, and construction.

From the world's longest water tunnel, the Delaware Aqueduct Tunnel in New York, to the 142 tunnels constructed for the Los Angeles Aqueduct in California, and what began in Chicago with the first raw water intake tunnel to the construction of Lake Mead Intake Tunnel No. 3 under the highest water pressure ever achieved, U.S. tunneling achievements have led the way.

REFERENCES

Arnold, A.B. 1966. *The California Aqueduct in Southern California*.

ASCE (American Society of Civil Engineers) Illinois Section. 2003. *150 Years of Engineering Excellence*. Palatine, IL: Aerodine Magazine.

Bellamy II, J.S. 1995. *They Died Crawling and Other Tales of Cleveland Woe*. Cleveland: Gray.

Burkholder, J.L. 1939. *The Great Aqueduct: The Story of the Planning and Building of the Colorado River Aqueduct.* Los Angeles: Metropolitan Water District of Southern California. pp. 34–35.

Chicago Bureau of Public Efficiency. 1917. *The Water Works System of the City of Chicago.* Chicago: Chicago Bureau of Public Efficiency.

Chicago DWS (Chicago Department of Water and Sewers). 1956. *Chicago Water System: A Description of Its System and Sanitary Protection.* Chicago: DWS.

Christenson, D. (ed). 1973. *Chicago Public Works: A History.* Chicago: Rand McNally.

Dunn, S.M. 1940. Pumping and related equipment used on the Mono Craters Tunnel Project. *Journal of the American Water Works Association* 32(1):1242–1258.

ENR *(Engineering News-Record).* 1908. The Blue Island Avenue Tunnel of the Chicago Water Works System. *Engineering News-Record.*

ENR *(Engineering News-Record).* 1933. Concrete water tunnel inspected after 25 years. *Engineering News-Record* (March 23):374–375.

ENR *(Engineering News-Record).* 1998. Tunnel repairs succeed following collapse under lake. *Engineering News-Record,* p. U8.

Foley, A. 2009. Life on the cutting edge: Dick Robbins. *Tunnels and Tunnelling International* (May):23–25.

Fulcher, B., Bednarski, J., Bell, M., Tzobery, S., and Burger, W. 2007. Piercing the mountain and overcoming difficult ground and water conditions with two hybrid hard rock TBMs. In *Rapid Excavation and Tunneling Conference: 2007 Proceedings.* Edited by M.T. Traylor and J.W. Townsend. Littleton, CO: SME.

Gruen, J.P. 1998. *Colorado River Aqueduct: Historic American Engineering Record.* Washington, DC: National Park Service.

Hanson, W.D. 2005. *San Francisco Water and Power: A History of the Municipal Water Department and Hetch Hetchy Water System,* 6th ed. San Francisco Public Utilities Commission.

Hennessey, B. 2012. *Hetch Hetchy.* Images of America series. Charleston, SC: Arcadia.

Jacobs Associates and Korbin, G. 2008. *Foothill Tunnel Condition Assessment 2007.* Prepared for San Francisco Public Utilities Commission.

Jacques, H.L. 1939. Mono Craters Tunnel Construction Problems. *Journal of the American Water Works Association* 32(1):43–56.

Koncza, K., Churchill, D.H., and Miller, G.L. 1980. *Lawrence Avenue Underflow Sewer System: Interim Report, Planning and Constructon.* EPA-600/2-80-014. Cincinnati: U.S. Environmental Protection Agency, Office of Research and Development.

Maupin, M.A., Kenny, J.F., Hutson, S.S., Lovelace, J.K., Barber, N.L., and Linsey, K.S. 2014. *Estimated Use of Water in the United States in 2010.* U.S.Geological Survey. Circular 1405.

MWD (Metropolitan Water District of Southern California). 1978. Fiftieth anniversary Metropolitan Water District. *Aqueduct* 46(1).

MWD (Metropolitan Water District of Southern California). 1995. *Inland Feeder Project—San Bernardino Mountains Section.* Geotechnical Data Report. November.

MWD (Metropolitan Water District of Southern California). 2005. Internal website. http://intramet.mwd.h2o/Resources/FacilityRefMan/ColoradoAquPP/3.1GenInfo.asp#content (accessed July 2015).

Nadeau, R.A. 1950. *The Water Seekers*. New York: Doubleday.

Parmley, W.C. 1927. *The Parmley System of Arch Construction—Catalogue E, Segmental Concrete Plain and Reinforced*. New York: Walter C. Parmley.

Pestonatto, A., Heinlein, S., Malkos, G., and Lindell, J. 2003. 79th Street Water Tunnel Extension. In *Proceedings of the Rapid Excavation and Tunneling Conference*. pp. 31–40.

Pikarsky, M. 1971. Sixty years of rock tunneling in Chicago. *Journal of the Construction Division and Management* 97(2):189–210.

Proctor, R.J. 1998. A chronicle of California tunnel incidents. *Environmental and Engineering Geoscience* IV(1):19–53.

Proctor, R.J. 2002. The San Fernando tunnel explosion, California. *Engineering Geology* 67:1–3.

SFPUC (San Francisco Public Utilities Commission). 2015. *Water System Improvement Program Regional Projects Quarterly Report, 3rd Quarter/Fiscal Year 2014–2015*. San Francisco: SFPUC.

Simpson, J.W. 2005. *Dam! Water, Power, Politics, and Preservation in Hetch Hetchy and Yosemite National Park*. New York: Pantheon Books.

Standiford, L. 2015. Water to the Angels: *William Mulholland, His Monumental Aqueduct, and the Rise of Los Angeles*. New York: HarperCollins.

Stassburger, A.G., Moller, D.W., Davis, J.A., and Farley, R.V. 1984. Helms pumped storage project design through construction. In *Proceedings, Energy 1984*. Pasadena, CA: American Society of Civil Engineers.

State of California. 1967. *Office Report Engineering Geology of the Sand Bernardino Tunnel*. State of California, The Resources Agency Department of Water Resources Southern District Design and Construction Branch.

State of California. 1974a. *California State Water Project: Volume II, Conveyance Facilities*. State of California, The Resources Agency, Department of Water Resources.

State of California. 1974b. *Final Geologic Report San Bernardino Tunnel*. State of California, The Resources Agency, Department of Water Resources, Division of Design and Construction.

Thompson, T.F. 1966. San Jacinto Tunnel—Case histories of tunnels in Southern California. In *Engineering Geology in Southern California*. Glendale, CA: Association of Engineering Geologists. pp. 105, 107.

USBR (U.S. Bureau of Reclamation). 2015. Reclamation: Managing Water in the West. Cachuma Project. www.usbr.gov/projects/Project.jsp?proj_Name=Cachuma+Project.

USNCTT (U.S. National Committee on Tunneling Technology). 1984. *Geotechnical Site Investigations for Underground Projects,* Vol. 2. Washington, DC: National Academy Press.

Water and Power Associates. 2015. Construction of the Los Angeles Aqueduct. http://waterandpower.org/museum/Construction_of_the_LA_Aqueduct.html (accessed July 2015).

Weymouth, F.E. 1937. *Colorado River Aqueduct*, 3rd ed. Los Angeles: Metropolitan Water District of Southern California.

Weymouth, F.E. 1939. *The Metropolitan Water District of Southern California: History and First Annual Report*. Los Angeles: Haynes Corporation.

Zavattero, J. 1978. *The Sylmar Tunnel Disaster*. New York: Everest House.

7

WASTEWATER TUNNELS

It can be said with confidence that modern sewage collection, transport, and treatment have provided great improvements to human life, health, and longevity.

Humankind has long realized that the diversion and disposal of human waste is necessary for public health; however, it was only more recently that the necessity for human waste treatment became understood, given the science behind the health risks associated with polluted water.

Wastewater tunnels have played a significant role in this regard in the United States, along with the benefits they have provided to our waterways, wildlife, and citizens.

EARLY PROBLEMS

Early sewer systems in the United States were developed on an as-needed basis. Most of these systems were designed and built using the common sense of the time, with little or no guidance from trained "professionals"— few such trained people existed in those times.

In the early 1800s, many new communities in the United States used sewer systems only to handle stormwater flows. Depending on the circumstances, human wastes were either thrown into the streets, often to the peril of people passing below (Figure 7.1); left in cesspools, often to leak into the surrounding ground; or deposited directly onto the ground.

Many of the early waste or wastewater sewers were relatively straightforward in design, providing a path for sewage to be conveyed from its point of origin to nearby bodies of water. Construction incorporated the use of a variety of unique materials and techniques, ranging from drainage ditches to hollowed-out logs of varying size; some had no grade (or worse, reverse grade). Figure 7.2 shows a wooden log sewer pipe from the mid-1700s bored from a hemlock tree (Payrow 1928) and logs believed to be from the late 1800s that were unearthed in 2007 by construction crews in Holly, Michigan (sewerhistory.org 2016). Pipes such as these were in use from the earliest colonial days and some are still in use today!

Designs were based on very little information and often did not have adequate capacity to handle future growth. Low-lying and poorly draining soil, common to cities established along bodies of water, exacerbated these problems. As cities became more populated, accumulation of waste from people, livestock, and industrial development overwhelmed primitive waste-handling methods and often threatened freshwater sources. Accounts of life in Chicago, Illinois, during the early 1800s described manure from barns being dumped into streets, garbage clogging the drainage ditches, and filth collecting in shallow bogs. Meanwhile, the death rate from cholera in Chicago reportedly reached nearly 5.5% in the 1850s (Chicago Sewers Collection 1878). Anyone who has taken the Seattle Underground Tour (Bill Speidel's Underground Tour 2015) will also have heard that prior to the Great Fire of 1889, it was often noted that the streets would "bloat deep enough with mud to consume dogs and small children," and twice each day, when the tides came in, the toilets would overflow causing raw sewage to spill out. Similar scenarios were playing out all over the country, and the need for trained civil and sanitary engineers was becoming more evident.

By the end of the nineteenth century, scientific research on waterborne diseases, such as cholera and typhoid fever, gave a greater understanding of the nature of disease and the potential consequences of having poorly designed systems (or no systems) to handle and transport human waste away from population centers and sources of drinking water. At the same time, the United States was rapidly urbanizing. In 1840, only 11% of all Americans lived in

Figure 7.1 | **Chamber Pot Being Emptied into the Streets**

Figure 7.2 | **Wooden Log Pipes**

Previous Page | **Hoisting Materials in Access Shaft of Chattahoochee Tunnel in Georgia**

urban areas, but by 1860 the percentage increased to 20% and by 1880 had risen to 28% (Burian et al. 1999). Disinfection notices, such as the one shown in Figure 7.3 from Trenton, Tennessee, were distributed in an effort to prevent widespread cholera outbreaks.

Improved public understanding of the connection between sanitary conditions and disease was not enough to prevent widespread cholera and typhoid fever epidemics in densely populated areas, especially after storms caused sewage to overflow into drinking water sources. In Chicago, for example, between 1860 and 1900, deaths from typhoid fever averaged 65 per 100,000 people per year (ASCE 2003). Given the inadequacy of the waste systems and the potential life-threatening consequences, cities around the country sought solutions to solve their sewage problems. Through the efforts of those individuals now considered to be pioneers in the field of sewerage design and construction, cities began developing their own systems, often through the process of trial and error. This time period, appropriately labeled the "Great Sanitary Awakening," led to a series of technological innovations and improvements that helped to form the basis of modern sewage systems. As part of these new techniques, some of the earliest wastewater tunnels in the United States came into existence.

EARLY UNDERGROUND SOLUTIONS, PROMOTERS, AND DRIVERS

As cities across the country recognized the need to more effectively handle and transport their sanitary sewage away from drinking water sources, a growing public demand led to the creation of many centralized wastewater systems.

Boston

Sewers in Boston, Massachusetts, have been in operation since before 1700. The first sewers were privately owned and served the purpose of draining water from cellars and low-lying areas into nearby surface waters. Problems arose because of the nuisance caused by constructing these sewers as well as disagreements over ownership rights. In 1709, the Massachusetts General Court passed an act regulating the construction of sewers. The act also provided the basis for distributing sewer costs and charging to use them (BWSC 2016).

In 1833, sanitary waste was allowed into the system, creating a "combined" sewer system. The practice proliferated throughout the city, and one year later, the city encouraged adding rainwater from roofs to the system to assist in flushing the sewers of sanitary waste. This flushing did not solve the problem, and health issues related to water contamination, such as cholera, typhoid, and dysentery, began to increase among Bostonians. In 1875, a study was initiated to address this problem, which led to the construction of the Boston Main Drainage System (BMDS; BWSC 2016).

The BMDS was the original backbone of Boston's early sewer system. Constructed between 1877 and 1884, the purpose of the BMDS was to intercept local sewers and carry the sanitary waste and rainwater runoff to an offshore disposal point. The system included 25 miles of intercepting sewers, a pumping station in Dorchester, the Dorchester Bay Tunnel, and an outfall pipe at Moon Island in Boston Harbor (neither the pumping station nor the tunnel are now in use). As Boston's population grew and its city limits expanded due to annexation and land reclamation, the drainage works were subsequently enlarged to accommodate this growth (BWSC 2016).

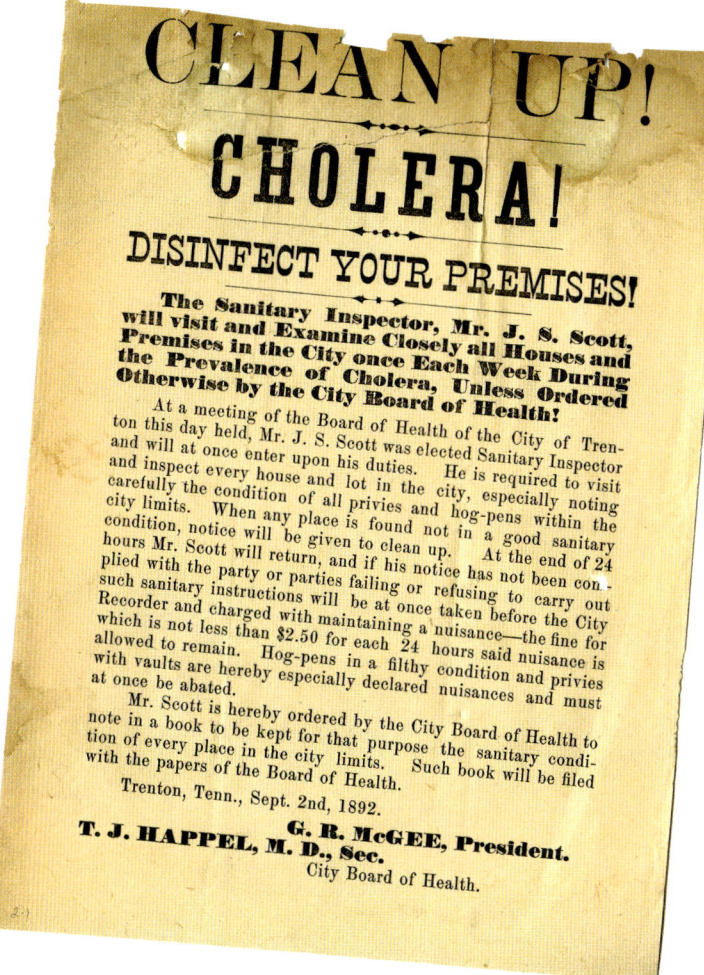

Figure 7.3 | **Cholera Notice, 1892**

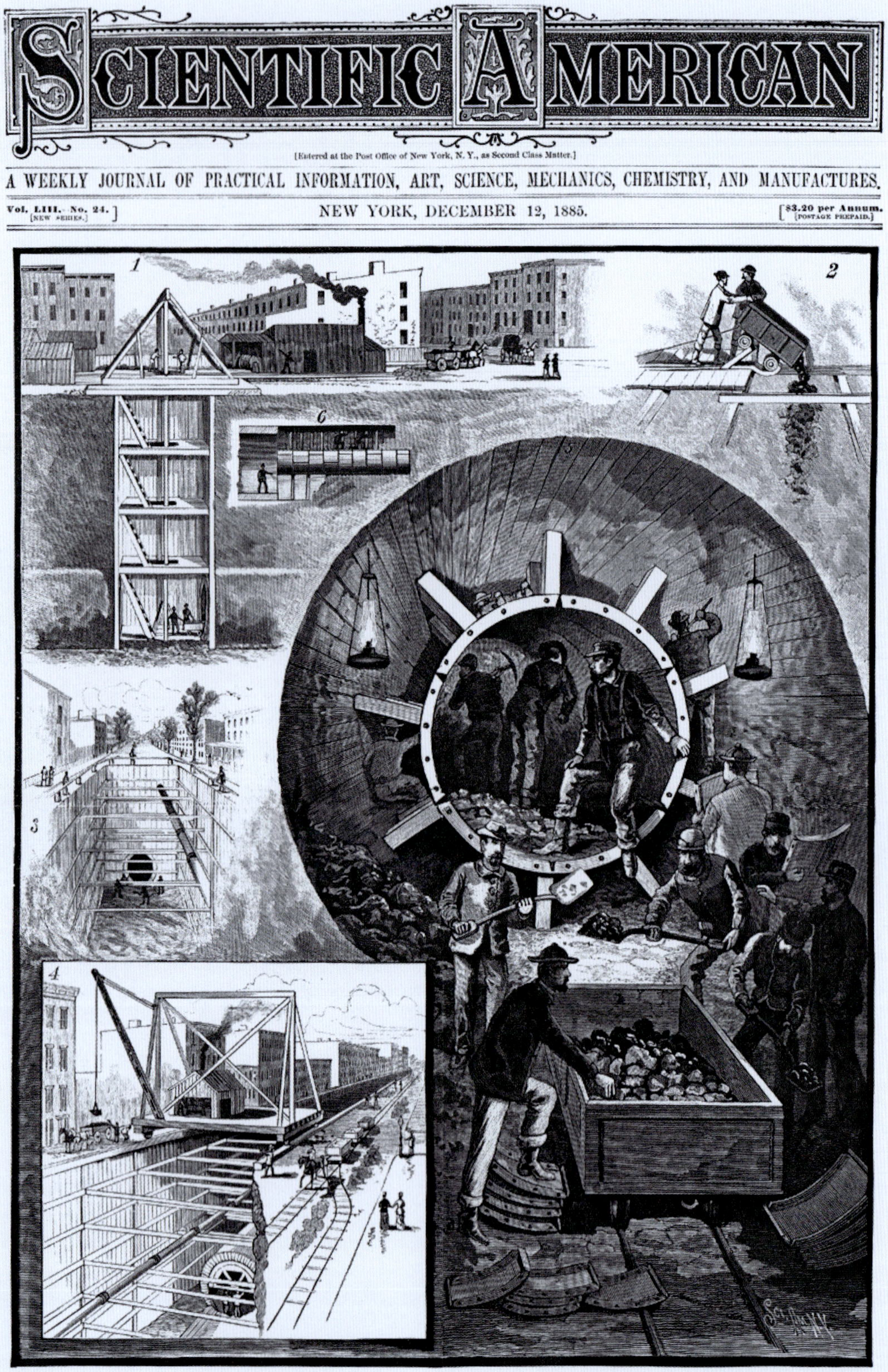

Figure 7.4 | **Tunneling Methods in the 1880s**

The BMDS disposed of a substantial portion of the city's waste; however, there were parts of Boston lying outside its service area. To address this, the Metropolitan Sewerage System was formed in 1889 to build one of the first regional sewerage systems in the country (BWSC 2016). The system soon became recognized as one of the best in the country, though it provided no treatment. It merely collected the wastewater and sent it out into the harbor. This resulted in the pollution of several clam beds located within the harbor. In 1933, all shellfish taken from the harbor required purification because of the worsening pollution.

New York City

By the mid-nineteenth century, the populations of America's urban centers had begun to explode. One particularly striking example was New York City, which doubled in population every decade from 1800 to 1880. With this increase in population density, the city's current methods of waste disposal began to prove both inadequate and unsafe, especially when located near drinking water sources. An article in a December 1885 issue of *Scientific American* made reference to this problem when discussing the reasons for a new sewer tunnel in Brooklyn. The article stated, "The necessity for the work is apparent from the fact that the present outlet…is discharged upon the lowlands at the head of Newtown Creek, making a nuisance greatly detrimental to public health" (*Scientific American* 1885). The work being referred to was a 12-foot-diameter brick sewer, which was being constructed by means of tunneling rather than open cut. Designers at this time determined that tunneling was the most feasible solution due to the required depth of the sewer grade lines (up to 60 feet below street level), potential danger from surface construction to heavy buildings on both sides of the street, and because tunneling caused much less disturbance to surface traffic (Figure 7.4).

Memphis

After a yellow fever epidemic swept through Memphis, Tennessee, in 1878, the newly created National Board of Health hired Civil War veteran and former drainage engineer for Central Park in New York City, George E. Waring Jr., to design and implement a better sewage drainage system. In Memphis, Waring designed a new system that separated sewage from regular stormwater runoff, an innovation that before then had not been implemented on a large scale in the United States (History.com 2016).

The approach of separating stormwater from sewage had been developed as early as 1842 in England by Edwin Chadwick and was supported by many because it seemed sensible to keep sanitary wastes and stormwater separate (Chadwick 1842). However, despite developments in Memphis, many cities across the country chose to combine their systems, as there was a widespread belief that combined sewers were more effective and less expensive to build.

Seattle

Prior to 1890, the city of Seattle, Washington, relied on a haphazard assortment of sewers and cesspools that, at best, drained into the surrounding water bodies. In 1891, faced with recurring threats of waterborne diseases, city engineer Benezette Williams designed Seattle's first centralized combined sewage system plan. This plan sought to remove as much city sewage as possible and transport it into the salt waters of Elliott Bay and Puget Sound, with more limited drainage into the freshwaters of Lake Washington and Lake Union. However, the hilly glacial terrain created a challenge for construction of such systems. To address this, it was determined that underground tunnels were the most feasible solution. The early tunnels of the nineteenth century were excavated using handheld spades and shovels, transporting the excavated material out of the tunnel by wheelbarrow. To support the excavation, workers used a combination of timber and steel to stabilize the surrounding soil prior to installing the final brick-and-mortar sewer structure. The earliest known record of a sewer tunnel during this time period was the Lake Union Sewer Tunnel. Completed in 1894, it consisted of approximately 5,700 feet of 72-inch internal-diameter sewer and transported sewage out into Elliott Bay on the west side of Seattle (Robinson et al. 2002).

By 1900, the population of the Seattle area had grown to more than 110,000, resulting in new areas being developed and the subsequent installation of additional raw sewage outfalls to the nearby lakes and rivers. It was apparent that improvements to the system were needed. Reginald H. Thomson, chief engineer for the City of Seattle, recognized this issue and sought to continue the work of Williams by designing and constructing additional tunnels to transport sewage away from the city's lakes and into the ocean. In the fall of 1904, a study was performed to determine a suitable location to place an outflow into Puget Sound. Discharging into Lake Washington was abandoned, and in 1911 the new sewer line was put into service. However, Thomson did not stop there. In planning the system for future growth, a brick sewer 12 feet in diameter was constructed across the north end of Seattle that went well beyond the needs of that time (in fact, some of the systems designed by Thomson are still in use today). The North Trunk Sewer project (Figure 7.5), involving almost 22 miles of new sewers, began in 1908 and took six years to complete. Because of the granular soils that flowed in the presence of groundwater, a compressed-air lock was installed and portions of the tunnel were excavated under air pressure of 18 to 22 psig (Robinson et al. 2002). Figure 7.6 shows an air lock used in the early 1900s to transition from standard atmospheric pressure to hyperbaric pressure at the tunnel face.

George E. Waring Jr. (1833–1898)

Considered by many as a founding father of the public works profession, Waring began his career in 1857 as an agricultural scientist and then served as drainage engineer for New York City's Central Park in the 1850s. He rose to the rank of Colonel during the Civil War while serving with the Missouri cavalry. After the war, he began a career as an agricultural drainage engineer. In 1880, he designed America's first "separate" sanitary sewage conveyance system for the Memphis, Tennessee, area— after the outbreak of two yellow fever episodes in the area in 1878 and 1879. He authored several works on agriculture, sewerage drainage, and street cleaning. In 1895, Waring became the street cleaning commissioner for New York City, during which time he created a model solid-waste program, providing for the physical removal of abandoned wagons and vehicles from city streets, the creation of a white-uniformed corps of street cleaners, and the recycling of certain components of the refuse stream.

Figure 7.5 | **North Trunk Sewer Tunnel in Seattle, 1913**

CHAPTER SEVEN | WASTEWATER TUNNELS

Figure 7.6 | Air Lock for Henderson Street Sewer Tunnel in Seattle, 1936

Although these systems provided measures to collect and transport sewage, they provided no means of treatment, leaving the water bodies where sewage was discharged heavily polluted. In fact, despite the fast-moving currents of Puget Sound, the plume of sewage from the North Trunk Sewer could be easily seen from the air and, on occasion, would deposit slime onto the nearby beaches. This resulted in health officials occasionally closing the beaches because of bacterial contamination.

Chicago

In the late 1840s, Chicago, Illinois, was growing rapidly. Given that the majority of the city was situated at the Lake Michigan water level, wastewater was unable to effectively drain out of the city. Cholera and dysentery epidemics occurred over several successive years; as a result, the city council resolved in 1855 to build a more comprehensive sewage system. Ellis Sylvester Chesbrough was appointed to the Board of Sewerage Commissioners because of his work on Boston's water distribution system and, in 1858, set out on the planning and design of what is now considered to be the first comprehensive sewer system in the United States. The main problems Chesbrough faced were moving wastewater out of the city, and keeping it from flowing into Lake Michigan and polluting the city's drinking water source. His plan was twofold: first, to build the sewer system aboveground, and then raise city buildings as much as 6 feet. This would allow the sewers to freely drain by gravity to the Chicago River. New brick sewer tunnels, generally 5 feet in diameter, were constructed on top of the existing streets and covered with dirt (Chicago Sewers Collection 1944).

Although these improvements effectively improved local sewage conveyance problems, the amount of sewage being conveyed into the Chicago River dramatically increased. Spring floods caused the river to overflow into Lake Michigan, endangering the quality of the city's drinking water. In 1863, Chesbrough proposed driving a 5-foot-wide brick-lined tunnel beneath Lake Michigan to construct a new intake for clean drinking water, 2 miles out from the sewer discharges at the mouth of the river. The tunnel was constructed from two shafts, one on the shore and one out in the lake at the intake crib. Miners excavated the tunnel in clay using hand tools, and masons followed 10 or 20 feet behind laying the brick tunnel lining. When the two tunnels met in 1866, they were found misaligned by a little over 6 inches (ASCE 2003).

After the Great Chicago Fire of 1871, new construction once again caused the population of Chicago to surge and exacerbated threats to public health from contaminated drinking water. In 1889, disease outbreaks and the public nuisance of the contaminated Chicago River led to the creation of the Sanitary District of Chicago. The strategy adopted by the Sanitary District was to address the problem by reversing the flow of the Chicago River, directing sewage away from Lake Michigan and eventually carrying it to the Mississippi River. To accomplish this feat,

Ellis Sylvester Chesbrough (1813–1886)

Chesbrough started out at an early age working as a railroad survey chainman. In 1855, he became engineer for the Chicago Board of Sewerage Commissioners and oversaw the construction of water and sewerage tunnels under Lake Michigan. In 1858, he published what became recognized by his peers as the first significant work on sewerage. This publication contributed to Chicago becoming the first big American city to design and install a comprehensive sewerage system. He was the eighth president of the American Society of Civil Engineers.

Figure 7.7 | **Rock Excavation for the Chicago Sanitary and Ship Canal, 1895**

the Chicago Sanitary and Ship Canal was constructed between 1892 and 1900 to connect the south branch of the Chicago River to the Des Plaines River (Figure 7.7). The channel is 28 miles long, more than 24 feet deep, and required excavation of in excess of 42 million cubic yards of rock and soil (Lanyon 2012).

With the reversal of the Chicago River, lake water could be diverted to aid in dilution of pollutants. The construction of intercepting sewers was undertaken to intercept outflows that would have discharged into Lake Michigan and to direct them to the Chicago River system (Figures 7.8 and 7.9). Given that the interceptors had to be lower than the sewers they were meant to intercept and had to slope downward to their destinations at the rivers, pump stations were required to pump the wastewater up to river level.

Cleveland

The Cuyahoga River and Lake Erie were two primary features that led Moses Cleveland to stake land at the mouth of the Cuyahoga in 1796, and the business district of the early city of Cleveland, Ohio, exploited the river where steamers, schooners, and canal boats exchanged imports and exports. The steel industry took off, and John D. Rockefeller began his oil empire on the shores of Lake Erie. Prosperity ensued, but polluted waters followed close behind.

At the time of the city's incorporation in 1836, civic leaders were satisfied with discharging raw sewage into Lake Erie and the Cuyahoga River simply to divert it away from public scrutiny. However, as people continued to settle in Cleveland, growing amounts of sewage began to mix with the same water that citizens drew from for drinking. The combination was deadly, causing waterborne illnesses that claimed hundreds of lives.

In 1881, Mayor Rensselaer Herrick declared Cleveland's riverfront "an open sewer through the center of the city" (NEORSD 2009). Despite a lack of public support, there began a series of public works to improve the quality of life in Cleveland, including the construction of a public water system and drainage sewers. In April 1882, the City Council appointed a special committee to plan for a comprehensive sewer system.

City officials thought it would be most efficient to construct a single interceptor sewer system, consisting of combined sewers, to carry sanitary sewage from one million people. These first "combined sewers" were scarcely more than drains, and simply carried sanitary sewage, industrial waste, and stormwater directly into nearby streams, the Cuyahoga River, and Lake Erie. As an increasing number of factories and oil refineries added to the river's vile condition, an outcry arose for better sewers. But nearly 40 years elapsed before a comprehensive system of sewers was adopted by the city.

While Cleveland's early sewers served to simply transport sewage away from the city's growing population, eventually these pipes became the conduit through which wastewater traveled to the plants for treatment. Along the way, the sewers fueled the development of outer-ring suburbs by providing them with access to Cleveland's wastewater treatment plants.

Figure 7.8 | **Calumet Intercepting Sewer Construction near Chicago, 1923**

Figure 7.9 | **West Side Intercepting Sewer Construction near Chicago, 1928**

By 1945, main interceptor sewers had been completed in the Easterly, Westerly, and Southerly districts to deliver sewage to three treatment sites. Meanwhile, the many suburbs that had evolved around Cleveland had constructed separate sewer systems and connected their sanitary sewers to the Cleveland combined sewer system.

Despite these improvements to Cleveland's sewer system, not enough was being done to adequately treat wastewater in a booming industrial city. The increased production and use of persistent toxic chemicals during and after World War II raised environmental concerns beyond those that accompanied the industrial and sewage pollution of earlier years. Something still had to be done to correct a steadily worsening situation.

A BURNING RIVER AND CREATION OF THE EPA

After enduring years of abuse, on June 22, 1969, the Cuyahoga River caught fire, galvanizing the nation and thrusting Cleveland into the national spotlight. Ironically, the 1969 fire was benign compared to at least 12 previous incidents—including a 1912 blaze that killed five men and a fire in 1952 that resulted in $1.5 million worth of damage to surrounding structures and water vessels (Figure 7.10). Comparatively, the 1969 fire caused just $85,000 in damage and no fatalities, but the event captured the public's imagination and ignited a growing environmental movement (NEORSD 2009). At one time, the pollution was considered an unfortunate but necessary side effect of American industrialization, a sign of economic prosperity and growth. Now, more than a century after the river's pollution was first noted, it became an international symbol of environmental neglect and abuse.

Figure 7.10 | **Cuyahoga River Fire in Cleveland, 1952**

Coupled with other environmental disasters at the time, such as the January 1969 oil spill off the Coast of Santa Barbara, California, which lasted 11 days and dumped more than 200,000 gallons of crude oil into the coastal waters, the burning of the Cuyahoga River spurred efforts to enact federal environmental legislation by attracting national media coverage to an already volatile subject.

In August 1969, *Time* magazine published an article titled "The Price of Optimism," detailing then Cleveland Mayor Carl Stokes' outrage over the incident. Stokes, a long-time advocate for environmental responsibility, criticized the federal government and vowed to fight for a cleaner river. He believed that cleaning the river was beyond the city's control and filed a formal complaint with the state of Ohio asking for government funds to help address the polluted waterway. Although help from the state never came, it was a precursor of things to come.

One year later, in 1970, the National Environmental Policy Act was passed in Congress, helping to establish the Environmental Protection Agency (EPA). The EPA was given the responsibility of protecting human health and managing environmental risks through the enforcement of regulations enacted in Congress. In 1972, Congress passed the Federal Water Pollution Control Act Amendments, which formed the basis for what would become the Clean Water Act of 1977. This act, which was supported by U.S. Representative Louis Stokes (brother of then former Cleveland mayor), gave the EPA the legal tools to help advance water pollution control by establishing the basic structure for regulating discharges of pollutants into the waters of the United States and regulated quality standards for surface waters (Pew 2008).

RESULTING MAJOR CONSENT DECREES AND TUNNELING PROGRAMS

Many of the earlier sewers prevalent in older cities throughout the United States were designed to carry both sewage and stormwater. When heavy flows of stormwater enter these combined sewers, control devices allow some of the untreated rainwater and sewage to overflow into area waterways, avoiding combined sewer backups and flooding. This combined sewer overflow (CSO) contains bacteria from human waste, industrial waste, and other pollutants swept from the ground's surface. This release of sanitary sewage into the environment can seriously compromise the quality of adjacent streams, rivers, lakes, and oceans (Figure 7.11).

Figure 7.11 | **Chicago Lakefront Beach Closed After CSO Events**

Presently, approximately 772 communities throughout the United States have combined sewer systems serving about 40 million people. In 1994, the EPA issued a policy requiring municipalities to make improvements to reduce or eliminate CSO-related pollution problems. In 2000, Congress amended the Clean Water Act to require municipalities to comply with this EPA policy. During negotiations with the EPA, several cities decided to construct a series of underground CSO storage tunnels to provide additional system capacity as a means to solve their pollution problems.

Chicago

After studying different options, the Metropolitan Water Reclamation District of Greater Chicago (MWRDGC, the modern successor of the Sanitary District of Chicago) selected the Tunnel and Reservoir Plan (TARP) in 1972 as the most cost-effective option to comply with federal and state water quality legislation. As illustrated in Figure 7.12, the TARP system consists of four large-diameter deep rock tunnel systems—Mainstream, Des Plaines, Calumet, and Upper Des Plaines—and three open reservoirs that serve the 375-square-mile sewer area consisting of Chicago and 51 surrounding suburbs (MWRDGC 2010). TARP was the nation's first comprehensive CSO control plan for a large urban area, aimed to eliminate CSOs by capturing and storing potential overflows until they could be properly treated at the nearby wastewater treatment plants.

More than 100 miles of TARP tunnels are in service in the Chicago area. These tunnels range from 5 to 33 feet in finished diameter and are 150 to 340 feet deep. The TARP tunnel system provides an overall CSO storage capacity of 2.3 billion gallons. TBM (tunnel boring machine) excavation was chosen as the primary method because it causes less rock disturbance, noise, and vibration than drill-and-blast methods. Figure 7.13 shows the bare dolomitic

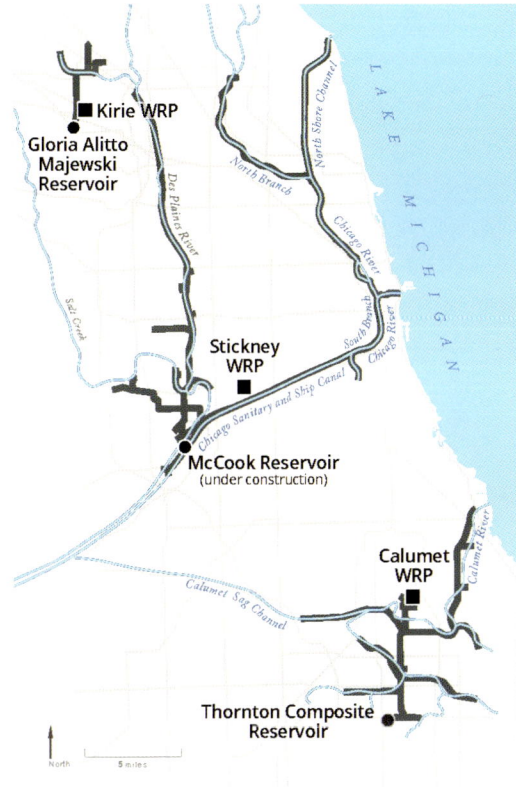

Figure 7.12 | **Chicago TARP System Map**

limestone rock surface left behind by a TBM on the Mainstream Tunnel. Many tunneling records were set over the years on the TARP project, including boring with the largest TBM ever used at the time. On the Mainstream Tunnel, TBMs performed beyond their then-proven capabilities—on average, three times faster than the 2 feet per hour that had been estimated in 1975 (ASCE 1986). The tunnel lining consisted of cast-in-place concrete poured in telescoping forms, which permitted the full circle of lining to be poured at once (Figure 7.14). MWRDGC was a trailblazer in rock tunnel boring, and the TARP projects undertaken were some of the largest and most ambitious for their time. Indeed, the MWRDGC tunnels are still some of the largest tunnels built in the United States. Lessons learned on these projects, both from a tunneling and sewer design perspective, have been invaluable to other designers and contractors who followed. Figure 7.15 shows the connection between the existing Mainstream Tunnel and the new McCook Reservoir Main Tunnel linking the Deep Tunnel System to the McCook Reservoir near Chicago in 2015.

Figure 7.14 | **Placement of Concrete Liner in TARP System, 1997**

Figure 7.13 | **Unlined TBM Bore in TARP Tunnel, 1981**

CHAPTER SEVEN | **WASTEWATER TUNNELS**　367

Figure 7.15 | Tunnel Connections near Chicago, 2015

Of the three reservoirs in the TARP system, the Gloria Alitto Majewski Reservoir (350 million gallons of storage) was the first to be completed after the tunnels in 1998. The Thornton Reservoir, dubbed the "Grand Canyon of the South Suburbs," was completed in September 2015, adding 7.9 billion gallons of storage (Figure 7.16). The McCook Reservoir, which is still under construction, will hold about 10 billion gallons when complete (CBS Chicago 2015). The success of the TARP system is evident by the dramatic improvements in the water quality of the Chicago River and other local waterways, in which fish populations have increased threefold (ASCE 2003). The system also continues to reduce flood damage as its storage capacity and conveyance are improved.

Milwaukee

In response to the Clean Water Act, the Milwaukee Metropolitan Sewerage District (MMSD) in Wisconsin initiated studies to improve its system and reduce sewer overflows. After extensive planning and public input, construction of Milwaukee's Deep Tunnel System began, consisting of approximately 28.5 miles of tunnel ranging in diameter from 17 to 32 feet (Figure 7.17).

Constructed in three phases, the entire system can store up to 521 million gallons of combined water and wastewater. The system is made up of several legs that run in different directions. These legs are all connected and flow by gravity to a main pump station at the Jones Island Water Reclamation Facility. Most of the tunnel system is constructed in bedrock at depths of up to 300 feet, with a 7-mile stretch placed approximately 135 to 175 feet underground. Figures 7.18 and 7.19 are photos of Milwaukee's Deep Tunnel System.

Out of all the combined sewage that has entered the regional sewer system since the tunnel started operating, MMSD has captured and cleaned 98.3%. Compared to the national goal of 85% for communities with combined sewers, this is a shining example of the benefits associated with underground wastewater

Figure 7.16 | **View from Connecting Tunnel into Thornton Reservoir, 2013**

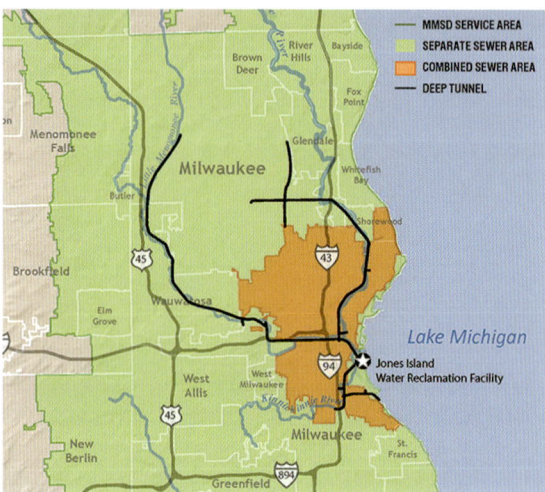

Figure 7.17 | **Milwaukee Deep Tunnel System Map**

CHAPTER SEVEN | WASTEWATER TUNNELS | 371

Figure 7.18 | **Blasting Preparation**

Figure 7.19 | **Two 22-Foot Tunnels Coming Together**

tunnels when used to mitigate sewage overflows into surrounding water bodies.

Like Chicago's MWRDGC, MMSD was also an early pioneer for the large-diameter CSO tunnel industry. Much research and modeling was performed to optimize these facilities, with many of the project's elements becoming industry standards for decades. As each of these early projects was built in the Midwest, success bred future success, and adversities led to innovation and still more success. Figure 7.20 shows workers inspecting a completed section of Milwaukee's Deep Tunnel Project that has prevented more than 100 billion gallons of pollution from entering Lake Michigan. A transition in a section of Milwaukee's deep tunnel is shown in Figure 7.21 where a 17-foot-diameter tunnel enters into a 32-foot-diameter tunnel.

Figure 7.20 | **Workers Inspecting Lining**

Figure 7.21 | **Transition from 17 to 32 Feet**

Cleveland

As a result of the Cuyahoga River fires and subsequent federal legislation, the Cleveland Regional Sewer District (later renamed the Northeast Ohio Regional Sewer District, or NEORSD) was created in 1972. NEORSD assumed ownership and operation of Cleveland's three wastewater treatment facilities and 200 miles of existing interceptors, and was charged with constructing several new "intercommunity" relief interceptors to further prevent suburban sanitary sewage from entering Cleveland's combined sewer system (Figure 7.22).

Today, NEORSD maintains more than 200 miles of large interceptor sewers and is in the process of constructing several deep underground storage tunnels to meet the Clean Water Act standards.

NEORSD's long-term plan for CSO control recommended constructing deep tunnel storage for combined wastewater. NEORSD has been building large-diameter sewer tunnels for more than 40 years and has completed many miles of CSO tunnels in that time in soil, sandstone, and shale. Since 1972, NEORSD's tunnel projects have drastically reduced CSO discharges in local waterways—from 9 billion gallons per year down to 4.1 billion gallons per year (in 2015).

Several years ago, NEORSD embarked on Project Clean Lake, a $3 billion, 25-year CSO-reduction project (Figure 7.23). By 2035, this program will have further reduced CSO pollution of Lake Erie and its tributaries to under half a billion gallons per year; a staggering 8.5 billion gallon reduction since 1972.

At the heart of Project Clean Lake is the construction of seven large-scale storage tunnels, ranging from 2 to 5 miles in length, up to 300 feet underground, and up to 24 feet in finished diameter. The 24-foot-diameter, roughly 3.5-mile long Euclid Creek Tunnel was the first tunnel completed under Project Clean Lake in 2015.

Figure 7.22 | **Heights/Hilltop Tunnel Construction in Cleveland, 1987**

A TBM used for the Euclid Creek Tunnel, nicknamed "Mackenzie," boasts an impressive machine transport system for conveying and installing precast concrete segments that comprise the lining of the tunnel (Figures 7.24 and 7.25). As the TBM advances, workers erect and connect these segments together, pushing out a finished product, one ring at a time. This "one-pass" technology is common in the industry in soft ground but not as frequently in rock, and it had never been used by NEORSD before. The adoption of this technology marked an advance over earlier two-pass construction projects in Cleveland, which had been subject to rock fall in shale, gas, and many other risks that were all eliminated or dramatically reduced with the precast concrete segmental lining (Figure 7.26). Figure 7.27 shows Mackenzie after breaking through the Euclid Creek Tunnel terminus shaft.

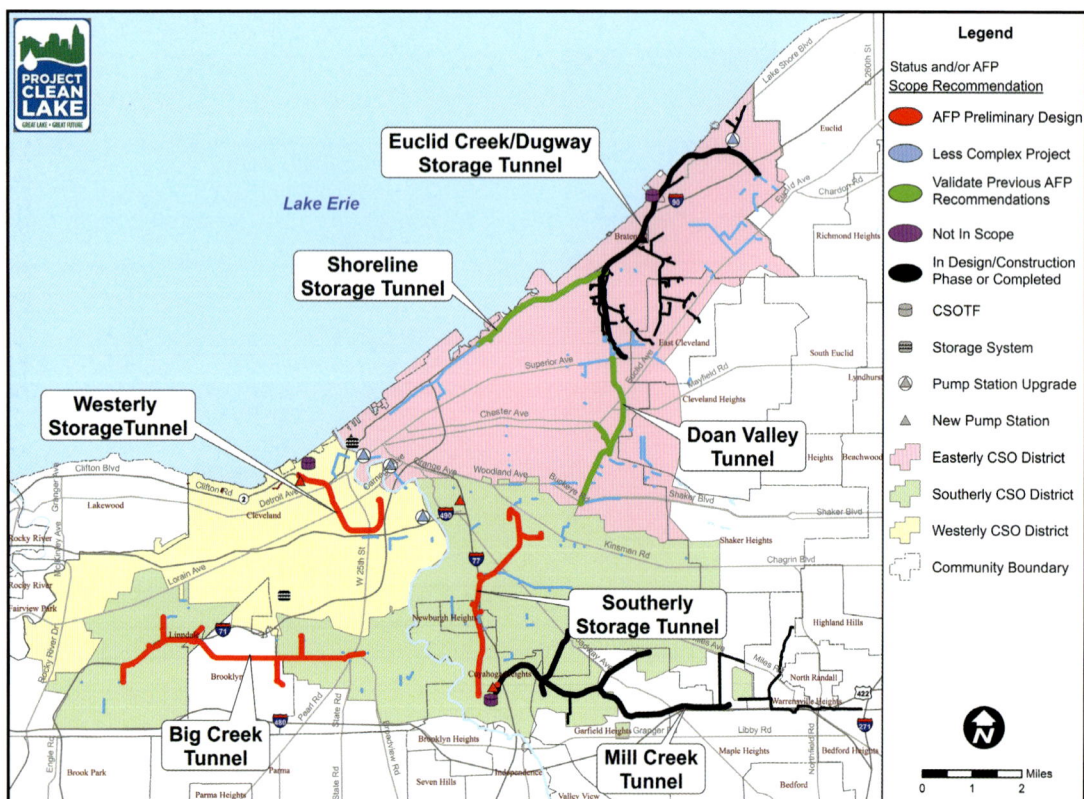

Figure 7.23 | **Cleveland "Project Clean Lake" System Map**

Figure 7.24 | **TBM Cutterhead for the Euclid Creek Tunnel**

Figure 7.25 | **The Launch of Mackenzie from the Mining Shaft**

Figure 7.26 | **Precast Segmental Lining**

The Euclid Creek Tunnel project also instituted new drop-shaft physical modeling, which built upon the drop structure work done in the late twentieth century by MWRDGC and MMSD. The NEORSD "Cleveland Baffle Drop" modeling has been adopted by many other agencies worldwide, and these organizations are further investigating and improving drop structure methods (and other sewer design elements) in turn.

The Euclid Creek Tunnel has the capacity to hold 62 million gallons of combined stormwater and wastewater. This project has won several national and international awards and was deemed quite successful by NEORSD. The success of this project bodes well for its remaining Project Clean Lake tunnels—and for continued improvements to water quality in the Cuyahoga River and Lake Erie.

Los Angeles

During the 1990s, several major sewers serving as the backbone for the City of Los Angeles, California, wastewater conveyance system showed signs of serious structural deterioration. The North Outfall Sewer was built in the 1920s and, over its 70-year life, had weakened due to corrosion of the mortar binding the clay tiles. Emergency repairs were required at various locations where the sewer roof collapsed, resulting in sinkholes.

In addition to deterioration, sewage flows had reached the design capacity of the aging sewer system during the dry season. During extended winter storms, sewers overflowed into city streets.

During the 1998 winter, El Niño storms caused millions of gallons of raw sewage to spill onto Los Angeles city streets. On September 14, 1998, the Regional Water Quality Control Board issued a Cease and Desist Order (CDO) requiring the City of Los Angeles to complete construction of several new sewer tunnels within a 7-year period.

Figure 7.27 | **Euclid Creek TBM Breakthrough**

Major CDO tasks included construction of the North Hollywood Interceptor Sewer, Eagle Rock Relief Sewer, North East Interceptor Sewer (NEIS), and East Central Interceptor Sewer (ECIS). These sewers are included in the system map shown in Figure 7.28.

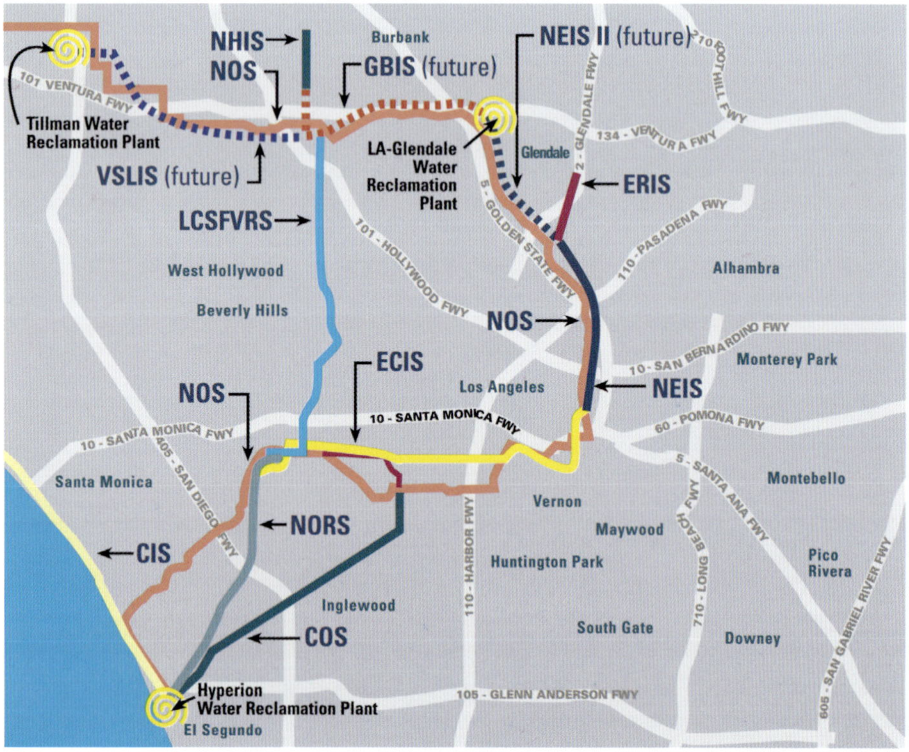

Upon issuance of the CDO, the City immediately accelerated action to design and construct the ECIS/NEIS within the tight schedule milestone dates established by the Regional Water Quality Control Board. The ECIS project consisted of approximately 11 miles of 11-foot inside-diameter sewer tunnel, excavated at depths ranging from 50 to 80 feet underground. The NEIS project consisted of a 5.3-mile long, 8-foot inside-diameter tunnel, divided into three segments identified as the Lower, Middle, and Upper Reaches, at depths ranging from 120 to 160 feet (Figure 7.29).

Design of the two systems was performed from 1998 through 2001. Various design challenges existed for both projects, including designing structures to convey sewage flows from existing large-diameter sewers into the new tunnels, crossing underneath the Los Angeles River as well as under active railroads, and specifying seismic features where the new sewer alignments crossed known earthquake faults. Construction was completed in 2004, successfully meeting the CDO milestone completion date originally set by the Regional Water Quality Control Board (Figures 7.30 and 7.31).

Figure 7.28 | **Map of Major Interceptor and Relief Sewers in the Los Angeles Area**

CHAPTER SEVEN | WASTEWATER TUNNELS | 375

Figure 7.29 | **Aerial Map of NEIS System**

Portland

The City of Portland, Oregon, began building CSO tunnels as part of the City's 20-year, long-term CSO Management Plan, which addressed the EPA requirements of the Clean Water Act of 1972 and CSO Control Policy of 1994. These tunnels improved the water quality of the Willamette River and Columbia Slough by controlling the frequency, volume, and location of CSOs (Figure 7.32).

There were three major elements to the Portland CSO control system: the Columbia Slough Consolidated Conduit (CSCC); the West Side CSO Tunnel and Swan Island Pump Station; and the East Side CSO Tunnel and Swan Island Pump Station, Phase 2 Pump Installation.

Constructed between 1996 and 2000, the CSCC consisted of a 2-mile-long storage tunnel with a 12-foot internal diameter and a 1.3-mile long conveyance tunnel with a 6-foot internal diameter located along an industrial corridor. The CSCC was designed to control flows from 13 Columbia Slough outfalls for up to the 10-year summer storm and cost approximately $70 million.

Figure 7.30 | **TBM Hole-Through into a Slurry Wall Shaft for the NEIS Tunnel**

Figure 7.31 | **Figure-Eight Slurry Wall Shaft for the NEIS Tunnel**

THE HISTORY OF TUNNELING IN THE UNITED STATES

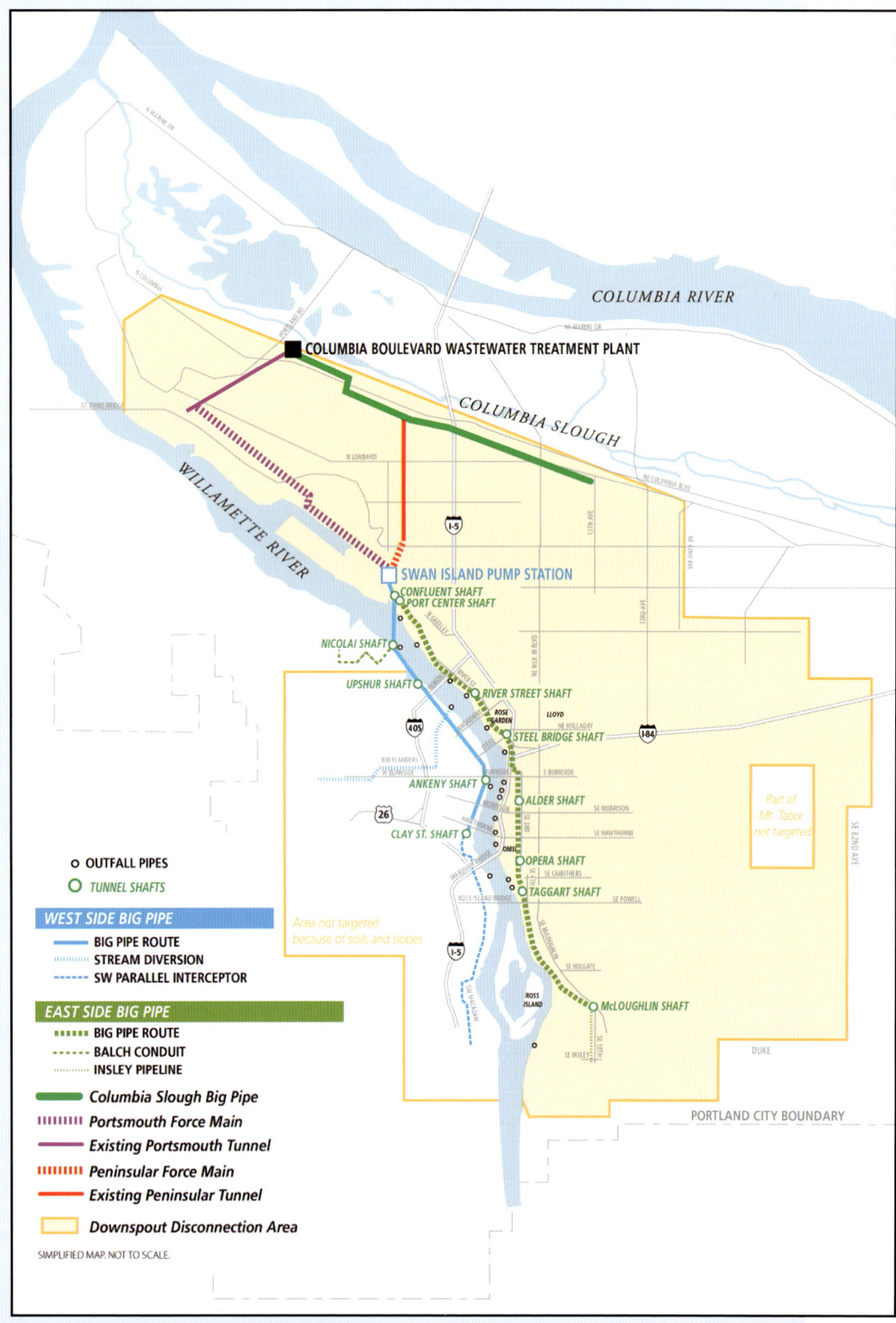

Figure 7.32 | **Portland CSO Control System Map**

The West Side CSO Tunnel was constructed between 2002 and 2006 through a heavy industrial and downtown waterfront setting along the Willamette River. Work included a 3.5-mile tunnel with 14-foot internal diameter, driven using two TBMs—one TBM traveling south along the riverfront and the other east under the Willamette River. Five shafts ranging from 100 to 140 feet deep were constructed to deliver captured CSO into the tunnel, with four outfalls remaining as overflow points to the river. The Swan Island Pump Station was constructed to lift and pump the captured CSO tunnel system conveyance flows north to the existing Columbia Boulevard Wastewater Treatment Plant. Figure 7.33 shows the final build-out of the Swan Island Pump Station for the West Side CSO Tunnel.

The East Side CSO Tunnel was constructed between 2006 and 2010 to control 14 Willamette River outfalls. This was a 6-mile-long tunnel with 22-foot internal diameter, driven using a 25-foot-diameter TBM (Figure 7.34). Seven large shafts ranging from 120 to 180 feet deep were needed for the project. Six shafts were used to convey flows down to the tunnel, with one shaft used solely for TBM launch and mining operations. Four outfalls remain open to the river as overflow points. The alignment was through areas of commercial, park and industrial waterfront, and large industrial use.

Through this underground tunnel system, the City of Portland has effectively removed 99% of the CSO volume previously released to the Columbia Slough and 94% of the volume released to the Willamette River. In addition, the number of CSO events was reduced from 50 per year to less than 4 (City of Portland BES 2016).

Washington, D.C.

Like the Cuyahoga and many other rivers in the United States, the water quality in the Potomac had deteriorated because of pollution, runoff, and sewer overflows by the time the Clean Water Act was passed. In an effort to curb this pollution, and as part of a federally mandated decree, the District of Columbia Water and Sewer Authority (DC Water) is currently in the process of

Figure 7.33 | Final Build-Out of Pump Station, 2005

constructing a series of deep underground storage tunnels through its Clean Rivers Project. In total, there will be approximately 13 miles of new 23-foot and 15-foot diameter concrete-lined storage and conveyance tunnels. The major tunneling projects associated with this program are the Blue Plains Tunnel, Anacostia River Tunnel, First Street Tunnel, Northeast Boundary Tunnel, and Potomac River Tunnel (Figure 7.35). TBM mining for the Blue Plains Tunnel was completed in June 2015. Figures 7.36 and 7.37 show "Lady Bird," the TBM responsible for boring the Blue Plains Tunnel. In addition to tunnel construction, the project involves multiple diversion structures and shafts to convey captured CSOs into the tunnels as well as provide access points for future maintenance. When complete, these tunnels will help to control CSOs to the Potomac River and its tributaries, the Anacostia River and Rock Creek. The Clean Rivers Project will reduce CSOs by 96% (DC Water 2016).

A LOOK TO THE FUTURE

Since the founding of the United States, great strides have been made in the way that we collect and dispose of wastewater, and perhaps tunneling is the biggest stride forward so far. This chapter is not by any means an exhaustive recounting of wastewater tunnel construction in the United States; there are many other municipalities that have undertaken or are just embarking on large CSO tunnel programs. Boston and Seattle are notable examples of such systems. The Detroit Metropolitan area also has a long history of sewer tunneling over the past 20 to 30 years. St. Louis, Missouri, and the Atlanta Metropolitan Region have also completed many miles of large-diameter CSO tunnels over the past 10 to 20 years and St. Louis is still building tunnels. Like Washington, D.C., other cities, including Indianapolis (Indiana), Pittsburgh (Pennsylvania), and Columbus and Akron (Ohio), are just embarking on (or have recently embarked upon) EPA-required tunnel programs at the time of this writing.

Figure 7.34 | **TBM Cutterhead Being Lowered into Shaft, 2007**

CHAPTER SEVEN | WASTEWATER TUNNELS   379

As the EPA continues to crack down on cities that are still dumping uncontrolled combined sewage into their local waterways, many more cities across the country will be required to develop strategies to solve their combined sewer and storm sewer issues. Many cities are looking for less expensive solutions to CSO abatement, including green infrastructure to capture stormwater and store it "in situ" before it can migrate to the sewers. Green infrastructure is ideal for non-point-source stormwater capture and it should be one tool in the "CSO toolbox." But when it comes to CSO abatement, neither green infrastructure or any other technology to date has proven as cost effective and expedient as deep storage and conveyance tunnels. Barring new technological breakthroughs or a relaxation of EPA regulations, CSO sewer tunneling is here to stay.

As the tunnel industry moves forward, it does so having learned from its predecessors of the tunnel systems such as those discussed herein, which have provided invaluable lessons and technical innovations that have shaped the tunnel industry of today: tunnels have become much safer and more cost effective with time; lining systems and TBM advancements make it possible to construct tunnels in a near "factory-like" environment; and TBM sizes now exceed 60 feet in diameter. Soil conditions that presented insurmountable risk are now routinely traversed with technology that is changing at a rapid rate. This technology will only make tunneling for sewer collection and disposal even more attractive in the years to come.

Figure 7.36 | **Blue Plains TBM Prior to Launch**

Figure 7.35 | **Washington, D.C., Clean Rivers Project Tunnel System**

Figure 7.37 | TBM Being Assembled in Construction Shaft of Blue Plains Tunnel, 2013

REFERENCES

ASCE (American Society of Civil Engineers). 1986. Boring Through TARP: 1986 Outstanding Civil Engineering Achievement. *Civil Engineering—ASCE.* 56(7).

ASCE (American Society of Civil Engineers), Illinois Section. 2003. *150 Years of Civil Engineering Excellence.* Palatine, IL: Aerodine Magazine.

Bill Speidel's Underground Tour. 2015. A Little History. http://undergroundtour.com/about/history.html.

Burian, S.J., Nix, S.J., Durrans, R., Pitt, R.E., Fan, C.-Y., and Field, R. 1999. Historical development of wet-weather flow management. *Journal of Water Resources Planning and Management,* 125(1):3–13.

BWSC (Boston Water and Sewer Commission). 2016. Sewer History. Accessed Jan. 11, 2016. www.bwsc.org/ABOUT_BWSC/systems/sewer/Sewer_history.asp.

CBS Chicago. 2015. Massive New Reservoir To Help Alleviate Chicago Area Flooding. Accessed Jan. 11, 2016. http://chicago.cbslocal.com/2015/09/01/massive-new-reservoir-to-help-alleviate-chicago-area-flooding/

Chadwick, E. 1842. Chadwick's Report on Sanitary Conditions (Excerpt from *Report...from the Poor Law Commissioners on an Inquiry into the Sanitary Conditions of the Labouring Population of Great Britain*) on The Victorian Web. Accessed Jan. 11, 2016. www.victorianweb.org/history/chadwick2.html.

Chicago Sewers Collection. 1878. Seven Days in Chicago [Box 1, Folder 15], Special Collections, Chicago Public Library.

Chicago Sewers Collection. 1944. "Chicago Sewer System" by A.J. Schafmayer in *Journal of the Western Society of Engineers* [Box 1, Folder 18], Special Collections, Chicago Public Library.

City of Portland BES (Bureau of Environmental Services). 2016. Controlling Combined Sewer Overflows. Accessed Jan. 13, 2016. https://www.portlandoregon.gov/bes/article/316721.

DC Water (District of Columbia Water and Sewer Authority). 2016. Clean Rivers Project. Accessed Jan. 11, 2016. https://www.dcwater.com/workzones/projects/cleanrivers.cfm.

History.com. 2016. George Waring. www.history.com/topics/george-waring.

Lanyon, R. 2012. *Building the Canal to Save Chicago.* Bloomington, IN: Xlibris.

MWRDGC (Metropolitan Water Reclamation District of Greater Chicago). 2010. Tunnel and Reservoir Plan. Accessed Jan. 11, 2016. www.mwrd.org/irj/portal/anonymous/tarp.

NEORSD (Northeast Ohio Regional Sewer District). 2009. What fueled the flames?: Reflecting on 1969 river fire 40 years later. *Environotes,* June. www.neorsd.org/enviro-2009-06-riverfire.php.

Payrow, H.G. 1928. Discussion. Historic Review of the Development of Sanitary Engineering in the United States During the Past One Hundred and Fifty Years: A Symposium. *Transactions of the American Society of Civil Engineers* 92(1):1287.

Pew, K.A. 2008. Interceptor and intercommunity relief sewers. In *Northeast Ohio Regional Sewer District: Our History and Heritage.* Edited by Kim Jones. Cleveland: Northeast Ohio Regional Sewer District.

Robinson, R.A., Cox, E., and Dirks, M. 2002. Tunneling in Seattle—A history of innovation. In *Proceedings of the North American Tunneling Conference.* Edited by L. Ozdemir. Exton, PA: Balkema.

Scientific American. 1885. The Knickerbocker Avenue Extension Sewer, Brooklyn, N.Y. *Scientific American* LIII(24):373.

Sewerhistory.org. 2016. The history of sanitary sewers: Pipes—wood. www.sewerhistory.org/photosgraphics/pipes-wood/.

Time. 1969. The price of optimism. *Time* (August 1).

8 INNOVATIONS IN TUNNELING

THE HISTORY OF TUNNELING IN THE UNITED STATES

Throughout the history of civilization, innovations to build projects faster, with higher quality, and at a lower cost or to address specific geological conditions have always been, and continue to be, very important in the planning, design, and construction of tunnel projects.

This has not changed over time; only the scope and size of the projects have changed. Important innovations of the past have allowed the tunneling industry to plan, design, and build the more complex and sophisticated underground structures required of a modern society.

Early civilizations, such as Ancient Egypt, Greece, and Rome, constructed tunnels and underground structures for religious purposes (burial caverns), water conveyance, and defensive purposes. These early tunnels were typically constructed in self-supporting geological formations. The excavation was achieved using rudimentary tools, such as bronze chisels and hammers. An early innovation in tunnel construction was the implementation of fire and cold water for rock excavation. As Bronze Age tools required a fissure in the rock formation to facilitate excavation, an ingenious method was developed to create cracks and fissures. A fire was started near the area to be excavated, and the rock was heated. The fire was removed and cold water was applied to rapidly cool the rock, thus cracking the rock formation. Unfortunately, this was not a healthy work environment.

A significant improvement was the development and use of iron handheld star drills (multiple-sided face) that were used to excavate a circular hole. The opening was then filled with black powder and ignited with a blasting cap. Multiple holes were typically set off at the same time to expand the impact of the blast. This technique was used in the late 1700s and first half of the 1800s. It was this development that helped to build the early canal and railroad tunnels, such as the Transcontinental Railroad from Omaha, Nebraska, to Sacramento, California.

In the mid-1800s, three major innovations were used within the tunneling industry throughout the world that made tunneling safer and less expensive. The introduction of pneumatic drills, nitroglycerin, and dynamite significantly impacted tunnel construction. The pneumatic drill was invented by Samuel Ingersoll in 1871 and incorporated steel drilling rods with replaceable bits (Figure 8.1). This enabled workers to drill more footage per day in a much less tiring and safer manner, as the pneumatic drill turned the drill steel and bit at a very high speed. All that was required of the worker was to control the direction of the drill and the depth of the hole.

Nitroglycerin was first invented in 1846 and used as an explosive material in tunnels in Europe and North America. The material was not very stable and required special handling to prevent premature detonations. The product was used to complete the final hard-rock tunnels and opencut passes in the High Sierras for the Transcontinental

Figure 8.1 | **Shaft Sinking in Rock Using Pneumatic Drills**

Previous Page | **TBM Cutterhead for the Atlanta Combined Sewer Overflow Tunnel**

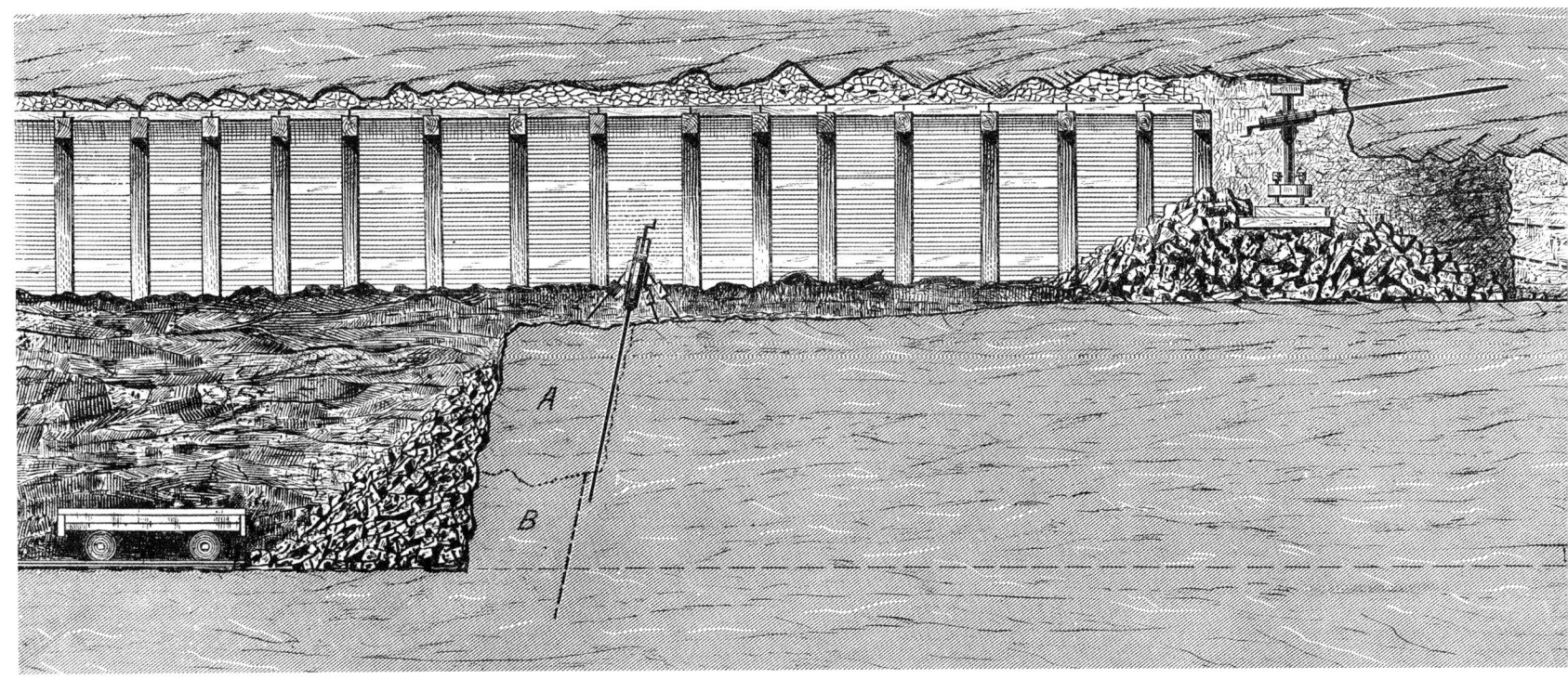

Figure 8.2 | **Vosberg Railroad Tunnel Construction Methods in 1887**

Railroad. In 1868, Alfred Nobel patented dynamite (U.S. Patent 78,317), a combination of nitroglycerin and silicate that is ignited with an electrical blasting cap. Dynamite is much safer to handle and results in a more productive explosion. This system of drilling holes at the leading face of tunnel construction (Figure 8.2), filling the holes with dynamite, then detonating the dynamite with an electrical charge is still used today.

The Industrial Revolution in Europe, bringing with it the invention of the early steam engine followed by the development of the steam locomotive, radically changed the face of transportation. As the railroad network grew to meet demands for rapid, direct goods shipment and passenger travel, so did the need for tunnels in areas where surface gradients were not amenable to track routing (Figure 8.3).

Up until the late 1800s, tunnels were typically constructed in self-supporting geological conditions because the technology was not available to safely support the soft ground conditions. With society requiring new tunnels for railroad and mass transit applications, a new tunnel construction system was developed in the United Kingdom and then brought to the United States. This was the circular self-advancing excavation shield designed for use in soft ground conditions. The shield system consisted of a steel cylinder with an internal excavation system and pneumatic or hydraulic cylinder shoving system that propelled off a cast-iron, and later a precast concrete, segmental final lining support system. This innovation was very important to the underground construction industry because it allowed for the building of more tunnels where they were required by society. Examples of these early shield tunnels were the original St. Clair River Tunnel between Port Huron, Michigan, and Sarnia, Ontario (Figure 8.4) and the Pennsylvania Railroad's Hudson River Tunnels in New York that are still in service today.

At the turn of the twentieth century, tunnel design became more of a science and was less reliant on trial and error. Tunnel design engineers began to use geotechnical investigation practices along with proven engineering formulas

Figure 8.3 | **Hudson & Manhattan Tubes in the Weehawken, New Jersey, Area**

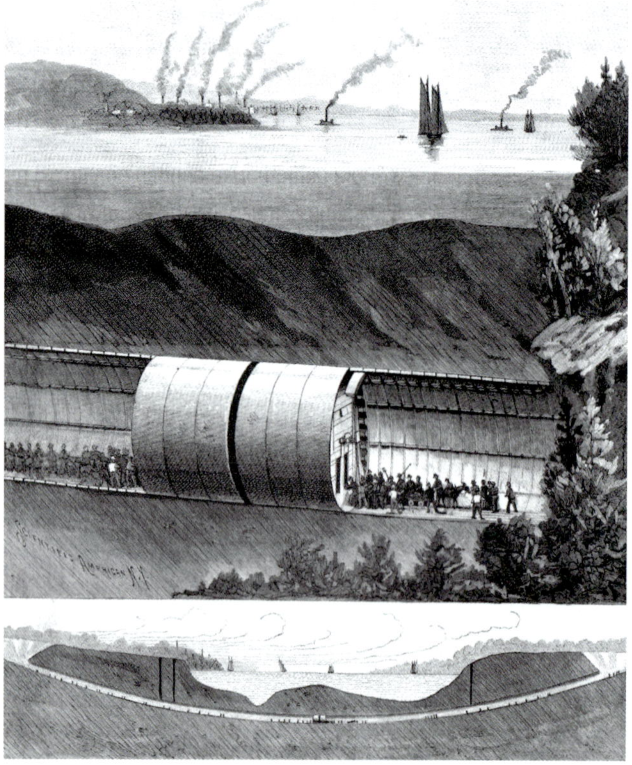

Figure 8.4 | **Beach Tunnel Shields at Work in the St. Clair River Tunnel**

and equations to design the support and the final lining of major underground projects. With the commitment to build underground subway systems in Boston and New York City in the late 1890s—which were already major metropolitan centers at the time—designers had to plan the proper route locations, then design tunnel systems that would last for decades. Consequently, scientific analysis entered the tunnel construction industry. This innovation was led by William Barclay Parsons, chief engineer for the New York Rapid Transit Commission, for the first mined tunnels in New York City. Parsons brought design innovation to the underground construction industry. Innovations in tunnel design have progressed concurrently with new techniques in tunnel construction. The design community has served the industry well for more than 100 years, as evidenced by the rarity of tunnel collapses in civil tunnel construction.

Soft-ground tunnels were designed with standard components and built using a shield and cast-iron linings or "tubbings." During tunnel excavation, there was an anticipated risk of water coming into the tunnel heading. Compressed air was often used to pressurize the tunnel by applying an air pressure greater than the force of the water pressure wanting to come into the tunnel. Compressed air had been in use since the mid-1800s (Eads Bridge [St. Louis] and Brooklyn Bridge caissons), but using it involved long-term health hazards for the people who worked in compressed-air environments.

For hard-rock (self-supporting) tunnels, the typical method of construction involved pneumatic drills, and later hydraulic ones, to drill holes for placing the dynamite. The tunnel heading was blasted by setting off electric blasting caps, igniting the dynamite. This process became known as *drill-and-blast*, a term still used today (Figure 8.5). During the 1950s and 1960s, innovations were made in the use of *mobile drilling gantries* (the industry term is *drill jumbo*), which supported many drills that could be moved in and out of large tunnel headings. The tunnel face would be simultaneously drilled by numerous drills at the same time. Workers became very adept at these new techniques. This process was used to construct the spillway and diversion tunnels for the Hoover Dam project, with this aspect of the project completed ahead of schedule and under budget. There are project management personnel in the industry today that "like to drill holes and blow things up!" and there is still a demand for this skill level in today's tunnel industry.

From the turn of the twentieth century until the mid-1960s, there was little innovation occurring in the industry. After World War II, the economy of the United States began to accelerate because of the pent-up demand for new infrastructure. Much of this work involved underground construction for new highway, railroad, subway, defense (missile silos and other underground structures), water, sewer, and hydroelectric tunnels. This led to a shortage of skilled personnel with conventional tunnel experience as society was demanding more and more underground structures. As often occurs in a commercially driven society, if a demand exists, someone will typically meet the request with an innovative idea or process. The age of mechanized tunneling was upon us.

It became necessary to build larger-diameter tunnels more quickly and safely, and the only way to accomplish this objective was to introduce automation/mechanization to the tunnel construction process. While tunnel boring machines (TBMs) had been invented earlier in the century, this modern transformation was initiated in 1953 when James S. Robbins introduced a rotating cutterhead TBM to the Oahe Dam project in South Dakota, achieving greater distances in a day than ever before possible (Figure 8.6). This was the beginning of new innovations in mechanical tunnel excavation that continue to this day whereby many tunnels are mechanically excavated in challenging geological or logistical conditions.

For noncircular or short tunnels where it is not economical or practical to set up a TBM and supporting equipment, other methods of tunnel excavation were developed, such as roadheaders and digger shields (Figure 8.7). Roadheaders are tracked machines that have an arm with a cutting head on the end that is used to cut strata, then gather the cuttings into an apron that puts the excavated material on a conveyor belt to the back of the machine. Digger shields are simple "cans" manufactured to a specific diameter for short soft-ground tunnels that require protection and support of the ground during excavation. Digger shields may be equipped with hydraulic excavators or roadheader booms. Primary tunnel supports are installed as the shield advances.

Primary support systems are methods used to support the rock or soil surrounding the tunnel before placement of the final (permanent) lining. These supports have advanced from cut heavy wood timbers, to tangential straight steel sets, to bent steel sets, to lattice girders encased in shotcrete as well as fabricated steel or precast concrete systems. For hard- and soft-rock tunnels, major innovations have occurred in rock-bolting technologies. Early rock bolts consisted of a steel rod with an expansion shell on the end that gripped only the rock at the end of drilled hole and were viewed as

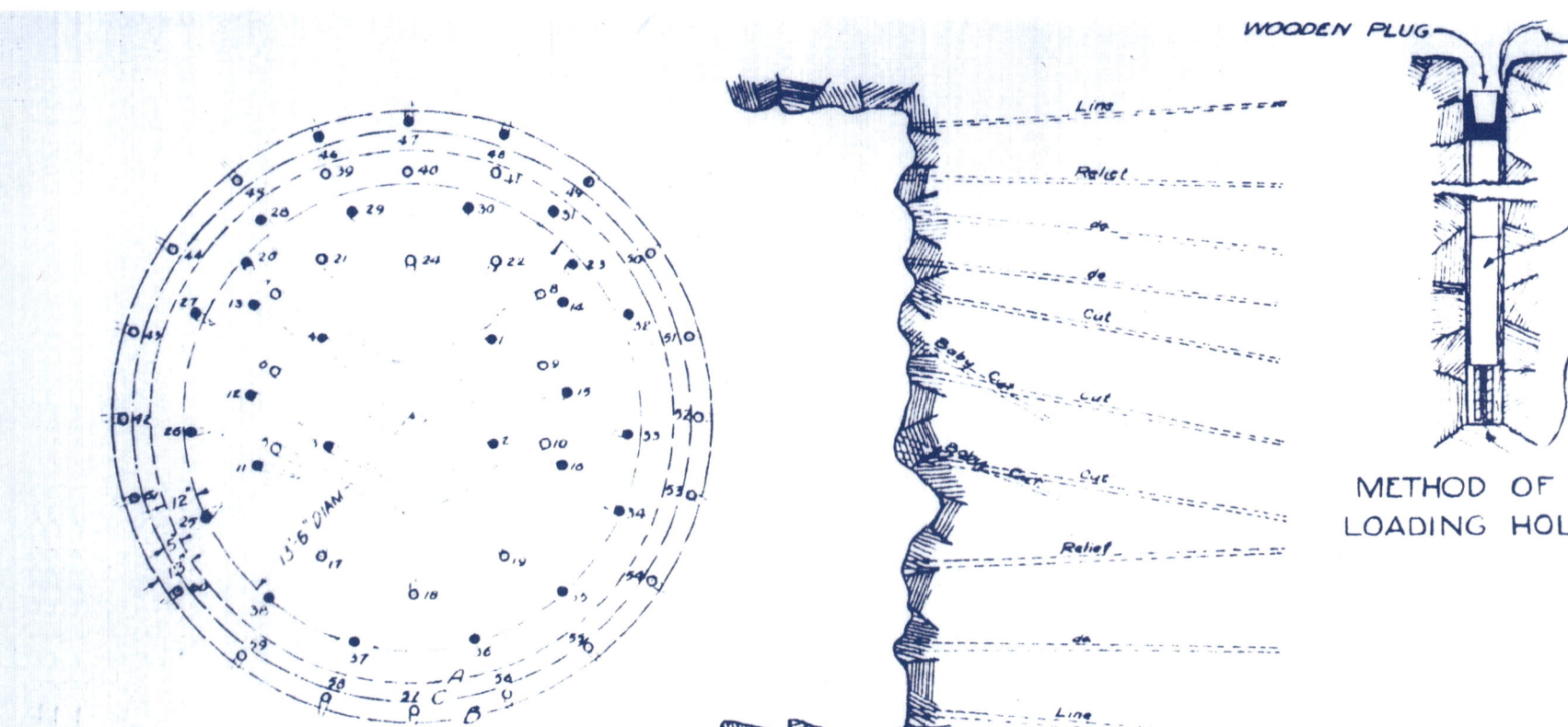

Figure 8.5 | **Typical Drill-and-Blast Tunnel Using a Multi-Level Drill Jumbo**

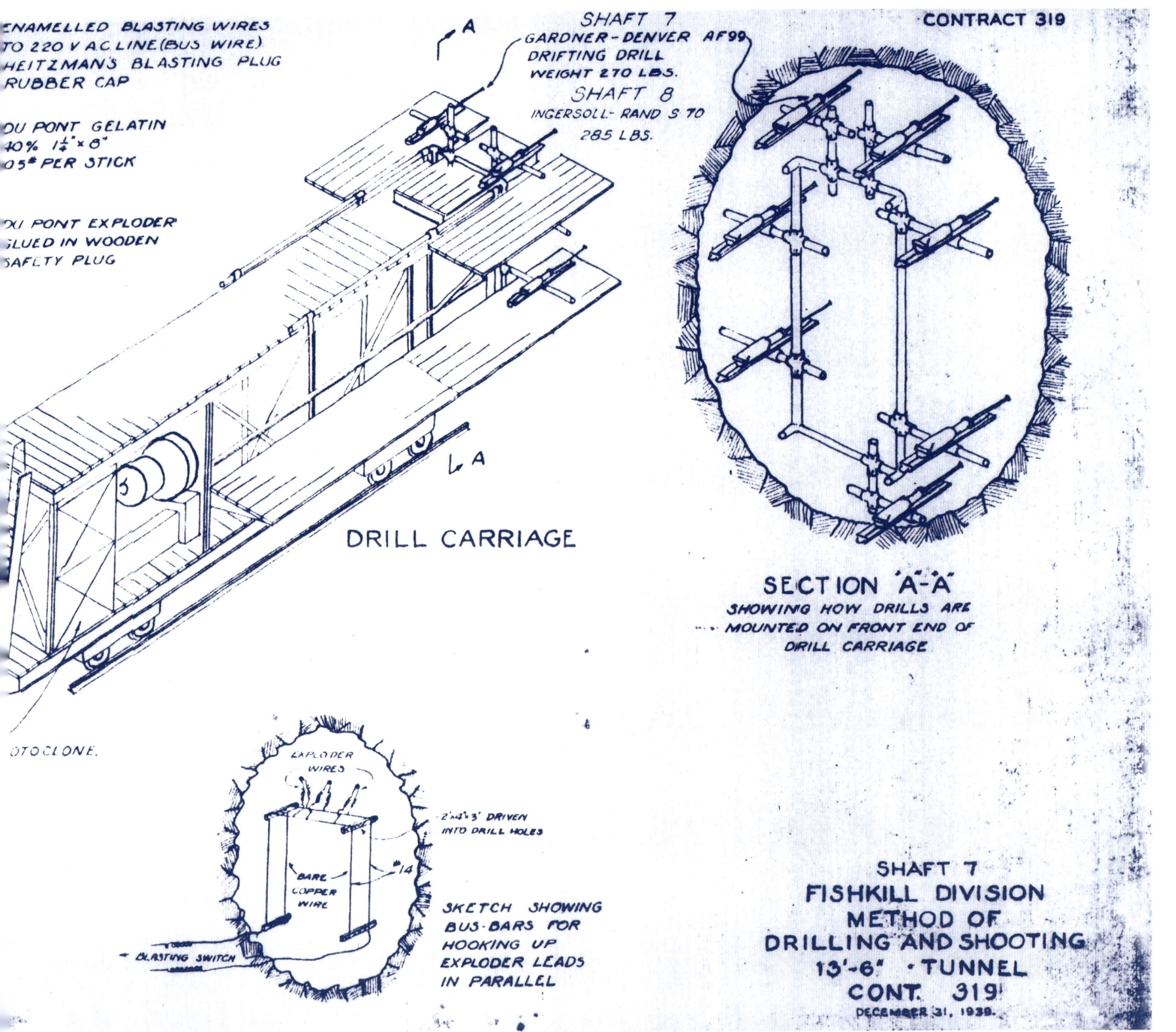

Figure 8.6 | **James S. Robbins in a Bored Tunnel Heading**

temporary support until a cast-in-place concrete tunnel lining was placed. These products have developed through many innovations that make them stronger and more efficient, can be installed faster and more safely, and are resistant to long-term corrosion and thus can be used as permanent support.

Circular tunnels have always been the most efficient method of tunnel construction, with the arching effect of the tunnel profile helping to support the surrounding strata. When the geology is soft ground, supports must be used during the excavation process to protect tunnel workers and the integrity of the opening. Regardless of the method of excavation, a circular support lining must be installed behind the mining machine to support the ground until either a permanent final lining is placed or the lining is used to perform as the permanent final lining.

In the initial soft-ground shield tunnel projects at the beginning of the twentieth century, the linings installed behind the shield excavation were typically cast iron and made to a specific diameter to fit the tunnel opening. These were initially cost-effective, but as foundry castings became more expensive and steel shortages developed during and after World War II, other methods of support were implemented. For primary tunnel support applications, a "ribs and lagging system" became prevalent as wood timbers cut to a specific size and length were installed between rolled wide-flange steel beams to create a continuous "barrel" that supported the ground (Figure 8.8). This was very cost-effective because the hardwood timber lagging and the steel-ring beams that formed the ribs were inexpensive and fast to install, and the components could be handled in the tunnel by the workers with no mechanical erection device required. This system is still used today for relatively short tunnels that are under approximately 16 feet in excavated diameter.

Because of the destruction of domestic steel manufacturing facilities during World War II, tunnel linings made of precast concrete were developed and implemented in Europe and Japan to rebuild their infrastructures and to regrow their respective economies. Precast concrete segmental tunnel linings were introduced to North America in

Figure 8.7 | **Fully Assembled Digger Shield for Metro Tunnels, Semi-Profile and Front View**

the 1970s but really took root in the United States in the 1980s on the Washington Metropolitan Area Transit Authority (WMATA) Expansion Program in Washington, D.C. (Figure 8.9), as well as in Baltimore. This allowed for the construction of soft-ground tunnels in very challenging geological conditions where both primary and one-pass precast tunnel linings were installed.

In the new millennium, one-pass precast concrete segmental tunnel linings have been a major innovation in the tunnel construction industry. The one-pass precast tunnel liner can be designed and manufactured to meet a variety of tunnel applications, from 10-foot-diameter tunnels for water and sewer up to 56-foot-diameter tunnels for highways. Innovations in the design and manufacture of precast tunnel linings combined with the advances in TBM technology have dramatically impacted the efficiency and capabilities of the tunnel construction industry.

Precast concrete tunnel linings can now be designed and manufactured to withstand more than 750 psi external water pressure acting on the lining and have concrete rheology and gasket sealing technologies designed to perform for more than 100 years. When integrated with the new hydraulic handling equipment built into today's TBMs, tunnels can now be excavated and supported faster and more safely than ever before.

Innovations are not always associated with bigger, faster machines. Improvements in tunnel construction safety programs have been dramatic since the conclusion of World War II. Skilled and productive workers are an asset that must be protected on the jobsite and in all aspects of life. The major tunnel construction companies have studied, developed, and implemented worker safety programs that prevent accidents and save lives (Figure 8.10). Every aspect of the tunnel construction process has been reviewed and changed where required to create a safe work environment for these very skilled people.

Tunnel project planning and financing, often referred to as "project delivery practices," is an area where one may think that innovation cannot occur, given that the process is well established and all projects are delivered using the same delivery process. The common approach is that the project is always planned, the money is allocated from the owner (federal, state, local, or private owner), the project is designed, and the project then goes out for bid and the low bidder builds the project. Although this is the most prevalent and commonly used delivery method within the underground industry, generally known as design–bid–build, there have been several innovations to this process throughout the last 150 years.

Figure 8.8 | **Steels Ribs and Timber Lagging as Initial Lining in the Fort Lawton Tunnel, Seattle**

Figure 8.9 | **Typical Precast Concrete Tunnel Lining for the WMATA Metro, Washington, D.C.**

Since 1990, innovations have occurred in how projects are procured, contracted, and ultimately delivered to control project costs and to shorten project completion times.

Although the public–private partnership is a recent innovation in project delivery, it is noteworthy that the first subway tunnels in both Boston and New York were built using this same mechanism. The tunnel construction industry seems to have innovated full circle in 125 years in this regard. This may infer that the industry is essentially linked to prevailing economic, political, and social events as well as the changing dynamics within our society. As this landscape tends to be cyclical, the construction industry correspondingly evolves and changes to meet the new challenges of the day.

EVOLUTION OF TUNNEL DESIGN

Figure 8.10 | **Green Cross of Safety Award in Construction**

From the early beginnings of society, there has always been some form of design associated with any construction. Design is, at its core principles, the result of trial and error, which is an assessment of what works successfully and what does not, and therefore fails in some way to meet the intent of the construction. Many theories, formulas, and approaches that have been derived from experimentation and academic study throughout the centuries have resulted in a common approach for undertaking design. Academia also adopts a trial-and-error approach, with successful theories being moved forward as a preferred approach until an improvement or alternative is derived that changes the in-vogue method of design.

As such, design for underground projects is no different. It is an industry that is very much based on and evolves through the consideration and application of lessons learned, or trial and error, for design and construction. Tunneling is both an art and a science, which means that design cannot be completed without an understanding and consideration of construction techniques and equipment applications that will act in combination with the appropriate applied design theory.

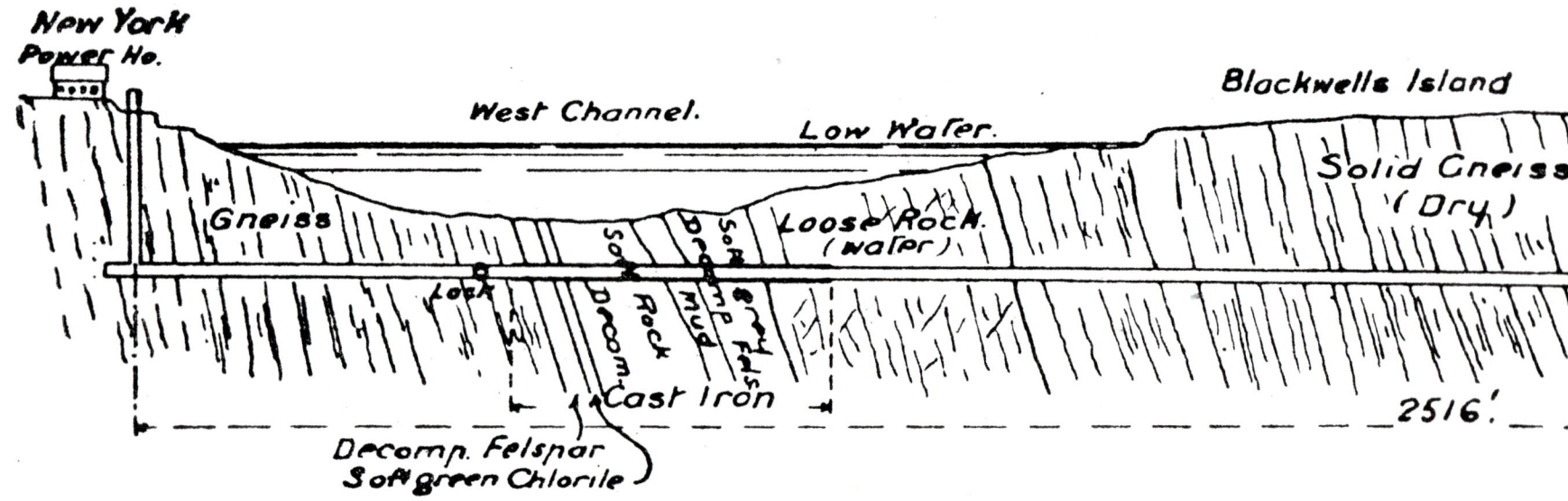

Figure 8.11 | **As-Built Geological Profile of the East River Tunnel in New York City in 1894**

A unique facet of this industry, and one of the greatest influences and considerations for all underground design, is the vital need to address the many risks that are imposed by Mother Nature, tunneling through the many varieties of ground conditions under all forms of infrastructure, while enabling society to continue as uninterrupted as possible (Figure 8.11).

Throughout the past 200 years, innovation in tunnel design and engineering has played a very important role in allowing the industry to build more complex tunnels to serve the needs of modern society. The role played by the major universities in Europe and North America was most important as they instituted scientific programs in the study of civil engineering practices, educating the industry about design and engineering practices that were successful and providing explanations for those that failed. This role of academia is still very important today in educating the new engineers and acting as a conduit for the implementation of innovative practices to the tunneling industry.

Before any underground project can be built, the engineer needs to know what geological matrix the tunnel will be constructed through so that a suitable construction system can be selected along with the proper initial and final lining (Figure 8.12). Rock mass classification systems have been in place for more than 100 years since Wilhelm Ritter (1879) attempted to develop an empirical approach to tunnel design with the objective of determining support requirements in advance of excavation.

One of the real leaders in the field of rock mass and soil classification for the purpose of tunnel support system design was Karl von Terzaghi, a civil engineer and geologist. Born in 1883 in Prague (now Czech Republic), he entered the Technical University in Graz, Austria, in 1900. He came to the United States in 1912, and in 1924 he published a paper stating many of his theories that transformed the field of geotechnical investigations.

Later in his career, Terzaghi consulted on the underground tunnel construction of the Chicago subway system and other major projects throughout the world. The ability to investigate and then determine how a specific soil or rock mass would react during excavation and construction was a major innovation.

Karl von Terzaghi (1883–1963)

Terzaghi was a pioneer in his field and is known as the "Father of Soil Mechanics." He graduated from Graz University of Technology in Austria with a degree in mechanical engineering in 1904. Following World War I, he relocated to Robert College in Istanbul where he continued his developmental work in soil mechanics, and in 1924 he published his magnum opus, *Erdbaumechanik*, which subsequently revolutionized this field of engineering. He later accepted a job offer from the Massachusetts Institute of Technology, setting up a soils laboratory and developing soils testing equipment, still in use today, with Arthur Casagrande as his assistant. In 1929, Terzaghi returned to Vienna and the Technische Hochshule and, using this as a base of operations, consulted on many projects throughout Europe. In 1938, he returned to Harvard University in the United States. Among his many achievements, Terzaghi is accredited with the establishment of the International Society for Soil Mechanics and Geotechnical Engineering and became its first president.

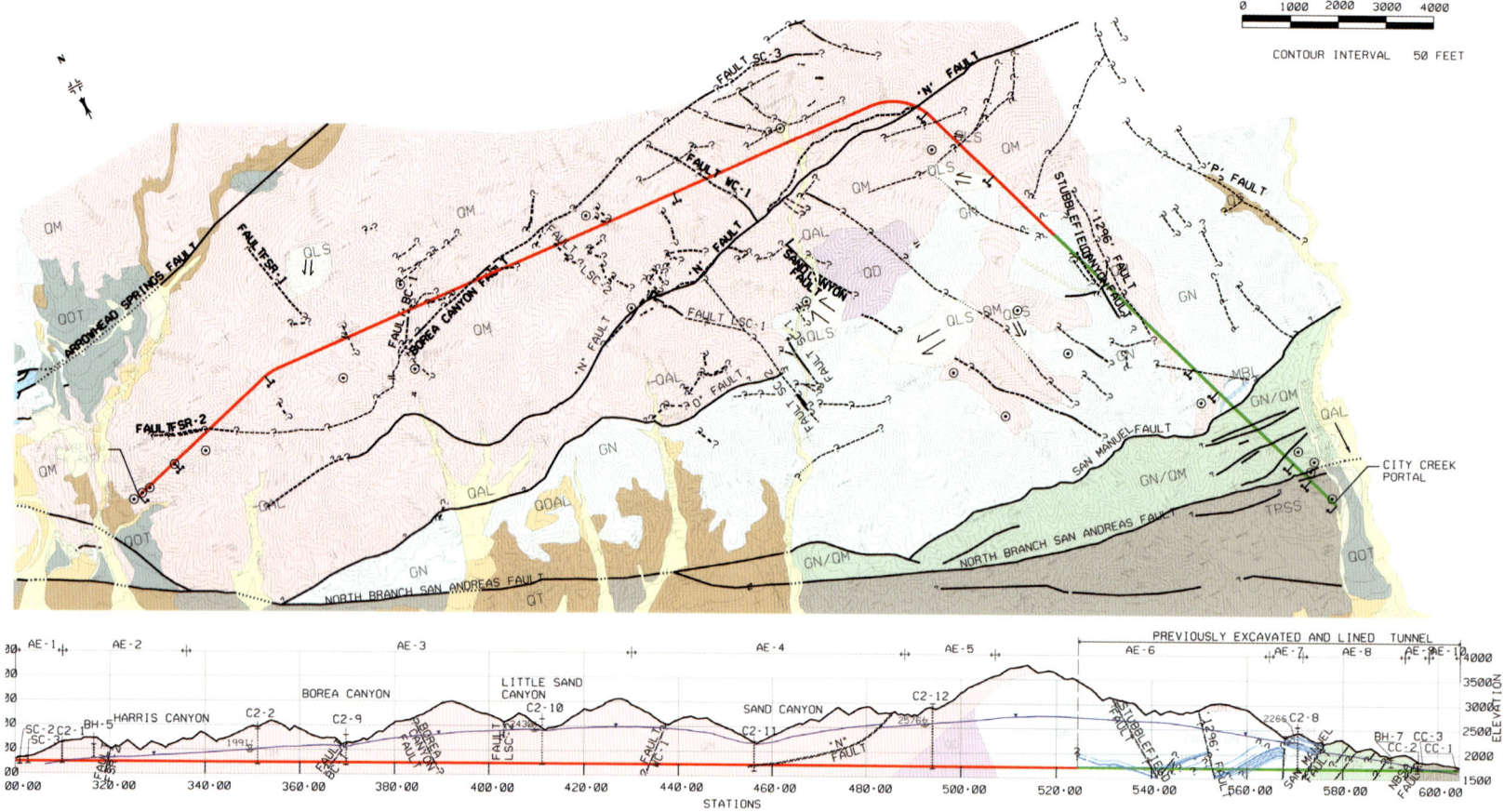

Figure 8.12 | **Geological Plan and Profile of the Arrowhead East Tunnel in San Bernardino, California**

Another innovation that made a major contribution to the tunnel industry was the development of the Rock Quality Designation (RQD) index by Don U. Deere in 1967. This provided a quantitative estimate of rock mass from geotechnical drilling programs. Deere's RQD index has been used extensively in North America since its inception. Various consultants and academics have attempted to relate the RQD index to Terzaghi's rock load factors and rock-bolt requirements in tunnels.

The evolution of geological core drilling, as referenced by Deere, has evolved into one of the most important tools for determining the geology that must be addressed when designing and constructing shafts and tunnels for a specific project. Many skilled geotechnical firms perform this service for the industry.

The tools of the design trade also changed from the 1970s to the 1980s and onward. We progressed from an era of floor-standing basic mechanical calculating machines, bulky logarithmic tables, and slide rules that were often referred to as "guessing sticks," to handheld calculators, later followed by development of the computers that we know today.

The approach to tunnel design largely followed an empirical methodology using hand calculations for two-dimensional (2-D) structures, typically in cross section, which took into account simplified and conservative assumptions for the weight of the ground and any associated groundwater that would likely apply load to the tunnel lining.

This approach, although resulting in comparatively thicker linings or greater amounts of reinforcement, performed well, but based on today's standards was less efficient. This changed with the use of computers in the process.

Commencing in the mid-1980s, the use of computers initially replaced the speed at which hand calculations could be performed, but this eventually extended to looking at the design with more efficient three-dimensional (3-D) theory while still remaining comparatively faster than the old methods (Figure 8.13). By 2000, there were several underground design software packages in common use depending on the type of ground conditions and the structure under design. This was a significant change, as design could be customized to each project to a greater extent, leading to cost efficiencies as the lining size, thickness, and reinforcement were optimized in comparison to the older methods.

Engineering drawing creation and production also changed with the advent of computerization and better printing and plotter equipment. Up until the late 1980s and early 1990s, most drawings were produced on a drafting table using either special ink drawing pens or pencils (Figure 8.14). Other drawing tools included scale rules, letter stencils for consistent fonts, compasses and dividers for developing curves and setting distances, and French curves for producing complicated compound curves. The drawing would be created on a specific tracing paper or a vellum stock that would allow reproduction when placed in a duplication machine. Paper prints or blueprints of the original could also be made one print at a time, which became very time-consuming when multiple identical documents were needed, such as bid documents for a specific project.

For public plans, executive committees, planning level drawings, or important purposes that needed higher quality and greater clarity, drawings were either drawn directly or duplicated onto a very high-grade linen stock (Figure 8.15). Different colors were introduced by hand-painting color wash, similar to water colors, in areas of the drawing that needed to be shaded (Figure 8.16). This process needed a skilled hand because any mistakes could not be corrected,

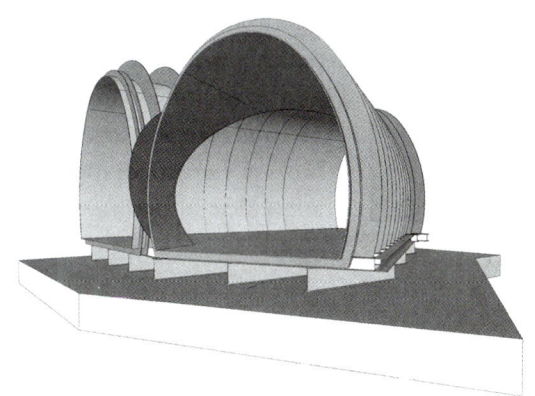

Figure 8.13 | **Rendering of a Fabricated Steel Wye for a Tunnel Bifurcation**

Figure 8.14 | **Typical Engineering Drafting and Printing Room, Circa 1910**

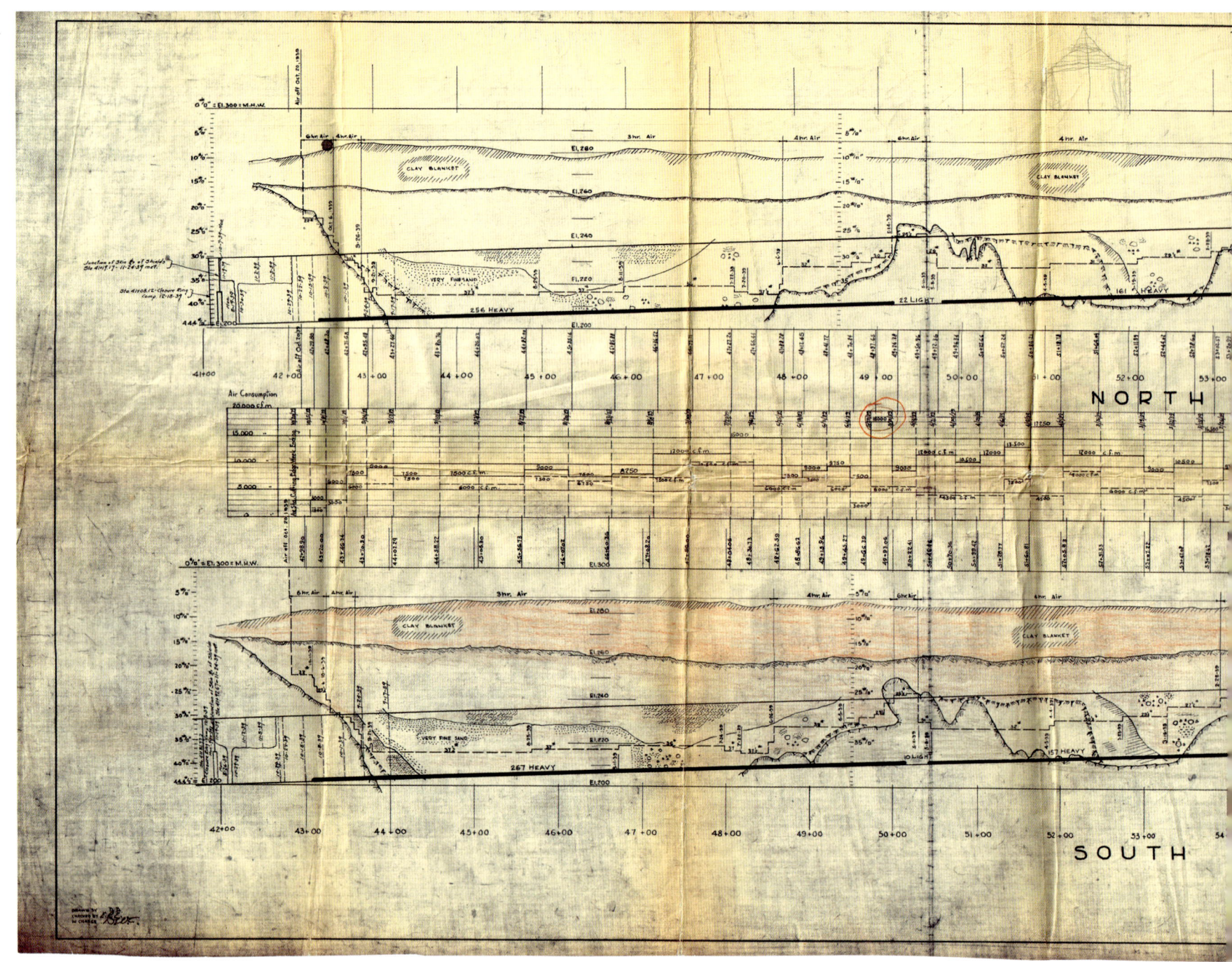

Figure 8.15 | As-Built Geological and Excavation Data from the Queens–Midtown Tunnel

CHAPTER EIGHT | INNOVATIONS IN TUNNELING

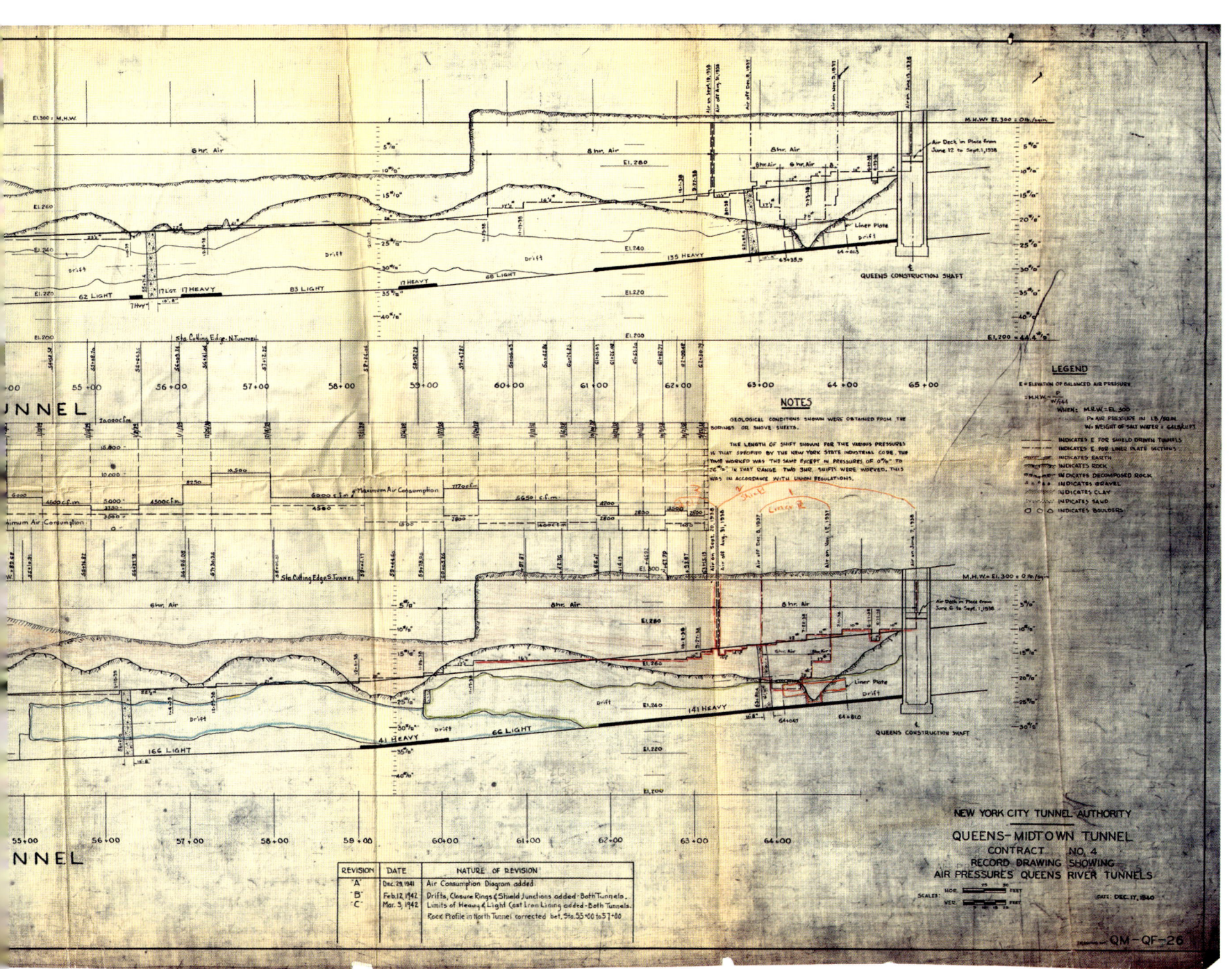

Figure 8.16 | **Steam Engine, Washington Navy Yard, Washington, D.C., Circa 1811**

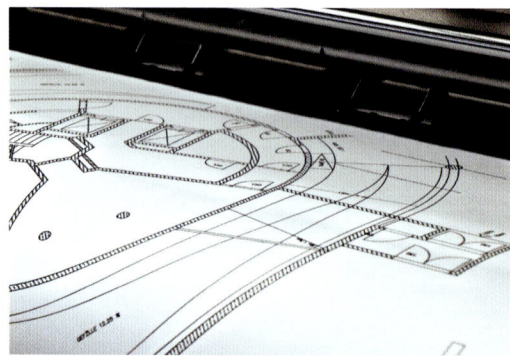

Figure 8.17 | **Modern-Day Electronic Drawing Plotting Equipment**

and the process was then repeated for each additional copy. There was indeed an art in achieving a continuous block of color without any blotches or runs from the color-washing process. By 2000, drawings were being sent to the plotter by computer, and the plotter produced the drawing with any colors desired. The art of manual drawing production and color washing was over, and drawings were now being produced with high efficiency (Figure 8.17).

What was also unexpectedly removed with this process was "thinking time" for design. Often the design engineer would send his or her first draft of a drawing to the draftsperson during which time the engineer could reevaluate his or her assumptions with sufficient time to make further design improvements and adjustments. Given that drawing production no longer took as much time to produce the first draft, thinking time for the engineer was removed or considerably reduced. The engineer needed a better way in which to develop the design with sufficient thought going into the design concepts. The evolution of the computer combined with comprehensive peer review using industry experts became the answer.

Throughout the 1970s and 1980s, computers were being used predominantly at universities for rudimentary structural calculations in design. This involved development of the actual programs as well as providing the data on punch cards, magnetic tape, and discs for data storage to be loaded into the computer each time they were used.

At the beginning of the 1990s, computers in turn evolved from having a colored font on a black screen display, cumbersome programming and printing capability, fairly basic calculating capability, and very limited memory capacity to full color displays by 1995, and significant calculating capability by 2000. In the space of 10 years, this innovation, together with the comparable evolution of the software, computers radically changed the way in which engineers were able to design tunnels.

In recent years, the industry has been moving toward the use of building information modeling (BIM) software. BIM is a new engineering tool that allows the design engineer to carry out the planning, design, and construction phases of a project regardless of size and complexity. BIM is a digital representation in three dimensions (3-D) of a construction project (Figure 8.18). This enables the designer to plan, schedule, and visualize a project before construction is initiated. The onboard rendering capabilities and ability to fly through a virtual completed facility from an underground perspective provides a capability that was only in the mind of design engineers looking at a 2-D drawing. As recently as the last two to three years, the capability has been enhanced to a BIM 5-D perspective, where time and costs can be derived from the 3-D model onscreen (Figure 8.19). Once the model is established and all data are plugged in, schedules and construction cost can be derived, validated, and assessed, making the entire process more efficient. The ability to see the construction envelope, complete with calculated loads and stresses, before the area is excavated can save millions of dollars in construction time and redesign (Figure 8.20). In addition, BIM adds to the safety of the project because hazardous situations can be identified in advance.

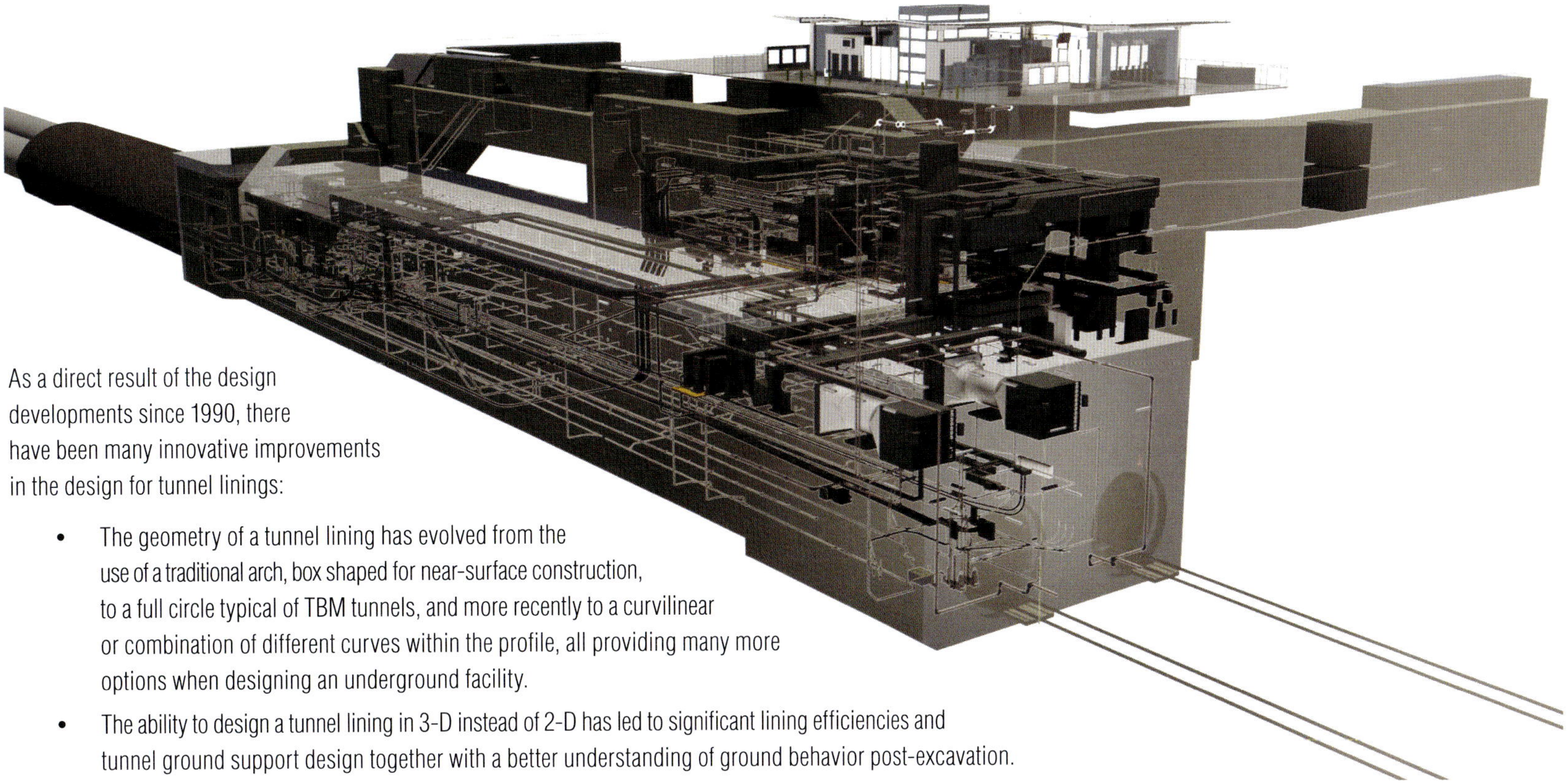

CHAPTER EIGHT | INNOVATIONS IN TUNNELING

As a direct result of the design developments since 1990, there have been many innovative improvements in the design for tunnel linings:

- The geometry of a tunnel lining has evolved from the use of a traditional arch, box shaped for near-surface construction, to a full circle typical of TBM tunnels, and more recently to a curvilinear or combination of different curves within the profile, all providing many more options when designing an underground facility.
- The ability to design a tunnel lining in 3-D instead of 2-D has led to significant lining efficiencies and tunnel ground support design together with a better understanding of ground behavior post-excavation.
- Reinforcement of tunnel linings is moving from traditional steel reinforcing bars to the use of steel and plastic fibers blended with the concrete mix, and fiberglass reinforcement for temporary conditions.
- The application of permanent sprayed concrete replacing traditional cast-in-place concrete has also seen some development, particularly for noncircular structures and difficult forming applications.

Figure 8.18 | **External BIM Image of a Complex Urban Metro Station**

Designing to contain and prevent infiltration of water has also seen significant development. Where the industry once used a combination of nominal low-pressure gaskets and internal caulking with asbestos and lead wool and oakum to seal leaks post-construction, it now relies on full membranes or high-pressure gaskets that are custom designed for the project and have the ability to seal against very high groundwater pressures.

With the recent trend to invest in maintaining infrastructure, together with resiliency of existing facilities, there is a renewed concern regarding the service life of tunnel structures. Design has kept pace with this requirement and continues to identify project-specific techniques for extending the design life of a tunnel.

Into the future, we see many further innovations in design that could come to fruition. Improvements in ground investigation techniques may lead to a better understanding of the ground conditions and behavior ahead of the tunnel face and help overcome the forces of Mother Nature during tunneling. Further enhancements of BIM are anticipated and potential integration with the many different structural and geotechnical analyses software programs that are commonly used within the industry to provide a one-stop underground design tool. Providing a seamless ability to profile, analyze, and evaluate

Figure 8.19 | **Internal BIM Image of a Modern Metro Tunnel Design**

existing tunnels to deliver an instantaneous health check on a regular basis is also not yet possible but is likely to be achievable in the future.

There is no question that underground design has seen substantial innovation, particularly in the last 25 years. There is recognition within the industry that this goes hand in hand with construction innovation given that one is inextricably linked to the other. History teaches us to consider and respect the lessons learned from both the design and construction of underground projects. When the envelope is pushed further for any aspect of the industry, we quickly learn and improve upon the challenges that were encountered. The ability to embrace state-of-the-art techniques, equipment, and materials has enabled projects to be successfully completed that would not have been possible or financially viable using the techniques of the past. It is this innovative mindset of the tunneling industry, considering both design and construction, that provides a solid foundation for the underground future.

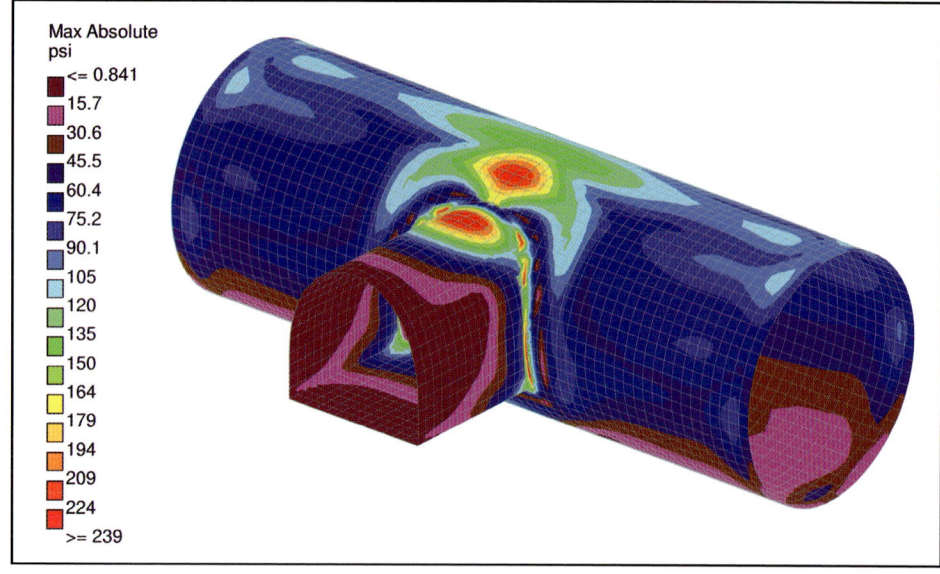

Figure 8.20 | **Finite Element Analysis of a Tunnel and Cross-Passage Connection**

INNOVATIONS IN PROJECT DELIVERY

From the time railroad and canal tunnel projects were initially planned and constructed in the early 1800s, a major issue has been how to finance the projects and what contracting model should be used to build them. The early railroad and canal tunnels were financed by the owner of the tunnels by selling additional shares or obtaining third-party financing. Construction was contracted to an outside company (Figure 8.21).

From 1863 to 1869, the U.S. government sponsored the construction of the Transcontinental Railroad. The Union Pacific and the Central Pacific Railroads were awarded the rights to and were paid for the line each built; however, each railroad had to determine the alignment, secure the financing, hire the contractor, and build the railroad to government specifications. The railroads were then paid for track miles built and accepted by the government. This is commonly referred to today as a design–bid–build method of project delivery.

In the mid-1890s, both Boston and New York City developed underground subway systems. At that time, the transportation systems in the cities were owned and operated by private companies that were typically owned by financiers. The project delivery method was the same for both Boston and New York City. The private transit company secured the rights of way, designed the project, secured financing from selling stock or by borrowing from lending institutions, hired the contractors, purchased the rolling stock, and operated the subway systems as stand-alone private entities. Today we call this the public–private partnership method of project delivery.

The industry has always looked for better ways to finance and deliver tunnel projects. Tunnel projects are becoming more complex and costlier, now exceeding $1.5 to $2.0 billion for individual contracts. More innovative and flexible contracting models and practices are being implemented in the United States to address risks and costs.

Contracting practices and procedures are a key determinant of successful project delivery with continued pressure to deliver on time and under budget, with minimal disputes and without claims or litigation. Given that underground projects tend to be complex undertakings with significant unknowns, the owner's contractual strategy and process is a critical consideration.

Specifics of U.S. Contracting Methods

Design-Bid-Build

Design–bid–build, or DBB, is the traditional method of public works procurement in which the owner contracts for design by an independent architect or engineer (or, in some cases, accomplishes the work in-house) for a set of design drawings and specifications that are the contractual basis of the contractor's bid. The architect or engineer (or the owner) certifies that the drawings and specifications meet all applicable requirements of building codes, engineering standards, and other design and safety requirements. This bid package is then made available for all qualified bidders to analyze, quantify, and provide a bid to the owner for construction of the work. For typical projects, DBB design and construction are accomplished by two independent, unrelated parties who are contracted separately to the owner, such as recent major subway construction in New York City (Figure 8.22) and where this is the owner's preferred contracting practice. The low responsive price from the contractor on bid day is generally accepted by the owner.

Design-Build

A growing trend is the use of design–build, or DB, in which the contracting agency identifies the required end result and design criteria. The prospective bidders then develop design proposals that optimize their construction abilities. The submitted proposals may be scored by the contracting agency on factors such as design quality, timeliness, management capability, and cost. These factors may be used to adjust the bids for the purpose of awarding the contract. A stipend may be provided by the owner for the submittal of a proposal.

The DB concept allows the contractor maximum flexibility (and additional responsibility) for innovation in the selection of design, materials, and construction methods. DB has been used for numerous infrastructure projects across the United States. Although not yet a "traditional" method in this country, it has decreased time to completion, and some projects show less cost growth than those using DBB, but this needs to be evaluated on a project-specific basis.

Construction Manager / General Contractor at Risk

The construction manager / general contractor at risk is a procurement method in which the design work is begun, either by the owner or a consultant, and a general contractor (GC) is engaged to work with the owner and designer to further develop the design and deliver the project. This GC is selected based on a combination of experience and price. The GC may not actually construct elements of the project but is responsible for delivery of the project at a guaranteed maximum price (hence, the term *at risk*).

The estimated cost for the work is developed by the contractor, and options to reduce cost, increase value, or shorten the schedule are evaluated with the owner,

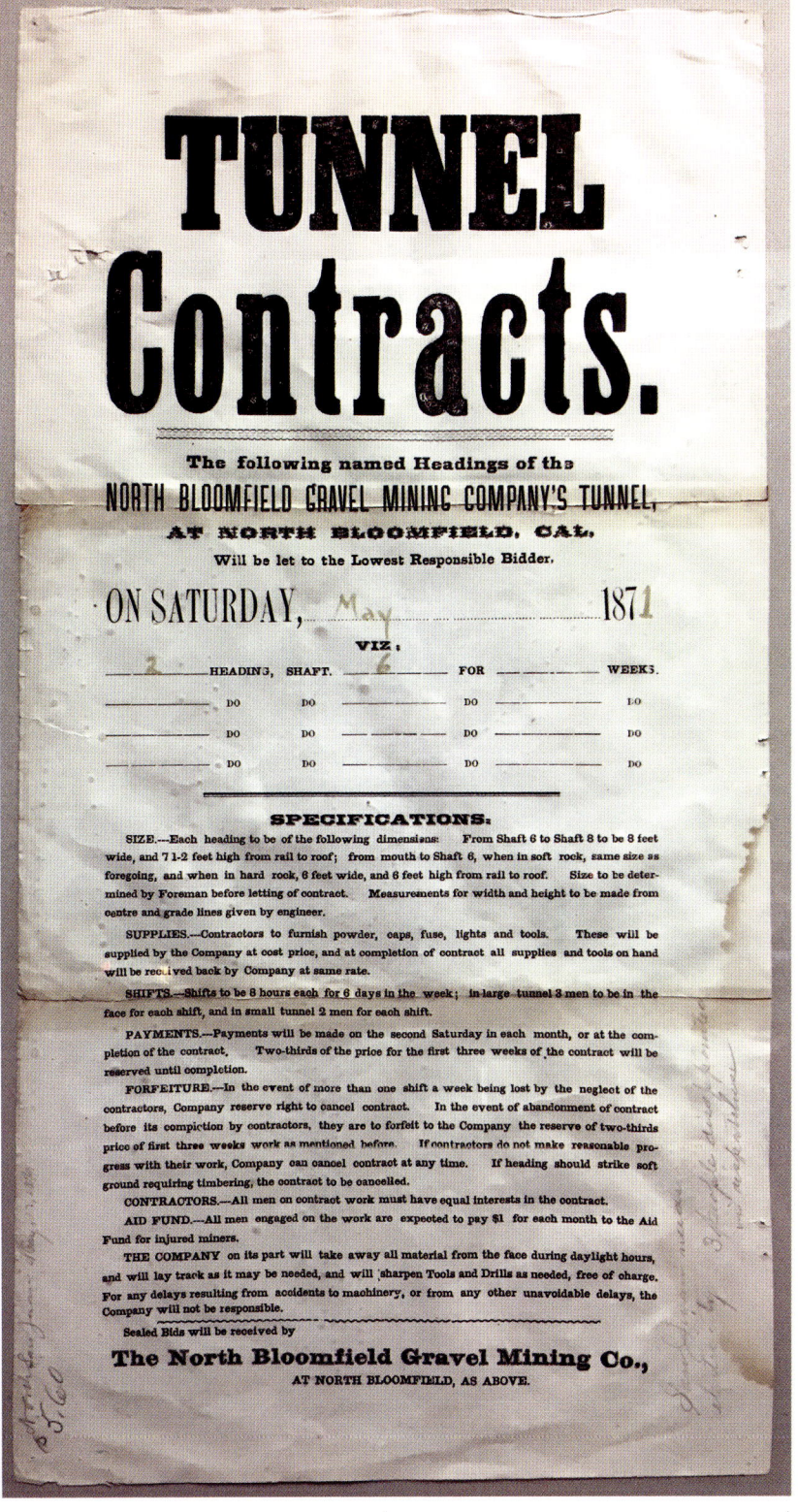

Figure 8.21 | **Call for Tunnel Contract Tenders, Circa 1871**

Figure 8.22 | **Second Avenue Subway Construction in New York City**

after which the contractor submits a guaranteed maximum allowable construction cost (GMACC), which the owner can accept or reject. If accepted, the GC can perform the work and is responsible for completing the work at or under the GMACC. If a particular subcontract ends up costing more than estimated, the GC is liable for the cost overrun.

Public–Private Partnership

A public–private partnership (PPP or P3) is a contractual arrangement between a public agency (federal, state, or local) and a private sector entity. Through this agreement, the skills and assets of each sector (public and private) are shared in delivering a service or facility for the use of the general public. In addition to the sharing of resources, each party shares the potential risks and rewards in the delivery of the service and/or facility.

Although P3s may look attractive on the surface to an owner, the success of P3s is very dependent on having highly skilled and motivated people with the authority to make binding decisions for all of the major participants in the P3. Before entering into a P3 tunnel project, one should study P3 tunnel delivery programs used throughout the world and analyze those that worked well and those that did not, because projects need to be a technical and financial success for all involved. For example, the recently completed PortMiami highway tunnel in Miami, Florida, was a very successful P3 tunnel project.

CONTRACTUAL INNOVATIONS IN THE UNITED STATES

Since 1955, contractually related improvements and innovations have included use of differing site condition clauses, alternative dispute resolution, Dispute Review Boards (DRBs), escrow bid documents, geotechnical baseline reports, partnering, and team alignment. The genesis of many of these methods can be traced to work begun in the early 1970s by the U.S. National Committee on Tunneling Technology, followed by subsequent committees and industry initiatives to minimize disputes, optimize the end product, and better coordinate the construction process.

The Basis for Contractual Disputes

The basis of many contractual disputes lies in the well-established, traditional, low-bid award process that forces contractors to submit the lowest responsive price to secure the work with an expectation that a profit can be realized at that price. Additional monies may be available as a result of changed or additional work directed by the owner as well as through claims for "changed conditions" to the work and conditions as provided for in the contract.

Addressing Contractual Disputes

Dispute Review Boards

DRBs are intended to assist the owner and contractor in resolving disputes quickly and reasonably. Such boards normally consist of three qualified and respected professionals selected jointly by the owner and contractor at the beginning of the work. These individuals are kept up to date on the progress of the work through regular meetings and written communications.

If the owner or contractor believe there is a potential dispute that could be resolved with the help of the DRB, they can submit their case to the DRB who will render an opinion on the merits of the issue. The opinion of the neutral third-party DRB is a benefit in resolving many potential disputes. It has also encouraged owners and contractors to resolve potential disputes directly, rather than submit to the DRB.

Escrow Bid Documents

With escrow bid documents, the process escrows the contractor's bid and supporting documents, storing and securing them in a proprietary and confidential manner for the duration of the contract. They are only used if there is a need to verify the bid basis to assist in the negotiation of contract changes or the resolution of disputes, claims, and controversies. Often, large subcontractor bid submissions are also escrowed.

Geotechnical Baseline Report

The geotechnical baseline report (GBR) sets out, as a contract document, the owner's anticipated subsurface conditions with an estimate of their impact on design and construction. This allows potential bidders to have a common understanding of the subsurface conditions affecting the work, resulting in more uniform bid prices with less contingency, and less exposure to claims and changed circumstances due to underground conditions. If conditions are materially different from the baselined conditions and values and the financial impact can be demonstrated, the contractor is entitled to additional compensation. In this way, the owner accepts the risk for conditions that are different from the baseline. As projects become larger in scope and cost, the importance of a proper and professional baseline report has gained in acceptance as a means to define a contractually based mutual understanding and to reduce project risks, conflicts, and controversies. A comprehensive GBR is now considered to be an essential contract document.

Partnering and Team Alignment

To reduce the potential for adversarial contractual relationships to develop between owner and contractor, *partnering* was developed by the U.S. Army Corps of Engineers in 1989. It involves building a mutual understanding of key goals and objectives, and aligning all parties to those goals and objectives, so that they can work more productively and collaboratively together for the benefit of all parties to the contract.

Partnering has been credited with a significant reduction of disputes and claims, an increase in job performance and participant satisfaction, and has improved communication and working relationships on construction projects. *Team alignment* is a similar process, led by the owner using partnering principles, and is more applicable to management of complex megaprojects.

Innovations in project delivery are an essential part of the tunneling industry. Project owners expect to have a degree of certainty in terms of final project cost as well as delivery by the expected date. In the case of public owners in the modern era, society expects delivery on time and on budget when public money is financing the project. Having the right project delivery approach is always a challenge for any form of project. This is arguably more critical for an underground project where construction is attempting to overcome the forces of Mother Nature.

Major tunnels are frequently complex projects and can be extremely challenging to plan, design, finance, build, and operate, but the tunneling construction industry is well equipped and continues to innovate in this area to achieve everyone's common goal of project success.

INNOVATIONS IN GROUND TREATMENT, STABILIZATION, AND GROUNDWATER CONTROL

The underground construction industry as we know it today, with its sophisticated engineering technology and equipment, is built on foundations first laid down nearly 200 years ago by a handful of visionary engineers. Their ingenuity and determination, even in the face of setbacks and defeat, led to remarkable advances over a relatively short period of time, making mass urban underground transportation and vastly improved utility lines a practical reality by the end of the nineteenth century. Other innovators would follow in their footsteps, developing dewatering and ground improvement techniques that, even though they were not the primary objective at the time, ultimately enhanced and significantly advanced tunneling practices, leading to greater application and project success.

The Birth of Subaqueous Tunneling

In 1835, a significant collapse of the 1.4-mile-long Kilsby Tunnel in the United Kingdom occurred when workers encountered a previously unknown seam of water-filled sand and gravel (Figure 8.23). This led Robert Stephenson, son of the famed steam locomotive designer George Stephenson, to direct the installation of a series of steam-driven pumping wells to remove the water ahead of the tunnel face. The project was successfully completed in 1837. This innovation made the difference between completing the tunnel or abandoning the project and became a milestone within the underground industry, enabling tunnels to be constructed through saturated ground conditions. Around the same era, other British engineers were turning their attention to another issue: how to tunnel effectively beneath a body of water. Two critical engineering advances were to make this a reality—the tunneling shield and the use of compressed air to offset groundwater pressures.

The tunneling shield was originally conceived and patented by Marc Isambard Brunel in 1818. Brunel used a variation of his original patent to excavate the Thames Tunnel in free air. The project began in 1825 but, due to delays, was not completed until 1843. Nevertheless, this was the first tunnel to be successfully constructed beneath a navigable river.

During the extended construction period of the Thames Tunnel, other British engineers were improving on Brunel's tunnel shield concept. In 1864, Peter W. Barlow patented a circular shield. Two years later, R. Morton developed a segmental cast-in-place liner with shield. But perhaps one of the most pivotal developments in early subaqueous tunneling had come earlier when Sir Thomas Cochrane submitted a patent for the use of compressed air for shaft and tunnel construction in 1830. It was civil engineer James Henry Greathead, who had served an apprenticeship with Barlow, who put all the elements together. He teamed up with Barlow and together they undertook the Tower Subway, the second River Thames subaqueous tunnel. The Barlow–Greathead shield was cylindrical and fitted with screws that enabled it to be slowly jacked forward while the cast-iron segmental lining was placed behind.

Figure 8.23 | **Kilsby Tunnel on the West Coast Main Line Railway in England**

Figure 8.24 | **Brooklyn Bridge Caisson Under Construction with Sandhogs**

It would be some years, however, before Greathead added compressed air to the procedure for the City and South London Railway Tunnel in 1886. This was the first use of compressed air for a shield-driven tunnel. It was the knowledge sharing of these innovations that radically changed tunneling in the United States. Up until this point in time, crossing under major water bodies had proved to be a significant barrier to tunneling.

Until construction of the Brooklyn Bridge, little was known about the use of compressed air for groundwater control for soft-ground tunneling or the dangers it posed for workers. This changed in 1872 when the tunnelers of New York City, known as "sandhogs," went on strike against the terrible conditions they had endured for two years to excavate the bridge foundations within compressed-air-filled caissons.

Thirteen years in the making, the Brooklyn Bridge, brainchild of engineering genius John Roebling, captured the public's imagination with its daring superstructure design. The approach to foundation construction was just as innovative. Massive wooden caissons, the size of half a city block and open at the bottom, were progressively sunk deeper into the banks of the East River. Compressed air was used to expel water from within the caissons. Sandhogs would enter the caisson via an air lock (Figure 8.24). The effects of being "in the air" were excruciatingly painful, causing weakness, crippling joint pain, abdominal cramps, blinding headaches, fever, and often delirium on returning

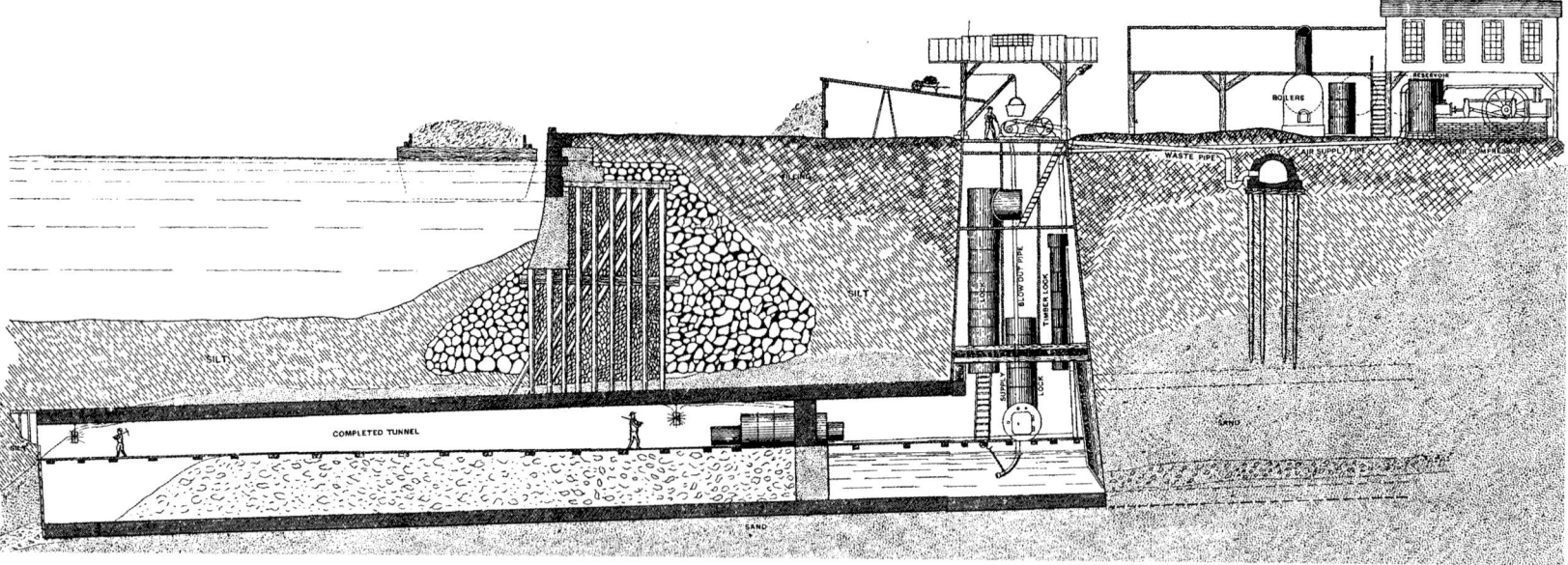

Figure 8.25 | **Original Construction Scheme for the Hudson & Manhattan Tunnels**

to the surface. The deeper the caissons went, the more the air pressure was increased, and more and more men succumbed to what ultimately became known as "caisson disease."

When two men working in the deeper New York caisson died within a week of each other and more workers became very seriously ill, the sandhogs organized and the entire belowground crew, some 200 men, went on strike. If they were putting their very lives on the line every day, they wanted $3 for a four-day work week. But under the threat of being fired, and with out-of-work men eager to take their places, the sandhogs capitulated after only three days. The strike was not a total loss. The negative publicity forced management to pay more attention to workers' lives. Digging was halted, original plans were modified, and the caissons were completed. It would be two years before the next major New York tunneling job commenced, and the sandhogs would once again play a central role in tragedy.

Hudson River Tunnel

In 1873, the Hudson River, which was too wide for a bridge to span, prevented a direct link to the west from New York City. Trains carrying passengers or freight had to either enter from the north or stop at the New Jersey shore and transfer to ferries or barges. Despite the ventilation difficulties associated with running a steam train through an underwater tunnel of such length, Colonel Dewitt Haskin, a railroad engineer, secured financing for the construction of twin tubes beneath the Hudson, connecting Jersey City with Manhattan (Figure 8.25). The work would be done under his newly formed Hudson River Railroad Company.

This was a landmark project and an ambitious one. A tunnel under a major U.S. river had never been attempted before. The water and the riverbed silts bearing on the excavation could not be held at bay by timbering (wood supports) alone. Haskin had followed the progress of the Brooklyn Bridge and reasoned that compressed air could work equally well for the Hudson Tunnel. However, the relatively new tunneling shield technology was not a part of his work plan.

Caissons were sunk in 1874, but legal injunctions delayed the work several times. Excavation finally began in 1879. Then, in July 1880, disaster struck and closed down the project completely. Soon after 28 men had entered the tunnel for their shift, the unmistakable screech of escaping air signaled a blowout in the roof of the temporary entrance between the working shaft and the tunnel. The seven men working in the entrance managed to escape back to the air lock, but the rest of the crew perished in the water that rapidly rushed in. It took three months for all the bodies to be recovered. Work resumed, but Haskin's financial backers had lost faith in the venture. Without their support, the company closed down and the work was abandoned in 1882. It was resumed in 1891, this time with tunnel shields and compressed air, under British contractor S. Pearson & Son, but this attempt ran out of funding and the project was once again abandoned. Of great importance during this attempt though was the development of the medical air lock by E.W. Moir, Pearson's project manager, which greatly improved worker safety.

The tunnels were eventually completed in 1905 by J.V. Davies and Charles M. Jacobs as part of what became known as the Hudson & Manhattan (H&M) Tunnels. Jacobs, who was British, had completed the Ravenswood gas tunnel under the East River in 1894, the first subaqueous tunnel in New York City. This was to become an integral part of what was the most ambitious tunnel undertaking of the day—excavation of multiple transportation tunnels under the city's Hudson River (known as the North River in the vicinity of New York and northeastern New Jersey) and the East River for the Pennsylvania Railroad Company (PRR).

The Pennsylvania Railroad Tunnels

All six of the proposed PRR tunnels in New York City under the East and North Rivers were constructed under compressed air. The distances to be tunneled, the river depth, and the material through which mining would take place made the PRR project unprecedented in size and scope.

The North River Division of the project, which included two single-track tunnels running 2.76 miles from Hackensack, New Jersey, to Manhattan, came under the engineering direction of Jacobs. The tunnels were driven simultaneously from both ends. The tunnel sections met on September 10, 1906, a year ahead of schedule, with a variation of just 1/16 inch. The event was marked by the ceremonial passing of a box of cigars between sections, and two days later, Jacobs led a party of PPR officials on foot through the entire tunnel length.

The contractor and shield designer for the four tunnels under the East River was S. Pearson & Son, which had worked on the abandoned H&M Tunnels. The tunnels were each about 6,000 feet in total length and involved 3,900 feet of subaqueous tunneling between deep shafts on either side of the river. For much of the crossing, cover above the tunnel crown to the river bottom was anticipated to be as little as 8 feet, bringing with it the danger of frequent blowouts of compressed air. Consequently, alternative methods were evaluated, including an extended test of the then little-known ground freezing technology. However, this idea was abandoned because of the time it would take to achieve freeze closure. Nevertheless, tunneling progress under compressed air was good, though marked by blowouts and, despite all precautions, 14 deaths from caisson disease. Work was eventually completed in 1909.

The Dawn of Pre-Drainage Dewatering

The next real advance that would eventually eliminate the need for compressed-air tunneling came about by chance in the United States. In the early 1920s, Thomas Moore, a trenching contractor turned trenching machine manufacturer, had leased a machine to a contractor working on a sewer line in Hackensack, New Jersey. Bulls liver silts (highly plastic, gelatin-like, silty wet soil) that are highly sensitive to excavation disturbance were encountered. The trench became unstable, and pipe-laying dwindled to just 48 feet per week.

Wellpoints, a recent invention, were installed in an attempt to lower the groundwater level, but they had previously only been successful in clean sands and rapidly clogged up in the silts. Moore agreed to take over the contract

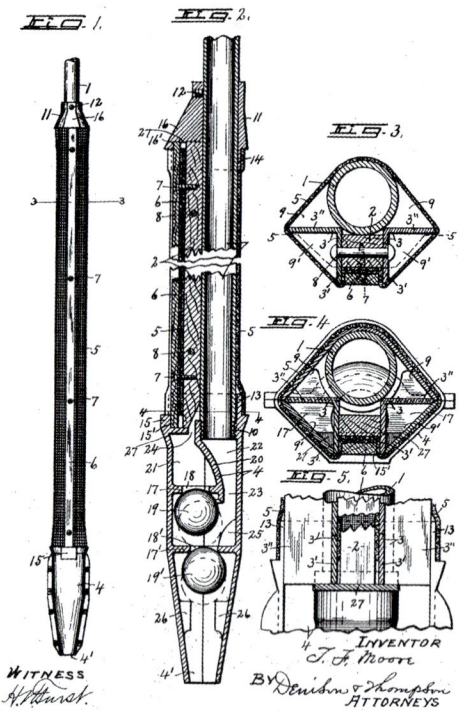

Figure 8.26 | **Early Moretrench Wellpoint Pre-Drainage Device**

and set to work to resolve the problem. Within two weeks, he had developed and field tested a workable dewatering device, and production increased to 60 feet of pipeline per day.

Moore's pre-drainage device, patented in 1924, was the first practical wellpoint system and revolutionized the manner that similar excavations were approached (Figure 8.26). For the tunneling industry, the timing could not have been better. Construction of New York City's subway system was under way, and by 1932, Moore's wellpoints had been used on 14 major cut-and-cover subway tunnel sections.

Beyond Wellpoints—Deep Wells and Eductors

With the adaptation of the eductor pump and the submersible electric pump for pre-drainage of soils in the mid-1950s, dewatering to much greater depths became more practical and opened up the range of tunneling-related applications to deep-bored tunnels. A new wave of subway construction beginning in the late 1960s saw deep wells, wellpoints, and the recently developed eductor wells put to use across the country.

Until the 1990s, large-diameter soft-ground tunneling in the United States had been performed for decades with what was commonly known as the open-face digger shield. The open-face digger shield used a digging arm to excavate the face and load the soil onto a conveyor belt in its invert. The typical open-face digger shield would have some breasting capability built in to restrain sloughing ground in the tunnel face and crown, and advance rates of hundreds of feet per day could be achieved in favorable mining conditions. However, the open-face digger shield was extremely sensitive to the presence of groundwater. A significant dewatering effort was, therefore, critical for safe tunneling, typically requiring a "picket fence" of wells installed on both sides of the tunnel alignment (Figure 8.27). With the aid of dewatering, hundreds of tunnels were successfully advanced using open-face digger shields. Still, there were many projects that experienced "daylighting" events, where a small amount of water would result in the loss of the tunnel face and a chimney of soil would rapidly find its way into the shield. The industry felt the impact of more than just a few major groundwater-related tunneling events while using open-face shields. Hence, this technique fell out of favor, and with advancement in TBM technology, open-face shields have largely been replaced by pressurized-face TBMs.

Figure 8.27 | **Dewatering System Alongside Tunnel Alignment**

The use of eductors was prominent in the early stages of construction for the World Trade Center in Manhattan in 1967. The twin H&M Tunnels, constructed under compressed air and lined with cast-iron segments, would be exposed during construction of the World Trade Center foundations. These tunnels would later serve as the Port Authority Trans-Hudson (PATH) tunnels to the twin towers and the 16-acre World Trade Center site. Stabilization of the H&M Tunnels was critical to prevent them from distorting during dewatering of the soft, compressible silts ahead of the mass excavation within the perimeter structural slurry wall. Precisely controlled eductor dewatering along either side of the tubes allowed the drawdown to occur at a slow and steady rate so as not to abruptly stress the site that was previously under water.

Ground Freezing

As tunnels and access shafts grow deeper and deeper, dewatering for dry construction becomes more difficult. For very deep work, ground freezing, which provides both groundwater control and excavation support, has been the method of choice since its widespread use on deep shafts for New York City Water Tunnel No. 3 in 1970.

Artificial ground freezing was first used in the coal mining valleys of South Wales for shaft sinking in the early 1860s. Patented by German scientist F.H. Poetsch in 1884, the technique's first use in the United States was in 1888 at the Chapin Mining Company in Iron Mountain, Missouri.

In Europe, where aggressive mining practices over the years had resulted in the need to go deeper underground, ground freezing for deep shaft sinking gained in popularity. In the United States, however, minerals could still be readily accessed at shallower depths. Consequently, the second deep-shaft freezing project was not commenced until 1954, at the Potash Company of America in Carlsbad, New Mexico. In the early days, artificial ground freezing was also limited to significant shaft sinking efforts that could warrant the assembly of the massive and costly freeze plants that took several months to set up. This was to change by the 1970s, with the development of mobile refrigeration plants (Figure 8.28). It was this development that made artificial ground freezing practical for civil tunneling projects.

Figure 8.28 | **Refrigeration Plant for Artificial Ground Freezing**

Figure 8.29 | Ground Freezing at Central Artery/Tunnel Project in Boston

Ground freezing was specifically developed for groundwater control and excavation support for deep vertical mine shafts sunk through difficult ground, and the technique remained largely within this niche for more than 80 years (until 1970). However, with the newly available, portable, compact freeze plants requiring little surface area to set up and operate, a few enterprising U.S. geotechnical contractors foresaw opportunities in the urban civil construction and tunneling markets and began to add ground freezing to their services.

The technique was still not widely known outside of the mineral and fossil fuel mining industries, and acceptance was initially very slow. Early ground freezing projects came as a result of contractors "selling" the technique as a more reliable alternate to conventional methods, particularly in challenging circumstances where rigid controls were needed. The introduction of liquid nitrogen as a coolant—costlier but much faster-acting than the usual brine—opened up the potential for emergency response work for sewer tunnel repair.

New York City's Water Tunnel No. 3, begun in 1970, represented the largest capital construction project in the city's history. The tunnel lies hundreds of feet belowground, within metamorphic granite. Many of the vertical access shafts rise through glacial deposits and water-bearing sands. Between 1988 and 1992, ground freezing was used for earth retention and groundwater control for construction of the overburden portions of four deep access shafts within very restricted urban environments in Brooklyn. Successful freezing to depths of up to 275 feet led to the technique being specified for five additional shafts in Midtown Manhattan during the second phase of the project.

The single-largest ground freezing project ever undertaken in the United States to date began in Boston in 1999 for the Central Artery/Tunnel Project, known as the "Big Dig". This method facilitated the jacking of three massive precast concrete box tunnels beneath the active rail lines serving Boston's

South Station. Not only were the tunnels located just a few feet below the tracks, which were required to remain in full service, but jacking was to take place through some of the most difficult subsurface conditions imaginable. The upper 20 feet of fill was laden with myriad obstructions, some "historic fill material" dating back to the 1800s. These included old granite seawalls, wharf structures, and abandoned brick structures as well as assorted building debris. Beneath the fill lay soft organic material and Boston blue clay. Groundwater was several feet below the surface.

The general contractor elected to use ground freezing to stabilize the entire volume of soil along each of the tunnel alignments so that each 38-foot vertical face could be cut and obstructions safely removed without the need for additional support (Figure 8.29). Chilled brine was recirculated through as many as 2,000 vertical freeze pipes for three to four months in advance of the jacking of each precast concrete tunnel. It would be 30 months from initiation of the freezing operation until the system was turned off after completion of the tunnel jacking operation. During this time, no interruption to rail traffic was experienced, and jacking was accomplished without incident.

These projects paved the way for ground freezing contractors to expand into other tunneling-related work. Today, this includes stabilization of mixed-face conditions, launch and retrieval box excavation, safe havens for TBM repair, and horizontal frozen canopies for soil stabilization above the tunnel alignment. Of note is the 2013 horizontal freezing for two of the five cross passages at the PortMiami Tunnel project, the first time in the United States that ground freezing has been used for this purpose. The ground freezing system operated for approximately six months during the freeze formation and excavation of the cross passages, which were constructed with a permanent waterproofing system and cast-in-place concrete liner.

Grouting in Soft Soils

The various grouting techniques used in tunneling applications are very different in methodology, and all were originally conceived in different parts of the world for various applications. In underground construction, the grouting plan and methods differ with the geological conditions being encountered. Methods suitable for soft soils include permeation, compaction, jet, compensation, and soil fracture grouting.

Permeation Grouting

Permeation grouting evolved as a natural progression from neat cement grouting of fine fissures, which was never more than partially successful. In 1896, Albert Francois, a Belgian mining engineer, developed what became known as the Francois cementation process specifically for mine shaft sinking. He brought the process to England in 1911 where it was used successfully at the Hatfield Colliery in South Yorkshire. In 1913, his process resolved a long-standing problem at the Thorne Colliery, also in South Yorkshire. Two shafts that were began in 1909 had come to a standstill at 500 feet below ground level because of water ingress through porous sandstone. Francois' "silicatization" of the sandstone—injection of sodium silicate and aluminum sulfate solution that he believed acted as a lubricant, followed by neat grout—proved a commercial success and established permeation grouting in the engineering community.

A critical advancement in permeation grouting came with the development of the tube-à-manchette grout pipe by Swiss engineer Ischy in the early 1930s. Grout placement via regularly spaced holes (ports) drilled in the pipe and covered with tightly fitting caps that act as check valves allow precise grout placement at specific locations (hole depths) and repeated injections at any port (Figure 8.30). Permeation grouting imparts increased strength to the soil and decreases permeability for water control.

412 THE HISTORY OF TUNNELING IN THE UNITED STATES

Figure 8.30 | **Chemical Permeation Grouting Operations for a Cross Passage**

Figure 8.31 | **Drilling for Compensation Grouting**

Today, sodium silicate and acrylate grouts are the most common permeation grouting materials, with sodium silicate being the most widely used. Prior to acrylates, the use of acrylamide grouts, particularly AM9, was widespread. Developed in the United States, AM9 is extremely controllable, a highly desirable requirement for the permeation grouting process. However, a number of events involving toxicity issues occurred, beginning in Japan in 1974 and culminating in the Hallandsås Tunnel in Sweden in 1997. At Hallandsås, acrylamides released into the groundwater caused serious health and environmental issues in the surrounding countryside, leading to a gradual withdrawal of AM9 from the global grouting marketplace.

Sodium silicate grouting to create stabilized canopies over the tunnel alignment, in conjunction with extensive dewatering, has enjoyed considerable success on major U.S. tunneling undertakings, including the Washington, D.C., Metro system.

Compaction Grouting—An American First

Compaction grouting as a remedial technique has its origins in the mudjacking technique developed in 1929 by John Poulter and used by the Iowa Department of Transportation for highway maintenance. In 1934, the Koehring mudjack machine was introduced, which allowed more viscous grouts to be injected to raise and support road slabs on grouted columns. Zero slump grouting as a method of controlled displacement and densification of loose soils to lift and re-level settled structures was developed in California in the late 1940s. Major improvements of the technique by James Warner in the early 1950s provided better grout quality, and he and fellow engineer Ed Graf, considered by many to be the founding fathers of compaction grouting, championed the technique within the engineering community. In fact, the term *compaction grouting* was coined by Graf in the first theoretical description of the technique in a 1969 journal article. Warner and Douglas R. Brown highlighted the characteristics of the technique in 1973. Thus the stage was set for compaction grouting to become a widely accepted technology and expand beyond slab jacking to ground improvement for structural foundations and deep excavations.

Jet Grouting

The earliest known patent for jet grouting was applied for in England in the 1950s. However, it is commonly acknowledged that the practical development of the technology took place in Japan in the 1960s before making its way to Europe during the 1970s and then to the United States in the early 1980s. The U.S. market was cautious, however, due to perceived legal risks, but acceptance began to grow, albeit slowly, thanks to specialty geotechnical contractors willing to embrace the new technology and prove its effectiveness for difficult ground conditions. Initially, U.S. jet grouting projects were primarily for structural underpinning, but as awareness and understanding of the technology increased within the engineering community, so did the range of applications. By the mid-1990s, jet grouting had become much more established, and the range of completed projects included several for groundwater control and tunnel/ground stabilization. One such project was for the Islais Creek tunnels, built between 1990 and 1997 in San Francisco.

The Islais Creek tunnel alignment passed through San Francisco Bay mud, well-known for its squeezing properties. The original design approach included using a compartmentalized breast-boarding shield with compressed air and back-grouting to overcome the effects of squeezing ground. However, existing structures, utilities, and active commuter lines above the tunnel alignment raised concerns about surface settlements and caving caused by squeezing of the soft soils. The general contractor, therefore, looked for an alternative to compressed air tunneling and found it in a Value Engineering proposal to stabilize the soils by full-face jet grouting to allow for successful open-face tunneling. Approximately 1,000 overlapping jet grout columns were installed for the project, which represents the first recorded application of full-face jet grouting for soft-clay tunneling worldwide.

In the years since, jet grouting has been successfully implemented on numerous tunneling projects across the United States for applications such as tunnel face stabilization, canopies, groundwater control, bottom seals, and soil stabilization behind the launch and retrieval shaft excavation support for TBM access and egress. Cross-passage excavations have also benefited from the use of jet grouting applications.

Figure 8.32 | **Surface Batch Plant Used for Compensation Grouting**

Compensation Grouting

Structural settlement remediation by compaction grouting methods gained steady acceptance and has remained the primary application since 1985 (Figures 8.31 and 8.32). Settlement of overlying structures induced by soft-ground tunneling was approached the same way as any other structure, with remediation taking place after the event. Then in the late 1970s, a paradigm shift in thinking by Wallace Hayward Baker brought about a historic development.

During construction of the Bolton Hills Tunnel, which is part of the northwest line of the Baltimore Region Rapid Transit System, the retrofit approach was specified for 40 vulnerable structures located above the tunnel alignment. Baker, however, offered an innovative alternative: inject compaction grout through pre-placed grout pipes as the tunnel shield passed beneath. This would immediately compensate for soil loss from the tunneling operation and limit movements nearer their source before the structures above were affected. This landmark project marked the beginning of international acceptance of concurrent compaction grouting techniques for ground loss compensation applications.

Soil Fracture Grouting

While the term *compensation grouting* has been applied to concurrent compaction grouting, in current tunneling practice it is typically performed with *fracture grouting*—the deliberate, locally confined and precisely controlled hydrofracturing of soils between the tunneling operation and overlying structures using multiple injections of medium- or high-viscosity grouts strategically injected through sleeve port pipes. Similar in intent to compensation grouting using compaction grouting techniques, fracture grouting is initiated at the tunneling shield if real-time monitoring indicates the potential for foundational soil loss.

Initially developed in France and widely used in Europe, the first successful application of soil fracture grouting to compensate for tunneling ground loss in North America was in 1993. The alignment of a new train tunnel beneath the St. Clair River between Port Huron, Michigan, and Sarnia, Ontario, included a large, sensitive research building and other structures at a commercial facility in Sarnia. With the crown of the tunnel just 22 feet below the building footings, maximum settlements of 4 inches were predicted to occur as the full-face earth pressure balance machine passed through the underlying soft clays. The project's design and construction management consultant drew up a performance specification for compensation grouting to control settlement to within acceptable limits and mitigate damage potential.

Although soil fracture grouting had not been previously performed in North America, the technique was accepted for this project based on European experience. The pre-condition grouting of the soils ahead of fracture grouting, together with extensive sophisticated instrumentation and real-time electro-level monitoring during the tunneling, were key to the success of the project. The beneficial effect of pre-conditioning minimized the amount of fracture grouting required, and settlements were within specified limits of 0.4 inches.

INNOVATIONS IN EXCAVATION

Drill-and-Blast

Since the beginning of civilization, humans have discovered ways to drill holes through solid materials. Stone Age artifacts more than 32,000 years old include perforated shells, ivory, bones, and teeth. Jumping ahead a few centuries,

Figure 8.33 | **Manually Drilling an "Upper" Blasthole**

Figure 8.34 | **Modern Multi-Boom Drill Jumbo**

Figure 8.35 | **Drilling and Blasting in a Niagara Water Tunnel, Circa 1892**

the Romans developed a drill that featured a crosspiece attached to a spindle by cords that ultimately wound and unwound around it. A downward push on the crosspiece rotated the spindle. This design, later made of iron or steel, remained popular until the end of the nineteenth century.

Underground excavation has advanced since 52 AD when the first long-driven tunnel was completed during the reign of Emperor Claudius. Pliny recorded it as a 3.5-mile bore that drained Fucine Lake in the Italian province of L'Aquila. This 6-foot-by-10-foot tunnel took more than 30,000 men 11 years to complete.

The history of underground rock excavation shows what humans are capable of when the need arises. The natural and most effective way to drill a hole in rock is to strike a steel chisel or drill bit with a hammer. The early tunnelers used this method to excavate the tunnels and caverns required for a developing infrastructure. Nonetheless, this was very labor intensive and dangerous, especially for the man holding the drill bit (Figure 8.33). We have come a long way to the present state of technology and equipment (Figure 8.34).

The first rock drill designers sought to duplicate the "turning of the drill steel by hand" technique. This is the reason that modern rock drill always rotates the steel to the left (counterclockwise), the same action that the hand driller used. In the early mechanical developments, they were forced to adopt a rudimentary technique where the entire drilling element was tied to a piston and reciprocated back and forth with it. It took nearly 50 years to devise a method for separating the two elements and evolving toward the original hammer principle similar to that used in hand drilling (Figure 8.35).

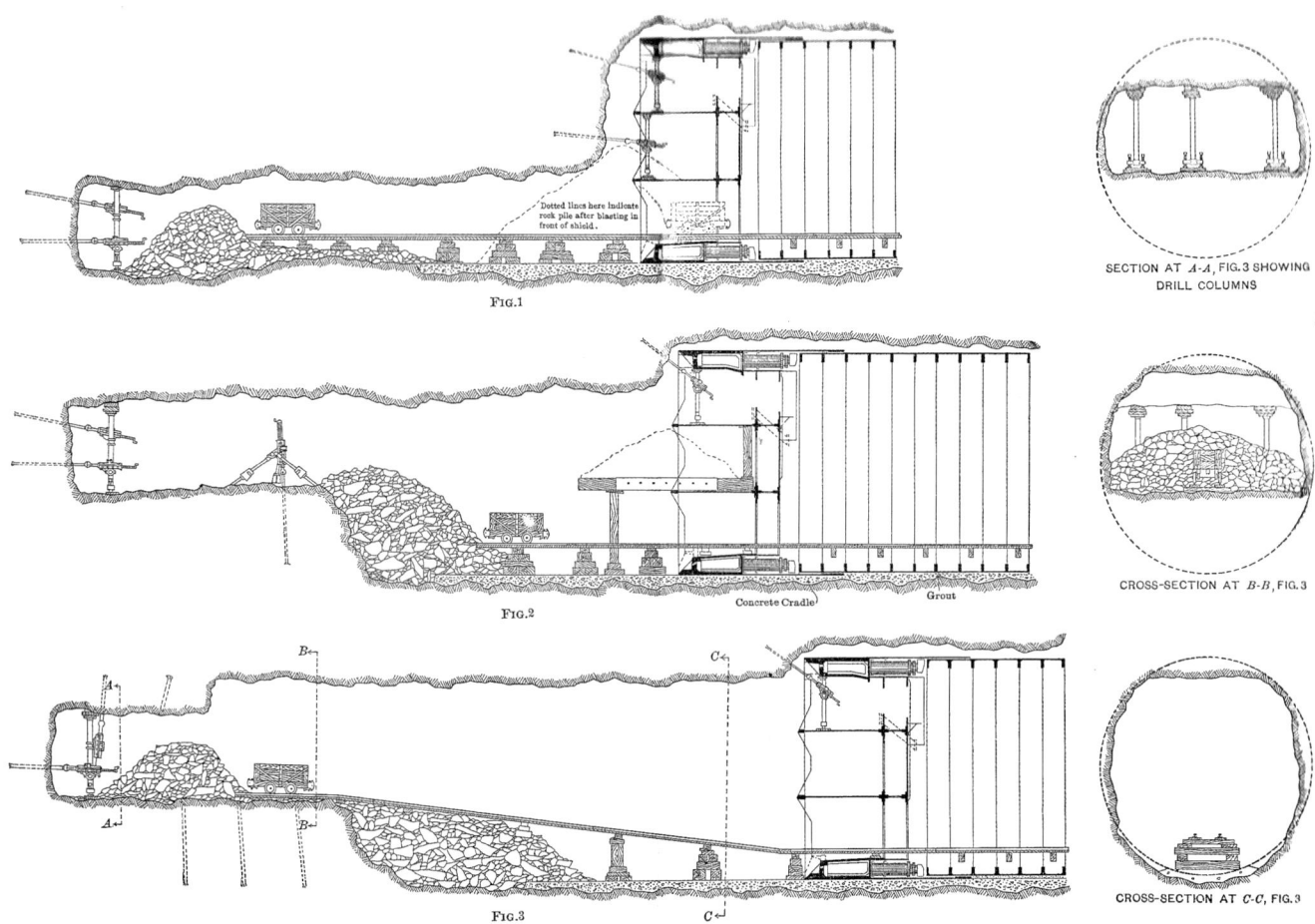

Figure 8.36 | **Early Pilot Tunnel and Enlargement, Circa 1905**

By 1874, the emerging rock drill industry was getting a good start. Its products were in considerable demand, as they almost immediately proved themselves to save time and money. In the 1940s, for example, the rock drill was rated number one in importance among labor-saving machines. The plain truth is that the first drills were cumbersome, costly, and unreliable (Figure 8.36). An example of the early difficulties of underground construction using this technique were the 18 years between 1855 and 1873 that were required to drive the nation's first major tunnel, the Hoosac railroad bore in Massachusetts. Drilling and blasting methods eventually improved with more efficient equipment and approaches to muck handling.

One of the limitations to increasing the drill penetration rate was the quality of the drill steel itself. The drill steel problem often continues to provide difficulties to this day as the industry strives for ever greater production rates and reliability in performance. As the striking power of drills is continually increased, the difficulties of finding a steel that will withstand the hammering effect and the rotational strain grow ever greater. This is comparable to the problems encountered when the first hydraulic drills were introduced in the 1970s.

Drill strings (steel rods) tended to weld themselves together under the strain imposed, or they failed prematurely. Metallurgists developed both a drill piston (striking bar) that could withstand the constant action of striking steel on steel and a drill steel that would be able to efficiently transfer the energy to the drill bit.

The development of drill steel, the detachable rock bit, and tungsten carbide were significant improvements in solving many of the issues of early drilling and in taking the industry to meet the demands of present day.

When comparing the expected productivity of the drills used to construct the world-class underground projects prior to 1950 with what can be done today, one is in awe with what they were able to accomplish more than 65 years ago (Figure 8.37). This includes such major drilling projects as Grand Coulee Dam, Hoover Dam, and even the Glen Canyon Dam in the western United States. During this time period, production was measured in inches drilled per minute.

The industry has witnessed the development of drifter drills with their power coming first from steam, then compressed air, and today's drills being powered hydraulically. Initially, drills were developed to be powered by compressed air, but to make them more powerful, the piston size was increased. This was not possible for drills used in underground construction or tunneling because of weight and space considerations. These drills had to be lighter weight while generating the power necessary to drill the hardest of materials. It was this necessity that drove the industry to find a more efficient and robust power source, which was brought to market in the 1970s with the introduction of the initial hydraulic drills.

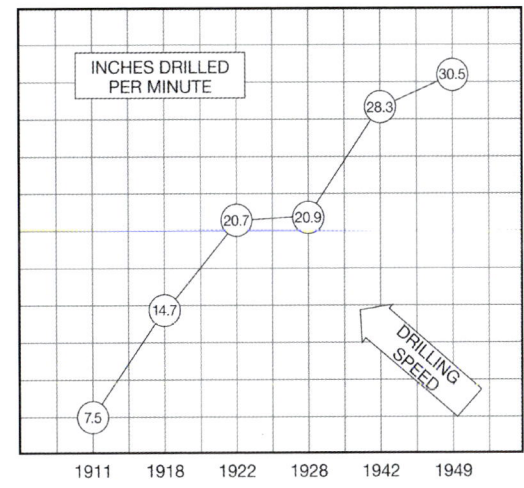

Figure 8.37 | **Inches Drilled per Minute over Four Decades**

With this innovation in drilling, rates of production that used to be measured in inches per minute were now being measured in feet or meters per minute (Figure 8.38). As rock drill productivity increased, it became necessary to redevelop all aspects of the underground excavation process (Figure 8.39). There have been many new advances in rock support with bolting and shotcrete, loading and haulage equipment, ventilation, surveying, and more. Key to these developments was the technology that to this day continues to make the drill-and-blast method competitive with any other means of excavation.

In the past, the combination of poorly placed holes together with black powder as an explosive resulted in either significant over-excavation (overbreak) or under-excavation (tights) that cost more time and money to address. Over the last few decades, computer-controlled equipment has been integrated into the drilling-and-blasting business. Modern drilling equipment uses computerized booms and drills that drill very accurate holes in precise patterns across the face with little operator interface (Figure 8.40).

Drilling has come a long way since the advent of what was considered the modern rock drill of the time in 1874. The modern rock drill is a product of the patient and persevering efforts of many innovators. From steam to compressed air to hydraulically driven drills, considerably more rock excavation has been accomplished. The inclusion of technology in underground excavation has increased the production that can be expected from a single driller to the point that it now requires just one person to operate a drilling platform consisting of multiple high-powered, laser-guided, programmable drills.

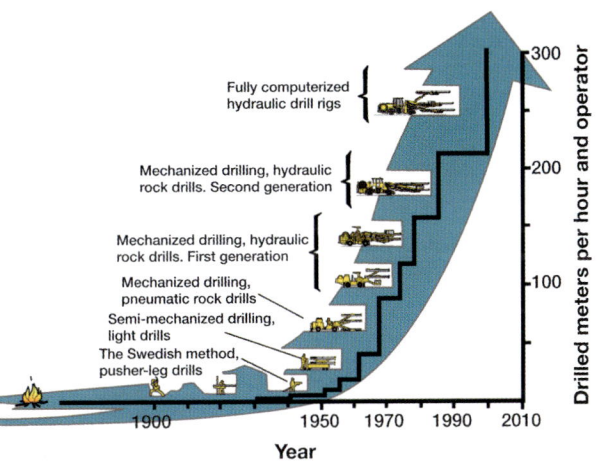

Figure 8.38 | **Increases in Drilling Productivity in the Past 100 Years**

Tunnel Explosives Technology

The need to construct tunnels in hard rock has challenged humankind for thousands of years. The previous section discussed the advancements and innovations that have occurred in the drilling of the holes for drill-and-blast tunneling. The second important aspect of this type of tunnel construction is the blasting, or fracturing, of the rock so that it can be easily removed from the tunnel and the tunnel advanced toward its final objective.

Figure 8.39 | **Modern Drill Jumbo Using a Laser Positioning and Drilling System**

418 THE HISTORY OF TUNNELING IN THE UNITED STATES

Figure 8.40 | **Modern Drill Jumbo, and Computerized Programmable Intelligent Drilling System**

In ancient times, tunnels in rock were advanced by building fires against the rock, which would cause it to expand and then spall off when doused with water. Hand tools were also used. The work was very slow, and many workers were harmed or killed. The advance rate of manually excavated tunnels in hard rock was only around 30 feet per year.

Black powder was the first explosive material developed and was initially used in fireworks by the Chinese. The drill-and-blast era of tunneling and mining began in 1627 when miners in Hungary blasted rock with black powder placed in holes bored in rock. As the use of black powder increased, the number of accidents also grew. Miners used straws and goose quills filled with black powder to initiate main charges. The actual burning time was not reliable, and charges sometimes detonated before workers could retreat to safe areas.

Edward Howard developed mercury fulminate in England in 1800. Mercury fulminate is a highly explosive substance used in detonators that could be attached to wires and initiated electrically from remote and safer locations.

In 1831, English merchant William Bickford developed fuses that could be attached to detonators containing fulminate-of-mercury charges (Figure 8.41). The use of detonator-fuse assemblies was safer than using black powder in goose quills, but even so, too many accidents still occurred because once the fuses were lit, there was no way to stop the blast if needed.

Having developed a process to make nitroglycerin in 1846, Italian chemist Ascanio Sobrero was astonished by its destructive power and concluded it was unsafe for any practical use. Despite his warnings, liquid nitroglycerin was used for commercial rock blasting, which caused some horrible accidents.

Alfred Nobel made history in Sweden in 1862 when he demonstrated that the sensitivity of nitroglycerin could be managed when it was absorbed into diatomaceous earth, which is essentially hollow plankton skeletons. Nobel packaged this mixture in paper cartridges and modern-day "dynamite" was born (Figure 8.42).

In the United States, nitroglycerin and fulminate of mercury fuses like the one shown in Figure 8.41 were first introduced to build the Hoosac railroad tunnel through the Hoosac range in western Massachusetts. Work on this 4.75-mile tunnel began in 1851 with an estimated cost of $2 million and ended in 1875 with a total cost of $21 million. With inflation, the cost today would be around a half billion dollars. Because of the extreme hardness of the rock, other important innovations in drilling technology, such as the use of star bits and pneumatic drilling methods, were also introduced to complete this very challenging tunnel.

Black powder and various forms of crude fuses were used in the United States to blast rock in a series of railway and subway tunnels built through the hard rock of the Palisades ridge in New Jersey and Manhattan. A map showing these tunnels as they existed around year 1900 is shown in Figure 8.43. In later years, explosives development became a science as chemists strove to develop more powerful and yet safer explosives.

Figure 8.41 | **Electric Fuse with Fulminate-of-Mercury Detonator**

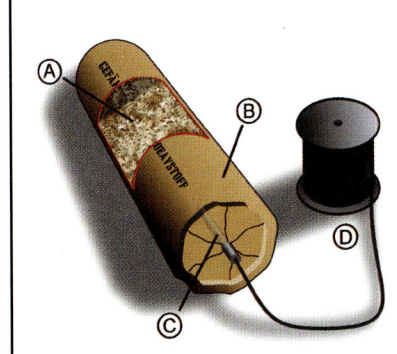

A. Sawdust (or other type of absorbent material) soaked in nitroglycerin

B. Protective coating surrounding the explosive material

C. Blasting cap–electric detonator

D. Wire connecting blasting cap to detonator source

Figure 8.42 | **Components of a Typical Dynamite Charge**

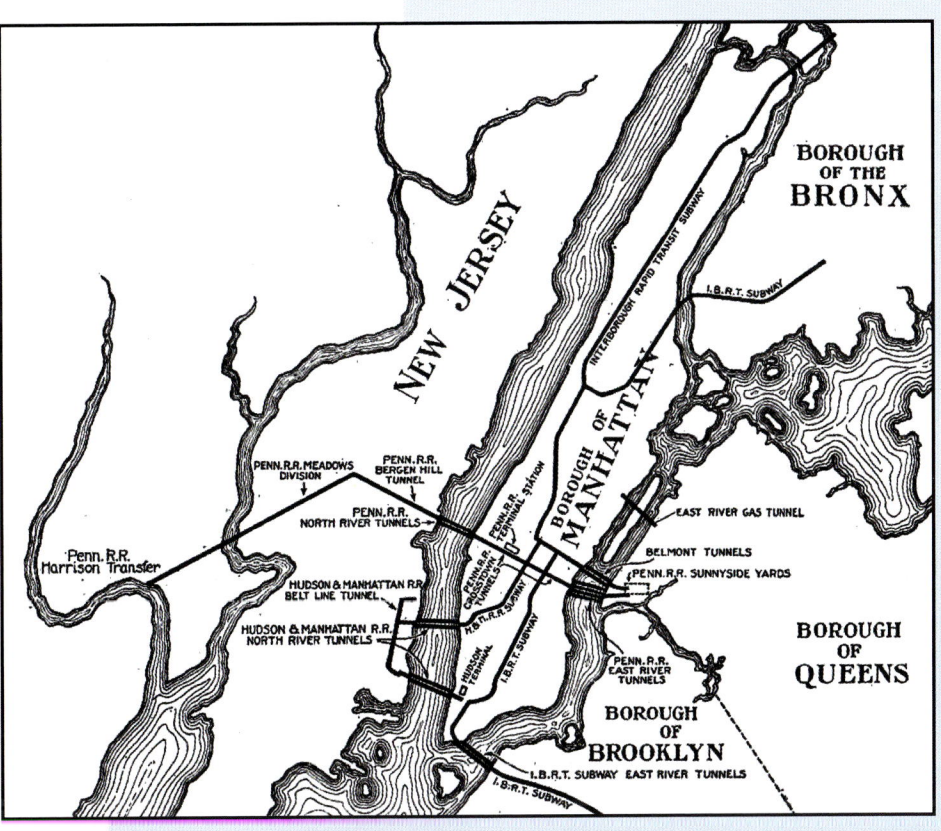

Figure 8.43 | **Location of Tunnels Excavated in the New York City Area, Circa 1900**

THE HISTORY OF TUNNELING IN THE UNITED STATES

Figure 8.44 | **Prepping Dynamite with Nonelectric Detonators**

Many historians believe that the invention of dynamite was one of the most important inventions of modern times because the tunnels needed to build the transcontinental railways across North America that fueled settlement of the West and the Industrial Revolution would not have occurred without it. In contrast to the ancient tunnels that typically advanced as low as 30 feet a year, tunnels blasted with dynamite could be advanced as rapidly as 30 feet a day. Electric detonators with delay-timing intervals were developed in the early 1900s and improved tunnel blasting results through better control of charge firing sequences. Detonators using plastic shock tubes (nonelectric) to transmit blast initiation signals were invented in the 1960s and 1970s under the leadership of Per-Anders Persson. Unlike electric detonators, these nonelectric devices are not affected by stray currents or radio waves. Because of their improved safety and ease of use, shock tube detonators are now employed in most tunneling operations (Figure 8.44).

In our modern era, scientists are accelerating improvements in explosives technology. Most of the explosives used today do not contain nitroglycerin or other self-explosive ingredients. Modern emulsion explosives are made from mixtures of nonexplosive fuels and oxidizers that provide great energy and safer handling characteristics. Emulsion explosives can be formulated so they have a putty-like consistency for packaging in paper or plastic cartridges, or made into slurries that can be pumped into holes. The slurry form is common in the mining and tunneling industries.

Figure 8.45 | **Drilling Blastholes for Sequential Excavation Rounds**

CHAPTER EIGHT | INNOVATIONS IN TUNNELING   421

Figure 8.46 | **Blast Round Initiation in the Tunnel Heading**

Like other innovations of our time, computers are being used to improve the efficiency of blasting operations. Electronic detonators with built-in computer circuits can be programmed to fire with extremely high accuracy, and control units can also communicate with all detonators to validate secure connections. For some tunnels, computer-controlled pump trucks for blasting materials are used to automate hole-charging operations and speed up the rate of the work.

Mechanical excavation methods have replaced the use of drill-and-blast methods in many tunnels. But there are locations where blasting will remain the most efficient choice due to the properties of rock and configuration of the excavation (Figures 8.45 and 8.46). With every new drilling and explosives product development, safety of the work improves, vibration and noise are better controlled, and tunneling costs are reduced.

Modern technology has improved the explosives used in underground construction to be very safe to handle and very productive in the form of energy released to fracture the rock strata (Figure 8.47). In urban areas where controlled blasting is a requirement, the industry has the abilty to design an explosives plan and detonation sequence that will break the rock but also minimize the potential vibration impacts to the surrounding area. Although this may require more rounds or cycles and increase the cost of the project, the objectives of the owner, the project's neighbors, and the contractor can be met using today's explosives technologies.

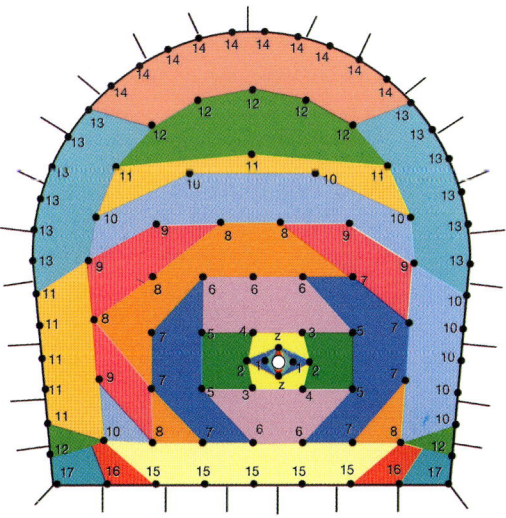

Figure 8.47 | **Typical Blast Detonation Sequence in a Tunnel Heading**

THE HISTORY OF TUNNELING IN THE UNITED STATES

With the recent expansion of the subway system in New York City, there were many projects that required controlled explosive practices. This work was performed in a safe and efficient manner by very skilled sandhogs, allowing for the construction of subway lines and stations in the areas where they were needed, regardless of the geological conditions encountered or the proximity to existing structures.

Tunnel Boring Machines and Shields

The introduction of mechanical excavation equipment was one of the major innovations to the tunneling industry. As the projects became larger, labor became more expensive and the availability of skilled, experienced workers became more difficult to obtain, so contractors started looking at mechanized tunnel excavation as a means to become more competitive and timely in the delivery of projects.

Mechanized tunneling was considered and experimented with as early as the mid-1800s by various individuals throughout the world. As was often the case with many of the early tunneling pioneers, they were not successful, because most of these early tunneling machines were underpowered due to limitations in mobile power technology at the time.

It was not until the advent of high-pressure hydraulics, higher-strength metals for gears and cutting tools, and improvements to the efficiency and power of electrical motors that tunnel excavation machines became the preferred tunneling method for the industry.

The first tunnel shield was invented by Marc Isambard Brunel in England in 1818. The features of Brunel's shield included the division of the face into small compartments, so each smaller cell could either be excavated or closed off ("breasted") separately to stabilize the tunnel face (Figure 8.48). The shield was thrust forward by screw jacks, and the tunnel lining was erected within the protection of the tail of the shield body. These last features are common to current tunnel shields and TBMs.

Figure 8.48 | **River Thames Tunnel Shield and Final Brick Lining**

Figure 8.49 | **Brunel Shield Tunnel Arrangement for the Thames Tunnel**

Brunel was appointed chief engineer of the Thames Tunnel Company. Two Brunel shields were used for tunneling beneath the River Thames in London (Figure 8.49). The first shield was used from 1825 to 1828 but did not complete the crossing because the company faced a lack of funds. The second shield, a rectangular one, was used from 1835 to 1843, when this tunnel was completed. The best rate of production was 14.1 feet in one week. Water and running sand often made the work very difficult. During the construction period, other engineers suggested that Brunel use compressed air to keep out the water and stabilize the sand at the face. A patent had been taken out in the United Kingdom by Lord Cochrane for this use of compressed air in tunneling and for design of an air lock. But Brunel believed that his shield was sufficient to protect the excavation. Several layers of brickwork and mortar were used for the final lining, which to this day are still intact and in remarkably good condition.

Other tunneling shields were designed and patented in the last half of the nineteenth century. Many of these designs, however, were never constructed and used underground. American Alfred Ely Beach patented a shield in 1870 that was used to excavate a tunnel for a pneumatically powered subway system beneath Broadway in New York City. Electrically powered subways were not available until years later. Beach's shield was the first to use a single cylindrical body and hydraulic jacks to provide thrust. Beach's shield design was also used for the first attempts to tunnel under the Hudson River starting in the 1880s (Figure 8.50).

Hand excavations of the face were used in the early soft-ground shields described above. The first mechanized shield with a rotating cutterhead was patented in the United Kingdom in 1876 by J.D. and G. Brunton. In 1886, Frank O. Brown of New York patented a shield that had an airtight inclined bulkhead near the face. A screw conveyor was mounted in the center of the shield and acted as a drilling tool and muck conveyor (Figure 8.51). If necessary in harder ground, hatches in the bulkhead could be opened for hand excavation

Figure 8.50 | **Silt Being Extruded Through the Doors as the Beach Shield Is Pushed Forward for the Original Hudson River Railroad Tunnels**

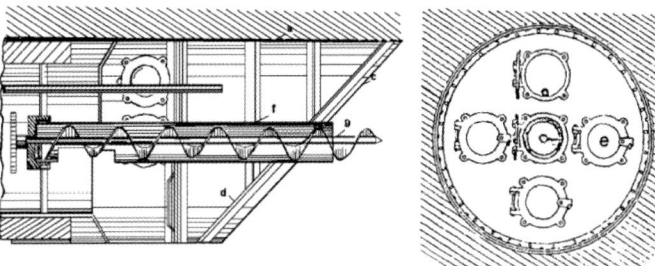

Figure 8.51 | **Brown's EPB Tunnel Shield Arrangement**

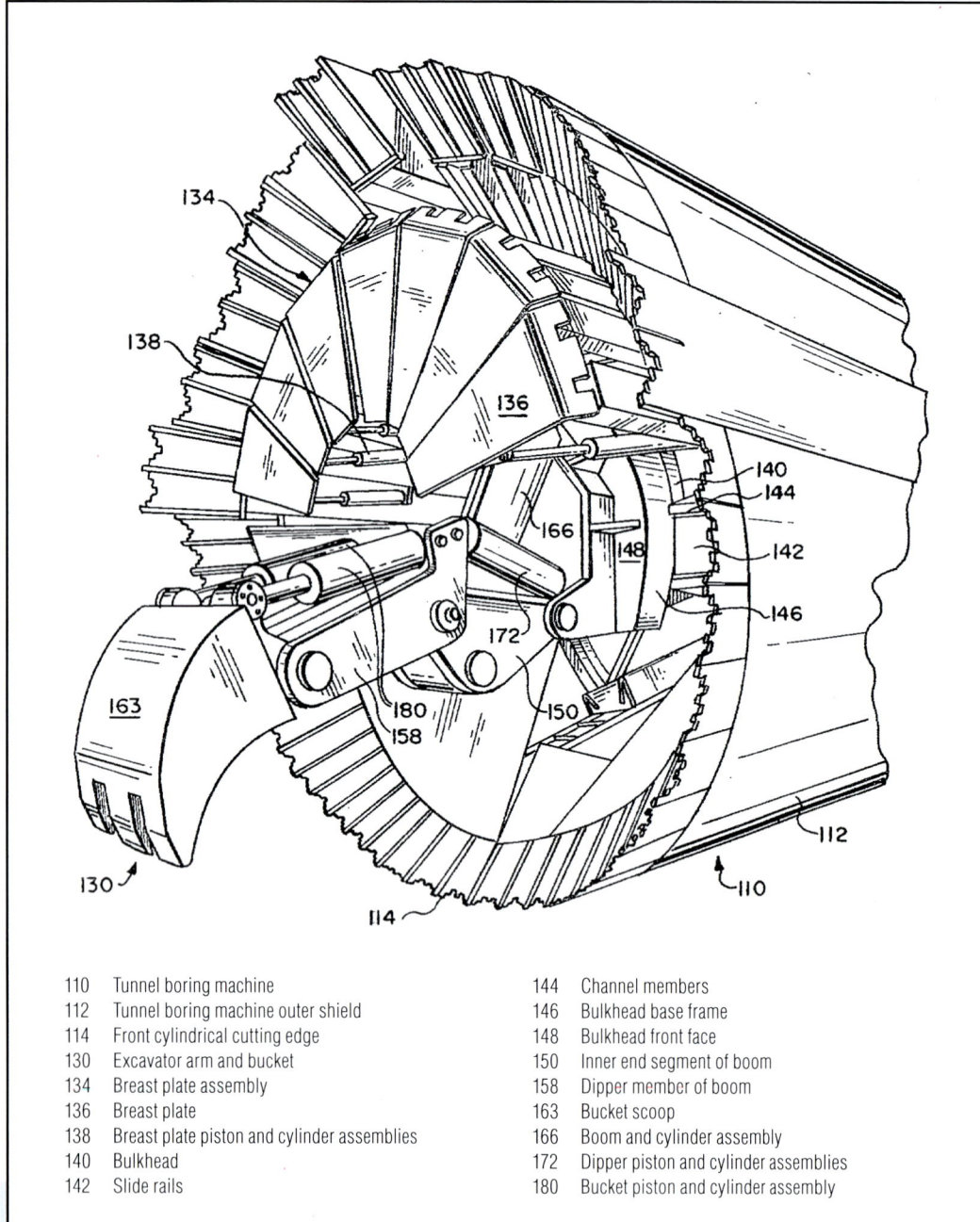

110	Tunnel boring machine	144	Channel members
112	Tunnel boring machine outer shield	146	Bulkhead base frame
114	Front cylindrical cutting edge	148	Bulkhead front face
130	Excavator arm and bucket	150	Inner end segment of boom
134	Breast plate assembly	158	Dipper member of boom
136	Breast plate	163	Bucket scoop
138	Breast plate piston and cylinder assemblies	166	Boom and cylinder assembly
140	Bulkhead	172	Dipper piston and cylinder assemblies
142	Slide rails	180	Bucket piston and cylinder assembly

Figure 8.52 | **Zokor "Big John" Digger Shield**

to augment the drill. This simple closed-face shield can be compared in some ways to modern earth pressure balance shields in operation today.

Mechanized shields with full-face rotary cutterheads gradually came into use and were effective. Such machines were designed and manufactured in North America by several pioneering companies and by other manufacturers from overseas. The Zokor "Big John" digger shield was typical of one of the largest made at the time (Figure 8.52).

Concurrent with the developments of full-face rotary cutterhead machines, other mechanized shields using a ripper/digger arm to excavate the face, known collectively as "digger shields," were produced in the United States. The Zokor shield had hydraulically extendable forepoling boxes in the upper half of the shield periphery. These boxes could be extended forward to hold loose material in the face and to form a protective hood. In addition, hinged breasting plates could be actuated to breast the center part of the face to reduce the unsupported excavated face area and minimize potential ground movement. A powerful rotating excavation bucket would dig around the entire perimeter of the shield's leading edge. The shield was thrust forward by pushing off precast concrete segments that were erected within the tail of the shield. Such digger shields achieved excellent advance rates in suitable ground conditions. They were considered particularly useful where boulders, pilings, foundation tie-backs, or other uncharted obstructions were to be encountered, because the face was accessible and such obstructions could be manually removed. Digger shields are still in use today.

Throughout the twentieth century, the use of compressed air became prevalent in soft-ground tunneling to hold back groundwater and running ground conditions as the complexity and scope of these tunnels increased to meet the needs of society. More stringent health and safety regulations came into effect to improve the safety of tunnel workers in compressed-air conditions, such as decompression times, medical locks, oxygen decompression, and other procedures. Although these actions improved worker safety, it became apparent that closed-face shields, with the face pressurized to control the ground conditions and with a pressure bulkhead to allow the remainder of the workings to be at atmospheric pressure, were a necessary improvement. This is the basis of the closed-face soft-ground TBMs that are used today.

In 1952, James S. Robbins convinced Mittry Construction Company to build a 26-foot-3-inch-diameter full-face rotary cutterhead TBM for use in shale at the Oahe Dam project in South Dakota. This machine was delivered in 1953 and achieved world records for advancing 61 feet in an eight-hour shift and 161 feet in a day. With this achievement, the TBM method had truly become a competitive method for rock tunneling.

Subsequent machines featured improvements in disc cutters for harder formations, better ground support, positive gripping and continuous steering, and improved hydraulic design. More machines followed, including a 25-foot-6-inch-diameter unit for the South Saskatchewan Dam in Canada (Figure 8.53).

James S. Robbins (1907–1958)

James S. Robbins graduated with a BSc in Mining Engineering from the Michigan College of Mines in 1927. During his early career, he worked at various mines for Arctic Circle Exploration Inc., rising to the position of general manager. In 1947, James launched his own consultancy, Mechanical Miner Company, developing continuous coal mining equipment. He founded James S. Robbins and Associates in 1952, the precursor to The Robbins Company, which was the first company dedicated exclusively to the design and manufacture of tunnel boring machines (TBMs).

The world's first rock TBM, known as the "Mittry Mole," was delivered in 1953 to Mittry Construction Company for the Oahe Dam diversion project near Pierre, South Dakota. The project included seven diversion tunnels and five power tunnels. Following the success of the Mittry Mole, additional machines were built for sewer tunnels in Pittsburgh and Chicago, and the Robbins name became synonymous with hard-rock TBMs.

Tragically, James was killed in an airplane accident in 1958 and the tunneling industry lost one of its leading TBM pioneers.

Figure 8.53 | **Early Robbins TBM for a Dam Diversion Tunnel**

Figure 8.54 | Robbins TBM at Mangla Dam, Pakistan

CHAPTER EIGHT | INNOVATIONS IN TUNNELING   427

In 1963, Robbins delivered a 37.7-foot-bore TBM, at that time the largest TBM in the world, for the Mangla Dam Project in Pakistan (Figure 8.54). A Goodman/Robbins continuous conveyor system, which was originally developed for coal mining, was adopted for use behind the TBM. This was the first use of a continuous conveyor system in TBM tunneling. Decades later, the TBM/continuous conveyor combination emerged again, allowing record-breaking performances and improved safety for many projects.

The Chicago Tunnel and Reservoir Plan (TARP) was a major water quality improvement project that involved more than 100 miles of rock tunnels. Many rock TBMs, up to 35 feet in diameter, were used on the project. In 1978, Jarva delivered a TBM with a 30-foot-bore diameter to the project constructors (Figure 8.55). This machine drove 5.4 miles of rock tunnel. Production rates were very good, as this TBM had a higher cutterhead rotation speed, as compared with other rock TBMs of that era.

Hard-rock TBMs continued to advance over the next 30 years. Improvements included better disc cutters, more robust machines with higher power, more versatile ground support equipment, and better ancillary equipment to allow higher TBM utilization rates. Backloading cutterheads allowed cutter inspections and changes to be carried out from within the safety of the cutterhead structure.

Figure 8.55 | **Robbins 30-Foot Hard-Rock TBM Used on TARP Tunnels in Chicago**

Figure 8.56 | **Hard-Rock TBM Assembly at the Niagara Tunnel Portal**

Figure 8.57 | **Completed TBM Assembly at Niagara Tunnel Portal**

There has always been close cooperation and a sharing of technology between the U.S. and Canadian tunneling industries. In 2006, Robbins delivered the 47.2-foot-bore TBM subassemblies to the Niagara Falls site (Figures 8.56 and 8.57). The project was designed to divert 8% of the Niagara River flow from the upstream side of the Horseshoe Falls to the Sir Adam Beck hydropower station in Niagara Falls, Ontario. This additional flow increased the power station's capacity by 28%. The machine had not been fully assembled in a workshop prior to delivery. Instead, the subassemblies were shipped separately and assembled onsite to form the finished TBM, saving shipping costs and six months on the project schedule. To date, it is the largest hard-rock TBM ever built, with a cutterhead driven by 6,330 HP (Figure 8.58).

Development of closed-face shielded TBMs continued around the world. These were basically two types of machines: slurry pressure balance (SPB) TBMs and earth pressure balance (EPB) TBMs. The EPB-type TBM has a pressurized bulkhead and an encased screw conveyor. The excavated ground is mixed with soil conditioners and fills the cutting chamber in front of the pressure bulkhead. The excavated material is allowed to discharge at a controlled rate through the screw conveyor. As the machine thrusts and cuts, the pressure of the mixed soil in the cutting chamber is increased to just balance the pressure of the in-situ ground and groundwater, thus the name "earth pressure balance." The SPB-type TBM has a pressure bulkhead that is pressurized with a slurry, typically bentonite. As the machine thrusts forward and cuts, the mixture of soil and slurry is pumped from the excavation chamber via steel pipes to a separation plant on the surface. The soil and slurry is processed by removing the solids, and the bentonite slurry is pumped back into the tunnel, creating a closed continuous loop, thus the name "slurry pressure balance."

Soft-ground TBMs continue to record remarkable achievements in terms of the extent to which this underground technology is applied. The second St. Clair River Tunnel Project, for example, was completed in 1995. It crossed the international boundary between the United States and Canada, and carried Canadian National Railway Company (CN) trains for international trade. In November 1993, a 31-foot-2-inch-diameter Lovat EPB TBM named "Excalibore" was launched to excavate a new, larger rail tunnel beneath the St. Clair River (Figure 8.59). Excalibore began boring and progressed with the EPB feature providing good control of surface settlement. The tunnel was successfully completed and open for rail traffic in early 1995. The EPB-type TBM design provided safe, efficient excavation in soft ground beneath the river. This new tunnel paralleled the existing original rail tunnel that had been completed in 1891. The first tunnel was excavated using Beach hydraulic shields and compressed air and supported with cast-iron segments (Figures 8.60 and 8.61). The original tunnel become obsolete as freight trains grew larger and international trade between Canada and the United States increased.

The East Side Access Project in New York City is a major tunneling program that has been ongoing since the 1970s. It consists of an immersed-tube tunnel beneath the East River, TBM and drill-and-blast tunneling in hard rock, ground freezing, and soft-ground tunneling with slurry TBMs. For these tunnels, two 22-foot Herrenknecht slurry TBMs

CHAPTER EIGHT | INNOVATIONS IN TUNNELING   429

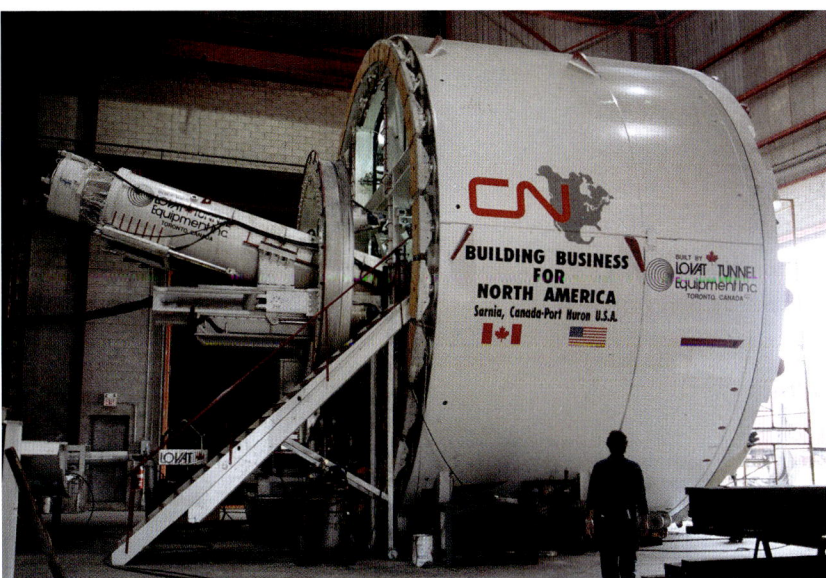

Figure 8.59 | **Soft-Ground Lovat TBM Used to Build the Second St. Clair River Tunnel**

Figure 8.58 | **View of Hard-Rock TBM Cutterhead in Tunnel**

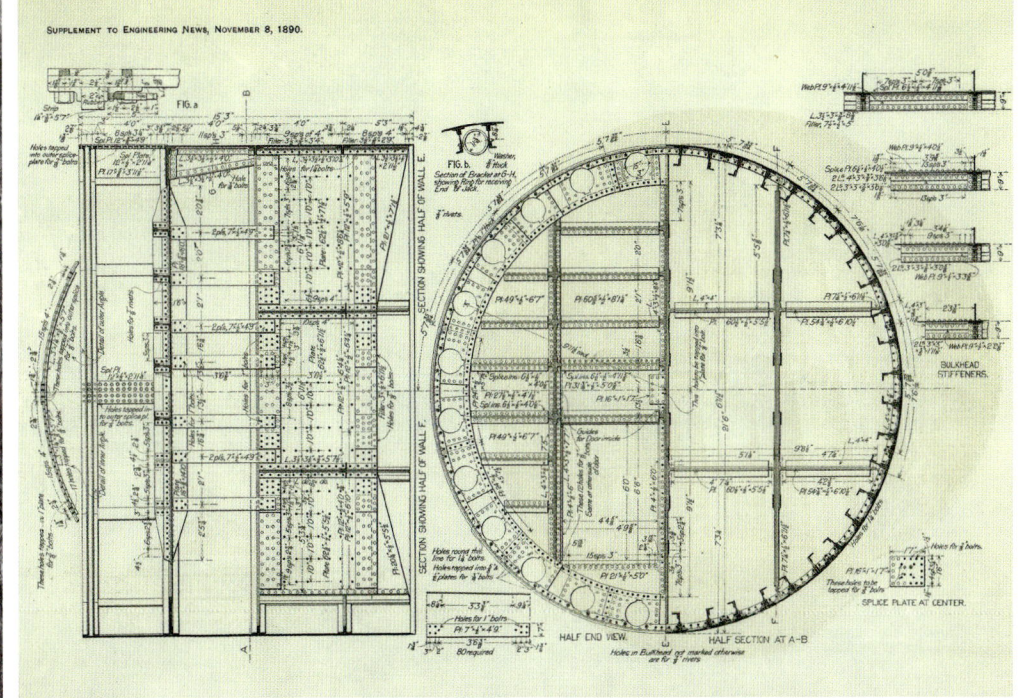

Figure 8.60 | **Original Beach Tunnel Shield Design for the First St. Clair River Tunnel**

were used. Two TBMs, named "Tess" and "Molina," are shown in Figure 8.62.

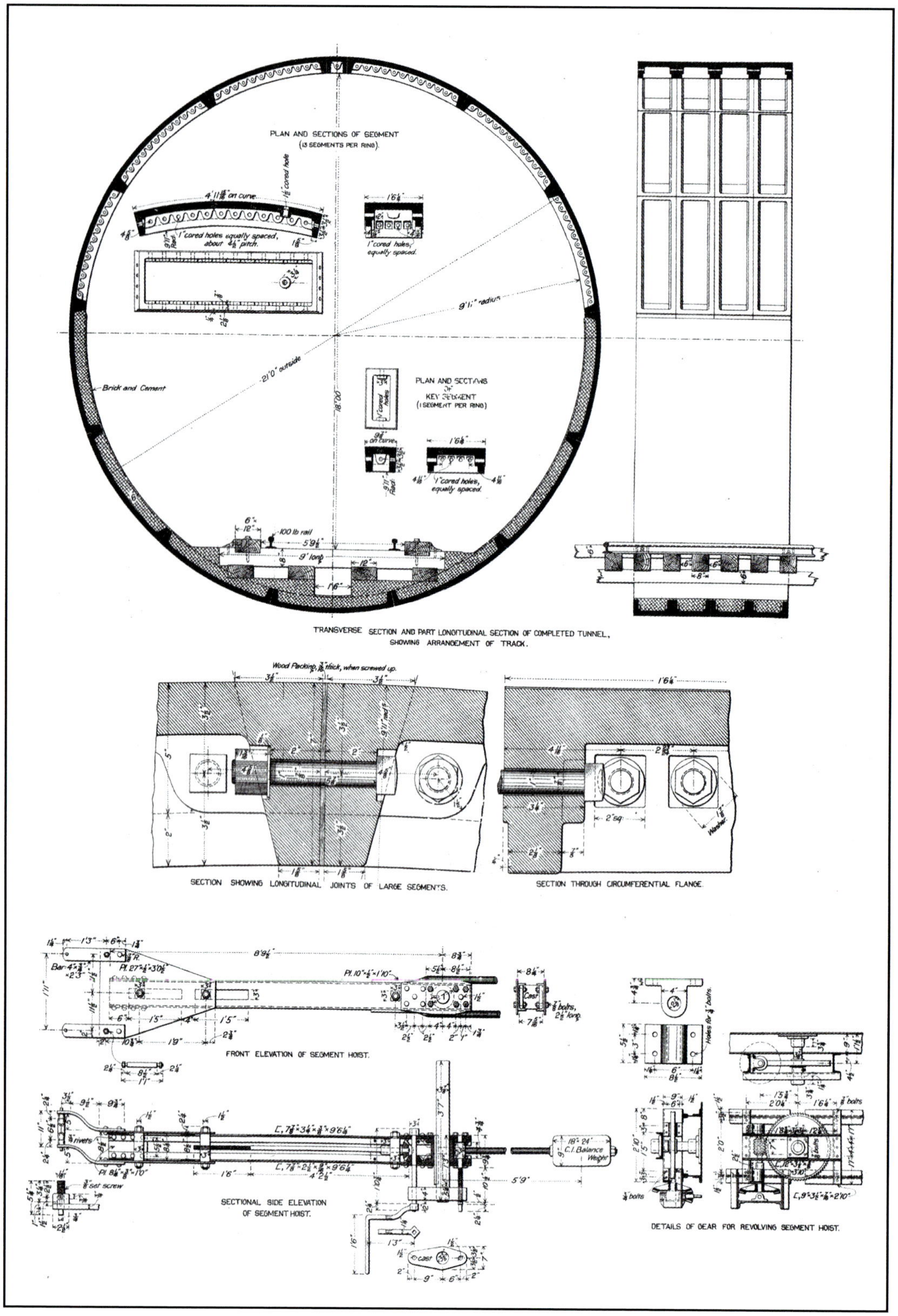

This project facilitates Long Island Rail Road (LIRR) commuter trains to proceed directly to Grand Central Terminal. The tunnels were connected into the surface LIRR tracks at Sunnyside Yard in Queens. Four tunnels make the final connection at Sunnyside. The tunnels were excavated under low cover under the busiest rail yard in the United States. Other challenges included the high water table, full-face hard-rock and mixed-face conditions with soft ground and boulders, and contaminated ground from past industrial activities.

An inside view of a slurry TBM is shown in Figure 8.63. Bentonite slurry is pumped into the cutterhead chamber and is mixed with the excavated soil. The slurry and mixed excavated material is discharged from the chamber at a controlled rate and pumped to a separation plant on the surface. The excavated material is separated and the slurry is returned to the TBM. The slurry and a compressed air "bubble" in the chamber control the pressure at the face and balance against the natural soil and groundwater pressure fluctuations. Slurry TBMs control the excavation face pressures in mixed-face conditions and keep contaminated soils and groundwater contained in the slurry circuit.

Figure 8.61 | Original Tunnel Lining Design for the First St. Clair River Tunnel

Figure 8.62 | **Herrenknecht Slurry TBMs Onsite in Queens, New York**

CHAPTER EIGHT | INNOVATIONS IN TUNNELING | 431

Road tunnels are another ideal application for TBMs, and such TBMs tend to be of large size to accommodate multiple lanes of traffic. The Port of Miami (now PortMiami) Tunnel was constructed between the Miami Interstate Highway System and the Port of Miami on Dodge Island. The intent was to reduce vehicular traffic going to the port through downtown streets. Twin tunnels, 4,200 feet long each, carry the roadway. The tunnels cross beneath Biscayne Bay in difficult soft-ground conditions.

The tunnels were constructed under a public–private partnership form of contract, with financing provided by private and public entities. The constructor was Bouygues of France, who was considered to have relevant experience tunneling in soft ground under a waterway, gained from its work on the Channel Tunnel and many other international projects of similar complexity. The PortMiami highway tunnel ground consisted of extremely porous and permeable marine deposits, with a hydrostatic head up to 120 feet. A 41-foot Herrenknecht EPB TBM was used to excavate the two tunnels (Figure 8.64).

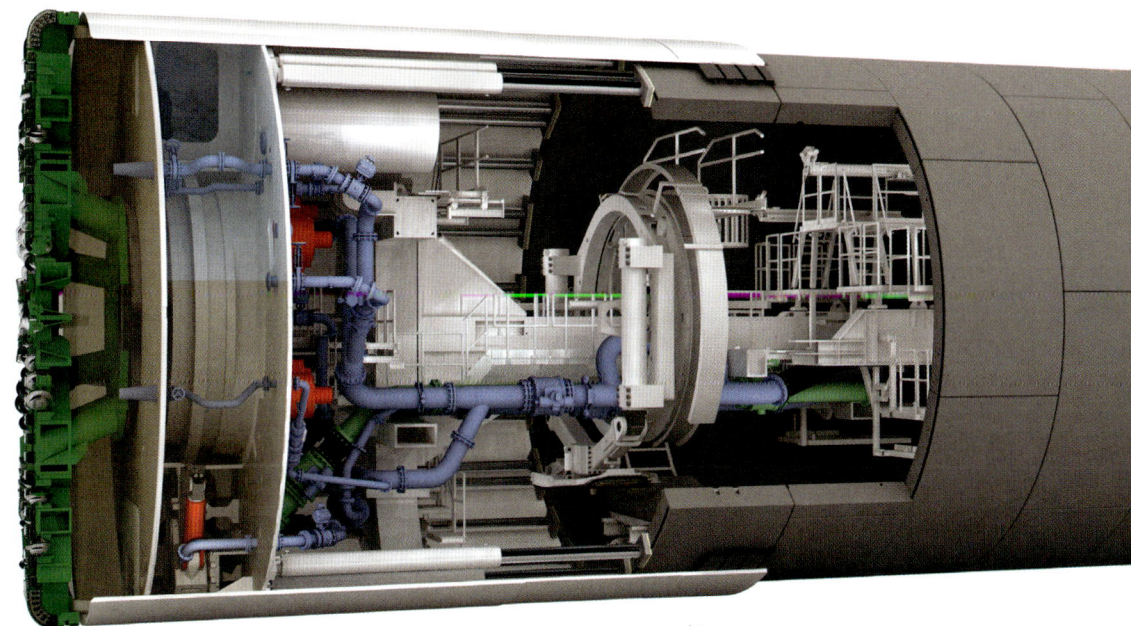

Figure 8.63 | **Herrenknecht Slurry TBM Design Interior Details with Precast Lining**

The Central Subway tunnels in San Francisco were constructed to extend the existing subway line 1.7 miles northward to a station at Chinatown. The tunnels were excavated in the downtown area beneath the water table in variable formations of clays, sands, and bedrock. Other project conditions included sharp curves, "potentially gassy" categorization (methane or other dangerous gases at potentially explosive concentrations), and close proximity while passing beneath existing Bay Area Rapid Transit (BART) and Muni Metro subway tunnels. Two 20-foot-4-inch-diameter Robbins hybrid soil/rock EPB TBMs were used to excavate these tunnels. The machines operated in closed mode in soft, water-bearing ground conditions and in open mode when in dry rock conditions.

The Alaskan Way Viaduct in Washington is an elevated roadway structure that carries State Route 99 (SR 99) along the waterfront in Downtown Seattle (Figure 8.65). The viaduct was damaged in the 2001 Nisqually earthquake. As a replacement to the viaduct, the tunnel option provided great environmental benefits along the waterfront, since the elevated roadway could be removed and the space used for other urban uses. One of the largest TBMs in the world was needed to excavate this roadway tunnel. A 57-foot-diameter EPB-type TBM was chosen for the work. This TBM, named "Bertha" was manufactured in Osaka, Japan, by Hitachi Zosen Sakai Works and at the time of manufacture was the world's largest diameter for a soft-ground TBM.

Figure 8.64 | **41-Foot-Diameter EPB TBM for the PortMiami Tunnel**

Because of the water level lowering in Lake Mead (Nevada) as a result of a continued drought in the western United States, a new tunnel, Intake No. 3, has been constructed to secure a long-term source of potable drinking water for the city of Las Vegas. The tunnel excavation started in July 2011 on land from a starter tunnel in the 600-foot-deep shaft and was excavated for 15,800 feet using 23.6-foot-outside-diameter precast segmental linings as the permanent lining. The tunnel linings were designed to seal to 250 psi of external water pressure and performed as designed. A Herrenknecht single-shield machine designed for hard-rock conditions and the potential for high-pressure water inflows was used. The TBM was equipped for pre-excavation drilling and high-pressure grouting capabilities under the premise of attempting to "grout off" the high-pressure water areas before the TBM reached these critical sections. On December 10, 2014, the contractor broke through into the project's previously constructed underwater intake structure that was open to the lake. The tunnel was flooded and is now currently providing the people of Las Vegas with additional fresh drinking water.

Figure 8.65 | **Subsurface View of the SR 99 Tunnel in Seattle, Washington**

Roadheaders

A roadheader consists of a cutting head connected to a hydraulically actuated boom attached to the main body of the machine with an apron or gathering bed under the boom to gather the rock cuttings onto a conveyor belt that discharges the cuttings at the back of the machine. The rock cuttings are then loaded and transported out of the tunnel. The machine is supported on a crawler undercarriage that is used to move and position the machine in the tunnel.

Roadheaders were initially designed for operating in coal mines. Beginning in the 1950s, they were first put to work in the European general tunneling industry in an attempt to mechanize tunnel excavation, thus steering away from the conventional drill-and-blast methodology. The first tunnel projects in North America using roadheaders were completed in the early 1970s.

Roadheader usage increased in the 1980s in tunnel applications with the introduction of heavier and more productive machines that were brought into the U.S. marketplace from Japan and Europe. With the need to mine big caverns and large flat-bottomed tunnels, roadheaders became the excavation method of choice in suitable geological conditions. Large cross-sectional water tunnels and caverns were successfully excavated in Kentucky, Texas, and California. Introduction of the new Austrian tunneling method / sequential excavation method (NATM/SEM) in Washington, D.C., in the early 1980s brought additional applications for roadheaders. They were used to excavate the multiple drifts associated with the SEM type of construction. Roadheaders, when properly sized for the project and geology to be encountered, have proven to be an effective method for tunnel excavation.

CHAPTER EIGHT | INNOVATIONS IN TUNNELING | 433

In the beginning, roadheaders had the reputation of only being capable of cutting soft rock with good stand-up time (similar to conditions found in a coal mine), but since 2005, major changes have occurred with the technology of rock cutting using roadheaders. Initially, the operating weight of these early machines did not exceed 30 tons; therefore, they had the ability to cut only soft rock below about 10,000 psi UCS (unconfined compressive strength).

Roadheaders are now offered with operating weights from 13 tons to 180 tons with a cutting power of up to 535 HP (Figures 8.66 and 8.67). The heavier the machine, the greater the reach and greater the cutting capacity. As these are mobile machines, mass is required to offset the forces encountered at the end of the cutter boom. Today, the heaviest and most powerful roadheaders are able to cut rock with up to 30,000 psi UCS, three times the strength of the original smaller machines. Today's sophisticated roadheaders can be

Figure 8.66 | **Heavy-Duty Roadheader**

Figure 8.67 | **Modern Compact Roadheader**

equipped with numerous options such as drill booms, remote control, automatic guidance systems, and pre-support spiling hammers, making them versatile tunnel excavation tools.

Roadheaders are an alternative means of tunnel excavation to the drill-and-blast method and use of TBMs for rock excavation in tunnels. This method of excavation allows complex underground geometries to be excavated. Excavation is more accurate when compared with drill-and-blast and not possible with circular TBMs unless used in combination with other excavation methods.

Roadheaders, however, are prone to significant tool wear during the excavation process, and tool replacement times slow production in highly abrasive ground.

Since 1975, there have been many projects in the United States where roadheaders have been successfully used. For example, more recent locations include

- New Irvington water tunnel, Sunol, California;
- New York East Side Access Project, New York;
- Caldecott Fourth Bore Project, between Oakland and Orinda, California;
- Dulles International Airport, Dulles, Virginia;
- Outfall Tunnel for Water Treatment Plant No. 4, Austin, Texas;
- Devil's Slide Tunnel, between the cities of Pacifica and Montara, California;
- Cumberland Gap, North Carolina
- Waller Creek Tunnel Project, Austin, Texas; and
- NATM Tunnel at Stanford Linear Accelerator, Menlo, California.

New Austrian Tunneling Method / Sequential Excavation Method

Not all tunnels need to be circular, and sometimes the length of the tunnel precludes the expense of a TBM. Shotcrete was invented in the United States at the turn of the twentieth century and is an efficient support system. But one problem with shotcrete is that it works well in compression but needs reinforcement for large-opening applications such as railroad, highway, and subway tunnels. In 1964, Austrian professor Ladislaus von Rabcewicz established the name "new Austrian tunneling method," or NATM, for a support system design that included shotcrete and lattice girders.

Widespread use in Europe and Australia ensued, and by the 1980s, its use became widely accepted with WMATA's construction of the Washington D.C. Metro system. Shotcrete and lattice girders (Figure 8.68) are very useful tools in weak soils and provide an excellent template for the addition of shotcrete and welded wire fabric, which effectively serves as the structural component of the ground support system.

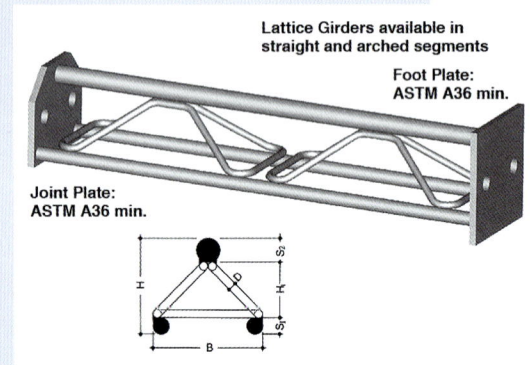

Figure 8.68 | **Typical Three-Strut Lattice Girder Fabrication Details**

Lattice girders provide enormous flexibility in creating complex tunnel shapes of compound curvature and vast cavern spaces (Figure 8.69). They also provide a unique ability to guide the proper depth of shotcrete placement in final lining applications. Girders are lightweight and come in customized segments for easy placement and erection in the tunnel space (Figure 8.70). They are easily adaptable to heading and bench excavation schemes, which are the hallmark of NATM. With the widespread use of lattice girders and shotcrete in the United States, the NATM name has changed to *sequential excavation method*, or SEM.

Figure 8.69 | **Typical Lattice Girder Installation Under a Pipe Canopy**

Figure 8.70 | **Lattice Girder Installation in a Large-Span Tunnel Heading**

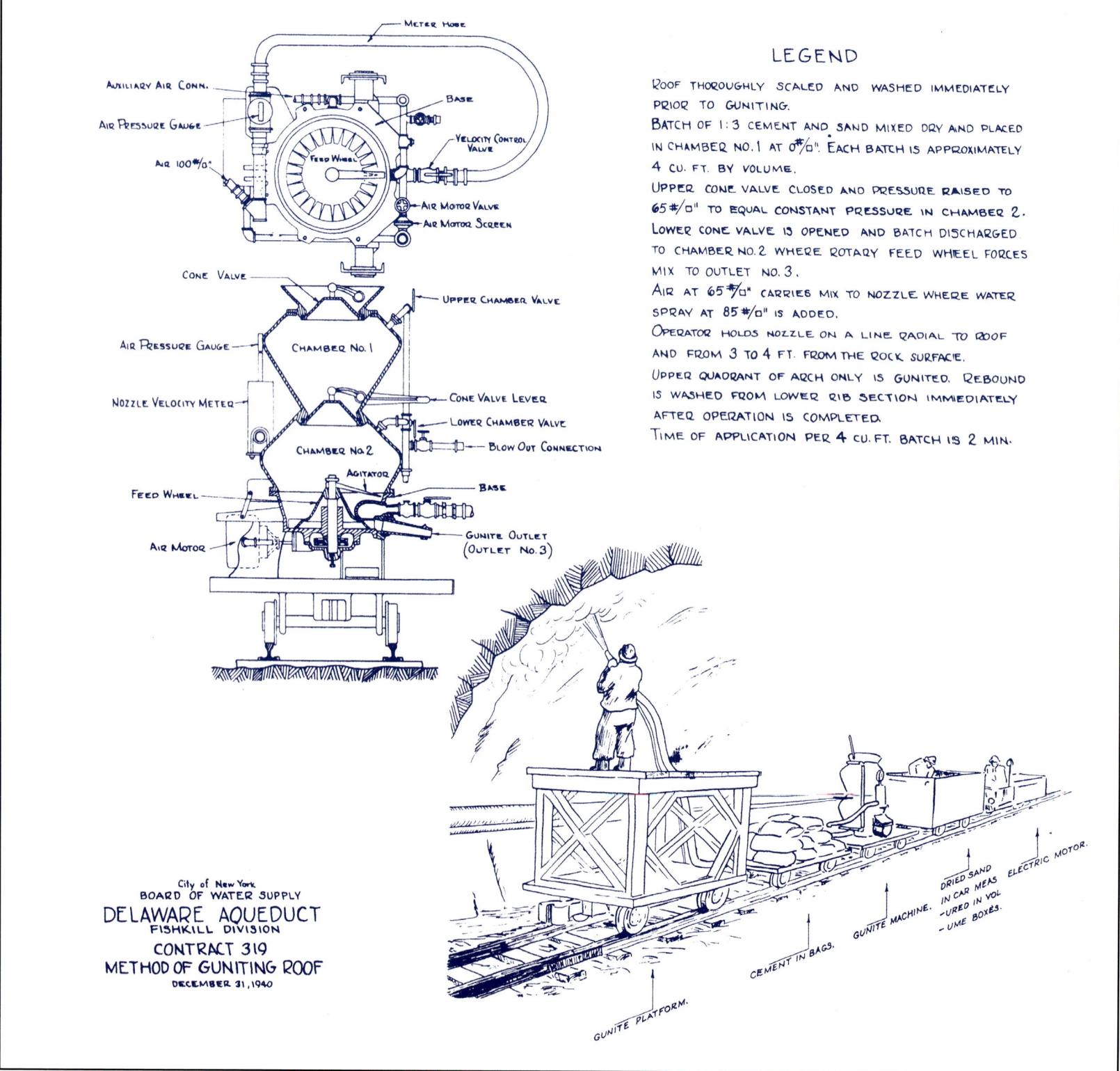

Figure 8.71 | **Vintage Drawing Illustrating Dry Mix Shotcrete Application**

INNOVATIONS IN SPRAYED SHOTCRETE TUNNEL LININGS

Sprayed concrete (shotcrete) has long been a key component for initial support during excavation of underground space in many ground types. Developed in Chicago in 1907 and then improved and commercialized in Allentown, Pennsylvania, machines and accessories for spraying a cementitious mixture onto a surface quickly became a versatile and useful construction tool (Figure 8.71). An early record (1917) indicates that mortar was sprayed in a small tunnel built for moving coal from a mine to an adjacent power plant in Western Pennsylvania, and the support and durability of this initial lining was noted.

Applications through the early 1970s further developed the use and acceptance for initial rock support in tunneling as the growth of rail and road building expanded in Europe and America. The "cement gun" machines and process were exported to Europe in the early 1920s and manufactured under license, in England, soon after. The first new innovation for the dry process was the development of a rotary machine in the early 1950s. This allowed for larger aggregates in mixtures and increased output. The rotor type of machine is very much in use in today's tunneling industry for dry shotcrete application. Dry shotcrete mixtures are designed to meet the capability of new types of dry spraying equipment and to meet higher volume and performance demands (Figure 8.72).

Wet mix shotcrete is also popular for tunneling applications for initial and final support requirements. Small, self-contained batching plants have been developed along with portable robotic spraying units.

Steel fibers for concrete reinforcement were adapted to initial shotcrete support in the early 1970s. Steel-fiber-reinforced shotcrete (SFRS) has been reviewed and standardized in ASTM International, American Concrete Institute, German Institute for Standardization, and other technical authority documents in most markets globally. This innovation provided solutions for many projects in difficult ground and logistical conditions. Although steel fibers have changed little, the understanding of performance, as shotcrete technology has evolved, provides abundant data for design of efficient lining types and thicknesses. Data exist to allow the designer to select steel fiber as a reinforcement component that will be compatible to the geology and geometry of the project. Spraying SFRS is common today. New variations and materials for fibers are also appearing, and testing data and performance history can provide direction for choice.

Figure 8.72 | **Modern-Day Dry Mix Shotcrete Machine**

The equipment innovations for applying sprayed shotcrete in modern tunnels, for support and final lining have evolved from product testing and experience. Fully automated robotics are not yet ready for the tunnel market; however, a properly trained nozzleman at the controls of a boom-mounted spray head or a qualified team in the basket of a manlift for hand spraying will produce quality and cost-effective results in a safe and efficient manner (Figure 8.73). Innovations in shotcrete admixtures, accelerators, steel fiber and plastic reinforcement, and shotcrete machine technologies have contributed to the underground construction industry in many ways and are important tools in the construction of major underground projects (Figure 8.74).

For difficult ground conditions where face support and crown pre-support is critical, a ground support system called "canopy pipes or spiling" was introduced to the U.S. tunneling market from Austria in the 1980s. Terminologies vary, but the general term for elements used to pre-support the crown include *rebar spiles, pipe spiles*, and *wood or steel poling plates* (Figure 8.75). Pipe spiles are used in NATM/SEM in conjunction with lattice girders. They provide exceptional strength and are capable of being driven far ahead of the excavation face at a slight angle from the last excavated round in the tunnel. The spacing of the

Figure 8.73 | **Robotic Arm Spraying Wet Shotcrete**

Figure 8.74 | **Remote-Controlled Robotic Arm Applying Shotcrete in a Cavern**

pipe spiles is adjusted depending on the soil conditions encountered. Lattice girders provide an ideal support structure for the pipe spiles to be driven across and into the soil mass ahead. A variation of this concept is a combination of "sawtooth" lattice girders with either rebar spiles or pipe spiles. On a basic repetitive girder design of increasing height and/or width, even the most complex geological conditions can be accommodated.

When extreme geological conditions are encountered, the use of pipe canopies has become the standard practice. Pipe diameters can become large. When the pipe wall thickness is increased, the pipe canopy can be drilled or driven more than 300 feet to provide exceptional starter tunnel space with a high confidence of safety and stability (Figure 8.76). Most applications of this type also use grouting of the pipe interior and permeation grouting through the pipe wall as part of the ground support system. The pipe spacing is also variable to meet the anticipated soil conditions.

With the innovations that have occurred in the various types of ground support systems, tunnels can now be supported safely and efficiently during the excavation process.

INNOVATIONS IN GROUND SUPPORTS

Initial supports are defined by the industry as a temporary support system installed at the time of excavation to support the geological conditions encountered. Initial support systems range from nothing at all if the rock is competent, up to heavy-duty support systems designed to hold up the face and the full weight of the overburden. There have been many innovations over the past 100 years in the area of initial steel support systems that have made tunneling safer and more productive.

Steel Sets

Prior to the 1920s, use of timber supports for both mining and tunneling was the normal practice. Some railroad tunnels and older mines still rely on the sturdy

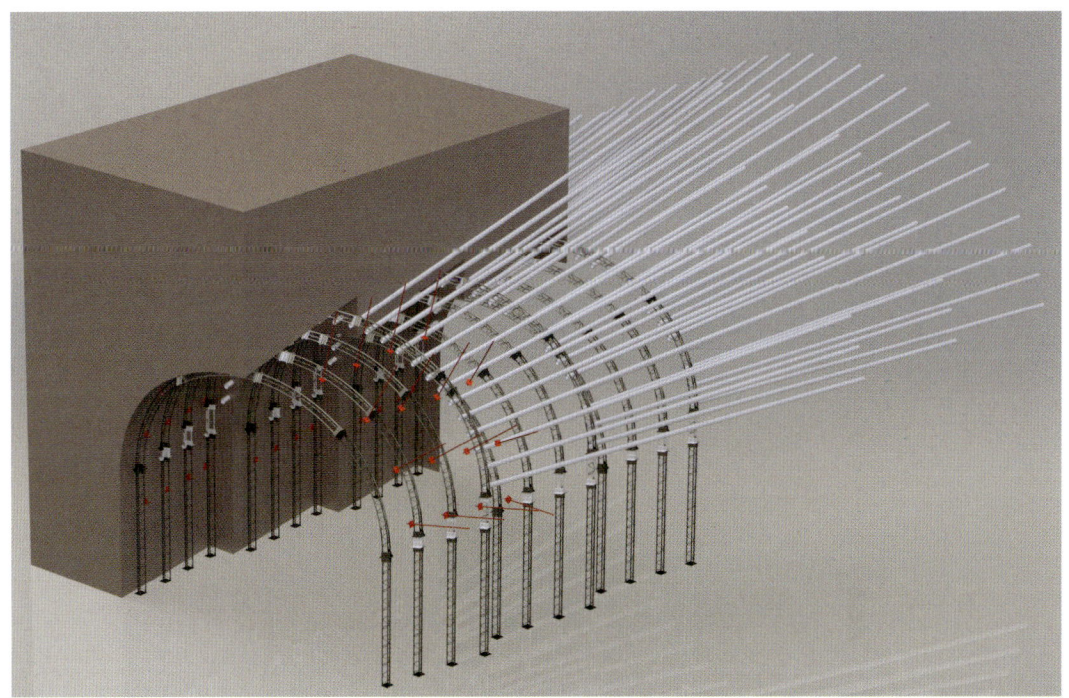

Figure 8.75 | **Typical Lattice Girder Steel Canopy Support System**

Figure 8.76 | **Typical Pipe Arch Canopy System**

Figure 8.77 | **Rendering of Typical Straight-Leg Horseshoe Steel Sets and Lagging**

Figure 8.78 | **Steel Rib-Wall Plate and Bench Legs**

and well-crafted timber support structures today. The tunnels constructed for the Transcontinental Railroad were supported with wood timbers that were typically sourced locally.

Introduced in the 1920s, structural steel support systems (commonly referred to as ribs, tunnel ribs, shaft ribs, or steel sets) have become a mainstay in underground construction. Structural steel supports are installed at or near the face of the tunnel heading and are principally used with some form of lagging, either wood, steel, or shotcrete. Steel ribs are a passive system that requires installation of blocking material (wood blocks, shotcrete, or concrete) to fill the gap between the outside flanges of the beam to the rock mass, thus integrating the beam to the forces acting on the tunnel opening.

Where higher anticipated ground loads are expected, the rib spacing can be reduced or the size of the structural support rib increased, or both, depending on various factors encountered. The rib geometry has a significant effect on the support capacity of a given rib set. Circular ribs provide the highest load-carrying capacity, while square sets provide the least (Figure 8.77).

As the use of steel rib sets expanded, the potential for larger underground tunnels, caverns, and access shafts developed. These applications included such projects as the Eisenhower–Johnson Tunnels in Colorado where the two tunnels cross the Continental Divide. In this area, 24-inch-deep wide-flange beams were supplied bent and later installed by the contractor to support the loading from the Rocky Mountains. Since each tunnel is different, tunnel engineers must design support systems with consideration to the steel section/profile required, the section component pieces to accommodate fabrication, shipping and installation in the tunnel, and the connection details necessary to withstand the ever-increasing rock and soil loads (Figure 8.78).

The primary reason for the use of steel rib sets becoming so widespread was due to the ability of specialty fabricators to bend increasingly large steel sections to meet the shape requirements selected by designers and contractors (Figure 8.79). Presently, beams up to 27 inches deep can be bent along the primary sectional axis.

Many types of steel profiles can be bent for tunnel use, including W sections, C sections, and TH sections. In addition, these sections can be supplied in varying steel grades to meet the requirements of the tunnel design engineer. In some instances of high rock stress, "yieldable" steel sets are used. These are typically of the TH section variety. In this application, the connection details allow for the steel to slip, or deform, in a controlled manner. These are special sections that are not commonly used in North American civil tunneling applications. These were used, however, on the recently completed Gotthard Tunnels in Switzerland where tunnel depths approached 2,650 feet of rock cover.

Steel Rock-Bolt Technology

Rock bolts of varying types (mostly mechanical and timber friction bolts) have been used for well over 100 years. In the United Kingdom, a North Wales slate quarry documented bolt usage as early as 1872. The first reported systemic use of rock bolts was at the St. Joseph lead mine in Missouri in the 1920s. Internationally at this time, bolt technology and usage became widespread. From Australia to the United States and across Northern Europe, rock bolting became a common method of temporary and permanent roof support in rock tunnels and caverns.

Figure 8.79 | **Circular Steel Ribs and Timber Lagging, and Rendering (inset)**

Modern steel rock bolts were originally developed and received major acceptance for use as a roof support in the coal and hard-rock mining industries in the late 1940s and 1950s. The U.S. Bureau of Mines was under a mandate to develop and test better roof control systems for safer and more productive means of mineral extraction. Roof falls were one of the major safety hazards associated with underground mining, and the development of steel bolting systems to support the ground surrounding the opening was a major innovation that was soon adapted to civil tunnel construction.

In the mining industry, the practice of installing steel bolts is called *roof bolting*, and in the civil construction industry, the practice is termed *rock bolting*. No matter the terminology, the practice is the same. A hole is drilled at a predetermined length and hole diameter into the substrate. Next, a steel rod is installed in the hole with mechanical or grouted anchorage to hold the bolt in place. Then a plate and nut for tensioning is installed against the rock for retaining tension applied by direct pull or by torqueing.

Bolts can be made of various metals, such as carbon steel, stainless steel, and also glass fiber reinforced polymer (GFRP) for use in a variety of applications inclusive of highly corrosive areas or in areas where high strength and long-term performance is required as part of a permanent ground support system. Bolts are used both as temporary and permanent support systems in tunneling and mining operations. There are three basic types of rock bolts:

1. **Mechanical bolts.** In this category, a threaded bar is used in conjunction with expansion anchor shells attached to the bolt (Figure 8.80). Upon insertion of the bolt bar into a drilled hole, the bar is spun to expand the shell, thereby applying pressure on the surrounding rock and creating a point from which resistive load can be generated, creating a tensioned bolt.

2. **Friction bolts.** For this type of bolt, there is no end piece that creates an anchor point (Figure 8.81). Instead, a friction bolt relies on nearly continuous contact of the bolt shaft with the borehole periphery. The bolt shaft is typically slit to allow circumferential compression as the bolt is inserted the length of the borehole. Alternately, the bolt shaft is hydraulically expanded to fill and compress against the bore hole.

3. **Grouted bolts.** This type of bolt uses high-strength grout, either resin or cementitious, to bond with and restrain a standard billeted rebar or cables with the borehole walls (Figure 8.82). It is essentially a concrete footing in a rock hole.

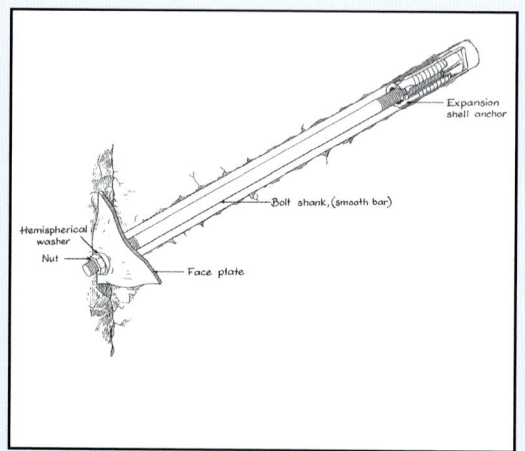

Figure 8.80 | **Mechanical Bolt Assembly**

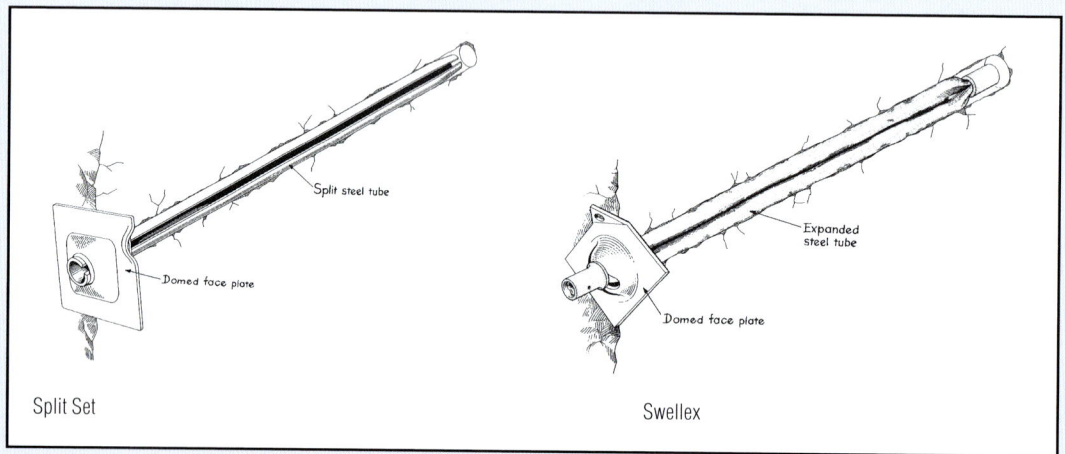

Figure 8.81 | **Friction Bolts and End Plates**

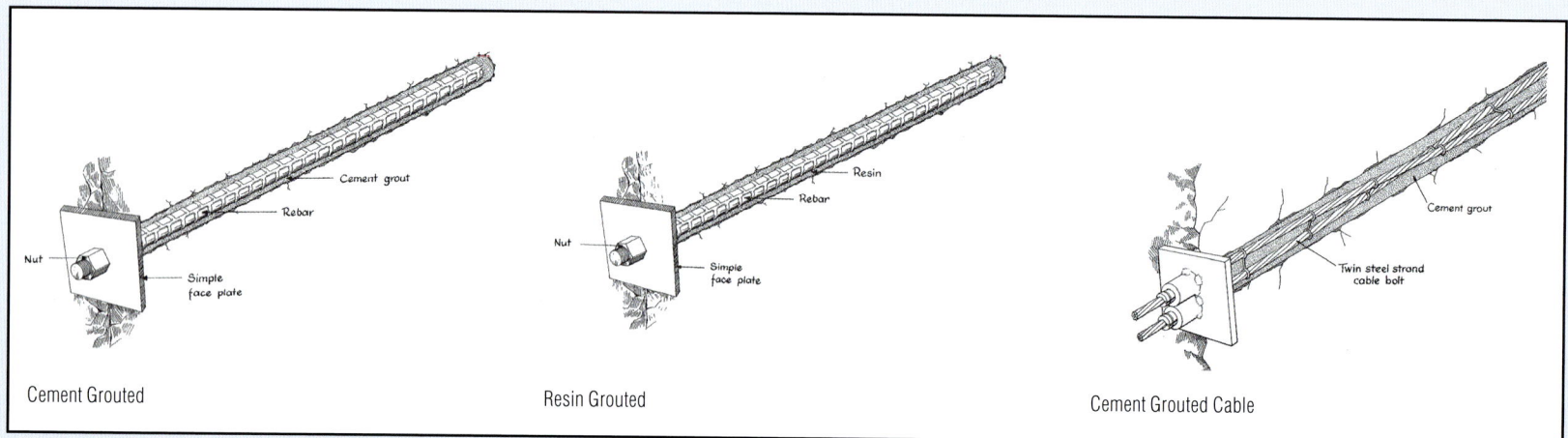

Figure 8.82 | **Grouted Bolt Assemblies**

Other Advances in Rock-Bolt Technology

In the 1960s, the use of grouted (mostly cementitious) rock bolts developed wide acceptance and use. The 1970s ushered in widespread use of "split sets" and Swellex friction-held rock bolts, where mechanical bolts were less effective due to certain geological conditions. As industry in general ushered in the age of plastics (Figure 8.83), the same occurred in the tunnel industry with the development of resins that significantly improved the adaptability of grouted bolts. Resins provide very high pullout strength and their designed viscosity allows for customized resins that can flow outward into adjacent rock masses, creating an increase in load-bearing capacity.

Combinations of the basic three types of bolt systems have developed with remarkable improvement and adaptability to specific rock conditions and load-bearing requirements.

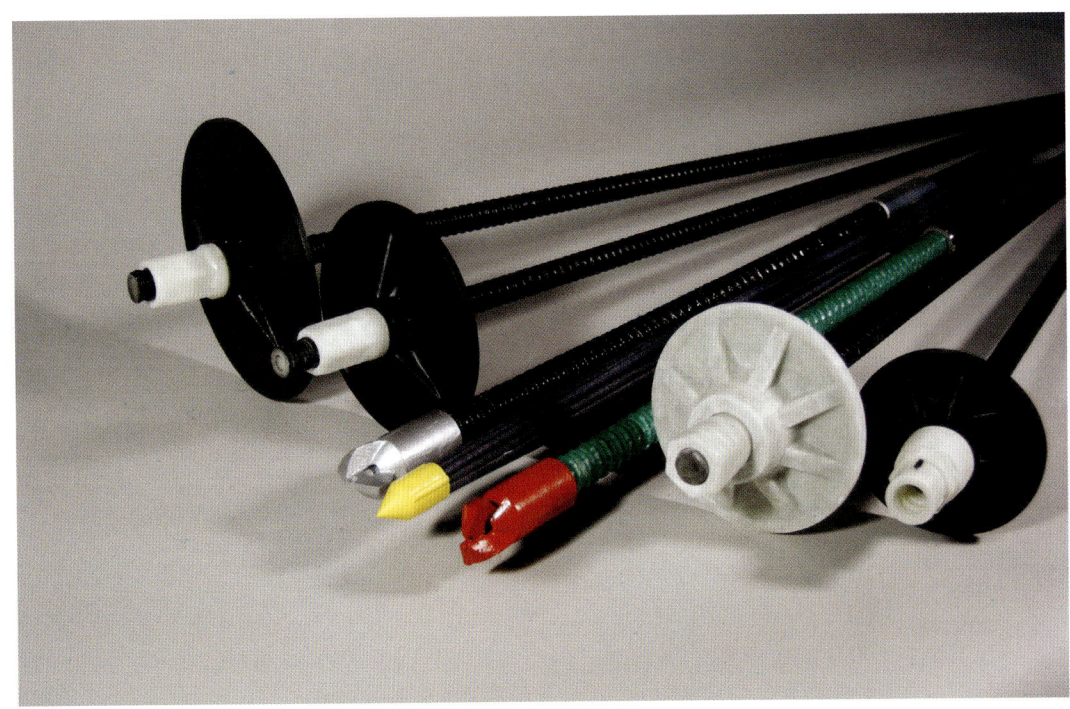

Figure 8.83 | **Various GFRP Bolts and End Hardware**

INNOVATIONS IN TUNNEL WATERPROOFING

Many innovations have enabled contractors to provide tunnels that are dry and thus require less maintenance over their service lives. One of the more important waterproofing innovations that has provided society with dry tunnels is the introduction of flexible membrane waterproofing systems to the U.S. tunnel industry.

Throughout history, there has always been the desire to have dry tunnels, especially in the wetter, more northern climates. Some of the first railroad tunnels were driven through mountainous terrain at high elevations where in winter conditions the water leaking into the tunnel would freeze. If the ice was not removed, the train would be damaged or, at worst case, the ice buildup could preclude the train from traveling through the tunnel.

In later years when the tunnels were electrified, the need for dry tunnels became even more important, because water, if not controlled and diverted, would infiltrate the electrical power system, causing damage and power outages.

Early attempts at waterproofing tunnels consisted more of a water control system where a metal sheeting (panning) or shield system would be installed at specific locations to direct the water to a drainage system. This worked to some degree in subway tunnels that did not freeze, but it was not effective in railroad and highway tunnels exposed to freezing temperatures. During winter months, railroads and highway departments employed ice removal crews with the sole purpose of removing the ice buildup in the operating tunnels.

Once concrete linings were installed in tunnels, other waterproofing methods, such as the installation of metal sheeting and drainage blankets, were adopted prior to the placement of the concrete to direct the water away from the middle of the tunnel to side drains. Alternatively, post-construction grouting behind the concrete tunnel lining was performed in an attempt to fill the space, creating a cementitious seal. This method had limited success as a long-term solution for creating "dry" tunnels.

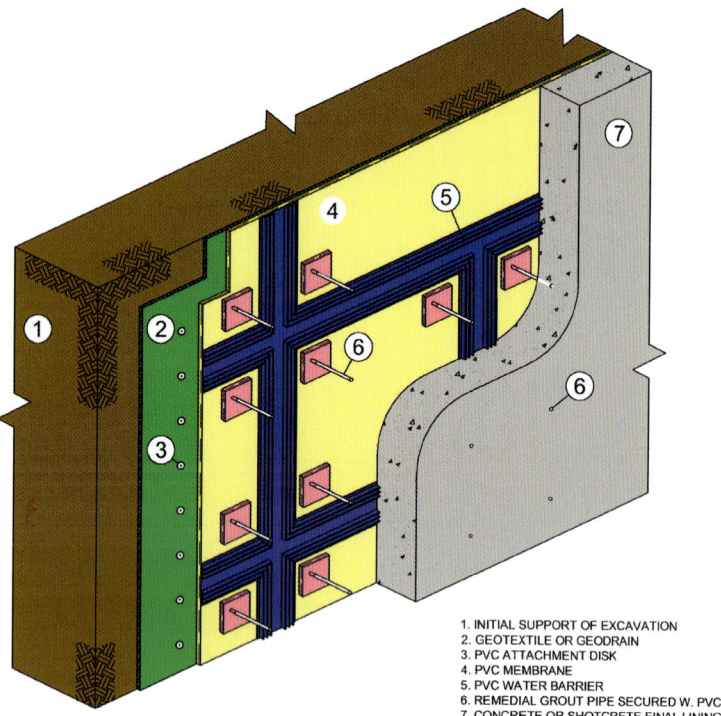

Figure 8.84 | **Typical Composite PVC Waterproofing Installation Detail**

Figure 8.85 | **Membrane Installed with PVC Discs Before Welding**

Other types of "engineered" waterproofing systems were attempted, such as rigid high-density polyethylene panels in areas where hydrocarbons were present, bentonite panels that expanded and sealed in the presence of water, and chemical spray-on application systems. These were only nominally successful and have so far had limited industry acceptance. The biggest drawback to these alternate systems is that the tunnel must first be dry when the material is applied and tunnel construction can often be a harsh, wet environment. To be successful, a waterproofing system needs to perform as an effective long-term seal, having been installed in such adverse conditions during construction.

Creating dry rock tunnels has always been a worldwide problem. Some European countries were leaders in these technologies and innovations. In the 1980s, a polyvinyl chloride (PVC) membrane tunnel waterproofing system was installed for the first time in the United States in portions of Washington, D.C.'s Red Line tunnels and Wheaton Station (Figure 8.84).

Based on WMATA's pre-design investigations, the flexible membrane PVC waterproofing provided a dry rock tunnel, which had been successfully applied mainly for mountainous highway tunnels in the Austrian Alps. At WMATA, electrical maintenance was a major cost item because of the corrosive nature of the groundwater in the area of the Wheaton project. WMATA management saw this waterproofing system as a solution to prevent additional long-term maintenance issues. This analysis proved to be invaluable since, after 30 years, the WMATA Wheaton Station and tunnels remain dry. This flexible membrane waterproofing system was implemented on many more WMATA developments and, to this day, continues to be a standard design for their projects.

The typical PVC waterproofing system is engineered to be flexible, long lasting, and, most importantly, tested for waterproofing integrity before the placement of the final lining cast-in-place concrete. The PVC sheets are typically installed circumferentially in the tunnel, then each seam is thermally welded with controlled heat to fuse the PVC sheets together. Each joint is then tested with compressed air to verify the integrity of the lining. One significant characteristic of the PVC membrane that made it applicable to underground waterproofing was that the membrane could be installed in wet conditions and the PVC sheets could be successfully welded even when wet, a common situation in tunnel construction (Figure 8.85).

One of the attributes of the PVC system is that the system can be easily repaired if it becomes damaged. The PVC system has evolved since 1985 to include a grid isolation system using water-barrier devices that allow for compartmentalization of leaks in the tunnel (Figure 8.86). If a leak does occur behind the final tunnel lining, this system allows for the area to be

Figure 8.86 | Completed Membrane Tunnel Waterproofing System

446 THE HISTORY OF TUNNELING IN THE UNITED STATES

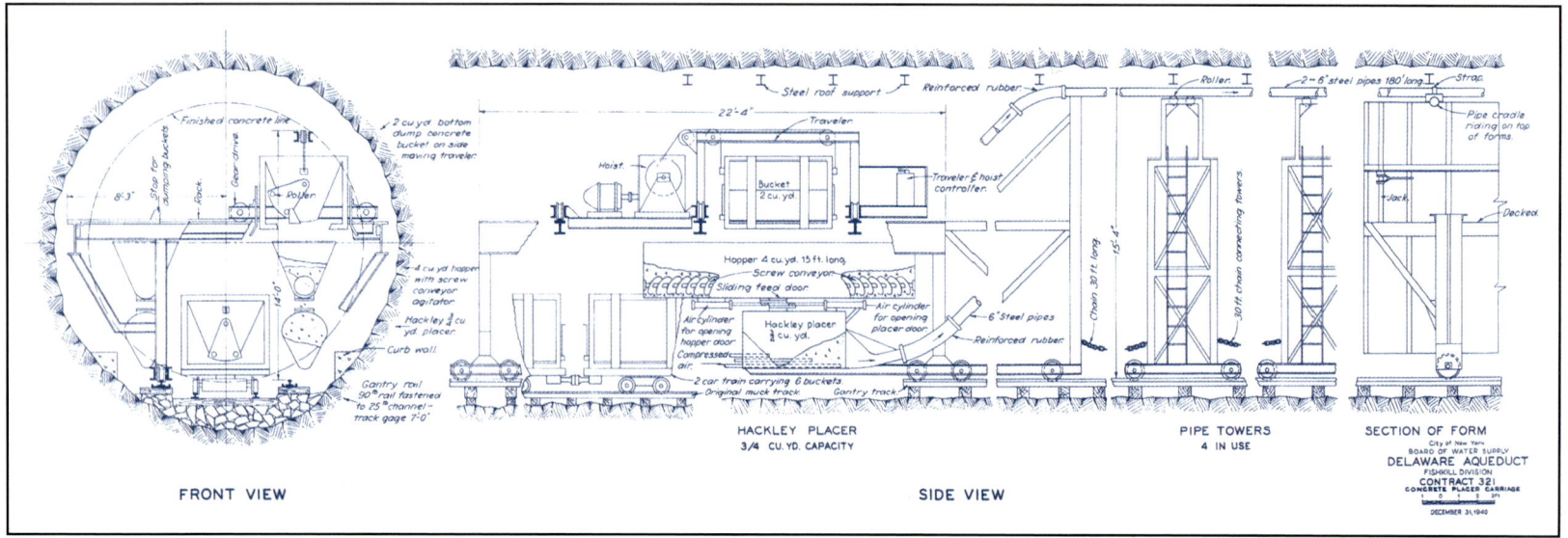

Figure 8.87 | **PVC Waterproofing System with Grid Isolation Panels in Frozen Ground**

Figure 8.88 | **Vintage Work Sequence Drawing for Cast-in-Place Concrete Tunnel Lining**

grouted in a small, local area with chemicals or micro-fine cement to seal off the leak. The grid isolation system has been successfully implemented on many projects in the United States to date.

New York City's Northern Boulevard project for New York City's East Side Access program is of significant industry importance because a PVC membrane waterproofing system was successfully installed in a tunnel at the same time the surrounding ground was frozen to provide support of the ground during construction (Figure 8.87). The PVC membrane was protected from the ice with a synthetic geotextile (fleece) layer. After the placement of the final concrete lining, the freezing operation was terminated, the surrounding strata thawed, and the tunnel was put into service as a dry, functioning structure.

The successful performance of the flexible membrane PVC systems in varying geological and tunnel end-use applications has made the waterproofing system applicable to a variety of tunnel project applications, such as Metro stations, cross passages, shafts and adits, open-cut retaining structures, and special structures. To date, flexible membrane waterproofing systems have been installed in more than 200 different tunnel projects and applications.

Cast-In-Place Concrete Linings

Since the time of Ancient Rome, concrete, in its semi-liquid state after mixing, has been placed using a form to retain the concrete into a desired final profile (Figure 8.88). The same is still true today in the tunnel industry, as it is industry practice to install a final concrete lining to facilitate the long-term final use of the tunnel.

Circular cast-in-place final concrete linings are typically associated with water and sewer tunnels, and flat-bottomed, cast-in-place, horseshoe-profiled tunnels are associated with transportation tunnels (Figure 8.89).

Most underground caverns rely on a permanent concrete lining for structural support. In the case of large-span Metro station caverns in rock or soil conditions, permanent support is often achieved with a cast-in-place concrete arch (Figure 8.90). This may also serve as the interior architectural finish as well as support intermediate slabs and service rooms.

The cast-in-place final concrete lining has multiple purposes. The primary purpose is to provide long-term structural support against the forces acting on the tunnel opening from the strata through which the tunnel was excavated. The cast-in-place final lining also assists in keeping water out of the tunnel and provides a mechanism for attaching

Figure 8.89 | **Circular Cast-in-Place Tunnel Lining Formwork (with Waterproofing)**

Figure 8.90 | **Semicircular Cast-in-Place Cavern Lining Formwork (with Waterproofing)**

Figure 8.91 | **Reinforced Concrete Cylinder Pipe Secondary Tunnel Liner for Water Transmission**

Figure 8.92 | **Reinforced Concrete Cylinder Pipe Secondary Tunnel Liner Installation**

Figure 8.93 | **Completed Devil's Slide Tunnel with Interior Stone Facade**

accoutrements associated with the ultimate use of the tunnel. To assist in strengthening the design of the final lining, steel reinforcement is placed within the final lining concrete pour. The reinforcement typically consists of deformed steel bars that bond with the cured concrete to provide tensile support to the final lining. Concrete works very well in compression, and the steel reinforcing rods provide tensile strength to the structure. A recent industry development is the use of high-strength steel fibers, thin wires about two inches long that create a reinforcement matrix. The use of plastic fibers are also gaining acceptance as a fire protection measure.

In some circular tunnels, there are forces acting equally around the cast-in-place final lining, thus putting the concrete lining in compression. In such cases, steel reinforcement is not required. In areas that are subject to earthquakes, special considerations need to be taken into account by the tunnel design engineer to deal with these seismic loads. This special loading condition is typically addressed by adding more reinforcement to the final cast-in-concrete design.

Many water transmission tunnels require a secondary liner to improve hydraulic characteristics. Steel, concrete, ductile iron, plastic, and fiberglass are some of the typical materials used. In some cases, a combination of materials may be required, including steel-plate-encased reinforced concrete pipe (Figures 8.91 and 8.92).

The California Department of Transportation recently completed the Devil's Slide highway tunnels south of San Francisco with special seismic design considerations (Figure 8.93). A project-specific custom steel concrete forming system was designed and built to cast the final lining for this project that had both structural and architectural requirements.

Modern concrete forming systems have increased the efficiency and the quality of concrete placement in major tunnel projects (Figure 8.94). The Devil's Slide project received many awards for its structural and eco-friendly design.

Precast Concrete Segmental Tunnel Linings

Throughout history, humankind has always had the desire to build tunnels in soft ground (loose, unconsolidated alluvial geological conditions that are typically located in river plains with water present) to cross a river too wide for a bridge or to build water and sewer tunnels to serve the needs of expanding metropolitan areas. It was the development of cast iron in the 1800s in conjunction with high-strength steel that provided the material to allow for the construction of tunnels in soft ground. Steel allowed for the development of the tunneling shield and liner plates, and cast iron was the basis for the development of cast tunnel linings, frequently called "tubbings" due to their noncorrosive properties.

The iron tubbings were cast at a nearby foundry to the specific diameter and width to match the requirements of a particular project. Pockets were formed into the cast-iron lining for weight reduction and assembly purposes. After molding and machining of the mating faces, the cast-iron pieces were bolted together in the shield. The joints were sealed with bitumen-impregnated wood layers installed

CHAPTER EIGHT | INNOVATIONS IN TUNNELING 449

Figure 8.94 | **Devil's Slide Final Lining Form at Plant Acceptance**

ring to ring and segment to segment. The inside face had a machined caulking groove where lead wool was hammered into this space in an operation called "chinking." This system has worked very well, as there are many tunnels built in this manner that are still in service today after more than 100 years of use, such as the Amtrak Hudson railroad tunnels and the PATH tunnels, both running under the Hudson River (Figure 8.95).

The cast-iron lining system was prevalent until the mid-twentieth century in Europe and the United States. In post–World War II Europe with limited steel manufacturing capabilities, the industry started to see the introduction of precast concrete segmental tunnel linings as an alternative to the cast-iron linings. In the United States, cast-iron tunnel linings were used in tunnel construction until the 1970s for such projects as WMATA's Blue Line in Washington, D.C., near the U.S. Capitol. For a short period of time, cast-iron linings were replaced with fabricated

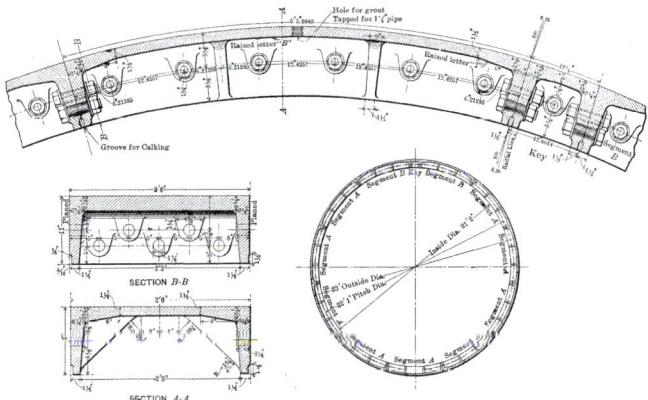

Figure 8.95 | **Typical Construction Details for Cast-Iron Tunnel Lining**

steel linings that were coated and installed in the tunnel. Fabricated steel linings were used for the San Francisco BART Tunnel from San Francisco to Oakland under the bay, sections of WMATA subway lines, and sewer tunnels in New York City.

Also in the 1970s, one-pass precast concrete segmental tunnel linings were being successfully used in the United Kingdom and West Germany. In the mid-1980s, one-pass precast concrete segmental tunnel lining projects were built in Washington, D.C., for WMATA's Navy Yard East, Navy Yard West, the Anacostia River Tunnel project, and the Baltimore Metro Lexington Avenue tunnel project. With the successful implementation of one-pass precast concrete segmental tunnel linings on these major projects combined with cost advantage over cast-iron and fabricated steel tunnel linings, one-pass precast concrete segmental tunnel linings quickly became the standard for soft-ground tunneling and especially for tunnels located below the water table (Figure 8.96).

Precast concrete segmental tunnel linings are based on the same concept as the old cast-iron tubbings. Segment sections are cast to a predetermined diameter and width, and taken into the tunnel where they are assembled inside a shield or TBM (Figure 8.97).

Primary Components of a One-Pass Precast Concrete Segmental Tunnel Lining

Concrete Admixtures

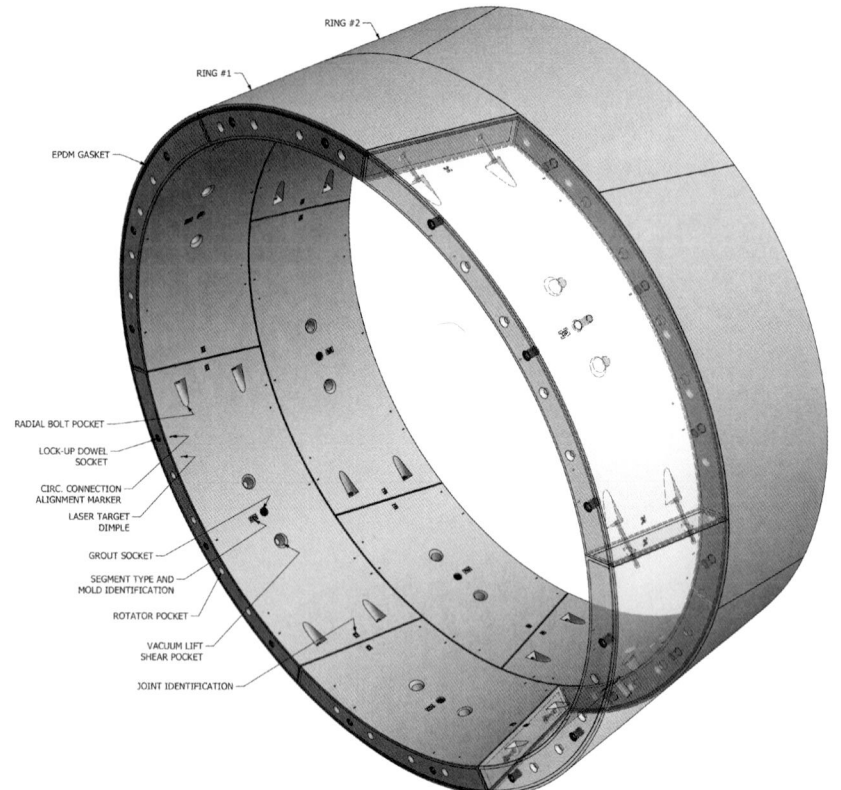

Figure 8.96 | **Isometric View of a Precast Concrete Tunnel Liner Ring**

One of the most important components in manufacturing high-quality, high-durability, one-pass precast segmental tunnel linings has been the innovations and advances in the design and control of the concrete with the use of various admixtures, all designed to perform a specific function in the manufacturing process. The major construction chemical manufacturers all have water-reducing agents and super-plasticizer admixtures that allow the precaster to reduce the water–cement ratio in the mix. These products increase the workability of the concrete without additional water that would lower the final compressive strength of the mix. If the concrete must be transported from the batch plant to a casting area, chemicals can be added that will retard the hydration process and will "awaken" the concrete when required.

Early concrete admixtures consisted of sugar or molasses added to concrete to improve the workability of the concrete. This was inexpensive but was more of an art than a science. In the mid-twentieth century, major chemical companies found this to be a large market potential and developed various chemical admixtures to control concrete behavior during the mixing and placement process.

Reinforcing Steel Options

Standard reinforcing steel options for precast segments include hand-tied, cut and bent deformed steel bars and machine cut, bent and welded deformed steel bars. These selections and design of the reinforcing materials are in response to temporary and long-term permanent loading conditions. Examples of the reinforcing steel cages for precast segmental tunnel liners are shown in Figure 8.98.

Bent high-strength steel fibers have become popular as the reinforcement material for precast concrete segmental tunnel liners (Figures 8.99 and 8.100). This is because of the improved long-term performance resulting from a thorough and even distribution of reinforcing materials in all portions of the segment. U.S. design codes are under review and further development is in progress to increase the use of steel and plastic fiber reinforcing materials.

Precast Concrete Segment Gasket Sealing Systems

With the development of precast concrete segmental tunnel linings, the need to have a reliable system for sealing between each segment and between each ring was essential. Early efforts consisted of bitumen-impregnated wood with a final seal of lead wool near the inside face of the segment, a technology transferred from the cast-iron tubbings era.

The precast concrete segmental lining must be watertight upon completion of the tunnel excavation because owners are looking for the tunnel to remain dry throughout its design operating life. The industry has advanced beyond these earlier practices with the development of extruded polymer rubber gaskets. Polymer or synthetic rubber was an innovation that changed society. It was developed during World War II as a replacement for natural rubber tires for military vehicles.

Polymer rubber is available in many grades, hardnesses, and types. The standard type used for tunnel sealing gaskets is EPDM (ethylene propylene diene monomer) because it has excellent resistance to ultraviolet rays when segments are stored outside. This compound is also very good in resisting common contaminants encountered in underground applications, in addition to its overall durability. Tunnel gaskets provide a watertight tunnel when adjacent gasket profiles attached to each segment are compressed and deformed to create a compression seal.

The precast concrete segment component supply business is international, as the requirements of tunneling projects are often very common throughout the world. Developed through meeting the requirements of rigorous international codes and standards, these essential sealing components are well proven in their applications. Gaskets are designed and manufactured by various companies throughout the world based on information provided to them on a project-specific basis. The most important factor is the depth of the tunnel and what external (and in some cases, internal) pressure will be acting on the gasket sealing system. Different gasket profiles have been developed for a wide variety of tunnel conditions (Figure 8.101). For higher-pressure applications or special conditions, it is typical to make the gasket profiles wider and taller.

Figure 8.97 | **Precast Tunnel Segments in Transit in a Water Tunnel**

Figure 8.98 | **Welded Wire Fabric Reinforcing Cage Placed in Segment Mold**

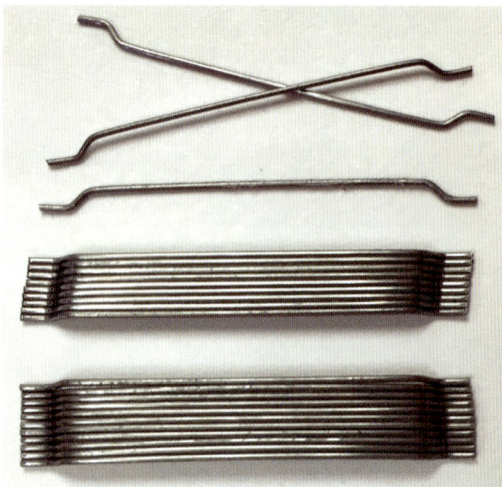

Figure 8.99 | **Hooked-End High-Strength Steel Fiber Reinforcement**

Figure 8.100 | **Steel Fiber Reinforcement Visible in Fresh Concrete**

Another important factor in gasket design is the need for exacting precast segment installation tolerances. This means that tunnel gaskets must seal to a predetermined leakage and pressure rating even if the precast segments are not perfectly installed in the tunnel. For such conditions, the gasket designer must consider the gap (distance from segment face to segment face) and offset (gaskets that are not perfectly aligned) (Figure 8.102).

The sealing gaskets are typically installed on segments at the precast manufacturing plant and shipped to the project site ready to be transported into the tunnel. Care must be taken to ensure that the gaskets and the precast elements are not damaged during transport and installation, otherwise the lining may not seal as designed.

Since 2000, the U.S. underground industry has been placing ever-higher demands on gasket application and performance, as tunnels are being located in more challenging and deeper locations with ever-increasing groundwater pressures. Current gasket designs offered by a variety of international manufacturers allow for pressures acting on the gaskets that can safely exceed 650 psi. Recent high-pressure gasket application projects in the United States include the Arrowhead Tunnels project (300 psi) and the Lake Mead Intake Tunnel No. 3 project (250 psi). Innovations in polymer materials and gasket design will allow the tunnel designers to meet the challenge of even higher pressure sealing requirements.

A new innovation in the marketplace of tunnel gaskets are anchored gaskets (Figure 8.103). These gasket profiles are placed in the casting mold and are cast into the precast concrete segment as permanent fixtures, eliminating adhesives and the associated environmental issues. Anchored gaskets have been successfully used on several recent tunnel projects in the United States.

Precast Concrete Segment Connection Systems

Once the precast segments are manufactured, the segmental lining components must be transported into the tunnel and assembled as a full-circle ring within the tail shield of the TBM. The standard mechanism to assemble precast segments is an erector arm that places the segments into a ring. Steel bolts are installed through holes in the segment and tightened in a threaded socket cast into the segment, as illustrated in Figure 8.104.

Figure 8.101 | **Standard Tunnel Segment Gasket Profiles**

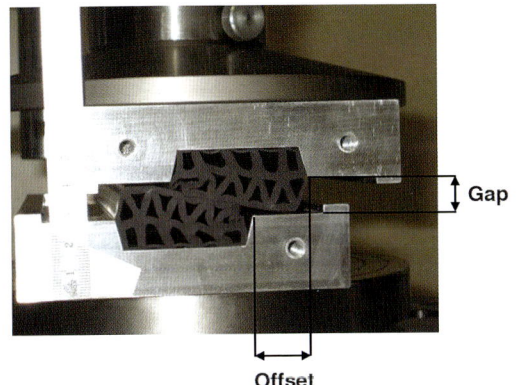

Figure 8.102 | **Gasket Gap and Offset Conditions Under Testing**

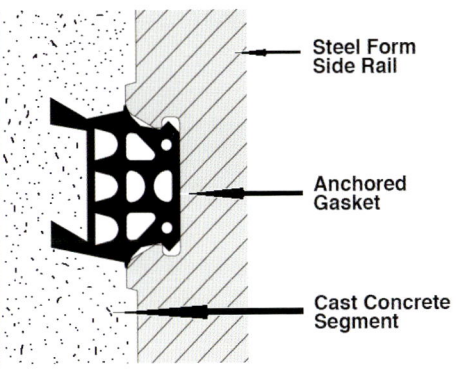

Figure 8.103 | **Anchored Gaskets Shown in the Steel Segment Form**

For corrosion protection and long-term service life, the steel connection bolts are either galvanized or made of stainless steel. There is some debate among engineers as to whether the bolts should remain in place after the ring has been erected, grouted, and stabilized against movement. The practice of removing these bolts is common in Europe but not in the United States.

One of the major innovations that has occurred in the design and manufacture of precast concrete segmental tunnel linings is the invention of high-strength plastic dowels as a replacement for steel bolt connector systems. Wood, bronze, and iron pins have been used since the time of the Greeks to align rock columns when in a state of compression. Machined wood dowels were used, with limited success, to connect the precast concrete segmental lining for the Boston Outfall Tunnel in the early 1980s. In the 1990s, some European manufacturers were providing plastic connecting/alignment devices for smaller 10-foot-diameter sewer tunnels.

Circumferential connecting dowels are a three-part system consisting of two female embedded sleeves that are cast into adjacent segments at predetermined locations and a center connecting dowel that is placed at the time of ring assembly and locks into

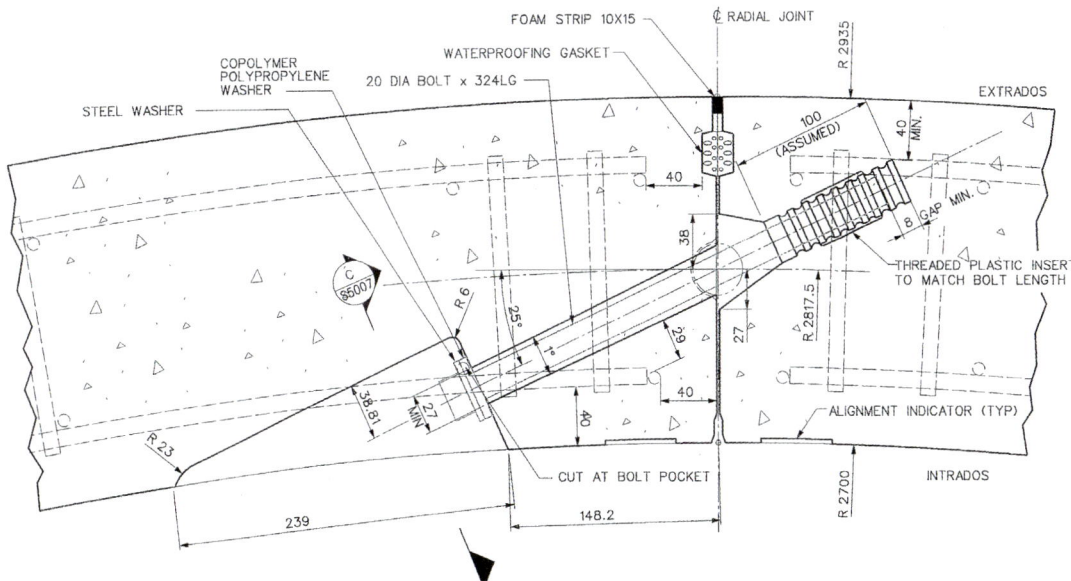

Figure 8.104 | **Standard Bolted Connection Detail Between Precast Segments**

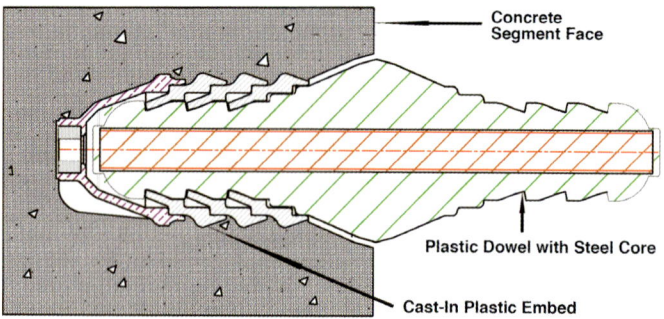

Figure 8.105 | **Circumferential Connecting Dowel Cross Section**

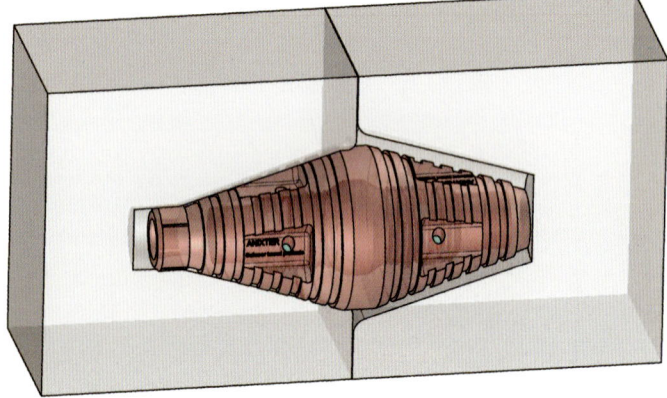

Figure 8.106 | **Typical Circumferential Joint Alignment/ Shear Dowel Detail**

Figure 8.107 | **Typical Precast Segment Casting Mold**

the two cast-in embeds when the TBM shove jacks push on the segment for final placement in the ring (Figure 8.105). The cast embedded sleeves and dowels have mating teeth that engage when compressed together, permanently locking the segment into its final location. To facilitate final placement, dowels typically have a larger-diameter center core section that helps guide the precast segment to its proper seating location and minimizes offsets and gaps that are detrimental to gasket performance.

Alignment dowel systems are plastic bicones installed in the circumferential face connection joints to help in segment alignment during segment ring assembly and to provide shear resistance for long-term tunnel lining performance. Alignment dowels are available from a variety of manufacturers and range in size from 2.4 inches to 4.8 inches in diameter. Alignment dowels come in varying grades of plastic and shear performance characteristics. For high-shear requirements, a steel rod (axis) is cast into the center of the plastic bicone (Figure 8.106). These dowels have no thread engagement for pull-out resistance.

Precast Concrete Segment Manufacturing and Delivery

Manufacturing precast concrete segmental tunnel linings to high dimensional tolerances (±0.5 mm) requires precision molds and handling equipment (Figure 8.107). In recent years, there have been many innovations on an international basis to modernize the process. To maintain tight casting tolerances over the life of a project where the number of casts may exceed 1,000 for a full ring of molds, heavy-duty precision steel molds must be used in the casting operation. After casting the freshly poured segment, it must be demolded in a manner that will not damage the new concrete. The current state-of-the-art technology makes use of vacuum lifting systems (Figure 8.108).

At the precast plant after the segments are demolded, they are cleaned and inspected. If anchored gaskets are used, the gaskets are inspected for complete embedment. If standard gaskets are used, they are glued into the gasket groove using the gasket manufacturer's specified adhesive. Segments are generally stacked for storage in a manner that includes one full ring with the key segment on top (Figure 8.109). Specific dunnage members are used to provide adequate spacing between segments for handling needs.

If compression packings are specified, they are applied at this time in the segment precast plant. Compression packings are either wood or membrane materials installed on the circumferential and/or the radial joint areas where adjacent segment faces are in contact. Compression packings are designed to distribute the contact forces that are applied to the segmental ring meeting faces during assembly and afterward, when TBM propel forces are applied.

A precast concrete segmental tunnel lining is a complex system of highly engineered materials and components that when designed and manufactured to stringent industry quality control procedures provides a tunnel lining that can meet its service requirements for more than 100 years (Figure 8.110).

INNOVATIONS DRIVE TUNNELING

The tunneling industry has a long-proven record of pushing the boundaries on what has previously been attempted to meet the ever-increasing demands of a developing society. This has only been achievable through the passion and drive of those engaged within the industry, in combination with a willingness to seek constant improvements to safety, quality, and operations, as well as to continuously innovate from the limitations on what has gone before.

In the decades since the invention of the tunneling shield, followed by the wholly American invention of the first practical dewatering wellpoint and a spectrum of ground improvement techniques, underground construction has evolved dramatically. What was once an arduous and hazardous occupation for tunnelers is now an automated, streamlined, and efficient process. We shall continue to see improvements and refinements, but nothing can surpass the ingenuity and grit of those early tunneling pioneers who built so much with so little while risking their lives for project success.

Identifying and mirroring the successes that continue to be achieved around the world enables us to apply global innovative technology to our home projects. Adopting new innovative technology is not without risk; however, history has proven that the benefits far outweigh these risks as the industry evolves and society reaps the rewards as the end user of this underground infrastructure.

Over the last 200 years, the U.S. underground industry has seen significant innovation applied to solve the problems of the day. All facets of the industry have looked to innovate—from design and safety, to project procurement, through to construction. Construction has seen significant improvements and innovations in the materials used, whether to achieve better quality or to become lighter, cheaper, quicker to install, more reliable, durable, or, most importantly, safer to use. The excavation methods, types of equipment, techniques, and capabilities for underground construction have also seen significant improvements.

Advancements in the tunneling construction industry are not always clear, as the achievements and benefits in applying technology are sometimes offset by attempting feats of engineering that have not previously been contemplated. From the outset of a project, society may not appreciate the scale, challenge, and risks that may exist during construction. Nevertheless, the industry has learned to respect and address project risks. Identification and use of innovative technologies will continue into the future with the aim of delivering challenging underground projects successfully while addressing the needs of a modern society.

Figure 8.108 | **Demolding a Segment Using Vacuum Lifting**

Figure 8.109 | **Typical Segment Ring Storage and Transport Arrangement**

Figure 8.110 | **Successful Tunnel Build—A Work of Art**

REFERENCES

Graf, E.D. 1969. Compaction grouting technique. *Journal of the Soils Mechanics and Foundations Division*, ASCE 95(SM5, September).

Ritter, W. 1879. *Die Statik der Tunnelgewolbe [Statics of Tunnel Vaults]*. Berlin: Springer.

Warner, J., and Brown, D.R. 1973. Compaction grouting. Paper 9908. *Journal of the Soil Mechanics and Foundations Division*, ASCE 99 (SM8, August).

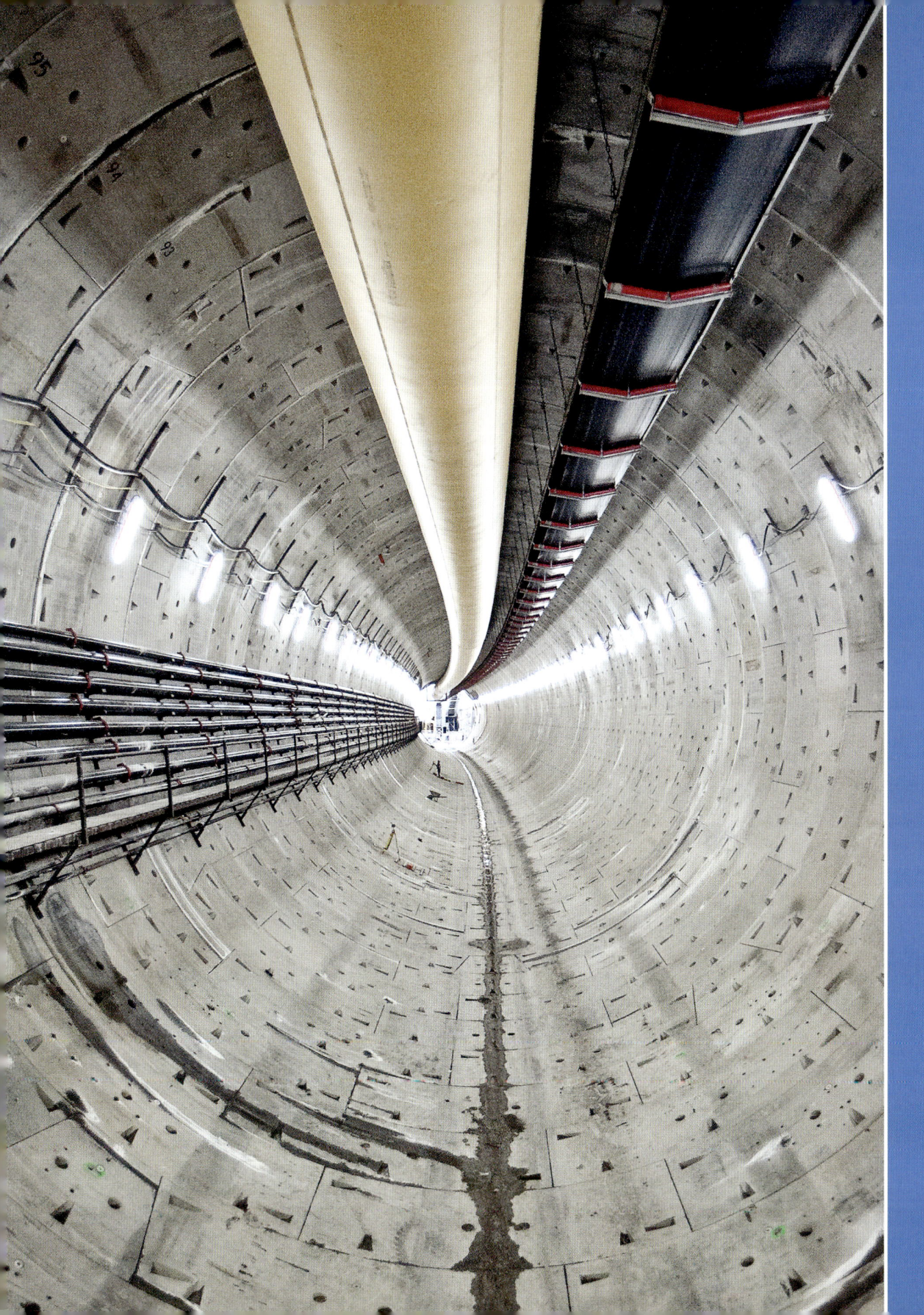

9 THE FUTURE OF TUNNELING

Our industry provides a glimpse into the future as we share some of the vision from the tunneling and underground industry regarding new and rehabilitated projects, technological improvements, and funding methods, as well as the general public's expectations for construction within their environment.

The tunneling industry expects the future of tunneling and underground construction in the United States to be "bright." Demands resulting from urbanization and environmental improvements will necessarily mean continued increasing use of underground space for conveyance, storage, and transportation, and to allow better uses for the surface. We anticipate an increased focus on urban underground infrastructure and greater interest in holistic use of underground space as urban areas grow—not only out or up, but also down. This will extend beyond traditional linear tunneling and will encourage us to look at our technologies for developing underground spaces of all shapes and sizes.

TUNNELING EXPECTATIONS

To expand this use of space, we must look not only at technology, but also at methods to procure and finance the work, educate and train our growing workforce, and gain acceptance by the general public.

Along with growth, we must continuously maintain and improve our existing infrastructure. It is well known that the importance of reliable infrastructure is a large part of the decisions that major industries make to keep or establish headquarters in the United States. Cities are dependent on reliable transport of goods, energy, water, and people. Not only will new tunnels be required for expansion and increased density, but there will also be considerable need for rehabilitation projects. In addition to megaprojects, we will see more use of less intrusive construction methods, such as trenchless technology and microtunneling, to address urban needs and to build projects with less inconvenience to the surrounding inhabitants.

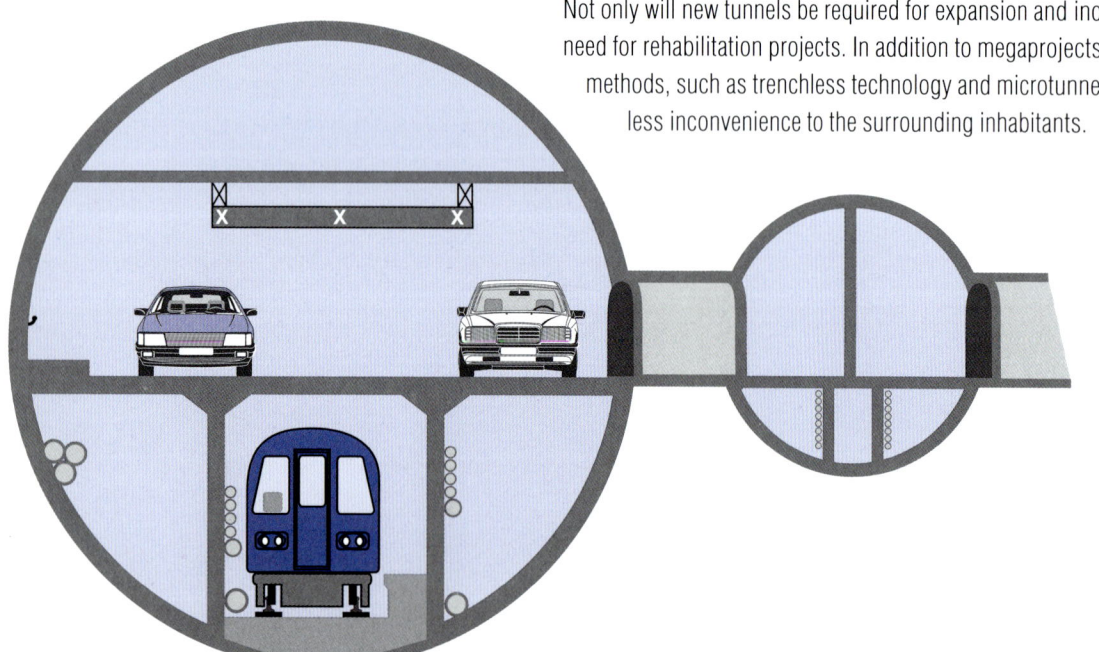

We believe market sectors will need to share technology. Tunneling and microtunneling industries, for example, will make mutually beneficial partnerships so there is a clear, continuing advancement in underground construction techniques and sharing of technological advances across industry boundaries. We will continue to see more applications for tunnel boring machine (TBM) technology from mining and oil-and-gas industry technologies being applied to tunneling programs.

We will not only build more tunnels and underground space with improved systems, but also rehabilitate and expand existing systems as more demand is placed on these aging facilities (Figure 9.1).

Figure 9.1 | **Multi-Use Tunnel**

Previous Page | **State Route 99 Tunnel in Seattle**

CHAPTER NINE | THE FUTURE OF TUNNELING

Urban Metro Systems—Efficient Use of Underground Space

Expansion and widening of freeways in the United States is no longer an attractive option when surface property costs are high and public acceptance cannot be obtained. Preserving space for metros (subways) and rail transit projects by allocation of a dedicated underground right-of-way is a clear solution. In fact, many of the existing systems are being expanded to allow increased ridership and service to more destinations.

Flood Control and Water Treatment Tunnels

Cities across the United States will use tunnels for flood control and protection of the environment. Projects such as the Portland (Oregon) East Side Combined Sewer Overflow Tunnel (Figure 9.2) and Boston's (Massachusetts) Deer Island Outfall Tunnel have improved local area water quality and marine habitats with vastly improved stormwater and sewage effluent treatment. More projects are on the horizon in cities with legal mandates for cleaner water and improved environments.

Megaprojects

Although not considered to be everyday construction, megaprojects (those on the order of $1 billion plus) have been and will continue to be essential in maintaining quality of life in urban areas. Recent examples include the Central Artery Project in Boston where an elevated highway system was placed underground to reduce traffic congestion, allowing more beneficial use of the surface above. Another megaproject, the PortMiami Tunnel in Florida, was constructed to remove freight traffic from already congested city streets and take traffic under water to the port facilities (Figures 9.3 and 9.4). Similarly, the State Route 99 (SR 99) Tunnel, currently under construction in Seattle, Washington, will replace an aging two-level highway viaduct underground, allowing for greater use of the waterfront area for business, pedestrians, and residences. The surrounding area will enjoy more unobstructed views and greatly reduced noise.

Figure 9.2 | **Pump Station for Combined Sewer Overflow Tunnels**

Figure 9.3 | **PortMiami Tunnel in Florida**

Figure 9.4 | **Construction of Port-Miami Tunnel Invert**

Highway Tunnels

Highway tunnels will be constructed for improved safety and traffic flow. Examples include tunnels to reduce roadway crossing of elements such as landslides, railroads, or other challenging conditions. Tunnels will also be upgraded for emergency egress and improved ventilation and operating systems.

Multi-Use Tunnels

Multi-use tunnels, such as the Stormwater Management and Road Tunnel (SMART) in Malaysia and the Silberwald in Moscow, should become more common in the United States. In the case of the Malaysia SMART, vehicular traffic is contained in two levels, with flood control on the lower level or entire tunnel, depending on the severity of the storm event (Figure 9.5). In the case of Silberwald, vehicular traffic is on the top level, and subway trains run below. Utilidors (utility corridors) for multiple installations and undergrounding of essential utilities also have potential, as do the tunnels, to protect vital infrastructure from weather, vandalism, or worse.

Water Supply, Storage, and Transport Tunnels

As the demand for reliable water supplies increases, tunnels and underground caverns will be relied upon even more for water storage, transport, and power generation. Recent North American examples include the Hollywood Water Quality Improvement Project in Los Angeles, California (Figure 9.6); Lake Mead's Intake Tunnel No. 3 in Nevada to tap water at shallower lake levels; and the Inland Feeder Project (Arrowhead West) in Southern California (Figure 9.7) to bring more reliable water supply from the north. Notable water supply tunnels currently under construction or in planning include the Delaware Aqueduct bypass tunnel and MetroWest Water Supply Tunnel in the Boston area.

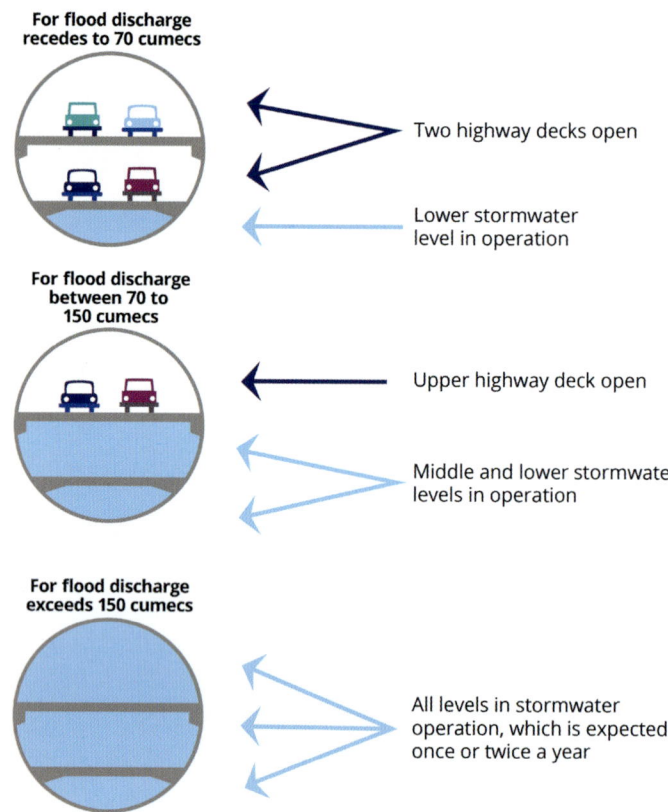

Figure 9.5 | **SMART Tunnel Concept**

Figure 9.6 | **Hollywood's Underground Water Storage**

Figure 9.7 | **Arrowhead West Tunnel at the Devil Canyon Portal**

Figure 9.8 | Basilica Cistern in Istanbul

Recycling of water is also becoming a viable option in drought-prone areas, with tunnels and pipelines needed to store and move water safely while reducing evaporation and infiltration losses. It is repeatedly said that the future often reflects the past, and this is true regarding water supply in arid climates, as water has been stored underground since ancient times. An early example in history is the beautiful, iconic Basilica cistern constructed by the Byzantines in sixth-century Istanbul, shown in Figure 9.8.

To respond to these increased future demands, the tunneling industry will face a number of challenges that it is already addressing, that is, continuously improving technology, public concerns about safety and impacts of construction, financing of projects, and training the workforce.

TUNNELING TECHNOLOGY

The underground industry has seen the growing application of tunnels, the increasing tunnel and cavern sizes, and construction in more difficult ground conditions—all with fewer disturbances to the surrounding environment. While the future demand is expected to be great, development of innovations will allow people and equipment to work under even more extreme underground conditions.

Projecting ahead from current trends, advances will be in the areas of mechanized tunneling, shaft drilling, geotechnical monitoring, and other technologies to improve performance and construction schedules.

Improved tunneling technology will allow construction of tunnels and underground space in more challenging ground conditions and at both deeper and shallower elevations. Tunnels are being constructed with increasingly larger cross sections and deeper levels below the surface, and at higher water pressures and under greater geologic stresses (Figure 9.9). TBM technology that is already under development will allow construction at higher groundwater pressures, with improvements in sealing systems as well as the replacement of cutting tools without worker exposure to high pressures. Larger-diameter TBMs allow fewer tunnel drives for bidirectional vehicular and rail traffic and other multi-use tunnels.

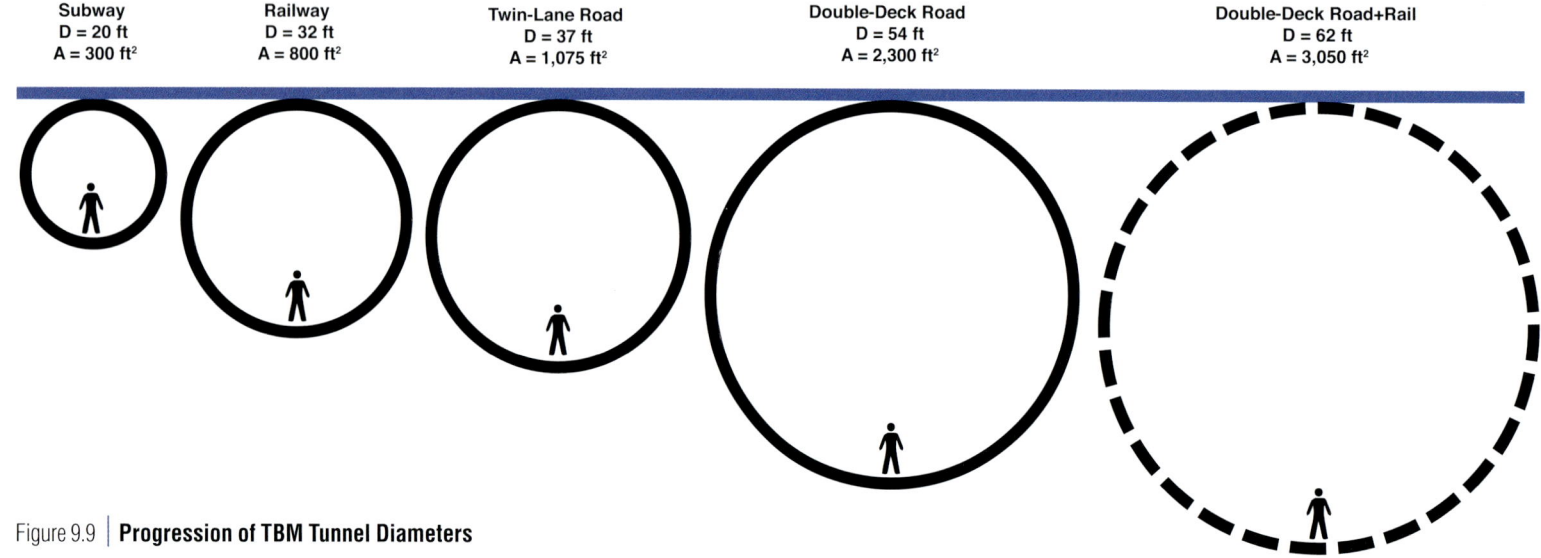

Figure 9.9 | **Progression of TBM Tunnel Diameters**

Mechanized Tunneling and TBM Technology

Many tunnels today are constructed under increasingly difficult ground conditions. The tunneling industry foresees TBM innovations in the following areas: multi-mode machines that can excavate in highly varying ground conditions, larger-diameter machines in excess of 60 feet in diameter, and machines capable of working under higher groundwater pressures than currently possible.

We envisage machinery that can operate in a wider range of ground and groundwater conditions. Machines will be able to handle uniform conditions along one tunnel alignment or adapt the configuration to offer a choice of methods for different reaches along a single alignment. This is a clear development trend toward multi-mode technologies that will continue for the coming years, presenting a significant benefit and will also provide greater potential for equipment reuse on subsequent projects where ground conditions were not considered during the manufacture of the original TBM.

A real tendency toward larger diameters for all machine types is clearly visible. Since about 2000, we have seen TBM tunnel diameters "grow" from 45 to 57 feet, an increase in face area of almost 60%. Figure 9.10 shows the Alaskan Way Viaduct TBM, which is the largest to date at 57 feet in diameter. There have been some temporary setbacks arising from unanticipated difficulties on some projects, but this has not brought further development to a stop. Careful review of lessons learned, leading to an effective resolution to these difficult experiences, often takes the industry forward. There are strong indications that the existing technical extrapolation models toward larger diameters have to be adjusted, and it is difficult to say where the ultimate size limit is for this technology. The interface between the machine and the ground becomes even more important as machine diameters increase. The real challenge for very large TBM diameters is less machinery oriented and more concerning the following issues: ground pressures, corresponding support of the tunnel face, and performance and longevity of the TBM cutterhead tools during excavation of the tunnel. As with all tunneling methods, the ground represents a significant factor in limiting development. To some extent, the intended purpose of the tunnel and final internal operation once in service is the key driver for larger tunnel diameters. We expect that in the future, technology will continue to evolve to meet this need.

High-pressure applications will experience an increasing demand as well. One or two decades ago, the pressure limitations were more involved with structural aspects and seal systems. Now the focus is on the question of safe worker access to the cutting chamber at these higher pressures. Technology to maintain cutterheads without human access under pressure will improve with advancing development in robotic solutions. At this time, it is difficult to say which of the two will be the way to the future. In any case, the future will see technologies that do not need as much exposure of personnel to a pressurized environment as they do today. However, there may be geometrical limits or minimum required TBM diameters that may pose significant challenges for such high-pressure applications.

For large-size projects, we see a trend toward "industrialized tunneling" that will increase. Tunneling operations will be considered a temporary underground factory that will ultimately produce a tunnel. What we already see today is an approach to having as much as possible of the construction completed while the TBM advances. Pressure will continue for projects to reduce the total time between project award and final completion. This has a tendency to move away from a sequential approach for different elements of construction work to a higher degree of simultaneous work, even if it slows down individual elements such as tunnel excavation.

The needs of the mining industry will support the development of partial face cutting machines for hard-rock conditions, and civil construction will benefit from this development.

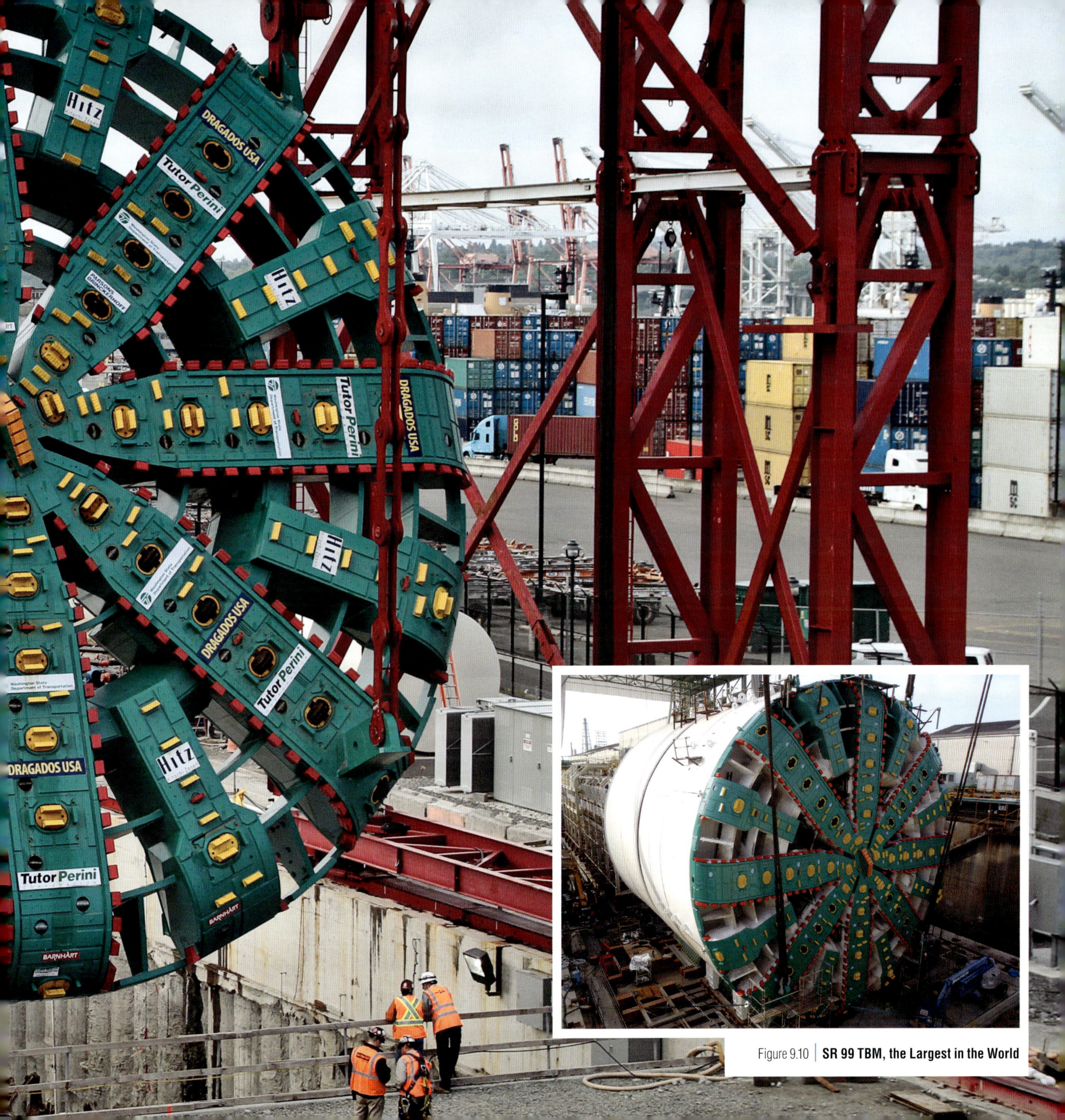

Figure 9.10 | SR 99 TBM, the Largest in the World

Noncircular TBMs, such as the binocular-shaped (Double O) tube used for the Shanghai Metro and the rectangle TBM made by China Railway Equipment Company (Figures 9.11 and 9.12), have been manufactured, but whether these types of noncircular TBMs will become more popular is very difficult to say. The technology has been available for almost 20 years, yet it has not been considered by the tunneling industry to have sufficient benefits to replace the traditional use of circular bore machines.

Longer tunnels will also be needed. The current longest is approaching 100 miles with future water and high-speed rail tunnels being planned over longer distances than previously contemplated. Improvements in ventilation, fire and life/safety systems, as well as TBM technology will make these tunnels more acceptable—especially in mountainous areas and under rivers, lakes, and oceans.

Specific technological improvements for TBMs will see continued advance in specialist areas, such as soil conditioners injected at the front of the TBM during excavation to stabilize the excavated ground; volume balance measurements for improved control and support of the ground when using pressurized-face TBMs; and having a better understanding for the management of soil abrasivity on the machine components, including tool wear. Other improvements include higher groundwater pressure designs for the TBM cutterhead, cutting tools, and TBM parts required to maintain during excavation.

Figure 9.11 | **Double O TBM**

Figure 9.12 | **Rectangle TBM**

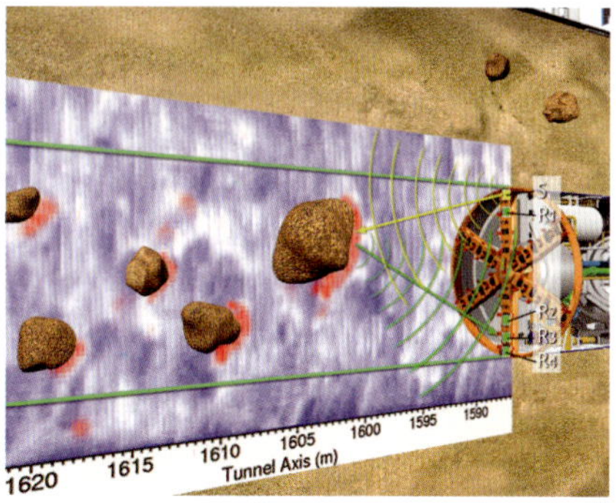

Figure 9.13 | **Ground Imaging**

Increased production rates will mean lower capital costs for tunnels. TBMs routinely make average advance rates of about 50 feet per day. Improvements with TBMs and precast segment design should allow much higher rates of production that will result in cost savings and reduced project schedules.

Expectations are also high for improvements in drilling and probing techniques in advance of the TBM. This includes data collection and analysis to accurately predict ground conditions ahead of the tunnel face, as well as better tools to image ahead (with methods such as ground-penetrating radar; Figure 9.13) and to non-destructively evaluate and effectively see behind tunnel linings.

Mechanized excavation of round tunnels are expected to be the most commonly used, but developments in other types of machines for shorter and shallower drives will be in wider use. "Shallow digger" types of machines are already on the market for short drives, including use for underpasses and utility installations (Figure 9.14). Increased use will allow more work in urban areas without cut-and-cover construction that usually requires significant excavation and disturbance at the surface. In addition, innovations for cross-passage connections between the larger bored parallel tunnels are being constructed with precast concrete segments. Systems for mechanized cross-passage construction using a sealed machine are expected to be available within the next few years. We have already seen pipe-jacked (pushed shield) systems used in cross passages to reduce pretreatment of ground conditions.

CHAPTER NINE | THE FUTURE OF TUNNELING

Figure 9.14 | **Herrenknecht Shallow Digger**

Design improvements in final structures through the use of more complex computational fluid dynamics models provide increased safety systems in addressing fire and other safety risks within transit and vehicular tunnels. Models accommodate not only smoke and fire properties, but extinguishing methods and analysis of passenger egress dynamics. Systems that include the use of safety refuges and escapeways are now becoming standard in new and rehabilitated tunnels.

Design and Monitoring Methods

Structural design efficiencies are expanding with increased use of computer models for more complex analysis. Software is used to model ground loading on the tunnel lining structure and for implementation of building information modeling, or BIM, for more efficient space planning. The models also promote more confidence in seismic designs and structural connections to allow more construction and acceptance of tunnels in seismically active regions.

With web-based collection and analysis, large-scale projects today include remote and real-time monitoring with the ability to access data at all times on multiple platforms (Figure 9.15). This reduces overall time for collection and distribution of data and allows earlier response times. Other automation includes TBM guidance systems to improve accuracy of the tunnel drives and reduce labor.

Instrumentation and remote monitoring developments, data recording, data processing, and data management of all involved activities will continue on a high level that supports operations, maintenance, management, quality control, and documentation. New use of instruments such as horizontal inclinometers and remote monitoring means lower cost for labor, data collection, presentation, and records management.

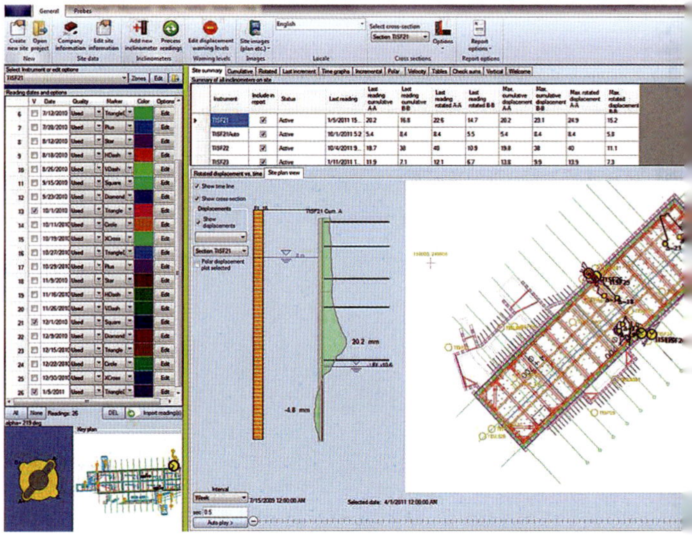

Figure 9.15 | **Remote Data Access**

Segments and Tunnel Linings

Along with mechanical tunneling innovations, developments in precast concrete tunnel linings will allow more rapid installation of the individual precast concrete segments that together form a complete tunnel lining ring, as well as reduced costs resulting from improvements in the materials used. Recent developments include tapered tunnel lining rings that can be rotated to allow better quality of the tunnel ring built to tighter tolerances in curved alignments. Further developments include geometries such as hexagonal shaped segments (Figure 9.16) that allow continuous excavation and ring installation (Harding and Chappel 2015).

Advances in precast segment concrete materials are being made using steel and plastic fiber reinforcement, in some cases replacing conventional steel-bar reinforcement, with significant labor savings. Strength improvements of unreinforced concrete will allow for thinner segments with increasing tunnel diameters. Segment connections have traditionally been bolted, with plastic doweled connections replacing bolts on the circle joints, providing better segment alignment and reduced corrosion. Similar reduction in bolted connections between radial segment joints will provide additional reduction in labor to allow faster ring erection times.

Shaft Sinking

In addition to tunneling, shaft sinking technology and inclined mechanical excavations will continue to improve for use in even more variable ground conditions. One example is development of the vertical shaft sinking machine (Figure 9.17). In this method, the shaft is filled with a bentonite slurry at the start of excavation. This is an advantage so that dewatering prior to excavation is not necessary. A slurry separation plant removes the excavated material from the slurry and recycles it back into the shaft excavation. The shaft concrete structure is lowered continuously while ring building of the shaft lining takes place at the surface. When the final depth has been reached, the machine is recovered. Subsequently, the shaft bottom is sealed by an underwater concrete plug and the annular gap is filled with grout, creating a frictional support locking the shaft in place. Once the slurry is pumped out, the shaft is ready for use. The technology has multiple applications for the future, especially in urban areas with soft ground and difficult groundwater conditions.

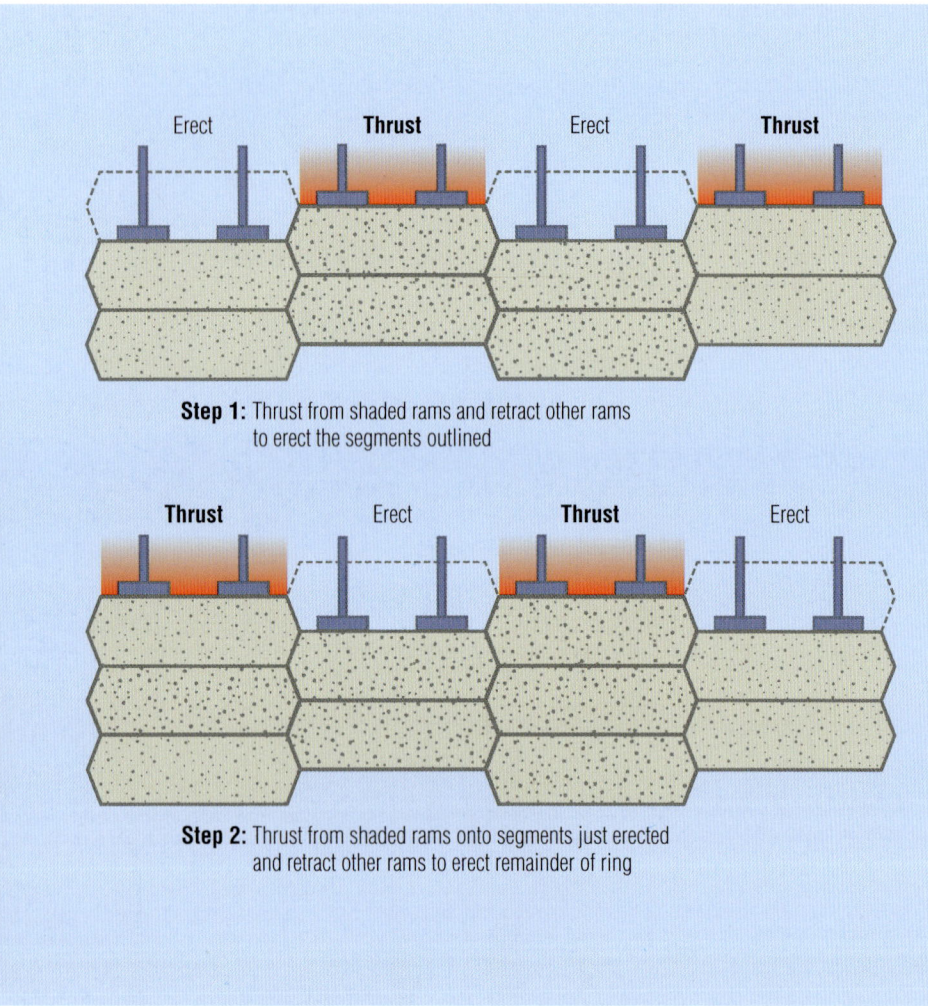

Figure 9.16 | **Hexagonal Segments**

Combining Temporary and Permanent Work

Designers and contractors are looking at more ways to incorporate initial and final support systems to gain efficiencies and more sustainable construction. Examples are common in Europe and Asia where slurry or secant pile walls are used for both the temporary and permanent support of excavation. This can be an overall time and materials savings, as there is no need for an initial support—for example, soldier piles and lagging—followed by a cast-in-place permanent structure. The systems also have advantages in providing a more watertight system to reduce dewatering during construction. Stiffer walls may also allow for the reduced use of temporary struts and their removal. Single-pass tunnel systems (initial and fixed support are the same) are already the standard in both soft- and-hard ground tunnels.

Long-Term Operations and Reduced Maintenance

Although tunnel capital (initial) costs are generally higher than bridges or elevated structures, tunnel costs are demonstrated to be lower comparatively than other construction when using a life-cycle cost analysis. This is because over the lifetime of the structure, the cost per year is generally less, as the tunnel service life is two to three times that of a bridge or viaduct, even though structural tunnel systems and facilities do require maintenance for active equipment (lighting and ventilation, roadway structure, rail, structural materials exposed to the environment, and so on). Further reducing the costs in the future will allow more consideration of tunnels when life-cycle costs are an important selection consideration in determining the best option for a project. Reduced maintenance can be achieved through improved materials, including use of precast concrete and other concrete with reduced permeability to produce greater durability, and better waterproofing methods and materials, including sprayed lining and membrane applications. Automated systems for inspection and reporting will be key for early detection of maintenance needs, reducing the costs of major rehabilitation. These systems will include technology for operations monitoring and maintenance, including remote monitoring for leakage and systems failures.

Traditional Methods

We will also continue to look back to previous technologies. As we make underground spaces of different sizes and shapes for our urban communities in the future, in hard-rock geology we will still need to consider old and new technologies. For example, the spaces needed for the future will require drill-and-blast excavation—this remains a practical solution in many applications. We need fundamental research and new concepts to make the supply, use, and potential impacts of explosives acceptable in an urban environment, especially where excavation requirements

Figure 9.17 | **Vertical Shaft Sinking Machine**

CHAPTER NINE | THE FUTURE OF TUNNELING

are not suitable for mechanized methods. This will include complex analysis of vibration effects on buildings of all ages, conditions, and building materials. Given recent successes in drill-and-blast excavations in New York City and Stockholm, Sweden, it is anticipated that continued development and improvements in this area will be made in the future in a manner that this form of excavation will remain attractive (Figures 9.18 and 9.19).

EXPANDING THE INDUSTRY AND WORKFORCE

As we produce increasingly complex and challenging projects, the industry will need to expand and train its workforce. This will drive planners, engineers, contractors, and craft labor to work collaboratively while technologies advance. Tunneling and underground construction traditions and culture will also adapt to the changing environment, the safety culture, and the use of technology required to tunnel through progressively more adverse conditions.

Continued expansion of the industry will include education to increase knowledge of the potential for using underground space. Such education and training is already underway with formation of organizations such as the International Tunnelling and Underground Space Association's (ITA's) Committee on Underground Space (ITACUS) and training programs to help leaders understand the role underground space can play in helping cities cope with rapid urbanization and the shortage of space. Making cities more resilient against natural hazards and the effects of climate change is also important. Outreach and education of decision makers through other organizations, such as the United Nations and the International Society of City and Regional Planners (ISOCARP), as well as through mainstream press publications and workshops, will go a long way toward the urban underground future. The demand for sustainable solutions that support growth must consider the economic environment, uncertainty about climate change, growing energy requirements, and the increasing importance of stewardship of the environment. These trends and demands will create a critical need for students graduating with degrees in tunneling and mining-related disciplines, and for continued professional and leadership development that will drive the industry in the future.

In response to global population expansion, rapid economic and societal growth within the developing world, and continued urbanization in the United States, engineering programs are recognizing the need for technical training with an emphasis on promoting sustainability, benefits to society, and economical solutions. Universities are developing more undergraduate and graduate programs to add faculty with broad experience from tunneling and mining industries, expand partnerships with industry to exchange knowledge, and teach business skills as well as traditional engineering courses. Mining and tunneling programs will require broader skills to be learned, including civil, mining, safety, oil and gas, life-cycle analyses, and "risk-based" approaches. As exciting as the field is, attracting students to engineering fields will be a challenge for universities and the industry.

Figure 9.19 | **East Side Access Excavation, New York**

Engineering and contracting firms have evolved over the years from smaller privately owned, and often family-owned, firms to global companies with more than 50,000 employees not uncommon. Contractor teams, especially on larger projects, are increasingly composed of diverse teams of design engineers from multiple disciplines as well as the more traditional civil backgrounds. As projects increase in size, we will continue to see the evolution of

Figure 9.18 | **New York City Second Avenue Subway, 72nd Street Station Construction**

firms through acquisition with competition on a global level for mega, multi-faceted projects. Even so, the majority of tunnel projects are smaller in scale, and there will always be a place for smaller firms with the initiative to do underground work.

Figure 9.20 | **Large-Diameter Tunnel Construction in Italy**

It is unavoidable that future underground projects will get larger in scope, construction time, and cost; however, there are some caveats with the advent of these larger projects (Figure 9.20). This consequence will stretch the capabilities of current contractor, engineer, and owner staff, not only in a technical sense, but also with respect to the management capability. We have seen throughout the history of tunneling that the nature of underground projects is such that they can take a very long time to develop, plan, design, and construct. Depending on the scale of the project, 5 years is not uncommon, and some of the megaprojects require 10 to 20 years from concept to final completion. If we assume that graduates from engineering and/or management schools will be active on a project level from their mid-20s through their mid-60s, a period of 40 years, then project management staff will have, at most, six to seven projects worth of experience to prepare them for their last (presumably most challenging) assignment. This is not sufficient experience for a project manager to have personally experienced enough challenges (problems, solutions, people, and so on) to be prepared for the scope of larger megaprojects that the underground industry will produce in the future.

A large project with its many challenges can provide a lifetime of experience upon which to draw throughout an individual's career. However, many tunnel projects are successful, providing few challenges, and are, therefore, not always sufficient preparation for a person to serve in a management or leadership role on a megaproject. Better composite blending of the leadership and management team targeting the anticipated project demands is essential.

In addition, the answer to part of this conundrum will lie in a better system of training and education, particularly of management practices for underground projects. Such management skills are not typically taught in undergraduate- or graduate-level engineering programs but in fact are learned through on-the-job training. A different and more suitable method of management training must be developed. If it is not, the megaprojects of the future, both aboveground and underground, will be curtailed and more likely suffer from time and cost overruns that will be considered management or engineering failures, and thus discourage future projects of size and complexity.

Training managers, however, will not be the only solution to successful megaprojects. Other systems in place to manage risk and expectations have been developed and continue to be implemented by large agencies sponsoring these projects. For example, *Guidelines for Improved Risk Management: On Tunnel and Underground Construction Projects in the United States of America* has recently been updated by the Underground Construction Association of SME for specific guidance on underground projects (O'Carroll and Goodfellow 2015).

Although management is paramount, underground construction involves a high percentage of labor, with major projects employing hundreds of workers from numerous trades within the United States in the associated labor crafts—laborers, operating engineers, electricians, and so on. To some extent, the labor forces and working "rules" have varied from city to city and have not always adapted to current tunneling practices, equipment, and improved technologies.

As projects become more complex with more sophisticated designs and technology used for construction, the labor forces will need to adapt to the construction methods, including ground support types, formwork use, excavated materials handling, reinforcing methods, concrete applications (precast, cast-in-place, and sprayed concrete), and more use of automatic controls and wireless communications (Figure 9.21).

These shifts will require more technical training in addition to the more traditional apprenticeship process. Also, for continued expansion of underground projects, the culture of safety must continue to permeate all aspects of the work. The "zero tolerance" for incidents is both a moral code and financial necessity in the current age of monitoring and reporting.

Figure 9.21 | **Modern TBM Operator's Cabin**

SAFETY IMPROVEMENTS

Safety improvements in design, construction methods, and equipment have been encouraging, including use of more automated equipment in the underground environment. Improvements in motor efficiency and nonpolluting equipment will allow for better working environments.

The tunnel industry has had its share of disasters that have helped shape tunnel safety as we know it today. The 1971 tunnel disaster where 17 tunnel workers died in a methane explosion near Sylmar, California, and the 1931 Hawk's Nest Tunnel disaster where more than 450 workers died from silicosis exposure from working in the tunnel at Gauley Bridge, West Virginia, are just two examples of major incidents. Both of these disasters were preventable and brought about legislation that protects workers today. More common injuries include those caused by falls from heights, being caught between moving and static objects or equipment, being struck by objects underground, and electrocution. Because of accidents and serious injury, the tunneling industry has come a long way in improving safety measures and standards.

On any jobsite now, one can see "Safety First" on hard hat stickers, reflective vests, and jobsite posters. New-hire safety orientation, safety meetings, and safety training are required on almost every jobsite. Owners and engineers who are planning the projects place a tremendous amount of emphasis on safety compliance and look for companies that highly value safety. But despite everyone's best efforts and continuous training, the construction industry as a whole tallies more accidents and injuries than any other industrial workforce (BLS 2010).

Since the inception of the Occupational Safety and Health Administration (OSHA) in 1970, construction-related accidents, injuries, and fatalities have greatly declined, but even with the decline, OSHA and the construction industry still strive to improve worker safety through intensive training and analysis of accidents. Many of the contractors within the tunnel industry have taken OSHA's lead by making training mandatory for management and field supervisors, as well as crews. These same companies continually push their management, field supervision, and safety staff to identify immediate hazards found on the jobsite and make the necessary corrections to provide a safe workplace.

Many safety professionals believe that the majority of their accident reports and root cause analyses indicate that human error is causing accidents and getting workers hurt. A variety of industry reports show that at least 90% of construction accidents are because of human error. The question many top managers ask is, "How can we keep our workers safe?" Merely telling a worker or an entire crew to work safely has never been the answer. Safety professionals will say that some of the problems point to workers who are undertrained, purposely ignore safety standards, and willfully take shortcuts. Behavior-based safety consists of observing the worker in his or her work environment, observing his or her actions, and then putting him or her "under the microscope" to determine whether or not the worker is safe through his or her actions, behavior, and decisions.

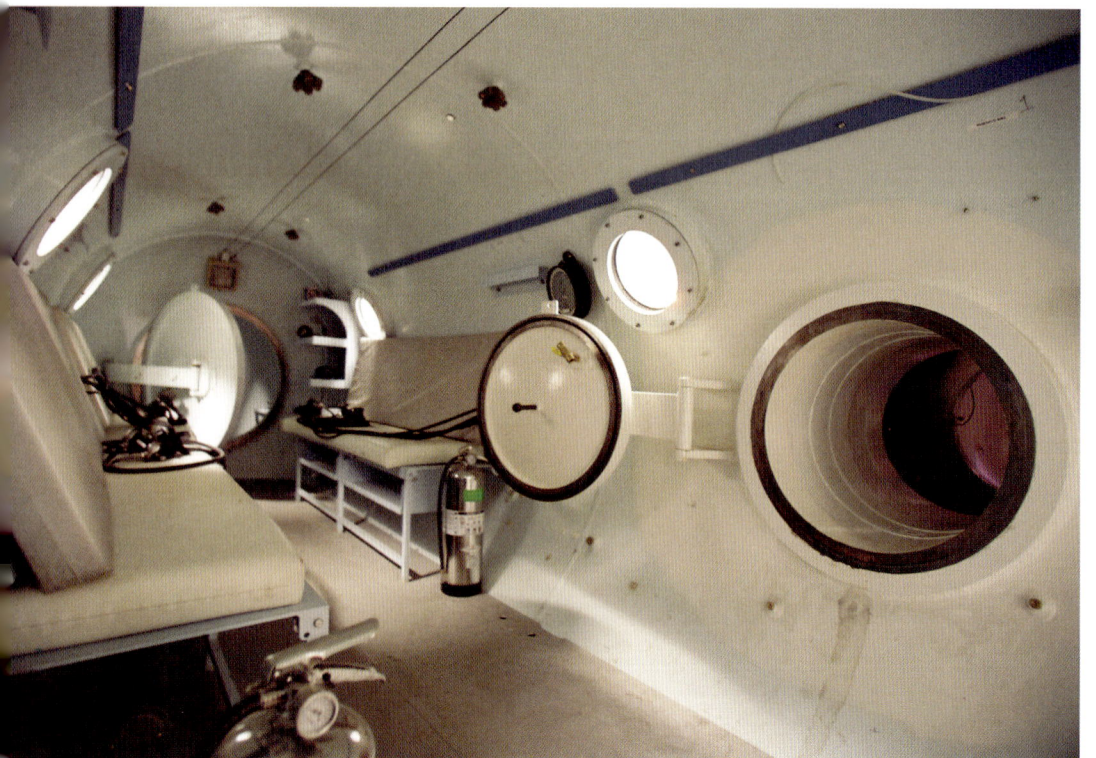

Figure 9.22 | **Decompression Chamber Used in Tunneling**

Although behavior-based safety has been very popular in the safety community, there are still those who focus on identifying and addressing the workplace hazards to maintain a safe environment. Construction companies rely on detailed work plans that enable engineers and field supervisors to map out each step of the work process. New technology in the tunnel industry has grown by leaps and bounds. New computer technology has given tunnel construction engineers the ability to design highly efficient structures, which minimizes time and material used in structure construction and still maintains its safety and integrity. Engineers are also using new technology to provide up-to-the-minute work schedules that can better pinpoint where worker-hours or costs are allocated, or to identify changes to the project's critical path. In safety, technology has also enhanced personal protective equipment for workers as well as improvements in medical safety and decompression techniques (Figure 9.22). New materials have been introduced that are designed to protect eyes, hands, and limbs in a better way. Safety glasses that are made of lighter-weight materials make for a more comfortable fit and are able to withstand greater impact (Figure 9.23). Steel-toed work boots have been designed with composite toe guarding, which provides a more comfortable fit and protects against penetration and impact. It is a synthetic fiber with very high tensile strength. Kevlar is now used in everything from cut-resistant gloves to work apparel that offers greater protection and comfort.

Figure 9.23 | **Train Crew in a Deep Tunnel**

Many companies conduct safety inspections and have their own way to process that information. This usually consists of management and safety staff making observations of unsafe conditions or unsafe behavior while working on site. These unsafe observations are noted, and the hazards are identified and then corrected. Some forward-thinking companies make time in their progress meetings, or even dedicate entire safety meetings, to review the safety inspections and ensure that corrections are made. The safety inspection forms are gathered and filed away to comply with company and OSHA recordkeeping requirements. It is within these records that the tunnel industry could use this valuable opportunity to take its safety to the next level by utilizing the information gathered in the safety records and leveraging technology to put that data to use in managing the risk on its jobsites.

Tunnel contractors who embrace computer-based technology are certainly years ahead of their competition. It is this same technology that will define the future of safety management. Being able to access and review the data collected in safety inspections can be the real crystal ball to see into the future of tunnel safety. Computerized data can be presented in many shapes and styles to maximize its effect. The information can be broken down by the second, minute, hour, or the day, week, month. The amount of information that is provided directly relates to the amount of data that can be calculated and used. This also applies to safety inspections.

Project safety data can be collated and sorted to represent all known hazards on the jobsite. The more detailed the data input, the better the information that can be analyzed and utilized. Eventually the information can show trends in unsafe conditions, and behaviors can then be identified and mitigated before an accident or injury can happen. This technology gives every safety professional a tool to foretell the future to identify the who, what, and where of the next accident or injury that will happen and hopefully prevent it.

There are other benefits to this safety document control. Imagine an OSHA inspector issuing a safety citation on a project site. Thorough safety records utilizing electronic document control can give the contractor an edge in managing the OSHA citation by mitigating the safety exposure to ensure that this type of citation will not become a second or third offense. This technology can also help companies break down safety observations into positive and negative ones. This can give the contractor the ability to see if the safety inspections are effective, well rounded, and show a complete safety picture of the job. Further, it can indicate if there are areas or aspects of the project that are not inspected at all.

Figure 9.24 | **Visualization Tools: Alaskan Way Viaduct vs. Underground Option Used During Environmental Phase**

Having the safety data available electronically makes it easier to share the information. Working with subcontractors is a prime example. Subcontractors may not have the sophistication to produce or provide such data; thus, providing them with valuable safety information can enable them to work safer and more productively. The prime contractor has the ability to show the subcontractor in black and white where there are schedule issues, which can be a big plus if they are part of the critical path. This information can also be used by other project stakeholders. Owners or clients can be quite demanding, especially in terms of requesting the latest and greatest safety information. With electronic data document control, owners and other clients can have their request satisfied with a keystroke.

Many noteworthy milestones in the tunnel construction industry can be directly related to new technology. These advances can be seen across the board, whether in surveying, estimating, project management, or safety. Technology is the future of tunnel safety, with seemingly endless and interesting opportunities.

Figure 9.25 | **Visualization of Construction Sequences**

PUBLIC ACCEPTANCE OF TUNNELING CONSTRUCTION AND USE OF UNDERGROUND SPACE

Although the benefits to the community of underground structures—metro systems, water and wastewater tunnels, traffic tunnels, tunneling in urban environments—are huge, the general public is increasingly less accepting of long-duration construction activities disrupting their daily lives and creating numerous challenges to be overcome. Even when the desire for a better future environment is there, reality sets in—construction activities mean jobsites in neighborhoods with associated noise, traffic, dust, and other temporary impacts. With large projects, these "temporary" impacts can last years. Progress in reducing impacts has been made, but future projects must also gain public support for acceptance before they can be realized. Positive experience and lessons learned from past projects will continue to be essential for future undertakings to be successful, and the message about these projects and how it is delivered to the public is a key factor in the process.

In the United States, environmental studies to analyze impacts of various alternatives are required to have input from the community and other stakeholders. Gathering consensus is difficult, with opponents often having a louder voice, and their concerns must be addressed. Politicians must also find ways to make decisions in the face of some opposition. Finding the political champions for these projects will be essential for future projects.

Community outreach to build consensus and prepare for acceptable construction phasing will build on the current momentum. Visualization tools to review options and demonstrate safety will advance, and there will be more use of the Internet for public outreach and consensus building during planning and environmental studies. Figure 9.24 shows the waterfront with and without the Alaskan Way Viaduct, allowing the public to visualize the area with several alternatives.

During construction, up-to-date on-line information for construction notifications and work progress available to the public will give people confidence in project compliance with agreements on working hours, noise levels, air quality, and other environmental mitigations. Figure 9.25 illustrates a construction sequence that can be tied to the calendar of events for a project. Even with all the web-based technology, there is still no substitute for face-to-face meetings. Personal outreach by the owner/agencies and contractors to local residents and business must continue on a daily basis. "Brick and mortar" visitor centers are extremely beneficial. Figure 9.26 shows a full-scale model of a highway tunnel and cars to demonstrate safety and technology features in a two-level highway tunnel, and Figure 9.27 shows displays in Amsterdam. These visitor centers promote the project's goals and educate the community on the need for the project, as well as the technical aspects of the design, and the technology used to build it.

Further reduction in community impacts will be needed to promote acceptance of major projects. This does not always mean performing work during off-peak hours with associated longer schedule impacts. More often now, the public is requesting more work to be performed over shorter durations, realizing that shortening the schedule reduces overall impacts, as well as saving overall costs. Full street closures for highway and metro work for long weekends (freeways) or weeks (metros) are becoming commonplace, rather than stretching out the inconveniences for weeks on end. Owners and contractors delivering to the promised schedules always goes a long way toward acceptance of the next project.

While it is acknowledged that projects cannot be "wished into place," methods to reduce disturbance and increase support for projects through education have been successful. Technology to reduce impacts, as well as innovative designs to move construction away from streets and neighborhoods, should also advance in the future. Exemplary projects—like Line 9 in Barcelona, Spain (Figure 9.28), where work for a new subway moved access shafts off streets to connect to tunnels under the streets—have been met with great enthusiasm.

Figure 9.26 | **Full-Scale Model of Tunnel and Cars in Versailles, France**

Figure 9.27 | **Amsterdam Metro Models of Subaqueous Tunnels and Stations**

Figure 9.28 | **Deep Station Entrance, Barcelona Line 9**

FUNDING AND CONTRACTING METHODS

Funding for multimillion- or multibillion-dollar projects and contracting methods to sustain multi-year projects is an ongoing challenge and will continue. The United States spends far less on underground infrastructure compared to Europe and Asia. Consider that more than 350 miles of heavy subway in 13 lines were built in Shanghai, China, from 1985 to 2015, while less than 40 miles were built in the entire United States during the same period.

Most major underground projects spend years in the planning phase, with funding taking as long to secure. This is due to long-range planning by agencies, requirements for alternatives analyses, and extensive environmental studies with public input (and opposition), and there are always shortages of funds for infrastructure. For example, in the United States, the time needed for planning and environmental studies, design, and construction is 15 years or more (Elioff et al. 2007). Projects currently under construction in Los Angeles and Seattle (Metro's Westside Purple Line Subway extension and Seattle's SR 99 Tunnel) began planning in 2007, with construction completion for the first phase of the Metro project expected in 2023 and the SR 99 Tunnel in 2018. Water and wastewater projects have similar program planning and construction schedules (Figure 9.29).

To initiate the planning and engineering process, projects need proponents with long terms in office to bring them to fruition and secure sufficient funding for the entire project. In addition, funding can be contingent on oversight and demonstrated progress toward maintaining cost and schedule. In the United States, we have even seen examples of projects getting sufficient funding to begin construction, only to be cancelled due to lack of confidence in being able to complete the project within budget.

Although standard methods of funding have been through taxes, user fees, bond measures, tolls, and so on, new sources of funding for projects have been underway, including public–private partnerships (called PPPs or P3s), including design–build–finance–operate–maintain. Conceptually these funding methods are designed to transfer responsibility of funding to the private sector when the public sector is unable to meet multimillion- or multibillion-dollar budgets.

Figure 9.29 | **Deep Shaft for Urban Combined-Sewer-Overflow Tunnel, Washington, D.C.**

The future of tunneling will involve more P3s to deliver these megaprojects, and to operate and maintain them over the long term. Governments no longer have the capacity or the political will to fund these megaprojects, so cities, counties, and states must use a mixture of local funds and creative financing to bridge the gap. The "traditional" alternative delivery methods in a P3 are availability payments or tolls. Availability payments require the owner to have some sort of funding source over the long term to pay back the concessionaire (design–build–finance–operate–maintain team), and the concessionaire must meet contract requirements on maintenance and lane availability to receive its annual payments. A toll P3 project is a bit riskier for the concessionaire. There is no guarantee that once the tunnel is built people will pay the toll to use it, so the concessionaire is betting that once the project is built and financed, the concessionaire will recoup their costs and make a profit over the agreed period of years by people paying tolls. In Europe we will see more of those types of funding systems, but in the United States we are more risk averse as a society.

While these funding methods are not unique to tunnels and underground projects, some notable innovations that look promising for the future are those used recently for the PortMiami Tunnel in Florida and by the State of Arizona Department of Transportation.

The PortMiami Tunnel is serving as a model for how large-scale infrastructure tasks serve the local and regional community while paying back their tax dollars, perhaps many times over. For this project, the State of Florida Department of Transportation (FDOT), Miami Dade County, and the City of Miami all collaborated with a concessionaire and two major private construction companies—Bouygues Civil Works Florida and Transfield Services Australia—to construct and operate the tunnel until 2044. Under the concession agreement, FDOT made payments to the concessionaire during the construction period, upon the achievement of contractual milestones. After construction, FDOT makes payments to the concessionaire, contingent upon actual use of the facility and service quality. The tunnel is to be returned to FDOT at the end of the contract in October 2044. The total cost of design and construction of the tunnel was set at $668.5 million. The state agreed to pay for 50% of the capital costs (design, construction) and all of the operations and maintenance, while the remaining 50% of the capital costs will be provided by Miami-Dade County and the City of Miami.

The economic benefits of the PortMiami Tunnel went beyond the thousands of construction jobs created and reduction of traffic congestion to the port. The tunnel is said to have generated billions of dollars in economic impact and contributes to annual income within the city as the transportation industry booms. Although some initially criticized the tunnel and its cost to taxpayers, the entire project actually came in $90 million under budget—a financial feat practically unheard of for infrastructure ventures of this size (Lilly and Associates 2015).

We should also see creative solutions to the funding gaps for operations and maintenance of existing tunnels. The Arizona Department of Transportation is currently exploring a design–build–operate–maintain contract to replace unsafe lighting at their I-10 Deck Park Tunnel in Phoenix. It is expected that this will involve more efficient LED lighting (and other efficiencies to cut costs), and agencies will look at the tunnels themselves as potential funding sources. Pedestrian and bike areas are popular community attractions, including the Mount Baker Ridge Tunnel in Seattle, Washington with its pedestrian and bike tunnels over the traffic lanes (Figure 9.30). Another experiment is the Los Angeles County Metropolitan Transportation Authority's (Metro's) "fan" experiment. Metro tested the effect of collecting energy from airflow in subway tunnels produced by trains pushing air through the tunnels (Chen 2014; Figures 9.31 and 9.32). It may be that future tunnels such as Metro Los Angeles' Sepulveda Tunnel proposal (both toll highway and rail transit) will have a new generation of hybrid tunnels that can help meet diverse stakeholders' needs while providing some future funding sources through tolls and fares. In Singapore, a study is currently being conducted regarding making reservations to use a toll road, much like one would make a reservation for a ferry or an airplane seat. This takes demand management to a whole new level and could especially be used at chokepoints like a tunnel or bridge.

In another example, the Colorado Department of Transportation will be implementing peak-period high-occupancy toll shoulder lanes (congestion pricing) to try to reduce morning (to the mountains) and afternoon (from the mountains) ski congestion along Interstate 70. There are rumors that this revenue stream may support a future P3 project in the corridor that could include another tunnel or expansion of the existing Eisenhower–Johnson Memorial Tunnel.

In addition to P3s, another prospect for the future is "green bonds." According to an article by Sarah Fister Gale (2015):

Figure 9.30 | **Bicycle Path in Mount Baker Ridge Tunnel, Seattle**

> *Green Bonds were originally created by the World Bank as part of the "Strategic Framework for Development and Climate Change" to help stimulate public and private investment in projects that mitigate the impact of climate change or help affected people adapt to it. They were designed in response to rising investor demand for triple-A-rated fixed income products that support environmental initiatives, explained Betsy Otto, global*

Figure 9.31 | **Workers Attaching Special Fans to the Tunnel Wall**

Figure 9.32 | **Close-Up of the Fans**

director of the Water Program for Water Resources Institute in Washington, D.C. "Green bonds have a lot of caché right now," she said. "They are drawing a lot of interest from investors who have set sustainability goals for their investments."

Initially, green bonds focused on major climate mitigation initiatives, like massive solar installations, off-shore wind projects and irrigation efforts in water-scarce agricultural communities. But as their popularity surged, they expanded to include a broader category of endeavors—including water and wastewater infrastructure projects, which align nicely with the goals of green bond programs.

The Metropolitan Water Reclamation District of Greater Chicago is another city group that has experienced great success with green bond offerings. The group issued a $297-million bond in December 2014, which included a variety of sustainably focused projects, including stream-bank stabilization efforts, construction of a phosphorus recovery facility and a capital improvements project.

FUTURE TUNNELING PROJECTS

Of course, we can't accurately predict the future, but there are a number of projects and uses for tunnels and underground space to look forward to now. Some future project examples are highlighted in the following sections.

Megaprojects

California High-Speed Rail

The California High-Speed Rail Authority is planning, designing, and currently building the first construction segment through the Central Valley between Fresno and north of Bakersfield (Figures 9.33 and 9.34). This will be the first high-speed rail system in the nation. The first segment of the system, now under construction, is scheduled to be completed for use as a 100-mile test track where initial train set prototypes will be tested with operating speeds of more than 220 miles per hour. Dynamic testing is scheduled to start in 2021 and a useable section slated to open in 2022, ahead of a fully planned operable segment in late 2024. The rail system will connect the major regions of the state, contribute to economic development and a cleaner environment, be a sustainable system, create jobs, and preserve agricultural and protected lands. The system will run from San Francisco to the Los Angeles basin in under three hours at speeds capable of more than 220 miles per hour. The system will eventually extend to Sacramento and San Diego, totaling 800 miles with up to 24 stations (CaHSRA 2015).

A series of tunnels will be necessary to connect the San Joaquin Valley to the Los Angeles area in the south and to the San Francisco Bay area at the north end. Although final alignments are not currently set, a number of short tunnel segments less than a mile in length are planned and tunnels several miles long are intended through the Tehachapi Mountains north of Los Angeles. Tunnels would be constructed as drill-and-blast or bored tunnels about 30 feet in diameter. Longer tunnels are planned between Palmdale and Burbank following the State Route 14 corridor though Santa Clarita or an alternative east corridor under the San Gabriel Mountains through Angeles National Forest. Individual tunnel lengths range from more than 12 miles to less than 4 miles with total length of tunneling varying from 18 to 21 miles depending on alignment alternatives. Additional tunnels through Pacheco Pass between Merced and Gilroy comprise several tunneling segments totaling approximately 12 miles and are currently planned to be under construction in 2018.

Figure 9.33 | **High-Speed Train in Station, 2015**

Figure 9.34 | **High-Speed Rail in Depressed Section of an Urban Area**

CHAPTER NINE | THE FUTURE OF TUNNELING

California WaterFix

The Sacramento–San Joaquin Delta (Delta) in California is an important ecosystem and source of water for much of the state (Figure 9.35). The Delta is located east of San Francisco and south of Sacramento and serves as the "hub" of California's water supply by conveying water mostly originating from Sierra Nevada snowmelt to water users in Central and Southern California as well as providing environmental flows to the Pacific Ocean through the San Francisco Bay. The Delta area contains farmland and waterways, and provides important habitat for fish and waterfowl. Over the years, water conveyance through the Delta has been controlled by a system of levies that are now aging and subject to seismic damage, and local plant and animal species have become endangered. The aging levee system is at risk to rising seas levels, which will also expose the area to changing salinity and other environmental issues. To improve the environment and to ensure reliable delivery of a critical water supply to farms, residences, and industrial uses, the state has proposed a number of alternative plans to convey water through the Delta. The project is currently in planning/environmental phases, with several alternatives under review. Alternatives include tunnels and pipelines—more than 30 miles total and over 30 feet in diameter.

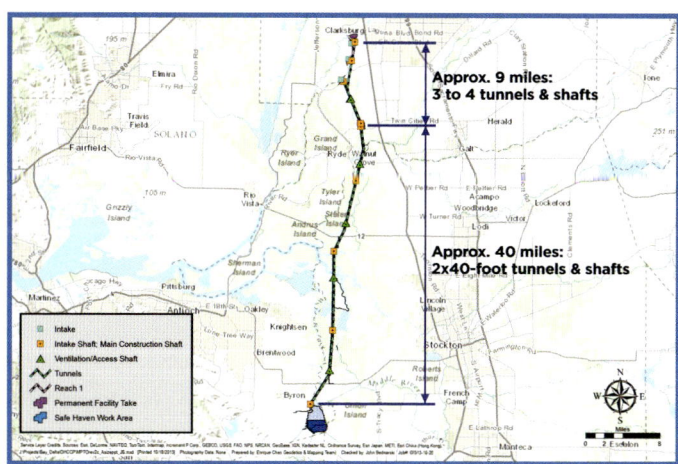

Figure 9.35 | **Bay Delta Tunnel**

Subway and Rail Expansion

With a growing population, traffic congestion, and increasing costs of gasoline and parking, ridership on San Francisco's BART (Bay Area Rapid Transit) heavy-rail system has been increasing. At the same time, the system has been "overstressed" with delays becoming more common. The transit agency has increased capacity with additional trains and system improvements, but a bottleneck remains at the transbay underwater crossing where four lines enter one tunnel, an immersed tube that was designed in the 1960s and built in the 1970s (Figure 9.36). Recent news is that the system is at a "tipping point" where steps will be taken to study, design, and construct a second bay crossing to increase capacity and reliability of the system (Figure 9.37). Michel Bernick (2015), former BART director, summarizes the situation well:

Figure 9.36 | **Immersed Tube—The Future as Depicted in the 1960s**

> Large-scale California transit projects nearly always have required a long period of gestation, in which concepts are presented, discussed, dismissed as impractical or too expensive, then set aside. But they may not be forgotten, and a combination of forces come together to achieve a type of tipping point in implementation. Traffic gridlock significantly worsens, or technological advances increase feasibility or reduce costs, or new funding becomes available, or usually a combination of these elements, combined with one or several project champions. This is the arc largely followed by the fourth bore of the Caldecott Tunnel, the Highway 4 widening, the BART extension to San Jose, and even for the basic BART system itself—which was first seriously discussed in the late 1940s. It may be that the second tube is reaching its tipping point.

Many other cities on both coasts have subway or other rail expansions in the planning phase. New York will be extending its Second Avenue Subway, to name one, with Los Angeles and other major cities also expanding and upgrading their existing systems.

THE HISTORY OF TUNNELING IN THE UNITED STATES

Highway Tunnels, Freeway Extensions and Replacements, and Underground Freeways

Tunnels have been used for rail transport in the United States since the 1800s, primarily to cross topographic features to allow transport by rail at operating grades. With the Federal-Aid Highway Act of 1956, freeways began to span the United States, with tunnels constructed where no alternative routes could be found. At the same time, freeways were built to traverse cities—with private commercial and residential properties in their paths taken by eminent domain. Many of these freeways were constructed at grade; however, many were also elevated to reduce the property taken and provide grade separation. Fifty plus years later, these structures are seen as detrimental to the environment, adding considerable noise, visual, and community impacts. Many have reached their design life and need replacement, especially in colder climates and seismic areas. Cities have debated and will continue to debate the alternatives for replacing these structures with tunnels as opposed to repairing or simply removing them.

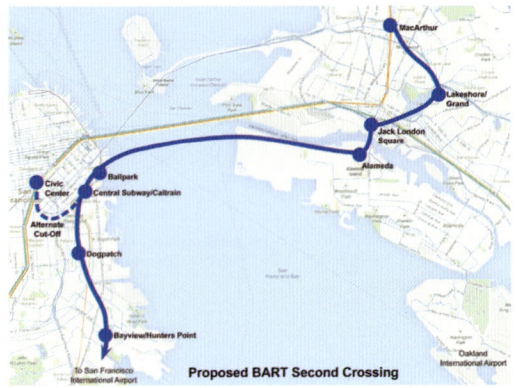

Figure 9.37 | **BART Second Tube Map**

Central Artery Project

The Central Artery Project in Boston, Massachusetts, is an example of replacement of a massive aerial viaduct structure (Interstate 93) with an underground roadway to reduce traffic congestion and use the surface for public (and green) space (Figure 9.38). The project, while often in the news during its multi-year construction, was visionary. Since its completion, traffic has flowed more freely, pedestrians again can cross overland to businesses and residences, and property values—and the city's tax basis—have increased. In addition to the beneficial outcome for the city, many technologies were advanced to allow construction in Boston's busy downtown area and in its water-bearing soils. For example, it was one of the first projects to broadcast real-time information on traffic conditions to alert drivers to detours and delays. Construction methods were advanced to limit movement of existing facilities such as active railroads, buildings, and roadways. One technological advance was the use of ground freezing under an active railroad to allow trains to continue operations while a precast concrete box tunnel structure was being installed near South Station. Use of slurry walls and ground improvement techniques were also advanced to new levels. Future projects of this nature will bring similar benefits and advancements of the industry.

Figure 9.38 | **Pedestrian Walk, over Central Artery**

Alaskan Way Viaduct

Another example of exemplary engineering practice is the replacement of the Alaskan Way Viaduct, an elevated freeway in Seattle, Washington, that was damaged in a 2001 earthquake. Numerous studies were done to assess replacement—either by tunnel, cut-and-cover, at grade, or with a similar aerial structure. After years of study and debate, the state elected to go with a single, double-decked tunnel and to have the city replace the viaduct space with revitalized waterfront property with quiet and unobstructed views. This will ultimately increase the land value and quality of life along Seattle's waterfront (Figure 9.39).

Figure 9.39 | **Existing Elevated Structure (left) To Be Replaced with Underground Roadway (right)**

CHAPTER NINE | THE FUTURE OF TUNNELING

California's 710 Gap Closure

New highway tunnels to reduce travel times are also on the drawing board in many states. In California, the state highway department is in planning and environmental phases to connect two existing freeways (the I-210 and I-710) via a large-diameter tunnel or tunnels. The SR-710 North Gap Closure, as it is known, could reduce travel times on the freeway systems, and thus improve air quality and reduce traffic on local surface streets (Figure 9.40). Other alternatives using subway and bus rapid transit are also being studied.

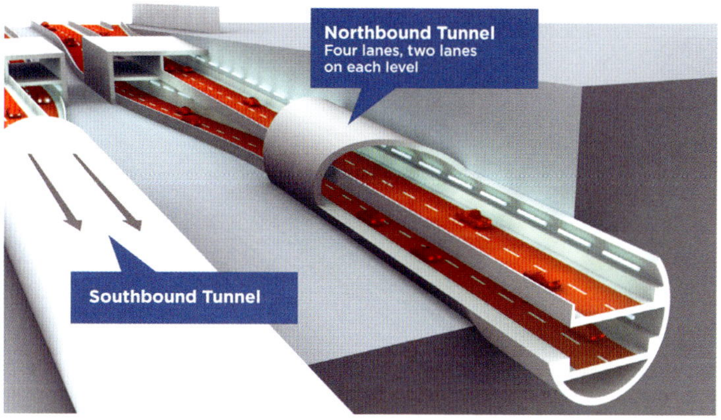

The Long Island Sound Link

The Long Island Sound Link is a proposed toll bridge or tunnel that would link Long Island to the south with Westchester County or Connecticut to the north across the Long Island Sound east of the Throgs Neck Bridge (Figure 9.41). The most recent proposal involves a tunnel between Rye, New York, on the mainland and Oyster Bay on the island. Feasibility studies for both bridges and tunnels have been conducted for numerous entry points. The highway tunnel would provide a link to improve enterprise on both sides, employment opportunities, and facility and goods movement—avoiding New York City congestion. The Sound is a minimum 6 miles wide at the proposed tunnel crossing point, but good portal points from which the tunnel might emerge make it necessary to make the total length of the tunnel 19 miles long.

Other Underground Projects

Other cities are taking note and studying underground concepts, including Toronto, Ontario, with its Gardiner Expressway; New York City's Gowanus Expressway; Detroit's Windsor Green Link; and Austin's I-35 Underground in Texas, among others. The future will bring more of these types of replacements for more efficient transportation and beneficial use of the surfaces above.

Rehabilitation and Retrofit Projects

Since Hurricane Sandy, New York City has had to implement major repairs and retrofitting to fix tunnel systems (subways and highway tunnels) damaged from the superstorm in 2012. Retrofits were done to replace electrical work as well as provide means to limit water inflows during storms. With expectations for global warming, more coastal cities will need to plan to prevent flooding and minimize impacts should major storm events occur.

Developing Uses—Freight Transport

Cargo and delivery tunnels are gaining renewed favor worldwide. Futurists are writing about "a tiny electric subway" that "will carry the packages to subscriber's homes or businesses, where itty-bitty forklifts will lift them from the subway" (Figure 9.42). These writers imagine that guidance systems will allow tunneling for several miles without "popping up for air," with construction being more or less difficult depending on the ground conditions (Barry 2009).

"CargoCap," described by ITA's Working Group 20, is a proposed safe and economical way to carry goods quickly and on time in congested urban areas by underground transportation pipelines (ITA 2012). This innovative

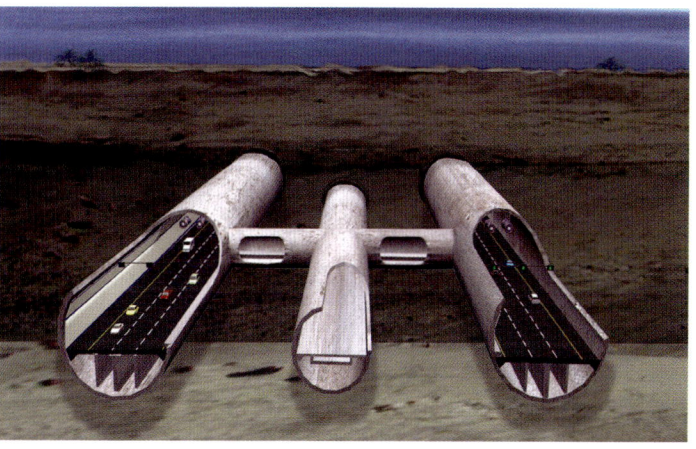

Figure 9.40 | **State Route 710 Gap Closure Tunnel Concept**

Figure 9.41 | **Rendering of the Long Island Sound Link**

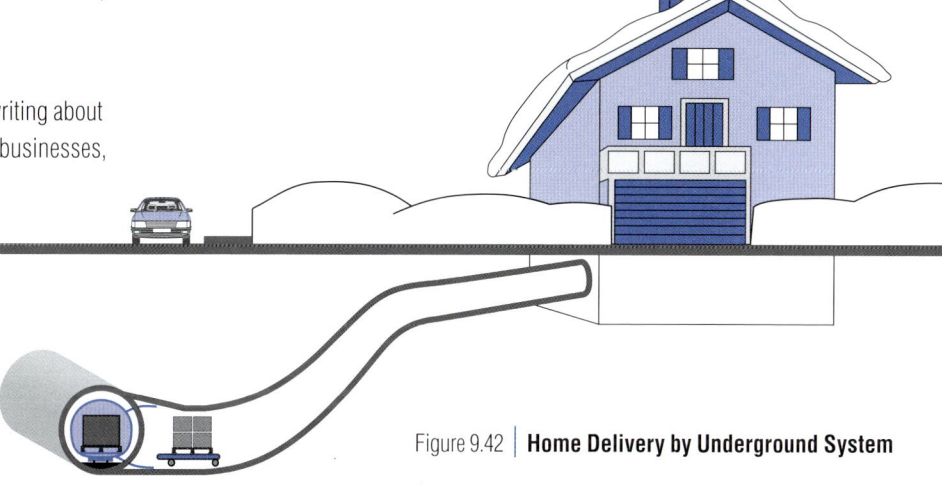

Figure 9.42 | **Home Delivery by Underground System**

Figure 9.43 | **CargoCap Model**

concept, which is currently under development, is the outcome of the interdisciplinary collaboration in research and development at the Ruhr University of Bochum, Germany. The system is independent from aboveground traffic congestion and weather conditions. According to the model displayed in Figure 9.43, each vehicle, the so-called cap, is designed for the transportation of two euro-pallets. Euro-pallets represent the majority of the general inner-European cargo transport, and therefore can be directed through pipelines with a diameter of 5.2 feet.

Since urban areas are becoming more crowded and polluted, it seems only a matter of time before other new ideas will be transferred into reality. When shops and offices can be supplied with goods via the underground, pollution and traffic congestion in cities can be greatly reduced, resulting in a better quality of life and an improvement of the work environment.

SUMMARY

The tunnel industry has evolved through a well-proven process of managing the many project risks and progressing from lessons learned that often result in innovation. This innovation pushes the boundaries on what has previously been possible, resulting in the many achievements and further application of underground solutions to meet society's needs. Looking into the future, the industry will continue to be challenged to develop underground space more efficiently through ground conditions that would not have been possible or contemplated in the past.

Tunnels and underground construction have played, and will continue to play, an important role for infrastructure to meet the future needs of space creation, transport, and improving the overall quality of life. Many large underground projects are currently on the horizon that will serve to help maintain our environment and bring forth much needed improvements and enhancements for transportation infrastructure. New technologies will assist in construction safety and quality, as well as cost savings to allow more possibilities in applying underground work.

Our history has shown that for more than 200 years, underground technology has benefited from experiences both within the shores of the United States and from the significant work of other nations around the world. This technology sharing is expected to continue into the future and enables the industry to go well beyond today's vision.

But in the future, we will also need to continue to address project financing in selecting underground projects and enabling a seamless progression into construction.

While we as an industry talk about sustainability and life-cycle design and engineering, we will need to do more to develop the analytical tools and approaches needed to understand life-cycle costs and performance more clearly, and relay the benefits to society to the elected representatives, decision makers, and the public so they will understand and support underground projects.

We also need our governments to have an enduring strategy in investment and support for infrastructure in general. For without this, society will be unable to prosper in the long term. History has shown that developing infrastructure, much of which was dependent on underground construction, has enabled current society to develop and sustain as the world's most successful and powerful nations. For this success to continue into the future, society needs to understand and embrace the importance of underground construction more than ever before. The challenges for underground construction will continue to increase, and our industry has never been better prepared to rise to the challenge.

The United States has benefited tremendously from our underground infrastructure, although our success story is not always well understood or appreciated. There is no doubt that, considering our underground heritage, our work has helped to make this world a better place. From our perspective, the United States has been, and will continue to be, one of the leading nations in developing underground facilities and infrastructure. We believe much of the future will continue to develop underground (Figure 9.44).

Figure 9.44 | **Light at the End of the Tunnel**

REFERENCES

Barry, K. 2009. Gear: Someday, a tiny subway will deliver your groceries. *Wired.* August 12.

Bernick, M. 2015. Is BART's second bay crossing at a tipping point? *San Francisco Chronicle*, March 20.

BLS (Bureau of Labor Statistics). 2010. *Economic News Release August 19, 2010: Census of Fatal Occupational Injuries Summary, 2009.* Washington, DC: BLS.

CaHSRA (California High-Speed Rail Authority). 2015. High-Speed Rail Program fact sheets. http://hsr.ca.gov/Newsroom/fact_sheets.html.

Chen, A. March 2014. Metro explores new green energy options: Placing a wind turbine in a subway tunnel. The Source: Transportation News & Views (blog). http://thesource.metro.net/2014/03/31/metro-explores-new-green-energy/.

Elioff, M.A., Miya, B.W., and Albino, J.H. 2007. Years of planning and implementation: Two major urban tunneling projects in Los Angeles, California, USA. In *Proceedings, World Tunneling Congress*, Prague, Czech Republic, May 5–10. pp. 51–57.

Gale, S.F. 2015. Green bonds: Are your projects a good fit? *WaterWorld* 31(3):14–19.

Harding, A., and Chappel, A. 2015. Segmental linings: A vision for the future. *Tunnelling Journal* (Feb/Mar):31–37.

ITA (International Tunnelling and Underground Space Association) Working Group 20. 2012. *Report on Underground Solutions for Urban Problems.* Lausanne, Switzerland: ITA.

Lilly and Associates. 2015. PortMiami Tunnel: The new standard in transportation infrastructure. www.infrastructureusa.org/portmiami-tunnel-the-new-standard-in-transportation-infrastructure/.

O'Carroll, J., and Goodfellow, B. 2015. *Guidelines for Improved Risk Management: On Tunnel and Underground Construction Projects in the United States of America.* Englewood, CO: UCA of SME.

U.S. CONSTRUCTED TUNNEL ARCHIVE

RAILROAD TUNNELS*

Name	Location	State	Length (ft)†	Year Constructed or Opened	Owner‡
Staple Bend Tunnel	Cambria County	PA	901	1833	Allegheny Portage Railroad
Wadesville Tunnel	Schuylkill	PA	1,477	1833	Danville & Pottsville Railroad
Girard (St. Clair) Tunnel	St. Clair	PA	800	1835	Danville & Pottsville Railroad
Harlem Tunnel (Old) (Murray Hill)	New York City	NY	844	1837	New York & Harlem Railway
Taft Tunnel	Lisbon	CT	300	1837	Norwich & Worcester Railroad
Black Rock Tunnel	Phoenixville	PA	1,932	1838	Reading Railroad
Elizabethtown Tunnel	Lancaster County	PA	900	1838	Pennsylvania Railroad
Howard Tunnel	York County	PA	275	1838	Northern Central Railway
Sherman's Tunnel	Sherman Station	PA	377	1838	Catawissa & Williamsport Railroad
Summit Tunnel No. 1	Summit Station	PA	1,050	1838	Catawissa & Williamsport Railroad
Buck Mountain Gravity Railroad Tunnel	Rockport	PA	135	1840	Buck Mountain Coal Company
Flat Rock Tunnel	West Manayunk	PA	937	1840	Reading Railroad
Harper's Ferry Tunnel	Jefferson County	WV	86	1840	Baltimore & Ohio Railroad
Rockport Tunnel	Carbon	PA	135	1840	Lehigh Valley Railroad
State Line (Canaan) Tunnel	Canaan	NY	580	1840	CSX Transportation
Doe Gully Tunnel	Morgan County	WV	1,207	1841	Baltimore & Ohio Railroad
Paw Paw Tunnel	Hampshire County	WV	250	1841	Baltimore & Ohio Railroad
Pulpit Rock Tunnel	Port Clinton	PA	1,637	1841	Reading Railroad
One Tunnel	Albany	NY	530	1842	Albany & West Stockbridge Railroad
William Henry Harrison Tunnel	Cleves	OH	1,782	1843	Cinncinati & Whitewater Railroad
Cobble Hill (Atlantic Avenue) Tunnel	Brooklyn	NY	256	1845	Long Island Rail Road
Phipps (Highland Street) Tunnel	Holliston	MA	92	1847	Boston & Worcester Railroad
Walpole Tunnel	Norfolk County	MA	201	1848	Boston Metro Transit Authority
Frankfort Tunnel	Franklin County	KY	435	1849	R.J. Corman Railroad
Greenfield Bridge Tunnel	Marriottsville	MD	419	1849	Springfield Terminal Railway
McGugin Tunnel	Washington County	PA	450	1850s	Pittsburgh & West Virginia Railway
Gwynedd Tunnel	Gwynedd	PA	500	Mid-1800s	North Pennsylvania Railroad
Tunnel on Section 25	Washington County	PA	1,335	Mid-1800s	Pittsburgh & Steubenville Railroad
Black Oak Ridge Tunnel	Dossett	TN	3,528	1850	CSX Transportation
Chetoogeta Tunnel (Old)	Tunnel Hill	GA	1,447	1850	Western & Atlantic Railroad
Everett's Tunnel	Piedmont	WV	350	1850	Baltimore & Ohio Railroad
Henryton Tunnel	Marriottsville	MD	430	1850	CSX Transportation
Mount Cobb Tunnel	Mount Cobb	PA	755	1850	Pennsylvania Coal Company
Murray's Tunnel	Cumberland	WV	250	1850	Baltimore & Ohio Railroad
Bellows Falls Tunnel	Windham County	VT	278	1851	New England Central Railroad
Breakneck Tunnel	Breakneck	NY	515	1851	New York Central & Hudson River Railroad
Carpenter's Tunnel	Western Pennsylvania	PA	450	1851	Pennsylvania Railroad
Chique's Tunnel	Lancaster County	PA	200	1851	Pennsylvania Railroad
Cruger's Tunnel No. 1	Cruger	NY	225	1851	New York Central & Hudson River Railroad
Cruger's Tunnel No. 2	Cruger	NY	61	1851	New York Central & Hudson River Railroad
Fort Montgomery Tunnel	Fort Montgomery	NY	350	1851	New York Central & Hudson River Railroad
Garretson's Tunnel	Poughkeepsie	NY	82	1851	New York Central & Hudson River Railroad

* This list is by no means exhaustive, but it tries to capture the most significant railroad tunnels in the United States with respect to age, length, and so forth.
† Length is usually the as-constructed length. Some tunnel lengths are approximate based on the accuracy of the source information.
‡ Tunnel ownership has changed multiple times. Where tunnels have been abandoned, the listing is typically the original constructor. Tunnels still in operation list the current operator, although some tunnels are operated by multiple railroads.

Previous Page | **Blue Plains Tunnel Screening Shaft During Tunnel Excavation, Washington, D.C.**

RAILROAD TUNNELS* (continued)

Name	Location	State	Length (ft)†	Year Constructed or Opened	Owner†
Garrison's Tunnel	Garrison	NY	603	1851	New York Central & Hudson River Railroad
Kelly's Tunnel	Poughkeepsie	NY	156	1851	New York Central & Hudson River Railroad
Laneys Tunnel	Glencoe	AL	1,000	1851	Alabama & Tennessee River Railroad
New Hamburg Tunnel	New Hamburg	NY	836	1851	Hudson River Railroad
Parkersburg - Linn Tunnel (No. 20)	Petroleum	WV	254	1851	Baltimore & Ohio Railroad
Parkersburg - Rodminer's Tunnel (No. 22)	Eaton	WV	338	1851	Baltimore & Ohio Railroad
Point Rock Tunnel	Columbia	PA	180	1851	Pennsylvania Railroad
Rodebaugh's Tunnel	Moreland County	PA	450	1851	Pennsylvania Railroad
Row Hook Tunnel	Row Hook	NY	72	1851	New York Central & Hudson River Railroad
Royersville Tunnel	Royersville	OH	1,050	1851	Iron Railroad
Spruce Creek Tunnel 1	Huntingdon County	PA	1,130	1851	Norfolk Southern Railroad
Whitehall Tunnel	Whitehall	NY	678	1851	Rensselaer & Sarratoga Railroad
Cumberland Mountain (Cowan) Tunnel	Franklin County	TN	2,228	1852	CSX Transportation
Eaton's Lower Tunnel	Wheeling	WV	265	1852	Baltimore & Ohio Railroad
Eaton's Upper Tunnel	Wheeling	WV	95	1852	Baltimore & Ohio Railroad
Glover's Gap Tunnel	Marion/Wetzel County	WV	300	1852	Baltimore & Ohio Railroad
Kingwood Tunnel (No. 1)	Preston	WV	4,137	1852	Baltimore & Ohio Railroad
Martin's Tunnel	Wheeling	WV	185	1852	CSX Transportation
McGuire's Tunnel	Preston County	WV	500	1852	Baltimore & Ohio Railroad
Shepherd Tunnel	Marchall County	WV	420	1852	Baltimore & Ohio Railroad
Cowan Tunnel	Radford	VA	3,302	1853	Norfolk Southern Railroad
East Barretts Tunnel	St. Louis	MO	630	1853	Missouri Pacific Railroad
Grant's Bend Tunnels (East & West)	Kenton County	KY	4,500	1853	CSX Transportation
Greenwood Tunnel (No. 2)	Greenwood	VA	538	1853	Blue Ridge Railroad
Ryan's Tunnel	Quakake	PA	180	1853	Catawissa & Williamsport Railroad
Tamaqua Tunnel	Tamaqua	PA	918	1853	Little Schuylkill Railroad
Tunnel on Section 13	Washington County	PA	351	1853	Pittsburgh & Steubenville Railroad
Tunnel on Section 17	Washington County	PA	1,315	1853	Pittsburgh & Steubenville Railroad
Welling Tunnel	Welling	WV	1,250	1853	Baltimore & Ohio Railroad
West Barretts Tunnel	St. Louis	MO	410	1853	Missouri Pacific Railroad
Allegheny (Summit) Tunnel	Gallitzin	PA	3,612	1854	Norfolk Southern Railway
Cork Run Tunnel (Berry Street)	Allegheny County	PA	2,371	1854	Port Authority of Allegheny County
Hollins Mill Road Tunnel	Lynchburg	VA	508	1854	Virginia & Tennessee Railway
Lofty Tunnel	Lofty	PA	1,100	1854	Catawissa Railroad
McLuney Tunnel	Zanesville	OH	1,165	1854	Cincinatti & Muskingum Valley Railway
Tunnel No. 1	Richland Furnace	OH	421	1854	Marietta & Cincinnati Railroad
Tunnel No. 2	Richland Furnace	OH	266	1854	Marietta & Cincinnati Railroad
Tunnel on Section 18	Washington County	PA	945	1854	Pittsburgh & Steubenville Railroad
Byers Tunnel	Byers	OH	413	1855	Marietta & Cincinnati Railroad
Doe Run Tunnel	West Union	WV	2,296	1855	Baltimore & Ohio Railroad
Huff's Summit Tunnel	Tuscarawas County	OH	1,010	1855	Cleveland & Pittsburgh Railroad
New Portage Tunnel	Gallitzin	PA	1,625	1855	Norfolk Southern Railway
Parkersburg - Bee Tree Tunnel (No. 21)	Wood County	WV	2,030	1855	Baltimore & Ohio Railroad
Pequabuck Tunnel	Plymouth	CT	3,580	1855	Pan Am Southern Railway
Pocono Tunnel	Monroe County	PA	497	1855	Delaware, Lackawanna, & Western Railroad
Scranton Tunnel	Luzerne County	PA	600	1855	Delaware, Lackawanna, & Western Railroad
Tunnel No. 1	Cooper County	MO	640	1855	Pacific Railway

RAILROAD TUNNELS* (continued)

Name	Location	State	Length (ft)†	Year Constructed or Opened	Owner‡
Brookville Tunnel (No. 3)	Albemarle	VA	864	1856	Chesapeake & Ohio Railway
Bulger Tunnel	Bulger	PA	300	1856	Pittsburgh & Columbus Railroad
Carr's Tunnel	Greensburg	PA	825	1856	Pennsylvania Railroad
Dinsmore Tunnel (No.4)	Washington County	PA	838	1856	Pittsburgh & Steubenville Railroad
Fergusson's Tunnel	Boyd County	KY	571	1856	Lexington & Big Sandy
Link Tunnel	New England	OH	260	1856	"Old Line" M&C Railroad
Nay Aug Tunnel - North	Dunmore	PA	755	1856	Delaware, Lackawanna, & Western Railroad
Rew Tunnel	Big Run	OH	313	1856	"Old Line" M&C Railroad
Sellersville Tunnel	Sellersville	PA	2,160	1856	North Pennsylvania Railroad
Tunnel No. 1	Jessamine County	KY	505	1856	Cincinnati Southern Railway
Tunnel No. 3 (Moonville)	Moonville	OH	250	1856	Marietta & Cincinnati Railroad
Tunnel No. 4 (King's or Mineral)	Mineral City	OH	362	1856	Marietta & Cincinnati Railroad
Van Gap Tunnel (Old)	Manuka Chunk	NJ	938	1856	Warren Railroad
Bell's Tunnel	Mansfield	PA	515	1857	Philadelphis & Reading Railroad
Big Tunnel No. 1	Lawrence County	IN	1,750	1857	CSX Transportation
Cady's Tunnel (No. 6)	Cady	VA	1,303	1857	Chesapeake & Ohio Railway
Coleman's Tunnel (No. 9)	Bath County	VA	853	1857	Chesapeake & Ohio Railway
Eastham's (Princess) Tunnel	Boyd County	KY	978	1857	CSX Transportation
Eaton Tunnel (No. 21, Old)	Eaton	WV	2,030	1857	Baltimore & Ohio Railroad
Finney Tunnel	Finney	PA	1,052	1857	Baltimore & Ohio Railroad
Fox Tunnel (No. 3)	Claysville	PA	694	1857	Baltimore & Ohio Railroad
Greenfield Tunnel (New)	Greenfield	WI	1,330	1857	Chicago, Milwaukee, St. Paul & Pacific Railroad
Jeddo Tunnel	Luzerne County	PA	900	1857	Lehigh Luzerne Railroad
Landis Ridge Tunnel	Bucks County	PA	2,142	1857	Norfolk Southern Railway
Mason's Tunnel (No. 8)	Bath County	VA	303	1857	Chesapeake & Ohio Railway
McClellan Tunnel (No. 4)	Taylorstown	PA	396	1857	Baltimore & Ohio Railroad
Parkersburg - Bonds Creek Tunnel (No. 13)	Cornwallis	WV	352	1857	Baltimore & Ohio Railroad
Parkersburg - Branch Tunnel No. 1 (Carr's)	Clarksburg	WV	2,708	1857	Baltimore & Ohio Railroad
Parkersburg - Branch Tunnel No. 12	Cornwallis	WV	577	1857	Baltimore & Ohio Railroad
Parkersburg - Branch Tunnel No. 14	Cornwallis	WV	182	1857	Baltimore & Ohio Railroad
Parkersburg - Branch Tunnel No. 15	Cornwallis	WV	478	1857	Baltimore & Ohio Railroad
Parkersburg - Branch Tunnel No. 16	Cornwallis	WV	220	1857	Baltimore & Ohio Railroad
Parkersburg - Branch Tunnel No. 17	Cairo	WV	452	1857	Baltimore & Ohio Railroad
Parkersburg - Branch Tunnel No. 18	Cairo	WV	963	1857	Baltimore & Ohio Railroad
Parkersburg - Brandy Gap Tunnel (No. 2)	Bristol	WV	1,086	1857	Baltimore & Ohio Railroad
Parkersburg - Butcher's Tunnel (No. 9)	Pennsboro	WV	855	1857	Baltimore & Ohio Railroad
Parkersburg - Calhoun's Tunnel (No. 7)	Pennsboro	WV	780	1857	Baltimore & Ohio Railroad
Parkersburg - Central Station Tunnel (No. 6)	West Union	WV	2,297	1857	Baltimore & Ohio Railroad
Parkersburg - Cunningham's Tunnel (No. 8)	Pennsboro	WV	588	1857	Baltimore & Ohio Railroad
Parkersburg - Farrell's Tunnel (No. 23)	Kanawha	WV	287	1857	Baltimore & Ohio Railroad
Parkersburg - Patterson's Tunnel (No. 10)	Ellenboro	WV	377	1857	Baltimore & Ohio Railroad
Parkersburg - Shannon Tunnel (No. 5)	Smithburg	WV	359	1857	Baltimore & Ohio Railroad
Parkersburg - Sherwood Tunnel (No. 4)	Sherwood	WV	846	1857	Baltimore & Ohio Railroad
Parkersburg - Silver Run Tunnel (No. 19)	Silver Run	WV	1,376	1857	Baltimore & Ohio Railroad

* This list is by no means exhaustive, but it tries to capture the most significant railroad tunnels in the United States with respect to age, length, and so forth.
† Length is usually the as-constructed length. Some tunnel lengths are approximate based on the accuracy of the source information.
‡ Tunnel ownership has changed multiple times. Where tunnels have been abandoned, the listing is typically the original constructor.
   Tunnels still in operation list the current operator, although some tunnels are operated by multiple railroads.

APPENDIX | U.S. CONSTRUCTED TUNNEL ARCHIVE

RAILROAD TUNNELS* (continued)

Name	Location	State	Length (ft)†	Year Constructed or Opened	Owner‡
Parkersburg - Teneriffe Tunnel (No. 11)	Cornwallis	WV	176	1857	Baltimore & Ohio Railroad
Parkersburg - Trough Tunnel (No. 3)	Long Run	WV	282	1857	Baltimore & Ohio Railroad
Perkasie Tunnel	Perkasie	PA	1,857	1857	Southeastern Pennsylvania Transportation Authority
Pitchers Tunnel	Francis	OH	1,670	1857	"Old Line" M&C Railroad
Ritner Tunnel	Lawrence County	IN	1,731	1857	CSX Transportation
Taylorstown Tunnel	Taylorstown	PA	479	1857	Baltimore & Ohio Railroad
Vineland Tunnel	Jefferson County	MO	800	1857	St. Louis, Iron Mountain & Southern Railroad
West Alexander Tunnel (No. 2)	West Alexander	PA	864	1857	Hempfield Railroad
Wheeling Tunnel (No. 1)	Wheeling	WV	448	1857	Baltimore & Ohio Railroad
Blue Ridge (Crozet) Tunnel (No. 5)	Nelson County	VA	4,273	1858	Chesapeake & Ohio Railway
Board Tree Tunnel	Marshall County	WV	2,350	1858	Baltimore & Ohio Railroad
Middle Tunnel	Pickens District	SC	616	1858	Blue Ridge & Atlantic Railroad
Saddle Tunnel	Oconee County	SC	475	1858	Blue Ridge & Atlantic Railroad
Stumphouse Mountain Tunnel	Walhalla	SC	5,865	1858	Blue Ridge & Atlantic Railroad
Burlington Tunnel	Burlington	VT	340	1860	New England Central Railroad
Madry Hill Tunnel	Lincoln County	TN	1,230	1860	Louisville & Cincinnati Railway
Kelly's Tunnel (No. 14)	Allegheny County	VA	460	1861	CSX Transportation
Lake's Tunnel (No. 13)	Allegheny County	VA	726	1861	CSX Transportation
Long Dock Tunnel	Hudson County	NJ	4,311	1861	Norfolk Southern Railway
Moore's Tunnel (No. 12)	Covington	VA	336	1861	CSX Transportation
Tunnel No. 1	Meridian	MS	428	1861	Vicksburg & Meridian Railroad
Oxford or Van Nest Gap Tunnel (Old)	Warren County	NJ	3,006	1862	Delaware, Lackawanna & Western Railroad
Brilzer Tunnel	Washington County	PA	377	1863	Pittsburgh & Steubenville Railroad
Vanderwarker Tunnel	Tunnel Station	OH	1,408	1863	"Old Line" M&C Railroad
Bow Ridge Tunnel (Old)	Westmoreland County	PA	630	1864	Pennsylvania Railroad
Gould Tunnel	Steubenville	OH	3,320	1864	Genesee & Wyoming Inc.
Mahanoy Tunnel	Mahanoy City	PA	3,410	1864	Reading Railroad
Panhandle (No. 42) Tunnel	Pittsburgh	PA	1,445	1865	Pittsburgh Light Rail
Donner Pass Tunnel 1 (Grizzly Hill)	Placer County	CA	514	1866	Union Pacific
Donner Pass Tunnel 2 (Emigrant Gap)	Placer County	CA	271	1866	Union Pacific
Junction Railroad Tunnel	West Philadelphia	PA	750	1866	Pennsylvania Railroad
Richmond Tunnel	Richmond	VA	600	1866	Richmond, Fredricksburg & Potomac Railroad
Turn Hole Tunnel	Jim Thorpe	PA	496	1866	Lehigh & Susquehanna Railroad
Webster Tunnel	Binghampton	NY	2,240	1866	Alany & Susquehanna Railroad
Washington Street Tunnel	Chicago	IL	1,605	1867	City of Chicago
Bank Lick Tunnel No. 1	Boone County	KY	181	1868	Louisville & Cincinnati Railway
Bank Lick Tunnel No. 2	Kenton County	KY	445	1868	CSX Transportation
Boyer Ridge	Henry County	KY	370	1868	CSX Transportation
Catoctin Tunnel	Point of Rocks	MD	494	1868	Baltimore & Ohio Railroad
Donner Pass Tunnel 3 (Cisco)	Placer County	CA	280	1868	Union Pacific
Donner Pass Tunnel 4 (Red Spur)	Placer County	CA	92	1868	Union Pacific
Donner Pass Tunnel 5 (Crocker's Spur)	Placer County	CA	128	1868	Union Pacific
Donner Pass Tunnel 6 (Summit)	Placer County	CA	1,659	1868	Union Pacific
Donner Pass Tunnel 7	Placer County	CA	100	1868	Union Pacific
Donner Pass Tunnel 8 (Black Point)	Placer County	CA	375	1868	Union Pacific
Donner Pass Tunnel 9 (Donner's Peak)	Placer County	CA	216	1868	Union Pacific
Donner Pass Tunnel 10 (Cement Ridge)	Placer County	CA	509	1868	Union Pacific
Donner Pass Tunnel 11 (Tunnel Spur)	Placer County	CA	577	1868	Union Pacific

RAILROAD TUNNELS* (continued)

Name	Location	State	Length (ft)†	Year Constructed or Opened	Owner‡
Donner Pass Tunnel 12 (Tunnel Spur 2)	Placer County	CA	342	1868	Union Pacific
Donner Pass Tunnel 13 (Lake Ridge)	Placer County	CA	870	1868	Union Pacific
Donner Pass Tunnel 14 (Alder Creek)	Placer County	CA	200	1868	Union Pacific
Donner Pass Tunnel 15 (Quartz Spur)	Placer County	CA	96	1868	Union Pacific
Dunleith Tunnel	Dunleith	IL	850	1868	Illinois Central Railroad
Eagle Creek Tunnel	Gallatin County	KY	637	1868	CSX Transportation
Lower Point of Rocks Tunnel	Frederick County	MD	788	1868	Baltimore & Ohio Railroad
Mill Creek Tunnel	Carroll County	KY	480	1868	Louisville & Cincinnati Railway
Ten-Mile Creek Tunnel	Grant County	KY	216	1868	Louisville & Cincinnati Railway
Transcontinental Tunnel No. 1	St. Mary's Creek	WY	215	1868	Union Pacific
Transcontinental Tunnel No. 2 (Echo)	Wahstch	UT	772	1868	Union Pacific
Transcontinental Tunnel No. 3	Ogden	UT	508	1868	Union Pacific
Transcontinental Tunnel No. 4	Ogden	UT	297	1868	Union Pacific
Cactoctin Tunnel	Frederick County	MD	494	1868	CSX Transportation
Woodbine Tunnel	Syksville	MD	240	1868	Baltimore & Ohio Railroad
Glenallen Tunnel	Bollinger County	MO	360	1869	Missouri Pacific Railroad
Hitchcock Tunnel	Garrett County	MD	400	1869	CSX Transportation
Housatiummit Tunnel	Fairfield County	MA	380	1869	New York & New England Railway
Johnson's Tunnel (No. 10)	Covington	VA	180	1869	Chesapeake & Ohio Railway
Van Gap Tunnel (New)	Manuka Chunk	NJ	938	1869	Warren Railroad
Packs Mountain Tunnel	Fayette County	WV	1,140	1870s	Chesapeake & Ohio Railway
Allegheny Tunnel (No. 16)	Allegheny County	VA	4,711	1870	CSX Transportation
American Flat Tunnel (No. 2)	Gold Hill	NV	566	1870	Virginia & Truckee Railroad
Barnesville Tunnel	Barnesville	OH	423	1870	Baltimore & Ohio Railroad
E Street Tunnel (No. 6)	Gold Hill	NV	587	1870	Virginia & Truckee Railroad
Glenn's Tunnel	Mansfield	PA	287	1870	Philadelphia & Reading Railroad
Homestead Tunnel (No. 3)	Gold Hill	NV	250	1870	Virginia & Truckee Railroad
Julia Tunnel (No. 5)	Gold Hill	NV	93	1870	Virginia & Truckee Railroad
Lake Shore Tunnel	Oil City	PA	956	1870	Lake Shore Railroad
Lake View Tunnel	Gold Hill	NV	367	1870	Virginia & Truckee Railroad
Oil City Tunnel	Oil City	PA	925	1870	Lake Shore & Michigan Southern
Red Hill Tunnel (No. 11)	Covington	VA	642	1870	Chesapeake & Ohio Railway
Stretchers Neck Tunnel	Lynchburg	VA	256	1870	Virginia Midland & Great Southern Railway
Yellow Jacket Tunnel (No. 4)	Gold Hill	NV	433	1870	Virginia & Truckee Railroad
Hannibal Tunnel	Marion County	MO	308	1871	Norfolk Southern Railroad
High View (also Bloomingburg, Shawangunk, & Wurtsboro) Tunnel	Mamakating	NY	3,855	1871	New York, Ontario & Western Railway
Hustler Tunnel	Omaha Trail	WI	875	1871	Chicago & North Western
Little Bend Tunnel (No. 22)	Fayette County	WV	750	1871	Chesapeake & Ohio Railway
Pinkerton Tunnel	Somerset County	PA	1,080	1871	Pennsylvania Railroad
Sand Patch Tunnel No. 1	Somerset County	PA	4,777	1871	Pittsburgh & Connellsville Railroad
Shepaug (Steep Rock) Tunnel	Washington County	CT	235	1871	New York, Newhaven & Hartford Railroad
Tunnel No. 1	Floyd County	IN	200	1871	Louisville, New Albany & St. Louis Air Line Railroad
Tunnel No. 2	Floyd County	IN	160	1871	Louisville, New Albany & St. Louis Air Line Railroad
Tunnel No. 3	Floyd County	IN	300	1871	Louisville, New Albany & St. Louis Air Line Railroad

* This list is by no means exhaustive, but it tries to capture the most significant railroad tunnels in the United States with respect to age, length, and so forth.
† Length is usually the as-constructed length. Some tunnel lengths are approximate based on the accuracy of the source information.
‡ Tunnel ownership has changed multiple times. Where tunnels have been abandoned, the listing is typically the original constructor.
  Tunnels still in operation list the current operator, although some tunnels are operated by multiple railroads.

APPENDIX | U.S. CONSTRUCTED TUNNEL ARCHIVE

RAILROAD TUNNELS* *(continued)*

Name	Location	State	Length (ft)†	Year Constructed or Opened	Owner‡
2nd Creek Tunnel (No. 19) (Old)	Greenbrier County	WV	1,550	1872	Chesapeake & Ohio Railway
Arnold Tunnel	Harrison County	IN	500	1872	Louisville, New Albany & St. Louis Air Line Railroad
Benford Tunnel	Fayette County	PA	406	1872	CSX Transportation
Big Tunnel No. 2	Floyd County	IN	4,328	1872	Louisville, New Albany & St. Louis Air Line Railroad
Blue Hole Tunnel (No. 25)	Raleigh County	WV	660	1872	CSX Transportation
Blue River Tunnel	Crawford County	IN	150	1872	Louisville, New Albany & St. Louis Air Line Railroad
Couchay Tunnel	Crawford County	IN	800	1872	Louisville, New Albany & St. Louis Air Line Railroad
Dade's Tunnel (No. 20)	Summer County	WV	1,150	1872	Chesapeake & Ohio Railway
Georgetown Tunnel No. 1	Harrison County	IN	230	1872	Norfolk Southern Railway
Georgetown Tunnel No. 2	Harrison County	IN	360	1872	Norfolk Southern Railway
Great Bend Tunnel (No. 21)	Summer County	WV	6,449	1872	Chesapeake & Ohio Railway
Lick Run Tunnel (No. 7)	Bath County	VA	252	1872	CSX Transportation
Navy Yard Tunnel	Washington	DC	1,550	1872	Baltimore & Potomac Railroad
Nesquehoning Tunnel	Carbon County	PA	3,805	1872	Lehigh Coal & Navigation Co.
Pope's Nose Tunnel (No. 24)	Raleigh County	WV	180	1872	Chesapeake & Ohio Railway
Ray's Hill Tunnel	Clearfield	PA	4,224	1872	Pennsylvania Railroad and Baltimore & Ohio Railroad
Shoo Fly Tunnel	Confluence	PA	307	1872	CSX Transportation
Virginia Avenue Tunnel	Washington	DC	3,788	1872	CSX Transportation
White Rock Tunnel (No. 18)	Greenbrier County	WV	402	1872	Chesapeake & Ohio Railway
Baltimore & Potomac Tunnel	Baltimore	MD	7,669	1873	Amtrak
Brookville Tunnel	Brookville	PA	754	1873	Allegheny Valley Railroad
Cazenovia Tunnel	Cazenovia	NY	1,631	1873	Syracuse & Chenango Railroad
Climax (Anthony's Neck) Tunnel	Climax	PA	517	1873	Allegheny Valley Railroad
Donner Pass Tunnel 0	Placer County	CA	711	1873	Union Pacific
Hawk's Mountain Tunnel	Sullivan County	NY	1,300	1873	New York & Oswego Midland
Lewis Tunnel (No. 15)	Allegheny County	VA	1,873	1873	Chesapeake & Ohio Railway
Long Point Tunnel	Lawsonham	PA	644	1873	Allegheny Valley Railroad
Stretchers Neck Tunnel (No. 23)	Fayette County	WV	1,600	1873	CSX Transportation
Tunnel No. 1 (Elroy-Sparta Trail)	Monroe	WI	1,694	1873	Chicago & North Western Railway
Tunnel No. 2 (Elroy-Sparta Trail)	Monroe	WI	1,694	1873	Chicago & North Western Railway
Tunnel No. 3 (Elroy-Sparta Trail)	Monroe	WI	3,810	1873	Chicago & North Western Railway
Union Tunnel	Baltimore	MD	3,410	1873	Amtrak
Caledonia Tunnel	Driftwood Junction	PA	425	1874	Allegheny Valley Bennett's Branch
Fort Ticonderoga Tunnel	Fort Ticonderoga	NY	439	1874	New York & Canada Railway
Harlem Tunnel (New)	New York City	NY	2,000	1874	New York & Harlem Railway
Hogback Tunnel	Centre County	PA	320	1874	Beech Creek Railroad
Muldroughs Hill Tunnel	Marion County	KY	1,900	1874	Cumberland & Ohio
Tunnel Hill Tunnel	Johnson County	IL	543	1874	Cairo & Vincennes Railroad
Wrays Hill Tunnel	Huntingdon County	PA	1,138	1874	East Broad Top Railroad & Coal Co.
Acheson Tunnel	Washington County	PA	1,299	1875	Pittsburgh & West Virginia Railway
Buncombe Tunnel	Lafayette	WI	350	1875	Chicago & North Western Railroad
Church Hill Tunnel (No. 1)	Richmond	VA	3,927	1875	Chesapeake & Ohio Railway
East Boston Tunnel (North & South)	Boston	MA	942	1875	Boston, Revere Branch & Lynn Railroad
Hoosac Tunnel	Western Massachusetts	MA	25,081	1875	Norfolk Southern Railroad
Musconetcong (Pattenburg) Tunnel	Hunterdon County	NJ	4,893	1875	Norfolk Southern Railway
One Tunnel	Emaus	PA	1,677	1875	Philadelphis & Reading Railroad
Port Henry Tunnel	Port Henry	NY	491	1875	New York & Canada Railway
Sideling Hill Tunnel 1	Fulton County	PA	830	1875	South Pennsylvania Railroad

RAILROAD TUNNELS* (continued)

Name	Location	State	Length (ft)†	Year Constructed or Opened	Owner‡
St. Louis Freight Tunnel	St. Louis	MO	4,880	1875	MetroLink Light Rail
Tunnel No. 3	Pulaski County	KY	1,067	1875	Cincinnati Southern Railway
Tunnel No. 4	Pulaski County	KY	1,165	1875	Cincinnati Southern Railway
Tunnel No. 12	Pulaski County	KY	872	1875	Cincinnati Southern Railway
Tunnel No. 18	Morgan County	TN	646	1875	Cincinnati Southern Railway
Tunnel No. 19	Morgan County	TN	360	1875	Cincinnati Southern Railway
Tunnel No. 20	Morgan County	TN	397	1875	Cincinnati Southern Railway
Tunnel No. 21	Morgan County	TN	261	1875	Cincinnati Southern Railway
Willsborough	Willsborough	NY	606	1875	New York & Canada Railway
Bergen Hill Tunnel No. 1 (North)	Hudson County	NJ	4,280	1876	New Jersey Transit
Burgin Tunnel	Buncombe County	NC	552	1876	Southern Railway
Greenfield Tunnel (Old)	Greenfield	WI	1,330	1876	Chicago, Milwaukee, St. Paul & Pacific Railroad
Hickory Tunnel	Washington County	PA	1,286	1876	Pittsburgh & West Virginia Railway
Mission Hill (No. 8)	Portero District	CA	927	1876	Southern Pacific Railroad
San Fernando Tunnel (No. 20)	Los Angeles County	CA	6,975	1876	Union Pacific
Summit Tunnel No. 2	Western Pennsylvania	PA	1,965	1876	Allegheny Valley Railroad
Tehachapi Tunnel No. 1	Kern County	CA	232	1876	Union Pacific
Tehachapi Tunnel No. 2	Kern County	CA	219	1876	Union Pacific
Tehachapi Tunnel No. 3	Kern County	CA	494	1876	Union Pacific
Tehachapi Tunnel No. 4	Kern County	CA	400	1876	Union Pacific
Tehachapi Tunnel No. 5	Kern County	CA	1,175	1876	Union Pacific
Tehachapi Tunnel No. 6	Kern County	CA	185	1876	Union Pacific
Tehachapi Tunnel No. 7	Kern County	CA	520	1876	Union Pacific
Tehachapi Tunnel No. 8	Kern County	CA	689	1876	Union Pacific
Tehachapi Tunnel No. 9	Kern County	CA	428	1876	Union Pacific
Tehachapi Tunnel No. 10	Kern County	CA	307	1876	Union Pacific
Tehachapi Tunnel No. 11	Kern County	CA	110	1876	Union Pacific
Tehachapi Tunnel No. 12	Kern County	CA	800	1876	Union Pacific
Tehachapi Tunnel No. 13	Kern County	CA	475	1876	Union Pacific
Tehachapi Tunnel No. 14	Kern County	CA	500	1876	Union Pacific
Tehachapi Tunnel No. 15	Kern County	CA	245	1876	Union Pacific
Tehachapi Tunnel No. 16	Kern County	CA	175	1876	Union Pacific
Tehachapi Tunnel No. 17	Kern County	CA	225	1876	Union Pacific
Tunnel No. 2	Lincoln County	KY	4,000	1876	Cincinnati Southern Railway
Tunnel No. 8	Pulaski County	KY	499	1876	Cincinnati Southern Railway
Tunnel No. 10	Pulaski County	KY	247	1876	Cincinnati Southern Railway
Tunnel No. 13	Scott County	TN	340	1876	Cincinnati Southern Railway
Tunnel No. 14	Scott County	TN	285	1876	Cincinnati Southern Railway
Tunnel No. 15	Scott County	TN	2,525	1876	Cincinnati Southern Railway
Tunnel No. 16	Morgan County	TN	1,084	1876	Cincinnati Southern Railway
Tunnel No. 17	Morgan County	TN	1,250	1876	Cincinnati Southern Railway
Tunnel No. 22	Morgan County	TN	699	1876	Cincinnati Southern Railway
Tunnel No. 23	Morgan County	TN	846	1876	Cincinnati Southern Railway
Tunnel No. 24	Morgan County	TN	2,021	1876	Cincinnati Southern Railway

* This list is by no means exhaustive, but it tries to capture the most significant railroad tunnels in the United States with respect to age, length, and so forth.
† Length is usually the as-constructed length. Some tunnel lengths are approximate based on the accuracy of the source information.
‡ Tunnel ownership has changed multiple times. Where tunnels have been abandoned, the listing is typically the original constructor. Tunnels still in operation list the current operator, although some tunnels are operated by multiple railroads.

APPENDIX | U.S. CONSTRUCTED TUNNEL ARCHIVE

RAILROAD TUNNELS* (continued)

Name	Location	State	Length (ft)†	Year Constructed or Opened	Owner‡
Tunnel No. 25	Morgan County	TN	238	1876	Cincinnati Southern Railway
Tunnel No. 26	Morgan County	TN	1,662	1876	Cincinnati Southern Railway
Beaverdam Tunnel	Penn View	PA	252	1877	Lewisburg & Tyrone Railroad
CNTP Tunnel 13	Scott	TN	751	1877	Cincinnati Southern Railway
CNTP Tunnel 14	Scott	TN	893	1877	Cincinnati Southern Railway
CNTP Tunnel 16	Morgan	TN	809	1877	Cincinnati Southern Railway
Poe Paddy Tunnel	Ingleby	PA	266	1877	Lewisburg & Tyrone Railroad
Port Perry Tunnel	Western Pennsylvania	PA	514	1877	Norfolk Southern Railway
Robertsville Tunnel	Robertsville	OH	520	1877	Wheeling & Lake Erie Railway Company
Tunnel No. 1	Altamont	CA	1,120	1877	Union Pacific
Tunnel No. 5	Pulaski County	KY	879	1877	Cincinnati Southern Railway
Tunnel No. 6	Pulaski County	KY	189	1877	Cincinnati Southern Railway
Tunnel No. 7	Pulaski County	KY	1,154	1877	Cincinnati Southern Railway
Tunnel No. 9	Pulaski County	KY	1,209	1877	Cincinnati Southern Railway
Tunnel No. 11	Pulaski County	KY	448	1877	Cincinnati Southern Railway
Tunnel No. 27	Roane County	TN	1,930	1877	Cincinnati Southern Railway
Washington Tunnel	Marin County	CA	1,358	1877	Sonoma & Marin Railroad
Bakerstown Tunnel	Bakerstown	PA	623	1878	Pittsburgh & Western Railroad
CNTP Tunnel 19	Morgan	TN	762	1878	Cincinnati Southern Railway
CNTP Tunnel 20	Morgan	TN	4,224	1878	Cincinnati Southern Railway
CNTP Tunnel 21	Morgan	TN	347	1878	Cincinnati Southern Railway
CNTP Tunnel 23	Morgan	TN	1,945	1878	Cincinnati Southern Railway
CNTP Tunnel 24	Morgan	TN	2,174	1878	Cincinnati Southern Railway
Sulphur Fork Tunnel	Henry County	KY	647	1878	Louisville & Cincinnati Railway
CNTP Tunnel 15	Scott	TN	2,533	1879	Cincinnati Southern Railway
Eastwood Tunnel	Jefferson	KY	1,049	1879	Louisville & National Railroad
Patton (Taswell) Tunnel	Crawford County	IN	769	1879	Norfolk Southern Railway
Raton Tunnel No. 1	Colfax County	NM	2,041	1879	BNSF Railway
Swannanoa Tunnel	Buncombe County	NC	1,808	1879	Southern Railway
Dillinger Tunnel	Emmaus	PA	1,793	1880s	Reading Railroad
Jefferson or Fish Trap Tunnel	Jefferson County	AL	1,200	1880s	Southern Railway
Tunnel No. 24	Toledo	OR	669	1880s	Portland & Western
Coast Line - Tunnel No. 2 (Summit or Wright)	Laurel	CA	6,028	1880	Southern Pacific Railroad
Coast Line - Tunnel No. 3 (Laurel or Glenwood)	Glenwood	CA	5,793	1880	Southern Pacific Railroad
Coast Line - Tunnel No. 4 (Mountain Charlie or Clems)	Clems	CA	910	1880	Southern Pacific Railroad
Coast Line - Tunnel No. 5 (Zayante)	Zayanite	CA	240	1880	Southern Pacific Railroad
Coast Line - Tunnel No. 7 (Summit)	Summit	CA	1,410	1880	Southern Pacific Railroad
Coast Line - Tunnel No. 8 (Felton)	Big Trees	CA	263	1880	Southern Pacific Railroad
Eagle Tunnel	Radcliff	OH	340	1880	Gallipolis, McArthur & Columbus Railroad
SPCR Summit Tunnel (No. 3)	Santa Cruz County	CA	5,793	1880	South Pacific Coast Railroad
Fairdale Tunnel	Harrison County	IN	570	1881	Norfolk Southern Railroad
Oak Street Tunnel	Hamilton	OH	1,600	1881	Cincinnati, Lebanon & Northern Railroad
Alpine Tunnel	Pitkin County	CO	1,825	1882	South Park & Pacific Railroad
Bonita Tunnel	Clinton	MT	896	1882	Montana Rail Link
Braswell Tunnel	Braswell	GA	750	1882	Southern Railroad
Campbells Tunnel	Radcliff	OH	896	1882	Gallipolis, McArthur & Columbus Railroad
Cow Creek Tunnel	Douglas County	OR	385	1882	Central Oregon & Pacific Railroad
Duncan Tunnel	Edwardsville	IN	4,295	1882	Norfolk Southern Railroad

THE HISTORY OF TUNNELING IN THE UNITED STATES

RAILROAD TUNNELS* (continued)

Name	Location	State	Length (ft)†	Year Constructed or Opened	Owner‡
Marengo Tunnel	Crawford County	IN	700	1882	Norfolk Southern Railroad
Nimrod Tunnel No. 4	Garnite County	MT	909	1882	BNSF Railway
The Little Tunnel	Berkshire County	MA	126	1882	Boston & Maine Railroad
Winslow Tunnel	Washington County	AR	1,702	1882	Arkansas & Missouri Railroad
Zeno Tunnel	Evans City	PA	350	1882	Pittsburgh & Western Railroad
Cowee Tunnel	Dillsboro	NC	700	1883	Great Smoky Mountain Railroad
Factoryville Tunnels (East & West)	Factoryville	PA	2,280	1883	Delaware, Lackawanna, & Western Railroad
Frisco - Meramec Highlands Tunnel	St. Louis County	MO	400	1883	St. Louis - San Francisco Railway
Glenwood Tunnel	Garfield County	CO	1,327	1883	Union Pacific
Meramac Highlands Tunnel	St. Louis County	MO	400	1883	St. Louis - San Francisco Railway
Mullan Tunnel	Mullan	MT	3,896	1883	Montana Rail Link
Schenley Tunnel	Pittsburgh	PA	2,872	1883	CSX Transportation
Weehawken Tunnel	Weehawken	NJ	4,156	1883	Hudson–Bergen Light Rail
Bozeman Pass Tunnel (Old)	Gallatin County	MT	3,652	1884	Northern Pacific Railway
Brook Tunnel	Somerset County	PA	1,690	1884	Hudson–Bergen Light Rail
Cal Park Tunnel	Marin	CA	1,100	1884	Northwestern Pacific Railroad
North Shore Railroad Tunnels - 1	Freedom	PA	400	1884	North Shore Railroad
North Shore Railroad Tunnels - 2	Freedom	PA	170	1884	North Shore Railroad
Peale Tunnel	Clearfield County	PA	1,277	1884	Beech Creek Railroad
Phoenixville (also Fairview) Tunnel	Phoenixville	PA	811	1884	Pennsylvania Railroad
Negro Mountain Tunnel	Somerset County	PA	734	1885	New York Central Railroad
Ray's Hill Tunnel	Bedford	PA	3,532	1885	Southern Pennsylvania Railroad
Fallsburg (also Neversink) Tunnel	South Fallsburg	NY	1,123	1886	Ontario & Western Railway
Grays Ferry Tunnel	Grays Ferry	PA	750	1886	CSX Transportation
Jenson Tunnel	Le Flore	OK	1,180	1886	Kansas City Southern Railway
Owensburg Tunnel	Lawrence County	IN	1,362	1886	Monongahela Railway
Clarity Tunnel	Quitaque	TX	742	1887	Fort Worth & Denver South Plains Railway
Midland Busk-Ivanhoe Tunnel (No. 16)	Pitkin County	CO	9,394	1887	Colorado Midland Railway
Midland Hagerman Pass Tunnel	Pitkin County	CO	2,161	1887	Colorado Midland Railway
Midland Mallon Tunnel (No. 17)	Hell Gate	CO	528	1887	Colorado Midland Railway
Midland Tunnel No. 1	Chaffee County	CO	211	1887	Colorado Midland Railway
Midland Tunnel No. 2	Chaffee County	CO	115	1887	Colorado Midland Railway
Midland Tunnel No. 3	Chaffee County	CO	165	1887	Colorado Midland Railway
Midland Tunnel No. 4	Ute Pass	CO	290	1887	Colorado Midland Railway
Midland Tunnel No. 6	Ute Pass	CO	264	1887	Colorado Midland Railway
Midland Tunnel No. 7	Ute Pass	CO	211	1887	Colorado Midland Railway
Midland Tunnel No. 8	Ute Pass	CO	317	1887	Colorado Midland Railway
Midland Tunnel No. 9	Elevenmile Canyon	CO	211	1887	Colorado Midland Railway
Midland Tunnel No. 10	Elevenmile Canyon	CO	250	1887	Colorado Midland Railway
Midland Tunnel No. 11	Elevenmile Canyon	CO	264	1887	Colorado Midland Railway
Midland Tunnel No. 12	Buena Vista	CO	53	1887	Colorado Midland Railway
Midland Tunnel No. 13	Buena Vista	CO	70	1887	Colorado Midland Railway
Midland Tunnel No. 14	Buena Vista	CO	100	1887	Colorado Midland Railway
Midland Tunnel No. 15	Buena Vista	CO	158	1887	Colorado Midland Railway

* This list is by no means exhaustive, but it tries to capture the most significant railroad tunnels in the United States with respect to age, length, and so forth.
† Length is usually the as-constructed length. Some tunnel lengths are approximate based on the accuracy of the source information.
‡ Tunnel ownership has changed multiple times. Where tunnels have been abandoned, the listing is typically the original constructor. Tunnels still in operation list the current operator, although some tunnels are operated by multiple railroads.

APPENDIX | U.S. CONSTRUCTED TUNNEL ARCHIVE

RAILROAD TUNNELS* (continued)

Name	Location	State	Length (ft)†	Year Constructed or Opened	Owner‡
Siskiyou Tunnel No. 2	Douglas County	OR	432	1887	Central Oregon & Pacific Railroad
Siskiyou Tunnel No. 3	Douglas County	OR	435	1887	Central Oregon & Pacific Railroad
Siskiyou Tunnel No. 4	Douglas County	OR	325	1887	Central Oregon & Pacific Railroad
Siskiyou Tunnel No. 5	Douglas County	OR	341	1887	Central Oregon & Pacific Railroad
Siskiyou Tunnel No. 6	Douglas County	OR	516	1887	Central Oregon & Pacific Railroad
Siskiyou Tunnel No. 7	Douglas County	OR	128	1887	Central Oregon & Pacific Railroad
Siskiyou Tunnel No. 8	Douglas County	OR	2,819	1887	Central Oregon & Pacific Railroad
Siskiyou Tunnel No. 9	Josephine County	OR	2,105	1887	Central Oregon & Pacific Railroad
Siskiyou Tunnel No. 13	Jackson County	OR	3,111	1887	Central Oregon & Pacific Railroad
Siskiyou Tunnel No. 14	Jackson County	OR	1,192	1887	Central Oregon & Pacific Railroad
Siskiyou Tunnel No. 15	Jackson County	OR	258	1887	Central Oregon & Pacific Railroad
Stewart Tunnel	Green	WI	1,260	1887	Illinois Central Railroad
Belmont Tunnel	Dawes	NE	698	1888	BNSF Railway
Oak Mountain Tunnel	Shelby County	AL	5,250	1888	Norfolk Southern Railroad
Stampede Pass Tunnel	Easton	WA	9,850	1888	Great Northern Railway
Winston Tunnel	Davies County	IL	2,493	1888	Chicago Great Western Railway
Cumberland Gap Tunnel	Middlesboro	KY	3,741	1889	CSX Transportation
Jacks Mountain Tunnel	Adams County	PA	540	1889	CSX Transportation
Mount Wood Tunnel	Wheeling	WV	1,203	1889	Wheeling Bridge & Terminal Railroad
Rexford Tunnel	Harrison County	OH	550	1889	Norfolk Southern Railway
Top Mill Tunnel	Wheeling	WV	587	1889	Wheeling Bridge & Terminal Railroad
Belden Tunnel - East	Gilman	CO	396	1890	Union Pacific
Chapline Hill Tunnel	Wheeling	WV	2,460	1890	Wheeling Bridge & Terminal Railroad
Indian Springs Tunnel	Indian Springs	IN	1,106	1890	Indiana Rail Road
Montgomery Tunnel No. 1	Montgomery	VA	663	1890	Norfolk Southern Railroad
Montgomery Tunnel No. 2	Montgomery	VA	663	1890	Norfolk Southern Railroad
Northfield Tunnel	Merrickville	NY	1,639	1890	New York, Ontario & Western Railway
Pando (Mitchel) Tunnel	Camp Hale	CO	242	1890	Union Pacific
Rock Creek Tunnel	Camp Hale	CO	530	1890	Denver & Rio Grande Western Railroad Co.
Tennessee Pass Tunnel (Old)	Eagle County	CO	2,200	1890	Denver & Rio Grande Western Railroad Co.
Bee Rock Tunnel	Appalachia	VA	47	1891	Louisville & Nashville Railroad
Pigeon Mountain Tunnel	Pigeon Mountain	GA	1,724	1891	Chattanooga Sothern Railroad
St. Clair Tunnel (Old)	Port Huron (to Sarnia, Ontario)	MI	6,028	1891	Grand Trunk Railway
Boulder Tunnel	Boulder-Wicks	MT	6,145	1892	Chicago, Milwaukee, St. Paul & Pacific Railroad
Dingess Tunnel	Mingo County	WV	3,327	1892	Norfolk Southern Railway
Garrison Tunnel	Garrison	MT	1,394	1892	BNSF Railway
Haskell Pass Tunnel	Kalispell	MT	1,400	1892	Great Northern Railway
Tunnel No. 0	Applegate	MT	6,145	1892	Great Northern Railway
Tunnel No. 10	Bonners Ferry	ID	395	1892	BNSF Railway
Big South Tunnel	Gallatin	TN	955	1893	CSX Transportation
CWR Tunnel No. 1	Fort Bragg	CA	1,112	1893	California Western Railroad
Little South Tunnel	Gallatin	TN	682	1893	CSX Transportation
Natural Tunnel	Scott	VA	838	1893	Norfolk Southern Railway
SPCR Summit Tunnel (No. 2)	Santa Cruz County	CA	6,208	1893	South Pacific Coast Railroad
Coast Line - Tunnel No. 9 (Rincon)	Rincon	CA	127	1894	Southern Pacific Railroad
Edgewater Tunnel	Fairview to Edgewater	NJ	5,280	1894	Delaware, Lackawanna & Western Railroad
Phantom Canyon Tunnels Nos. 1 & 2	Fremont	CO	515	1894	Florence & Cripple Creek Railroad
Van Buren Street Tunnel	Chicago	IL	1,541	1894	City of Chicago

RAILROAD TUNNELS* *(continued)*

Name	Location	State	Length (ft)†	Year Constructed or Opened	Owner‡
Coast Line - Tunnel No. 6 (Cuesta)	Cuesta	CA	3,610	1895	Southern Pacific Railroad
Conococheague Mountain Tunnel	Perry County	PA	200	1895	Path Valley Railroad
Howard Street Tunnel	Baltimore	MD	7,392	1895	CSX Transportation
Harper's Ferry Tunnel No. 2	Harper's Ferry	MD	885	1896	CSX Transportation
Rocheport Tunnel	Howard	MO	243	1896	Missouri-Kansas-Texas Railroad
Cumberland Gap Tunnel	Cumberland Gap	KY	4,750	1897	CSX Transportation
Elberta Slant Tunnel	Eureka	UT	100	1897	Tintic Railroad
Estelle Mining Company Tunnel No. 1	Estelle	GA	600	1897	Estelle Mining Company
Estelle Mining Company Tunnel No. 2	Estelle	GA	250	1897	Estelle Mining Company
Estelle Mining Company Tunnel No. 3	Estelle	GA	200	1897	Estelle Mining Company
Estelle Mining Company Tunnel No. 4	Estelle	GA	350	1897	Estelle Mining Company
Estelle Mining Company Tunnel No. 5	Estelle	GA	500	1897	Estelle Mining Company
Estelle Mining Company Tunnel No. 6	Estelle	GA	400	1897	Estelle Mining Company
Estelle Mining Company Tunnel No. 7	Estelle	GA	400	1897	Estelle Mining Company
Falls Cut Tunnel	Fairhope	PA	517	1897	CSX Transportation
Tunnelton Tunnel	Lawrence County	IN	1,750	1898	CSX Transportation
East Dubuque Tunnel	East Dubuque	IL	850	1899	CN Rail
Franklin Tunnel (No. 3)	Martinez	CA	5,596	1899	BNSF Railway
Little Tunnel	Lawrence County	IN	1,700	1899	CSX Transportation
Tunnel Springs Tunnel	Tunnel Springs	AL	840	1899	Alabama Railroad
Valley Division Tunnel No. 1	Martinez	CA	1,230	1899	BNSF Railway
Valley Division Tunnel No. 2	Martinez	CA	300	1899	BNSF Railway
Valley Division Tunnel No. 3 (Franklin)	Martinez	CA	5,596	1899	BNSF Railway
Whitehall Tunnel	Allegheny County	PA	1,630	1899	CSX Transportation
Glen Alum Tunnel	Wharncliff	WV	1,302	Late 1800s, early 1900s	Norfolk Southern Railroad
Goodview Tunnel	Bedford	VA	986	Late 1800s, early 1900s	Norfolk Southern Railroad
Gordon Tunnel	McDowell County	WV	1,271	Late 1800s, early 1900s	Norfolk Southern Railroad
Hardy Tunnel	Bedford	VA	757	Late 1800s, early 1900s	Norfolk Southern Railroad
Roderfield Tunnels (Middle, East, & West)	McDowell County	WV	924	Late 1800s, early 1900s	Norfolk Southern Railroad
Twin Branch Tunnels 1 & 2	McDowell County	WV	3,100	Late 1800s, early 1900s	Norfolk Southern Railroad
Vaughan Tunnel	McDowell County	WV	1,113	Late 1800s, early 1900s	Norfolk Southern Railroad
Willet Hollow Tunnel	Roane County	TN	1,540	1900s	Franklin Industrial Minerals Railroad
Adams Tunnel (No. 11)	Jefferson County	OH	630	1900	Norfolk Southern Railway
Brocks Gap Tunnel	Birmingham	AL	900	1900	CSX Transportation
Cascade Tunnel (Old)	Wellington	WA	13,728	1900	Great Northern Railway
Coast Line - Tunnel No. 12	Sudden	CA	811	1900	Southern Pacific Railroad
Coen Tunnel (No. 10)	Jefferson County	OH	856	1900	Norfolk Southern Railway
Everett Tunnel (No. 16)	Everett	WA	2,370	1900	BNSF Railway
Fellows Tunnel (No. 12)	Jefferson County	OH	1,038	1900	Norfolk Southern Railway
Martin Creek Tunnel	Chelan County	WA	1,492	1900	Great Northern Railway

* This list is by no means exhaustive, but it tries to capture the most significant railroad tunnels in the United States with respect to age, length, and so forth.
† Length is usually the as-constructed length. Some tunnel lengths are approximate based on the accuracy of the source information.
‡ Tunnel ownership has changed multiple times. Where tunnels have been abandoned, the listing is typically the original constructor.
   Tunnels still in operation list the current operator, although some tunnels are operated by multiple railroads.

APPENDIX | U.S. CONSTRUCTED TUNNEL ARCHIVE

RAILROAD TUNNELS* (continued)

Name	Location	State	Length (ft)†	Year Constructed or Opened	Owner‡
Raton Tunnel No. 2	Colfax County	NM	2,787	1900	BNSF Railway
Seaboard Air Line Tunnel	Richland	SC	600	1900	Seaboard Airline Railway
Sherril Cove Tunnel	Blue Ridge Parkway	NC	590	1900	Federal Highway Administration for National Park Service
Spruce Creek Tunnel 2	Huntingdon County	PA	1,075	1900	Norfolk Southern Railroad
Warren Tunnel (No 15)	Jefferson County	OH	384	1900	Norfolk Southern Railway
Willow Valley Tunnel	Martin County	IN	1,160	1900	CSX Transportation
Aspen Tunnel	Aspen	WY	5,941	1901	Union Pacific
CS&CC Tunnel No. 1	Cripple Creek	CO	325	1901	Colorado Springs & Cripple Creek Railway
CS&CC Tunnel No. 2	Cripple Creek	CO	185	1901	Colorado Springs & Cripple Creek Railway
CS&CC Tunnel No. 3	Cripple Creek	CO	278	1901	Colorado Springs & Cripple Creek Railway
CS&CC Tunnel No. 4	Cripple Creek	CO	521	1901	Colorado Springs & Cripple Creek Railway
CS&CC Tunnel No. 5	Cripple Creek	CO	263	1901	Colorado Springs & Cripple Creek Railway
CS&CC Tunnel No. 6	Cripple Creek	CO	200	1901	Colorado Springs & Cripple Creek Railway
CS&CC Tunnel No. 7	Cripple Creek	CO	270	1901	Colorado Springs & Cripple Creek Railway
CS&CC Tunnel No. 8	Cripple Creek	CO	170	1901	Colorado Springs & Cripple Creek Railway
CS&CC Tunnel No. 9	Cripple Creek	CO	268	1901	Colorado Springs & Cripple Creek Railway
Hermosa Tunnel No. 1	Albany County	WY	1,650	1901	Union Pacific
Moulton Falls Tunnel	Clark County	WA	330	1901	Chelatchie Prairie Railroad
Tunnel No. 0	Multnomah County	OR	471	1901	Portland & Western
Tunnel No. 1	Multnomah County	OR	4,105	1901	Portland & Western
Welch Tunnel	McDowell County	WV	1,335	1901	Norfolk Southern Railroad
Coal & Iron Tunnel No. 1	Glady	WV	1,800	1902	Coal & Iron Railway
Coal & Iron Tunnel No. 2	Glady	WV	1,000	1902	Coal & Iron Railway
Deere Creek Tunnel	Clearfield County	PA	1,080	1902	New York Central Railroad
Fulton Tunnel	Clearfield County	PA	2,695	1902	New York Central Railroad
Karthaus Tunnel	Clearfield County	PA	1,440	1902	New York Central Railroad
Mount Airy Tunnel	Frederick County	MD	2,757	1902	CSX Transportation
Shawsville Tunnel	Clearfield County	PA	1,725	1902	New York Central Railroad
Tunnel No. 3	Astoria	OR	193	1902	Portland & Western
Union Dam Tunnel	Milford Mill	MD	810	1902	CSX Transportation
Antler Tunnel No. 1	McDowell County	WV	600	1903	Norfolk Southern Railroad
Antler Tunnel No. 2	McDowell County	WV	613	1903	Norfolk Southern Railroad
Argyle Tunnel (No. 3)	Osage County	MO	1,223	1903	Chicago, Rock Island & Pacific Railroad
Carlin Rail Tunnel	Carlin	NV	1,584	1903	Union Pacific
Clinton Tunnel	Clinton	MA	1,080	1903	Central Massachusetts Railroad
Cooks Spring Tunnel	St. Clair County	AL	740	1903	Norfolk Southern Railroad
Craighead Tunnel	Washington County	PA	1,125	1903	Pittsburgh & West Virginia Railway
CWR Tunnel No. 3	Casper	CA	1,000	1903	California Western Railroad
Eugene Tunnel	Cole County	MO	1,667	1903	Chicago, Rock Island & Pacific Railroad
Freeburg Tunnel	Osage County	MO	700	1903	Chicago, Rock Island & Pacific Railroad
Hanna Tunnel (No. 17)	Harrison County	OH	1,505	1903	Norfolk Southern Railway
Hardwick Tunnel	St. Clair County	AL	960	1903	Alabama & Tennessee River Railroad
Hatfield Tunnel No. 1	Pike County	KY	990	1903	Norfolk Southern Railroad
Hatfield Tunnel No. 2	Pike County	KY	997	1903	Norfolk Southern Railroad
Ilchester Tunnel	Ellicott City	MD	1,405	1903	CSX Transportation
Moffat Road Tunnel No. 12	Boulder County	CO	429	1903	Union Pacific
Moffat Road Tunnel No. 13	Boulder County	CO	312	1903	Union Pacific
Moffat Road Tunnel No. 14	Boulder County	CO	434	1903	Union Pacific

RAILROAD TUNNELS* (continued)

Name	Location	State	Length (ft)†	Year Constructed or Opened	Owner‡
Moffat Road Tunnel No. 15	Jefferson County	CO	444	1903	Union Pacific
Moffat Road Tunnel No. 16	Jefferson County	CO	698	1903	Union Pacific
Moffat Road Tunnel No. 26	Gilpin County	CO	295	1903	Union Pacific
Penobscot Mountain Tunnel	Nuangola	PA	2,684	1903	Wilkes-Barre & Hazleton Railway
Roper Tunnel	Trussville	AL	800	1903	Alabama & Tennessee River Railroad
Tunnel No. 1	Eureka County	NV	874	1903	Union Pacific
Tunnel No. 2	Elko County	NV	1,887	1903	Union Pacific
Tunnel No. 3	Elko County	NV	2,473	1903	Union Pacific
Tunnel No. 4	Elko County	NV	3,917	1903	Union Pacific
Tunnel No. 18	Boulder County	CO	238	1903	Union Pacific
Tunnel No. 20	Boulder County	CO	460	1903	Union Pacific
Tunnel No. 21	Boulder County	CO	667	1903	Union Pacific
Tunnel No. 22	Boulder County	CO	180	1903	Union Pacific
Tunnel No. 23	Boulder County	CO	1,553	1903	Union Pacific
Tunnel No. 24	Boulder County	CO	812	1903	Union Pacific
Unionville Tunnel	Monroe County	IN	500	1903	Indiana Rail Road
Vale Tunnel	Kansas City	MO	446	1903	Chicago, Rock Island & Pacific Railroad
Wabash Tunnel	Pittsburgh	PA	3,650	1903	Pittsburgh & West Virginia Railway
Big Sandy Tunnel No. 1	Wayne County	WV	2,627	1904	Norfolk Southern Railroad
Big Sandy Tunnel No. 3	Wayne County	WV	1,840	1904	Norfolk Southern Railroad
Big Sandy Tunnel No. 4	Wayne County	WV	2,060	1904	Norfolk Southern Railroad
Big Sandy Tunnel No. 5	Wayne County	WV	1,190	1904	Norfolk Southern Railroad
Coast Line - Tunnel No. 28 (Chatsworth)	Chatsworth	CA	599	1904	Southern Pacific Railroad
Cooper Tunnel	Cooper	WV	698	1904	Norfolk Southern Railroad
Conway Tunnel	Conway	AR	1,250	1904	Union Pacific
Cotter Tunnel	Marion County	AR	1,034	1904	Missouri & Northern Arkansas Railroad
Crown Avenue Tunnel	Scranton	PA	4,747	1904	Lackawanna & Wyoming Valley Railroad
Divide Tunnel	Paudling County	GA	730	1904	Seaboard Airline Railway
Gallitzin Tunnel	Gallitzin	PA	3,612	1904	Pennsylvania Railroad
Hemphill Tunnel No. 1	Capel	WV	864	1904	Norfolk Southern Railroad
Hemphill Tunnel No. 2	Capel	WV	1,142	1904	Norfolk Southern Railroad
Indigo Tunnel	Little Orleans	MD	4,350	1904	Western Maryland Railway
Knobley Tunnel	Mineral County	WV	1,448	1904	Western Maryland Railway
Pyatt Tunnel	Marion County	AR	660	1904	Missouri & Northern Arkansas Railroad
Simi Valley Div. Santa Susana Tunnel (No. 26)	Chatsworth	CA	7,369	1904	Union Pacific
Simi Valley Div. Chatsworh Tunnel (No. 27)	Chatsworth	CA	924	1904	Union Pacific
Spellacy Tunnel (No. 16)	Harrison County	OH	856	1904	Wheeling & Lake Erie Railway Company
Williamson Tunnel	Mingo	WV	678	1904	Norfolk Southern Railroad
Crest Tunnel	Boone County	AR	3,500	1905	Union Pacific
Cricket Tunnel	Boone County	AR	2,657	1905	Missouri & Northern Arkansas Railroad
Dorsey Tunnel	Ellicott City	MD	1,022	1905	CSX Transportation
Great Northern Tunnel, King Street Station Tunnel (No. 17)	Seattle	WA	5,142	1905	BNSF Railway
Guernsey Tunnel No. 1	Guernsey	WY	3,400	1905	BNSF Railway
Guernsey Tunnel No. 2	Guernsey	WY	1,900	1905	BNSF Railway

* This list is by no means exhaustive, but it tries to capture the most significant railroad tunnels in the United States with respect to age, length, and so forth.
† Length is usually the as-constructed length. Some tunnel lengths are approximate based on the accuracy of the source information.
‡ Tunnel ownership has changed multiple times. Where tunnels have been abandoned, the listing is typically the original constructor.
   Tunnels still in operation list the current operator, although some tunnels are operated by multiple railroads.

RAILROAD TUNNELS* (continued)

Name	Location	State	Length (ft)†	Year Constructed or Opened	Owner‡
Lookout Mountain Tunnel	Chattanooga	TN	250	1905	Norfolk Southern
Nay Aug Tunnel - South	Dunmore	PA	755	1905	Delaware, Lackawanna & Western Railroad
Ridgetop (Baker) Tunnel	Robertson County	TN	4,621	1905	CSX Transportation
West Vivian Tunnel	Kimball	WV	680	1905	Norfolk Southern Railroad
Chicago Freight Tunnels	Chicago	IL	316,800	1906	Chicago Tunnel Company
Davis Tunnel	Ellicott City	MD	497	1906	CSX Transportation
First Street Tunnel	Washington	DC	4,033	1906	Amtrak
Gleason Canyon Tunnel	White Pine County	NV	310	1906	BHP Nevada Railroad
Kessler Tunnel	Allegheny County	MD	1,843	1906	Western Maryland Railway
Knobbley Mountain Tunnel	Mineral County	WV	4,106	1906	Baltimore & Ohio Railroad
Quemahoning Tunnel	Somerset County	PA	5,919	1906	Pittsburgh, Westmoreland & Somerset Railroad
Shoshone Tunnel - East	Glenwood Springs	CO	310	1906	Union Pacific
Shoshone Tunnel - West	Glenwood Springs	CO	240	1906	Union Pacific
Stickpile Tunnel	Allegheny County	MD	1,706	1906	Western Maryland Railway
Tunnel No. 1	Fremont	CA	4,317	1906	Union Pacific
Tunnel No. 1	Seward	AK	714	1906	Alaska Railroad Corporation
Tunnel No. 3	San Francisco	CA	2,345	1906	Union Pacific
Tunnel No. 4	San Francisco	CA	3,547	1906	Union Pacific
Tunnel No. 32	Spring Garden	CA	7,334	1906	Union Pacific
Western Pacific Railroad Tunnels	Niles Canyon	CA	4,950	1906	Union Pacific
Alligator Rock Tunnel (No. 6)	Hood	WA	657	1907	BNSF Railway
Blueslide Tunnel	Pende Oreille	WA	1,093	1907	Pend Oreille Valley Railroad
Bow Ridge Tunnel (New)	Westmoreland County	PA	630	1907	Pennsylvania Railroad
Burton Tunnel	Orange County	IN	2,127	1907	Southern Railway
Clearfield Tunnel 1	Cranberry	PA	967	1907	Franklin & Clearfield Railroad
Dotsero Cutoff Tunnel No. 36	Grand County	CO	229	1907	Union Pacific
Dotsero Cutoff Tunnel No. 37	Grand County	CO	134	1907	Union Pacific
Dotsero Cutoff Tunnel No. 38	Grand County	CO	109	1907	Union Pacific
Dotsero Cutoff Tunnel No. 39	Grand County	CO	294	1907	Union Pacific
Dotsero Cutoff Tunnel No. 40	Grand County	CO	63	1907	Union Pacific
Eggleston Tunnel No. 1	Eggleston	VA	1,195	1907	Norfolk Southern Railroad
Eggleston Tunnel No. 2	Eggleston	VA	925	1907	Norfolk Southern Railroad
Fallbridge Subdivision Tunnel No. 1	Cape Horn	WA	2,382	1907	BNSF Railway
Fallbridge Subdivision Tunnel No. 2	Drano	WA	122	1907	BNSF Railway
Fallbridge Subdivision Tunnel No. 3	Blum	WA	416	1907	BNSF Railway
Fallbridge Subdivision Tunnel No. 4	Severson	WA	267	1907	BNSF Railway
Fallbridge Subdivision Tunnel No. 5	Owl Rock	WA	394	1907	BNSF Railway
Fallbridge Subdivision Tunnel No. 6	Alligator Rock	WA	657	1907	BNSF Railway
Fallbridge Subdivision Tunnel No. 7	Hewitt	WA	966	1907	BNSF Railway
Fallbridge Subdivision Tunnel No. 8	Anderson	WA	755	1907	BNSF Railway
Fallbridge Subdivision Tunnel No. 9	Woldson	WA	392	1907	BNSF Railway
Fallbridge Subdivision Tunnel No. 10	Jackson County	WA	575	1907	BNSF Railway
Fallbridge Subdivision Tunnel No. 11	Lyle	WA	648	1907	BNSF Railway
Fallbridge Subdivision Tunnel No. 12	Wishram	WA	385	1907	BNSF Railway
Pembroke Tunnel	Giles County	VA	299	1907	Norfolk Southern Railroad
Pocahontas Tunnel No. 1	Pocahontas	VA	201	1907	Norfolk & Western Railroad
Pocahontas Tunnel No. 2	Pocahontas	VA	366	1907	Norfolk & Western Railroad
Rover Tunnel	Rover	ID	507	1907	St. Maries River Railroad

RAILROAD TUNNELS* (continued)

Name	Location	State	Length (ft)†	Year Constructed or Opened	Owner‡
Solway Tunnel	Knox County	TN	2,300	1907	CSX Transportation
Tunnel No. 1 - Mays Hill Tunnel	Cranberry	PA	967	1907	Franklin & Clearfield Railroad
Tunnel No. 2	Clarion	PA	2,176	1907	Franklin & Clearfield Railroad
Tunnel No. 3	Clarion	PA	1,726	1907	Franklin & Clearfield Railroad
Tunnel No. 13	Plymouth	WA	700	1907	BNSF Railway
Vail Tunnel	Ione	WA	810	1907	Pend Oreille Valley Railroad
Bergen Hill Tunnel (No. 1) (North)	New York City	NJ	5,920	1908	Pennsylvania Railroad
Bergen Hill Tunnel (No. 2) (South)	Jersey City	NJ	4,280	1908	New Jersey Transit
Joralemon Street Tunnel	New York City	NY	2,170	1908	New York City Transit Authority
Milwaukee Road Tunnel No. 2 (Red)	Ringling	MT	403	1908	Chicago, Milwaukee, St. Paul & Pacific Railroad
Milwaukee Road Tunnel No. 3 (Canyon)	Townsend	MT	180	1908	Chicago, Milwaukee, St. Paul & Pacific Railroad
Milwaukee Road Tunnel No. 4 (Eagle Nest)	Francis	MT	370	1908	Chicago, Milwaukee, St. Paul & Pacific Railroad
Milwaukee Road Tunnel No. 5 (Josephine)	Francis	MT	334	1908	Chicago, Milwaukee, St. Paul & Pacific Railroad
Milwaukee Road Tunnel No. 6 (Deer Park No. 1)	Deer Park	MT	493	1908	Chicago, Milwaukee, St. Paul & Pacific Railroad
Milwaukee Road Tunnel No. 7 (Deer Park No. 2)	Deer Park	MT	222	1908	Chicago, Milwaukee, St. Paul & Pacific Railroad
Milwaukee Road Tunnel No. 8 (Deer Park No. 3)	Deer Park	MT	163	1908	Chicago, Milwaukee, St. Paul & Pacific Railroad
Milwaukee Road Tunnel No. 9 (Lombard)	Lombard	MT	207	1908	Chicago, Milwaukee, St. Paul & Pacific Railroad
Milwaukee Road Tunnel No. 10 (Fish Creek)	Butte	MT	426	1908	Chicago, Milwaukee, St. Paul & Pacific Railroad
Milwaukee Road Tunnel No. 11 (Pipestone Pass)	Butte	MT	2,290	1908	Chicago, Milwaukee, St. Paul & Pacific Railroad
Milwaukee Road Tunnel No. 12 (Blacktail No. 1)	Butte	MT	2,490	1908	Chicago, Milwaukee, St. Paul & Pacific Railroad
Milwaukee Road Tunnel No. 13 (Blacktail No. 2)	Butte	MT	560	1908	Chicago, Milwaukee, St. Paul & Pacific Railroad
Milwaukee Road Tunnel No. 14 (Garrison)	Garrison	MT	1,975	1908	Chicago, Milwaukee, St. Paul & Pacific Railroad
Milwaukee Road Tunnel No. 15 (Nimrod)	Clinrton	MT	1,157	1908	Chicago, Milwaukee, St. Paul & Pacific Railroad
Milwaukee Road Tunnel No. 17 (Nine Mile)	Soudan	MT	170	1908	Chicago, Milwaukee, St. Paul & Pacific Railroad
Milwaukee Road Tunnel No. 18 (Cyr)	Alberton	MT	285	1908	Chicago, Milwaukee, St. Paul & Pacific Railroad
Milwaukee Road Tunnel No. 20 (St. Paul Pass)	Mineral County	MT	8,777	1908	Chicago, Milwaukee, St. Paul & Pacific Railroad
Milwaukee Road Tunnel No. 21 (Dry Creek)	Roland	MT	790	1908	Chicago, Milwaukee, St. Paul & Pacific Railroad
Milwaukee Road Tunnel No. 22 Moss Creek)	Roland	MT	1,516	1908	Chicago, Milwaukee, St. Paul & Pacific Railroad
Milwaukee Road Tunnel No. 23 (Small Creek 1)	Roland	MT	279	1908	Chicago, Milwaukee, St. Paul & Pacific Railroad
Milwaukee Road Tunnel No. 24 (Small Creek 2)	Roland	MT	377	1908	Chicago, Milwaukee, St. Paul & Pacific Railroad
Milwaukee Road Tunnel No. 25 (Loop 1)	Adair	MT	966	1908	Chicago, Milwaukee, St. Paul & Pacific Railroad
Milwaukee Road Tunnel No. 26 (Loop 2)	Adair	MT	683	1908	Chicago, Milwaukee, St. Paul & Pacific Railroad
Milwaukee Road Tunnel No. 27 (Clear Creek 1)	Falcon	MT	470	1908	Chicago, Milwaukee, St. Paul & Pacific Railroad
Milwaukee Road Tunnel No. 28 (Clear Creek 2)	Falcon	MT	178	1908	Chicago, Milwaukee, St. Paul & Pacific Railroad
Milwaukee Road Tunnel No. 29 (Deer Creek 2)	Kyle	MT	217	1908	Chicago, Milwaukee, St. Paul & Pacific Railroad
Milwaukee Road Tunnel No. 30 (Deer Creek 2)	Kyle	MT	221	1908	Chicago, Milwaukee, St. Paul & Pacific Railroad
Milwaukee Road Tunnel No. 31 (Glade 1)	Kyle	MT	332	1908	Chicago, Milwaukee, St. Paul & Pacific Railroad
Milwaukee Road Tunnel No. 32 (Glade 2)	Kyle	MT	638	1908	Chicago, Milwaukee, St. Paul & Pacific Railroad
Milwaukee Road Tunnel No. 33 (Kyle)	Kyle	MT	462	1908	Chicago, Milwaukee, St. Paul & Pacific Railroad
Milwaukee Road Tunnel No. 34 (Stetson 1)	Avery	MT	462	1908	Chicago, Milwaukee, St. Paul & Pacific Railroad
Milwaukee Road Tunnel No. 35 (Stetson 2)	Avery	MT	416	1908	Chicago, Milwaukee, St. Paul & Pacific Railroad
Milwaukee Road Tunnel No. 36 (Stetson 3)	Avery	MT	552	1908	Chicago, Milwaukee, St. Paul & Pacific Railroad
Milwaukee Road Tunnel No. 37 (Herrick)	Pocono	MT	515	1908	Chicago, Milwaukee, St. Paul & Pacific Railroad
Milwaukee Road Tunnel No. 40 (Benewah)	Ramsdell	MT	363	1908	Chicago, Milwaukee, St. Paul & Pacific Railroad

* This list is by no means exhaustive, but it tries to capture the most significant railroad tunnels in the United States with respect to age, length, and so forth.
† Length is usually the as-constructed length. Some tunnel lengths are approximate based on the accuracy of the source information.
‡ Tunnel ownership has changed multiple times. Where tunnels have been abandoned, the listing is typically the original constructor.
  Tunnels still in operation list the current operator, although some tunnels are operated by multiple railroads.

RAILROAD TUNNELS* (continued)

Name	Location	State	Length (ft)†	Year Constructed or Opened	Owner‡
Milwaukee Road Tunnel No. 41 (Watte)	Sorrento	MT	2,559	1908	Chicago, Milwaukee, St. Paul & Pacific Railroad
Milwaukee Road Tunnel No. 43 (Rock Lake)	Rock Lake	MT	756	1908	Chicago, Milwaukee, St. Paul & Pacific Railroad
Milwaukee Road Tunnel No. 44	Rock Lake	MT	704	1908	Chicago, Milwaukee, St. Paul & Pacific Railroad
Milwaukee Road Tunnel No. 45 (Boylston)	Yakima	WA	1,973	1908	Chicago, Milwaukee, St. Paul & Pacific Railroad
Otisville Tunnel	Orange County	NY	5,314	1908	Norfolk Southern Railroad
Tunnel No. 2	Pinnacle	MT	2,230	1908	BNSF Railway
Tunnel No. 3	Pinnacle	MT	510	1908	BNSF Railway
Tunnel No. 5A (Vista)	Whitefish	MT	840	1908	BNSF Railway
Tunnel No. 19	Mineral County	MT	474	1908	Chicago, Milwaukee, St. Paul & Pacific Railroad
Tunnel No. 21	Shoshone County	ID	790	1908	Chicago, Milwaukee, St. Paul & Pacific Railroad
Tunnel No. 22	Shoshone County	ID	1,516	1908	Chicago, Milwaukee, St. Paul & Pacific Railroad
Tunnel No. 23	Shoshone County	ID	279	1908	Chicago, Milwaukee, St. Paul & Pacific Railroad
Tunnel No. 24	Shoshone County	ID	377	1908	Chicago, Milwaukee, St. Paul & Pacific Railroad
Tunnel No. 25	Shoshone County	ID	996	1908	Chicago, Milwaukee, St. Paul & Pacific Railroad
Tunnel No. 26	Shoshone County	ID	683	1908	Chicago, Milwaukee, St. Paul & Pacific Railroad
Tunnel No. 27	Shoshone County	ID	470	1908	Chicago, Milwaukee, St. Paul & Pacific Railroad
Tunnel No. 28	Shoshone County	ID	178	1908	Chicago, Milwaukee, St. Paul & Pacific Railroad
Tunnel No. 29	Shoshone County	ID	217	1908	Chicago, Milwaukee, St. Paul & Pacific Railroad
Clearfield Tunnel 2	Clarion	PA	2,176	1909	Franklin & Clearfield Railroad
Clearfield Tunnel 3	Clarion	PA	1,726	1909	Franklin & Clearfield Railroad
Coosa Tunnel	Shelby County	AL	2,460	1909	Norfolk Southern Railroad
East River Tunnels (four tunnels)	New York City	NY	15,796	1909	Amtrak/Long Island Rail Road/Metro North Railroad
East Side Tunnel	Providence County	RI	5,080	1909	Providence & Worcester Railroad
Feather River Tunnel No. 1	Plumas County	CA	4,317	1909	Union Pacific
Feather River Tunnel No. 2	Plumas County	CA	408	1909	Union Pacific
Feather River Tunnel No. 4	Plumas County	CA	2,410	1909	Union Pacific
Feather River Tunnel No. 5	Plumas County	CA	2,922	1909	Union Pacific
Feather River Tunnel No. 6	Plumas County	CA	2,583	1909	Union Pacific
Feather River Tunnel No. 7	Plumas County	CA	4,406	1909	Union Pacific
Feather River Tunnel No. 8	Plumas County	CA	8,856	1909	Union Pacific
Feather River Tunnel No. 15	Plumas County	CA	3,115	1909	Union Pacific
Feather River Tunnel No. 17	Plumas County	CA	325	1909	Union Pacific
Feather River Tunnel No. 18	Plumas County	CA	162	1909	Union Pacific
Feather River Tunnel No. 19	Plumas County	CA	172	1909	Union Pacific
Feather River Tunnel No. 20	Plumas County	CA	294	1909	Union Pacific
Feather River Tunnel No. 21	Plumas County	CA	408	1909	Union Pacific
Feather River Tunnel No. 22	Plumas County	CA	306	1909	Union Pacific
Feather River Tunnel No. 23	Plumas County	CA	1,260	1909	Union Pacific
Feather River Tunnel No. 24	Plumas County	CA	617	1909	Union Pacific
Feather River Tunnel No. 25	Plumas County	CA	183	1909	Union Pacific
Feather River Tunnel No. 27	Plumas County	CA	365	1909	Union Pacific
Feather River Tunnel No. 28	Plumas County	CA	609	1909	Union Pacific
Feather River Tunnel No. 29	Plumas County	CA	588	1909	Union Pacific
Feather River Tunnel No. 30	Plumas County	CA	538	1909	Union Pacific
Feather River Tunnel No. 31	Plumas County	CA	687	1909	Union Pacific
Feather River Tunnel No. 32 (Spring Garden)	Plumas County	CA	7,344	1909	Union Pacific
Feather River Tunnel No. 33	Plumas County	CA	1,266	1909	Union Pacific

THE HISTORY OF TUNNELING IN THE UNITED STATES

RAILROAD TUNNELS* (continued)

Name	Location	State	Length (ft)†	Year Constructed or Opened	Owner‡
Feather River Tunnel No. 34	Plumas County	CA	305	1909	Union Pacific
Feather River Tunnel No. 36	Plumas County	CA	765	1909	Union Pacific
Feather River Tunnel No. 37	Plumas County	CA	6,002	1909	Union Pacific
Grant's Bend Tunnel (East)	Kenton County	KY	2,250	1909	CSX Transportation
Hogan's Tunnel, Shafter Subdivision	Wells	NV	5,676	1909	Union Pacific
Huddleston Tunnel	Bedford	VA	560	1909	Norfolk Southern Railroad
Milwaukee Road Tunnel No. 16 (Beavertail)	Clinton	MT	924	1909	Chicago, Milwaukee, St. Paul & Pacific Railroad
Milwaulkee Road Tunnel No. 16.5 (Bonner)	Missoula County	MT	636	1909	Chicago, Milwaukee, St. Paul & Pacific Railroad
Milwaukee Road Tunnel No. 19 (Dominion Creek)	Mineral County	MT	422	1909	Chicago, Milwaukee, St. Paul & Pacific Railroad
Milwaukee Road Tunnel No. 46 (Thorp #1)	Thorp	WA	496	1909	Chicago, Milwaukee, St. Paul & Pacific Railroad
Milwaukee Road Tunnel No. 47 (Thorp #2)	Thorp	WA	1,239	1909	Chicago, Milwaukee, St. Paul & Pacific Railroad
Milwaukee Road Tunnel No. 48 (Easton)	Easton	WA	203	1909	Chicago, Milwaukee, St. Paul & Pacific Railroad
Milwaukee Road Tunnel No. 49 (Whittier)	Whittier	WA	527	1909	Chicago, Milwaukee, St. Paul & Pacific Railroad
Natron Cutoff Tunnel No. 17	Lane County	OR	2,076	1909	Union Pacific
Natron Cutoff Tunnel No. 18	Lane County	OR	1,145	1909	Union Pacific
Natron Cutoff Tunnel No. 21	Lane County	OR	461	1909	Union Pacific
Park City Tunnel	Barren County	KY	400	1910s	CSX Transportation
Catasauqua Tunnel	Catasauqua	PA	735	1910	Crane Railroad
Dalecarlia Tunnel	Bookmont	MD	340	1910	Baltimore & Ohio Railroad
Ely's Peak Tunnel	St. Louis County	MN	450	1910	Duluth, Winnipeg & Pacific Railway
Michigan Central Railway Tunnel	Detroit	MI	8,450	1910	Canadian Pacific Railway
Natron Cutoff Tunnel No. 22	Westfir	OR	1,199	1910	Union Pacific
Natron Cutoff Tunnel No. 23	Westfir	OR	654	1910	Union Pacific
North River Tunnels (Twin)	Weehawken to Manhattan	NJ	29,150	1910	Amtrak / New Jersey Transit
Oregon Trunk Subdivision Tunnel No. 2 (Oakbrook)	Sherar	OR	654	1910	BNSF Railway
P&SW Tunnel	Columbia County	OR	710	1910	Portland & Southwestern Railroad
Tunnel No. 4	West Glacier	MT	227	1910	BNSF Railway
Big Savage Tunnel	Wellersburg	PA	3,295	1911	Western Maryland Railway
Borden Tunnel	Frostburg	MD	957	1911	Western Maryland Railway
Brush Tunnel	Corriganville	MD	914	1911	Western Maryland Railway
Cornelius Pass Tunnel	Portland	OR	4,100	1911	State of Oregon
CWR Tunnel No. 2	Willits	CA	795	1911	California Western Railroad
Huger Main Tunnels 1 & 2	Roanoke County	WV	724	1911	Norfolk Southern Railroad
Nada Tunnel	Powell County	KY	900	1911	Dana Lumber Company
Oregon Trunk Subdivision Tunnel No. 1 (Wishram)	Moody	OR	818	1911	BNSF Railway
Oregon Trunk Subdivision Tunnel No. 3 (Dant)	Dant	OR	528	1911	BNSF Railway
Oregon Trunk Subdivision Tunnel No. 4 (Davidson)	Kaskela	OR	586	1911	BNSF Railway
Oregon Trunk Subdivision Tunnel No. 5 (Gateway)	Gateway	OR	544	1911	BNSF Railway
Peninsular Tunnel (Penn)	Portland	OR	5,425	1911	Union Pacific
Redwood Peak Tunnel	Alameda County	CA	3,400	1911	Sacramento Northern Railway
Roseville Tunnel	Sussex County	NJ	1,024	1911	Delaware, Lackawanna & Western Railroad
Wayne Tunnel	Cascade County	MT	560	1911	BNSF Railway
Bellingham Subdivision Tunnel No. 18	Skagit County	WA	1,123	1912	BNSF Railway
Bellingham Subdivision Tunnel No. 19	Whatcom County	WA	141	1912	BNSF Railway

* This list is by no means exhaustive, but it tries to capture the most significant railroad tunnels in the United States with respect to age, length, and so forth.
† Length is usually the as-constructed length. Some tunnel lengths are approximate based on the accuracy of the source information.
‡ Tunnel ownership has changed multiple times. Where tunnels have been abandoned, the listing is typically the original constructor.
   Tunnels still in operation list the current operator, although some tunnels are operated by multiple railroads.

RAILROAD TUNNELS* (continued)

Name	Location	State	Length (ft)†	Year Constructed or Opened	Owner‡
Bellingham Subdivision Tunnel No. 20	Whatcom County	WA	327	1912	BNSF Railway
Bellingham Subdivision Tunnel No. 21	Whatcom County	WA	713	1912	BNSF Railway
Cartwright Tunnel	McKenzie County	ND	1,458	1912	Great Northern Railway
Donner Pass Tunnel 15	Placer County	CA	1,904	1912	Union Pacific
Donner Pass Tunnel 16	Placer County	CA	777	1912	Union Pacific
Donner Pass Tunnel 17	Placer County	CA	1,648	1912	Union Pacific
Donner Pass Tunnel 18	Placer County	CA	1,000	1912	Union Pacific
Donner Pass Tunnel 19	Placer County	CA	377	1912	Union Pacific
Donner Pass Tunnel 20	Placer County	CA	1,248	1912	Union Pacific
Donner Pass Tunnel 21	Placer County	CA	1,210	1912	Union Pacific
Donner Pass Tunnel 22	Placer County	CA	984	1912	Union Pacific
Donner Pass Tunnel 23	Placer County	CA	843	1912	Union Pacific
Donner Pass Tunnel 24	Placer County	CA	300	1912	Union Pacific
Donner Pass Tunnel 25	Placer County	CA	771	1912	Union Pacific
Donner Pass Tunnel 26	Placer County	CA	149	1912	Union Pacific
Donner Pass Tunnel 27	Placer County	CA	855	1912	Union Pacific
Donner Pass Tunnel 28	Placer County	CA	3,208	1912	Union Pacific
Donner Pass Tunnel 29	Placer County	CA	1,009	1912	Union Pacific
Donner Pass Tunnel 30	Placer County	CA	780	1912	Union Pacific
Donner Pass Tunnel 31	Placer County	CA	443	1912	Union Pacific
Donner Pass Tunnel 32	Placer County	CA	769	1912	Union Pacific
Donner Pass Tunnel 33	Placer County	CA	1,331	1912	Union Pacific
Grant's Bend Tunnel (New)	Kenton County	KY	2,250	1912	CSX Transportation
Milwaukee Road Tunnel No. 4 (Eagle Nest)	Townsend	MT	387	1912	Chicago, Milwaukee, St. Paul & Pacific Railroad
Ruston Tunnel (No. 1)	Tacoma	WA	325	1912	BNSF Railway
Sand Patch Tunnel No. 2	Somerset County	PA	4,475	1912	CSX Transportation
Tunnel 10.2	Boundary County	ID	479	1912	BNSF Railway
Verde Canyon Railroad Tunnel	Yavapai County	AZ	680	1912	Clarkdale Arizona Central Railroad
Alray Tunnels Nos. 1 & 2	San Bernardino County	CA	847	1913	BNSF Railway
Cambridge Tunnel	Guernsey County	OH	690	1913	CSX Transportation
Coal Dale Tunnel	McDowell County	WV	3,015	1913	Norfolk & Western Railroad
Donner Pass Tunnel 34	Placer County	CA	410	1913	Union Pacific
Fort Mason Tunnel	San Francisco	CA	1,500	1913	State Belt Railway of California
Milwaukee Road Tunnel No. 1 (Sage Creek)	Hoosac	MT	2,014	1913	Chicago, Milwaukee, St. Paul & Pacific Railroad
Milwaukee Road Tunnel No. 3 (Lacey)	Belt Creek	MT	481	1913	Chicago, Milwaukee, St. Paul & Pacific Railroad
Milwaukee Road Tunnel No. 4 (Amphitheater Tunnel)	Salem	MT	720	1913	Chicago, Milwaukee, St. Paul & Pacific Railroad
Milwaukee Road Tunnel No. 5 (Belt Creek)	Waltham	MT	690	1913	Chicago, Milwaukee, St. Paul & Pacific Railroad
National Tunnel	Doundary County	PA	623	1913	Montour Railroad
Nelson Bennett Tunnel (No. 2)	Tacoma	WA	4,391	1913	BNSF Railway
Old Tunnel	Kendall County	TX	920	1913	Fredericksburg & Northern Railway
Peacock Tunnel	Bishop	PA	600	1913	Montour Railroad
Tunnel 10.1	San Bernardino County	ID	610	1913	BNSF Railway
Big Four Tunnels 1 & 2	Huger	WV	821	1914	Norfolk Southern Railroad
Carothers Tunnel	Paw Paw	WV	996	1914	CSX Transportation
Corliss Street Tunnel	Pittsburgh	PA	420	1914	Pittsburgh Railways
Graham Tunnel	Allegheny County	MD	1,592	1914	CSX Transportation
Great Falls Line - Tunnel No. 6 (Red Coulee)	Cascade County	MT	780	1914	Chicago, Milwaukee, St. Paul & Pacific Railroad

RAILROAD TUNNELS* *(continued)*

Name	Location	State	Length (ft)†	Year Constructed or Opened	Owner‡
Greer Tunnel	Thompsonville	PA	235	1914	Montour Railroad
Island Mountain Tunnel	Trinity County	CA	4,313	1914	Southern Pacific Railroad
Meredith Tunnel	Marion County	WV	1,200	1914	Baltimore & Ohio Railroad
Milwaukee Road Tunnel No. 50 (Snoqualmie)	Snoqualmie Pass	WA	12,144	1914	Chicago, Milwaukee, St. Paul & Pacific Railroad
Randolph Tunnel	Hansrote	WV	1,015	1914	CSX Transportation
Stuart Tunnel	Hansrote	WV	3,350	1914	CSX Transportation
Clinchfield Line - 1st Rocky Tunnel	Marion County	NC	716	1915	CSX Transportation
Clinchfield Line - 2nd Rocky Tunnel	Marion County	NC	757	1915	CSX Transportation
Clinchfield Line - 4th Rocky Tunnel	Marion County	NC	1,688	1915	CSX Transportation
Clinchfield Line - 1st Washburn Tunnel	McDowell County	NC	770	1915	CSX Transportation
Clinchfield Line - 2nd Washburn Tunnel	McDowell County	NC	363	1915	CSX Transportation
Clinchfield Line - 3rd Washburn Tunnel	McDowell County	NC	915	1915	CSX Transportation
Clinchfield Line - Bald Knob Tunnel	Clinchport	VA	453	1915	CSX Transportation
Clinchfield Line - Blue Ridge Tunnel	Mitchell County	NC	1,865	1915	CSX Transportation
Clinchfield Line - Brush Creek Tunnel	Washington County	VA	304	1915	CSX Transportation
Clinchfield Line - Buffalo Creek Tunnel	Dickenson County	VA	352	1915	CSX Transportation
Clinchfield Line - Byrd Tunnel	McDowell County	NC	341	1915	CSX Transportation
Clinchfield Line - Caney Fork Tunnel	Dickenson County	VA	412	1915	CSX Transportation
Clinchfield Line - Click Tunnel	Hawkins County	TN	608	1915	CSX Transportation
Clinchfield Line - Clinch Mountain Tunnel	Scott's County	VA	4,135	1915	CSX Transportation
Clinchfield Line - Clinchport Tunnel	Clinchport	VA	637	1915	CSX Transportation
Clinchfield Line - Free Hill Tunnel	Washington County	TN	472	1915	CSX Transportation
Clinchfield Line - Goff Tunnel	Dickenson County	VA	784	1915	CSX Transportation
Clinchfield Line - Hewitt Tunnel	Dickenson County	VA	596	1915	CSX Transportation
Clinchfield Line - Hills Mills Tunnel	Russell County	VA	1,040	1915	CSX Transportation
Clinchfield Line - Holston Tunnel	Sullivan County	TN	154	1915	CSX Transportation
Clinchfield Line - Honeycutt Tunnel	McDowell County	NC	1,688	1915	CSX Transportation
Clinchfield Line - Indian Ridge Tunnel	Washington County	TN	1,023	1915	CSX Transportation
Clinchfield Line - Kendrick's Tunnel	South Fork Holston River	TN	502	1915	CSX Transportation
Clinchfield Line - Lower Bridle Path Tunnel	McDowell County	NC	1,618	1915	CSX Transportation
Clinchfield Line - Lower Pine Ridge Tunnel	McDowell County	NC	2,211	1915	CSX Transportation
Clinchfield Line - Marion Tunnel	Marion	NC	1,073	1915	CSX Transportation
Clinchfield Line - Martin Tunnel	Wakenva	VA	387	1915	CSX Transportation
Clinchfield Line - McClure Tunnel	Dickenson County	VA	331	1915	CSX Transportation
Clinchfield Line - Perkins Tunnel	Dickenson County	VA	496	1915	CSX Transportation
Clinchfield Line - Petitt Tunnel	Dickenson County	VA	379	1915	CSX Transportation
Clinchfield Line - Pool Point Tunnel	Elkhorn City	KY	642	1915	CSX Transportation
Clinchfield Line - Quinn's Knob Tunnel	McDowell County	NC	545	1915	CSX Transportation
Clinchfield Line - Red Ridge Tunnel	Dickenson County	VA	1,359	1915	CSX Transportation
Clinchfield Line - Rinehart Tunnel	Dickenson County	VA	617	1915	CSX Transportation
Clinchfield Line - Russell Tunnel	Russell County	VA	448	1915	CSX Transportation
Clinchfield Line - Sandy Ridge Tunnel	Dickenson County	VA	7,854	1915	CSX Transportation
Clinchfield Line - Sensabough Tunnel	Hawkins County	TN	348	1915	CSX Transportation
Clinchfield Line - Shannon's Tunnel	Dickenson County	VA	820	1915	CSX Transportation

* This list is by no means exhaustive, but it tries to capture the most significant railroad tunnels in the United States with respect to age, length, and so forth.
† Length is usually the as-constructed length. Some tunnel lengths are approximate based on the accuracy of the source information.
‡ Tunnel ownership has changed multiple times. Where tunnels have been abandoned, the listing is typically the original constructor. Tunnels still in operation list the current operator, although some tunnels are operated by multiple railroads.

RAILROAD TUNNELS* (continued)

Name	Location	State	Length (ft)†	Year Constructed or Opened	Owner‡
Clinchfield Line - Short Branch Tunnel	Dickenson County	VA	913	1915	CSX Transportation
Clinchfield Line - Skaggs Hole Tunnel	Dickenson County	VA	519	1915	CSX Transportation
Clinchfield Line - Snipe's Tunnel	McDowell County	NC	637	1915	CSX Transportation
Clinchfield Line - Spartanburg Tunnel	Spartanburg	SC	742	1915	CSX Transportation
Clinchfield Line - Speedy Tunnel	McDowell County	NC	288	1915	CSX Transportation
Clinchfield Line - Speer's Ferry Tunnel	Speer's Ferry	VA	1,116	1915	CSX Transportation
Clinchfield Line - Squirrel Branch Tunnel	Dickenson County	VA	668	1915	CSX Transportation
Clinchfield Line - Starne's Tunnel	Hill	VA	517	1915	CSX Transportation
Clinchfield Line - State Line Tunnel	Dickenson County	VA	1,523	1915	CSX Transportation
Clinchfield Line - Sykes Mill Tunnel	Dickenson County	VA	752	1915	CSX Transportation
Clinchfield Line - Towers Tunnel	Dickenson County	VA	921	1915	CSX Transportation
Clinchfield Line - Town's Tunnel	Dickenson County	VA	1,098	1915	CSX Transportation
Clinchfield Line - Twin Tunnel No. 1	Scott County	VA	308	1915	CSX Transportation
Clinchfield Line - Twin Tunnel No. 2	Scott County	VA	236	1915	CSX Transportation
Clinchfield Line - Upper Bridle Path Tunnel	McDowell County	NC	927	1915	CSX Transportation
Clinchfield Line - Upper Pine Ridge Tunnel	McDowell County	NC	1,600	1915	CSX Transportation
Clinchfield Line - Vance Tunnel	Mitchell County	NC	527	1915	CSX Transportation
Coos Bay Tunnel No. 13	Lane County	OR	2,496	1915	Central Oregon & Pacific Railroad
Coos Bay Tunnel No. 14	Lane County	OR	471	1915	Central Oregon & Pacific Railroad
Coos Bay Tunnel No. 15	Lane County	OR	2,143	1915	Central Oregon & Pacific Railroad
Coos Bay Tunnel No. 16	Lane County	OR	624	1915	Central Oregon & Pacific Railroad
Coos Bay Tunnel No. 17	Douglas County	OR	1,200	1915	Central Oregon & Pacific Railroad
Coos Bay Tunnel No. 18	Douglas County	OR	1,556	1915	Central Oregon & Pacific Railroad
Coos Bay Tunnel No. 19	Douglas County	OR	4,202	1915	Central Oregon & Pacific Railroad
Coos Bay Tunnel No. 20	Coos County	OR	874	1915	Central Oregon & Pacific Railroad
Coos Bay Tunnel No. 21	Coos County	OR	478	1915	Central Oregon & Pacific Railroad
Hayden Mountain Tunnel	Blount County	AL	1,800	1915	CSX Transportation
Mansion Tunnel	Campbell	VA	929	1915	Norfolk Southern Railroad
Ostrander Tunnel	Kelso	WA	1,165	1915	BNSF Railway
Robinson Creek Tunnel	Douglas County	KY	700	1915	CSX Transportation
Steinway Tunnel	New York City	NY	7,920	1915	New York City Transit Authority
Big Four Tunnels 1 & 2	Huger	WV	821	1916	Norfolk Southern Railroad
East Brady Tunnel	Phillipston	PA	2,468	1916	Allegheny Valley Railroad
Fishhook Tunnel	Avery	ID	415	1916	Chicago, Milwaukee, St. Paul & Pacific Railroad
Kennerdell Tunnel	Kennerdell	PA	3,350	1916	Allegheny Valley Railroad
Leesville Tunnel	Bedford	VA	824	1916	Norfolk Southern Railroad
Strawbridge Tunnel	Lafayette County	WI	650	1916	Chicago & North Western Railroad
Tunnel No. 1	Mineral County	MT	686	1916	Chicago, Milwaukee, St. Paul & Pacific Railroad
Wood Hill Tunnel	Rockland	PA	2,068	1916	Allegheny Valley Railroad
King's No. 6 Tunnel	Corbin	KY	748	1917	CSX Transportation
Hermosa Tunnel No. 2	Albany County	WY	1,650	1918	Union Pacific
Stanford Tunnel	Judith Basin County	MT	1,450	1918	BNSF Railway
Second Kingwood Tunnel	Preston	WV	4,202	1919	Baltimore & Ohio Railroad
Nehalem Divide Tunnel	Scappoose	OR	1,712	1920	Portland & Southwestern Railroad
Elk Rock Tunnel	Clackamas County	OR	1,396	1921	Portland & Willamette Valley Railroad
Garner Tunnel	Healy	AK	508	1921	Alaska Railroad Corporation
Moody Tunnel	Healy	AK	262	1921	Alaska Railroad Corporation
Donner Pass Tunnel 35	Placer County	CA	737	1924	Union Pacific

RAILROAD TUNNELS* *(continued)*

Name	Location	State	Length (ft)†	Year Constructed or Opened	Owner‡
Donner Pass Tunnel 36	Placer County	CA	325	1925	Union Pacific
Donner Pass Tunnel 37	Placer County	CA	410	1925	Union Pacific
Donner Pass Tunnel 38	Placer County	CA	921	1925	Union Pacific
Donner Pass Tunnel 39	Placer County	CA	279	1925	Union Pacific
Donner Pass Tunnel 40	Placer County	CA	315	1925	Union Pacific
Donner Pass Tunnel 41 (Summit)	Placer County	CA	10,325	1925	Union Pacific
Donner Pass Tunnel 42	Placer County	CA	892	1925	Union Pacific
Natron Cutoff Tunnel No. 3 (Summit)	Klamath/Lane County	OR	3,655	1925	Union Pacific
Natron Cutoff Tunnel No. 4	Lane County	OR	698	1925	Union Pacific
Tunnel No. 39	Elko County	NV	1,081	1925	Union Pacific
Tunnel No. 41	Elko County	NV	1,887	1925	Union Pacific
Tunnel No. 42	Elko County	NV	2,473	1925	Union Pacific
Tunnel No. 43	Elko County	NV	5,681	1925	Union Pacific
Jefferytown Tunnel	Enlow	PA	558	1926	Montour Railroad
Natron Cutoff Tunnel No. 5	Lane County	OR	964	1926	Union Pacific
Natron Cutoff Tunnel No. 6	Lane County	OR	566	1926	Union Pacific
Natron Cutoff Tunnel No. 7	Lane County	OR	3,164	1926	Union Pacific
Natron Cutoff Tunnel No. 8	Lane County	OR	671	1926	Union Pacific
Natron Cutoff Tunnel No. 9	Lane County	OR	1,144	1926	Union Pacific
Natron Cutoff Tunnel No. 10	Lane County	OR	466	1926	Union Pacific
Natron Cutoff Tunnel No. 11	Lane County	OR	779	1926	Union Pacific
Natron Cutoff Tunnel No. 12	Lane County	OR	360	1926	Union Pacific
Natron Cutoff Tunnel No. 13	Lane County	OR	875	1926	Union Pacific
Natron Cutoff Tunnel No. 14	Lane County	OR	212	1926	Union Pacific
Natron Cutoff Tunnel No. 15	Lane County	OR	150	1926	Union Pacific
Natron Cutoff Tunnel No. 16	Lane County	OR	2,213	1926	Union Pacific
CN Tunnel No. 2	Pope County	IL	6,840	1927	CN Rail
Holston Tunnel	Sullivan County	TN	154	1927	CSX Transportation
Chetoogeta Tunnel (New)	Tunnel Hill	GA	1,519	1928	CSX Transportation
Chumstick Tunnel	Chelan County	WA	2,601	1928	BNSF Railway
CN Tunnel No. 1	Johnson County	IL	2,550	1928	CN Rail
CN Tunnel No. 3	Pope County	IL	800	1928	CN Rail
Dotsero Cutoff Tunnel No. 34	Tabernash	CO	420	1928	Union Pacific
Dotsero Cutoff Tunnel No. 35	Gore Canyon	CO	157	1928	Union Pacific
Moffat Road Tunnel No. 1	Rollinsville	CO	376	1928	Union Pacific
Moffat Road Tunnel No. 2	Rollinsville	CO	516	1928	Union Pacific
Moffat Road Tunnel No. 3	Rollinsville	CO	369	1928	Union Pacific
Moffat Road Tunnel No. 4	Rollinsville	CO	174	1928	Union Pacific
Moffat Road Tunnel No. 5	Rollinsville	CO	585	1928	Union Pacific
Moffat Road Tunnel No. 6	Rollinsville	CO	536	1928	Union Pacific
Moffat Road Tunnel No. 7	Rollinsville	CO	208	1928	Union Pacific
Moffat Road Tunnel No. 8	Rollinsville	CO	753	1928	Union Pacific
Moffat Road Tunnel No. 10	Rollinsville	CO	1,572	1928	Union Pacific
Moffat Road Tunnel No. 11	Rollinsville	CO	238	1928	Union Pacific
Moffat Road Tunnel No. 17	Rollinsville	CO	1,730	1928	Union Pacific

* This list is by no means exhaustive, but it tries to capture the most significant railroad tunnels in the United States with respect to age, length, and so forth.
† Length is usually the as-constructed length. Some tunnel lengths are approximate based on the accuracy of the source information.
‡ Tunnel ownership has changed multiple times. Where tunnels have been abandoned, the listing is typically the original constructor.
  Tunnels still in operation list the current operator, although some tunnels are operated by multiple railroads.

RAILROAD TUNNELS* (continued)

Name	Location	State	Length (ft)†	Year Constructed or Opened	Owner‡
Moffat Road Tunnel No. 19	Rollinsville	CO	1,055	1928	Union Pacific
Moffat Road Tunnel No. 25	Gilpin County	CO	639	1928	Union Pacific
Moffat Road Tunnel No. 27	Rollinsville	CO	643	1928	Union Pacific
Moffat Road Tunnel No. 29	Rollinsville	CO	78	1928	Union Pacific
Moffat Road Tunnel No. 30	Rollinsville	CO	257	1928	Union Pacific
Moffat Road Tunnel No. 31 - Moffat	Rollinsville	CO	32,736	1928	Union Pacific
Moffat Road Tunnel No. 33	Rollinsville	CO	1,215	1928	Union Pacific
Moffat Road Tunnel No. 42	Rollinsville	CO	463	1928	Union Pacific
Swede Tunnel	Winton	WA	788	1928	BNSF Railway
Winton Tunnel No. 14	Chelan County	WA	4,059	1928	BNSF Railway
Cascade Tunnel No. 15 (New)	King County & Chelan County	WA	41,183	1929	BNSF Railway
Gray Summit Tunnel	Franklin County	MO	1,600	1929	Union Pacific
Labadie Tunnel	Franklin County	MO	590	1929	Union Pacific
Vasper Tunnel	Vasper	TN	1,566	1930s	CSX Transportation
Gateway Subdivision Tunnel No. 1	Keddie	CA	700	1931	BNSF Railway
Gateway Subdivision Tunnel No. 2	Keddie	CA	491	1931	BNSF Railway
Gateway Subdivision Tunnel No. 3	Keddie	CA	621	1931	BNSF Railway
Gateway Subdivision Tunnel No. 4	Keddie	CA	472	1931	BNSF Railway
Gateway Subdivision Tunnel No. 5	Moccasin	CA	275	1931	BNSF Railway
Gateway Subdivision Tunnel No. 6	Greenville	CA	1,109	1931	BNSF Railway
Hardy Tunnel No. 1	Cascade County	MT	494	1931	BNSF Railway
Hartman Tunnel	Frederick County	MD	215	1931	CSX Transportation
Hoover Dam Tunnels No. 1 Through 5	Lake Mead	NV	16,000	1931	U.S. Government
Big Bend Tunnel	Talcott	WV	6,500	1932	Chesapeake & Ohio Railroad
Grimms Bridge Tunnel	Grimms Bridge	OH	1,042	1933	Pittsburgh Coal Company
Dotsero Cutoff - Sweetwater Tunnel	Dotsero	CO	1,115	1934	Union Pacific
Dotsero Cutoff - Yarmony Tunnel	Eagle County	CO	647	1934	Union Pacific
Knapps Hill	Chelan County	WA	788	1936	Washington State Department of Transportation
Rutgers Street Tunnel	New York City	NY	5,479	1936	New York City Transit Authority
Little Switzerland Tunnel	Blue Ridge Parkway	NC	542	1938	Federal Highway Administration for National Park Service
Rough Ridge Tunnel	Blue Ridge Parkway	NC	150	1938	Federal Highway Administration for National Park Service
Twin Tunnel North	Blue Ridge Parkway	NC	300	1938	Federal Highway Administration for National Park Service
Twin Tunnel South	Blue Ridge Parkway	NC	401	1938	Federal Highway Administration for National Park Service
Sideling Hill Tunnel 2	Fulton County	PA	6,662	1940	South Pennsylvania Railroad
Wild Acres Tunnel	Blue Ridge Parkway	NC	330	1940	Federal Highway Administration for National Park Service
Craggy Flats Tunnel	Blue Ridge Parkway	NC	400	1941	Federal Highway Administration for National Park Service
Craggy Pinnacle Tunnel	Blue Ridge Parkway	NC	245	1941	Federal Highway Administration for National Park Service
Devil's Courthouse Tunnel	Blue Ridge Parkway	NC	665	1941	Federal Highway Administration for National Park Service
Begich Peak Tunnel	Begich Peak	AK	5,300	1943	State of Alaska Department of Transportation & Alaska Railroad Corporation
Tunnel 3.7	Coram	MT	325	1943	BNSF Railway
Tunnel 3.8	Flathead County	MT	2,281	1943	BNSF Railway
Tunnel 3.9	Flathead County	MT	30	1943	BNSF Railway
Whittier Tunnel (Anton Anderson Memorial Tunnel)	Prince William Sound	AK	13,728	1943	State of Alaska Department of Transportation & Alaska Railroad Corporation
Blue Ridge Tunnel (New)	Nelson County	VA	4,250	1944	CSX Transportation
Bozeman Tunnel (New)	Gallatin County	MT	3,015	1945	Montana Rail Link
Pinnacle Ridge Tunnel	Blue Ridge Parkway	NC	813	1945	Federal Highway Administration for National Park Service
Tennessee Pass Tunnel (New)	Tennessee Pass	CO	2,200	1945	Union Pacific
Lickstone Ridge Tunnel	Blue Ridge Parkway	NC	402	1946	Federal Highway Administration for National Park Service

RAILROAD TUNNELS* (continued)

Name	Location	State	Length (ft)†	Year Constructed or Opened	Owner‡
Bear Pen Gap Tunnel	Fremont	VA	2,500	1947	CSX Transportation
Big Witch Tunnel	Blue Ridge Parkway	NC	348	1947	Federal Highway Administration for National Park Service
Bunches Bald Tunnel	Blue Ridge Parkway	NC	255	1947	Federal Highway Administration for National Park Service
Altamont Tunnel	Aspen	WY	5,900	1949	Union Pacific
Gaynor Tunnel (No. 14.7)	Chelan County	WA	674	1949	BNSF Railway
Elkhorn Tunnel	Elkhorn	WV	7,100	1950	Norfolk Southern
Boysen Dam Tunnel	Shoshone	WY	7,920	1951	BNSF Railway
Cabinet Tunnel	Clark Fork	ID	380	1951	BNSF Railway
Casper Subdivision Tunnel No. 2.5	Boysen	WY	1,300	1951	BNSF Railway
Casper Subdivision Tunnel No. 3	Boysen	WY	950	1951	BNSF Railway
Casper Subdivision Tunnel No. 4	Boysen	WY	1,060	1951	BNSF Railway
Casper Subdivision Tunnel No. 5	Boysen	WY	2,960	1951	BNSF Railway
Lodgeville Tunnel	Harrison County	WV	3,236	1952	CSX Transportation
White Sulphur Tunnel (No. 17)	White Sulphur Springs	WV	300	1952	CSX Transportation
Front Street Tunnel	Ketchikan	AK	274	1954	Alaska Railroad Corporation
Rattlesnake Mountain Tunnel	Blue Ridge Parkway	NC	395	1958	Federal Highway Administration for National Park Service
Grassy Knob Tunnel	Blue Ridge Parkway	NC	770	1961	Federal Highway Administration for National Park Service
Frying Pan Tunnel	Blue Ridge Parkway	NC	577	1962	Federal Highway Administration for National Park Service
Nemo Tunnel	Morgan County	TN	2,174	1962	Cincinnati Southern Railway
Tanbark Tunnel	Blue Ridge Parkway	NC	780	1962	Federal Highway Administration for National Park Service
Ferrin Knob Tunnel No. 1	Blue Ridge Parkway	NC	561	1963	Federal Highway Administration for National Park Service
Ferrin Knob Tunnel No. 2	Blue Ridge Parkway	NC	421	1963	Federal Highway Administration for National Park Service
Ferrin Knob Tunnel No. 3	Blue Ridge Parkway	NC	375	1963	Federal Highway Administration for National Park Service
Fork Mountain Tunnel	Blue Ridge Parkway	NC	389	1963	Federal Highway Administration for National Park Service
Pine Mountain Tunnel	Blue Ridge Parkway	NC	1,434	1963	Federal Highway Administration for National Park Service
Young Pisgah Ridge Tunnel	Blue Ridge Parkway	NC	412	1963	Federal Highway Administration for National Park Service
Buck Springs Tunnel	Blue Ridge Parkway	NC	462	1964	Federal Highway Administration for National Park Service
Little Pisgah Tunnel	Blue Ridge Parkway	NC	576	1964	Federal Highway Administration for National Park Service
Moab Tunnel	Moab	UT	7,022	1964	Union Pacific
Flathead Tunnel	Lincoln County	MT	37,012	1972	BNSF Railway
North Bonneville Tunnel (No. 1.5)	Chelan County	WA	1,503	1977	BNSF Railway
Thistle Pass Tunnel	Thistle	UT	3,009	1983	Union Pacific
St. Clair Tunnel (New)	Port Huron (to Sarnia, Ontario)	MI	6,129	1994	Canadian Pacific Railway
Portage Lake Tunnel	Prince William Sound	AK	440	1998	State of Alaska Department of Transportation & Alaska Railroad Corporation
East Side Access Tunnels	New York City	NY	55,000	Projected Completion 2022	Metropolitan Transportation Authority
			1,955,225	Feet	
			370	Miles	

* This list is by no means exhaustive, but it tries to capture the most significant railroad tunnels in the United States with respect to age, length, and so forth.
† Length is usually the as-constructed length. Some tunnel lengths are approximate based on the accuracy of the source information.
‡ Tunnel ownership has changed multiple times. Where tunnels have been abandoned, the listing is typically the original constructor.
   Tunnels still in operation list the current operator, although some tunnels are operated by multiple railroads.

APPENDIX | U.S. CONSTRUCTED TUNNEL ARCHIVE

TRANSIT TUNNELS*

Name	Location	State	Length (ft)†	Year Constructed or Opened	Owner‡
Boston Green Line	Boston	MA	23,760	1897 (Streetcar), 1924 (Rapid Transit), 1954 (Revere Extension)	Massachusetts Bay Transportation Authority
Boston Orange Line	Boston	MA	10,032	1901 (Streetcar)	Massachusetts Bay Transportation Authority
Mount Washington Transit Tunnel	Pittsburgh	PA	3,500	1902	Port Authority of Allegheny County
Boston Blue Line	Boston	MA	11,088	1904 (Streetcar), 1924 (Rapid Transit), 1954 (Revere Extension)	Massachusetts Bay Transportation Authority
149th Street Tunnel	New York City	NY	641	1905	Metropolitan Transit Authority
Selby Avenue Trolley Tunnel	Minneapolis	MN	1,500	1907	Twin City Rapid Transit
Joralemon Street Tunnel	New York City	NY	8,888	1908	Metropolitan Transit Authority
Uptown Hudson Tunnels	Jersey City	NY/NJ	11,000	1908	Port Authority of New York & New Jersey
Downtown Hudson Tunnels	Jersey City	NY	11,300	1909	Port-Authority Trans-Hudson
North River Tunnels	New York City	NY	6,100	1910	Port-Authority Trans-Hudson
Boston Red Line	Boston	MA	45,936	1912 (Streetcar)	Massachusetts Bay Transportation Authority
East Side Trolley Tunnel	Providence	RI	2,000	1914	Unknown
Steinway Tunnel	New York City	NY	29,568	1915	Metropolitan Transit Authority
Lexington Avenue Tunnel	New York City	NY	1,283	1918	Metropolitan Transit Authority
Twin Peaks Tunnel	San Francisco	CA	11,986	1918	San Francisco Municipal Transportation Authority
Clark Street Tunnel	New York City	NY	5,900	1919	Metropolitan Transit Authority
Cincinnati Subway System	Cincinnati	OH	11,616	1920	Cincinnati Subway
Montague Street Tunnel	New York City	NY	7,009	1920	Metropolitan Transit Authority
Sunset Tunnel	San Francisco	CA	4,232	1928	San Francisco Municipal Transportation Authority
53rd Street Tunnel	New York City	NY	5,589	1932	Metropolitan Transit Authority
Cranberry Street Tunnel	New York City	NY	8,487	1932	Metropolitan Transit Authority
Chicago Subway	Chicago	IL	47,520	1943	Chicago Transit Authority
Tandy Center Subway	Fort Worth	TX	3,696	1963	Leonard's M&O Subway
Bay Area Rapid Transit	San Francisco	CA	195,360	1972	Bay Area Rapid Transit
Berkeley Hills Tunnel	San Francisco	CA	16,900	1973	Bay Area Rapid Transit
Transbay Tunnel	San Francisco Bay	CA	19,008	1974	Bay Area Rapid Transit
Washington, DC, Metro System	Washington	DC	167,080	1976–2014	Washington Metropolitan Area Transit Authority
Baltimore Subway	Baltimore	MD	30,624	1983	Maryland Transit Authority
Center City Commuter Connection Tunnel	Philadelphia	PA	9,504	1984	Southeastern Pennsylvania Transportation Authority
Buffalo Subway	Buffalo	NY	21,965	1985	Niagara Frontier Transportation Authority
Mount Lebanon Tunnel	Pittsburgh	PA	3,000	1985	Port Authority of Allegheny County
63rd Street Tunnel	New York City	NY	3,140	1989	Metropolitan Transit Authority
Bus Tunnel	Seattle	WA	13,728	1990 (Bus), 2007 (Bus/Transit)	Sound Transit
Los Angeles Subway	Los Angeles	CA	86,592	1993	Los Angeles County Metropolitan Transportation Authority
Blue Line Tunnel	Minneapolis	MN	17,952	2004	Metropolitan Council
Boston Silver Line	Boston	MA	5,280	2004 (Bus)	Massachusetts Bay Transportation Authority
Tren Urbano	San Juan	PR	4,900	2004	Puerto Rico Department of Transportation & Public Works
Beacon Hill Tunnel	Seattle	WA	10,560	2006 (Transit)	Sound Transit

* This list is by no means exhaustive, but it tries to capture the most significant transit tunnels in the United States with respect to age, length, and so forth.
† Length is usually the as-constructed length. Some tunnel lengths are approximate based on the accuracy of the source information.
‡ Tunnel ownership may have changed.

THE HISTORY OF TUNNELING IN THE UNITED STATES

TRANSIT TUNNELS* (continued)

Name	Location	State	Length (ft)†	Year Constructed or Opened	Owner‡
North Shore Connector	Pittsburgh	PA	6,336	2012	Port Authority of Allegheny County
University Link Tunnel	Seattle	WA	33,264	2012 (Transit)	Sound Transit
No. 7 Line Extension	New York City	NY	5,280	2015	Metropolitan Transit Authority
Second Avenue Subway	New York City	NY	44,880	2017	Metropolitan Transit Authority
Northgate Link	Seattle	WA	36,100	Projected Completion 2021	Sound Transit
East Link	Seattle	WA	4,000	Projected Completion 2023	Sound Transit
			974,292	Feet	
			185	Miles	

HIGHWAY TUNNELS*

Name	Location	State	Length (ft)†	Year Constructed or Opened	Owner‡
Breeden Tunnel	Mingo	WV	347	1833	West Virginia Department of Transportation
Murray Hill Tunnel	New York City	NY	1,600	1834	New York City Department of Transportation
Vernon Tunnel	Vernon	CT	108	1849	Connecticut Department of Transportation
LaSalle Street Tunnel	Chicago	IL	1,890	1871	City of Chicago
Twin Branch Tunnel	McDowell	WV	190	1890	West Virginia Department of Transportation
FRCO 67-317 Tunnel	Fremont	CO	312	1894	Colorado Department of Transportation
FRCO 67-318 Tunnel	Fremont	CO	218	1894	Colorado Department of Transportation
River Bend Tunnel	Kanawha	WV	316	1900	West Virginia Department of Transportation
TELL-8-TUN Tunnel	Teller	CO	243	1900	Colorado Department of Transportation
Fort Mason Road Tunnel	San Francisco	CA	1,245	1901	California Department of Transportation
Lower Gold Camp Tunnel No. 1	El Paso	CO	315	1901	Colorado Department of Transportation
Lower Gold Camp Tunnel No. 2	El Paso	CO	184	1901	Colorado Department of Transportation
Third Street Tunnel	Los Angeles	CA	1,245	1901	California Department of Transportation
Kennedy Tunnel	Oakland	CA	1,040	1903	Alameda & Contra Costa County
Happy Hollow Road Tunnel	Dearborn	IN	120	1907	Indiana Department of Transportation
Missionary Ridge Tunnel No. 1	Chattanooga	TN	300	1909	Tennessee Department of Transportation & Bridge
ASARCO Tunnel	Pierce	WA	325	1912	Washington State Department of Transportation
Dornan Drive Tunnel	Contra Costa County	CA	741	1912	California Department of Transportation
Solano Avenue Tunnel	Alameda	CA	464	1912	California Department of Transportation
Dingess Tunnel	Mingo	WV	3,331	1914	West Virginia Department of Transportation
East Side Trolley Tunnel	Providence	RI	2,000	1914	Rhode Island Public Transit Authority
Oneonta Tunnel	Columbia River Gorge	OR	125	1914	Oregon Department of Transportation
Stockton Street Tunnel	San Francisco	CA	911	1914	California Department of Transportation
Mitchell Point Tunnel	Columbia River Gorge	OR	385	1915	Oregon Department of Transportation
Fort Cronkite Tunnel	Marin County	CA	2,690	1918	California Department of Transportation
Mosier Twin Tunnels	Columbia River Gorge	OR	350	1921	Oregon Department of Transportation
Liberty Tunnels	Pittsburgh	PA	11,778	1924	Pennsylvania Department of Transportation
Second Street Tunnel	Los Angeles	CA	1,734	1924	California Department of Transportation
Harmon Tunnel	Madison County	IA	150	1925	Iowa Department of Transportation
Lincoln Tunnel - South	New York City	NY	8,003	1925	Port Authority of New York & New Jersey
Armstrong Tunnel	Pittsburgh	PA	1,298	1927	Allegheny County Department of Public Works
Beaucatcher Tunnel	Asheville	NC	1,039	1927	North Carolina Department of Transportation
Holland Tunnel - East	New York City	NY	8,371	1927	Port Authority of New York & New Jersey
Holland Tunnel - West	New York City	NY	8,558	1927	Port Authority of New York & New Jersey
Posey Tunnel	Alameda	CA	4,436	1928	California Department of Transportation
B-15-E Tunnel	Larimer	CO	95	1929	Colorado Department of Transportation
Bachmann Tunnel No. 1	Hamilton	TN	1,035	1929	Tennessee Department of Transportation & Bridge
Bachmann Tunnel No. 2	Hamilton	TN	1,035	1929	Tennessee Department of Transportation & Bridge
Sepulveda Tunnel	Los Angeles	CA	645	1929	California Department of Transportation
Big Oak Flat Road Tunnel No. 1	Mariposa County	CA	366	1930	Federal Highway Administration for National Parks Service
Big Oak Flat Road Tunnel No. 2	Mariposa County	CA	222	1930	Federal Highway Administration for National Parks Service
Big Oak Flat Road Tunnel No. 3	Mariposa County	CA	2,083	1930	Federal Highway Administration for National Parks Service
Bunker Road Tunnel	Marin	CA	2,350	1930	Federal Highway Administration for National Parks Service
Detroit-Windsor Tunnel	Wayne	MI	5,160	1930	Detroit & Canada Tunnel Corporation

* This list is by no means exhaustive, but it tries to capture the most significant highway tunnels in the United States with respect to age, length, and so forth.
† Length is usually the as-constructed length. Some tunnel lengths are approximate based on the accuracy of the source information.
‡ Tunnel ownership may have changed.

HIGHWAY TUNNELS*(continued)

Name	Location	State	Length (ft)†	Year Constructed or Opened	Owner‡
Missionary Ridge Tunnel No. 3	Chattanooga	TN	1,415	1930	Tennessee Department of Transportation & Bridge
Mount Carmel Tunnel	Zion National Park	UT	5,808	1930	Federal Highway Administration for National Parks Service
Cape Creek Lane Tunnel	Lane County	OR	714	1931	Oregon Department of Transportation
Cave Rock Tunnel - East	Douglas County	NV	153	1931	Nevada Department of Transportation
Figueroa Street Tunnels	Los Angeles	CA	1,751	1931	California Department of Transportation
Middle Matilija Tunnel	Ventura	CA	131	1931	California Department of Transportation
South Matilija Tunnel	Ventura	CA	205	1931	California Department of Transportation
Elk Creek Tunnel	Douglas County	OR	1,112	1932	Oregon Department of Transportation
Fort Columbia Tunnel	Pacific	WA	800	1932	Washington State Department of Transportation
Newcastle Tunnel	Placer	CA	550	1932	California Department of Transportation
Lyle Tunnel (14/215)	Klickitat County	WA	389	1933	Washington State Department of Transportation
Wawona Tunnel	Mariposa County	CA	4,233	1933	California Department of Transportation
Going-to-the-Sun Road Tunnels	Glacier	MT	590	1934	National Park Service
Sumner Tunnel	Boston	MA	5,653	1934	Massachusetts Turnpike Authority
Deer Creek Tunnel	Pierce	WA	508	1935	Federal Highway Administration for National Parks Service
Seymour Peak Tunnel (123/106)	Seattle	WA	510	1935	Washington State Department of Transportation
Soledad Canyon Road Tunnel	Los Angeles	CA	442	1935	California Department of Transportation
Breakneck Ridge Tunnels	Putnam	NY	1,166	1936	New York Department of Transportation
Chapim Mesa Road Tunnel	Mesa Verde	CO	1,478	1936	Federal Highway Administration for National Parks Service
Grizzly Dome Tunnel	Plumas	CA	390	1936	California Department of Transportation
Knapps Hill Tunnel	Chelan	WA	788	1936	Washington State Department of Transportation
Rim Rock Drive Tunnels	Fruita	CO	1,051	1936	Federal Highway Administration for National Parks Service
Rimrock Tunnel	Yakima	WA	577	1936	Washington State Department of Transportation
Tooth Rock Tunnel	Cascade Locks State Park	OR	811	1936	Oregon Department of Transportation
Yerba Buena Tunnel	San Francisco	CA	1,791	1936	California Department of Transportation
Arch Cape Tunnel	Clatsop	OR	1,228	1937	Oregon Department of Transportation
Arch Rock Tunnel	Butte	CA	261	1937	California Department of Transportation
Caldecott Tunnel No. 1	Contra Costa County	CA	3,610	1937	California Department of Transportation
Caldecott Tunnel No. 2	Contra Costa County	CA	3,610	1937	California Department of Transportation
Elephant Blue Tunnel	Plumas	CA	1,187	1937	California Department of Transportation
Fishhook Tunnel	Shoshone	ID	415	1937	Civilian Conservation Corps
Lincoln Tunnel - Center	New York City	NY	8,216	1937	Port Authority of New York & New Jersey
Tunnel No. 1	Skamania	WA	130	1937	Washington State Department of Transportation
Tunnel No. 2	Skamania	WA	408	1937	Washington State Department of Transportation
Tunnel No. 3	Skamania	WA	261	1937	Washington State Department of Transportation
Tunnel No. 4	Skamania	WA	257	1937	Washington State Department of Transportation
Tunnel No. 5	Skamania	WA	212	1937	Washington State Department of Transportation
Waldo Tunnel	Marin County	CA	2,000	1937 & 1954	California Department of Transportation
Presido Tunnel	San Francisco	CA	1,299	1938	California Department of Transportation
Scotts Bluff Tunnels	Summit Road	NE	505	1938	Federal Highway Administration for National Parks Service
Sideling Hill Tunnels	Fulton County	PA	13,564	1938	Pennsylvania Turnpike Commission
MacArthur Tunnel	San Francisco	CA	1,300	1939	California Department of Transportation
Rocky Butte Tunnel	Portland	OR	370	1939	Oregon Department of Transportation
Salt Creek Tunnel	Lane County	OR	914	1939	Oregon Department of Transportation

* This list is by no means exhaustive, but it tries to capture the most significant highway tunnels in the United States with respect to age, length, and so forth.
† Length is usually the as-constructed length. Some tunnel lengths are approximate based on the accuracy of the source information.
‡ Tunnel ownership may have changed.

HIGHWAY TUNNELS*(continued)

Name	Location	State	Length (ft)†	Year Constructed or Opened	Owner‡
U.S. 6 Tunnel No. 5	Clear Creek	CO	411	1939	Colorado Department of Transportation
U.S. 6 Tunnel No. 6	Clear Creek	CO	588	1939	Colorado Department of Transportation
Allegheny Tunnel No. 1	Somerset	PA	6,070	1940	Pennsylvania Turnpike Commission
Blue Mountain Tunnels	Franklin	PA	8,678	1940	Pennsylvania Turnpike Commission
Burnside Tunnel	Multnomah	OR	230	1940	Oregon Department of Transportation
Cornell Tunnel No. 1	Portland	OR	497	1940	Oregon Department of Transportation
Dennis L. Edwards Tunnel	Vernonia	OR	772	1940	Oregon Department of Transportation
Kittatinny Tunnel No. 1	Franklin	PA	4,727	1940	Pennsylvania Turnpike Commission
Laurel Hill Tunnel	Westmoreland	PA	4,541	1940	Pennsylvania Turnpike Commission
Mount Baker Ridge Tunnel - South	King	WA	1,466	1940	Washington State Department of Transportation
Queens–Midtown Tunnel - North	Manhattan & Queens, New York	NY	6,414	1940	Triborough Bridge & Tunnel Authority
Queens–Midtown Tunnel - South	Manhattan & Queens, New York	NY	6,272	1940	Triborough Bridge & Tunnel Authority
Ray's Hill Tunnel	Bedford	PA	3,532	1940	Pennsylvania Turnpike Commission
Sunset Tunnel	Washington County	OR	772	1940	Oregon Department of Transportation
Tuscarora Tunnel No. 1	Franklin & Huntingdon	PA	5,326	1940	Pennsylvania Turnpike Commission
West Burnside Tunnel	Portland	OR	230	1940	Oregon Department of Transportation
Angeles Forest Highway Tunnel	Los Angeles	CA	475	1941	California Department of Transportation
Bankhead Tunnel	Mobile	AL	3,389	1941	Alabama Department of Transportation
Cornell Tunnel No. 2	Portland	OR	247	1941	Oregon Department of Transportation
Colonial Parkway Tunnel	Williamsburg	VA	1,190	1942	Federal Highway Administration for National Parks Service
L-06-P Tunnel	Ouray	CO	165	1942	Colorado Department of Transportation
Anton Anderson Memorial Tunnel	Prince William Sound	AK	12,600	1943	State of Alaska Department of Transportation & Alaska Railroad Corporation
Lincoln Tunnel - North	New York City	NY	7,482	1945	Port Authority of New York & New Jersey
Wind River Canyon Tunnel No. 1	Fremont	WY	330	1947	Wyoming Department of Transportation
Wind River Canyon Tunnel No. 2	Fremont	WY	225	1947	Wyoming Department of Transportation
Heroes Tunnel - Northbound (West Rock)	New Haven	CT	1,200	1949	Connecticut Department of Transportation
Heroes Tunnel - Southbound (West Rock)	New Haven	CT	1,200	1949	Connecticut Department of Transportation
Rock Tunnel	Greenlee County	AZ	397	1949	Arizona Department of Transportation
Angeles Crest Tunnel No. 1	Los Angeles	CA	644	1950	California Department of Transportation
Angeles Crest Tunnel No. 2	Los Angeles	CA	475	1950	California Department of Transportation
Brooklyn-Battery Tunnel - North	New York City	NY	9,117	1950	Triborough Bridge & Tunnel Authority
Brooklyn-Battery Tunnel - South	New York City	NY	9,117	1950	Triborough Bridge & Tunnel Authority
Chatsworth Tunnel	Los Angeles	CA	475	1950	California Department of Transportation
John H. Wilson Tunnel No. 1	O'ahu	HI	2,813	1950	Hawaii State Department of Transportation
John H. Wilson Tunnel No. 2	O'ahu	HI	2,775	1950	Hawaii State Department of Transportation
Lake Keechelus Tunnel	Seattle	WA	500	1950	Washington State Department of Transportation
Missionary Ridge Tunnel No. 2	Chattanooga	TN	932	1950	Tennessee Department of Transportation & Bridge
Pali Tunnel No. 1	O'ahu	HI	1,500	1950	Hawaii State Department of Transportation
Pali Tunnel No. 2	O'ahu	HI	1,557	1950	Hawaii State Department of Transportation
Stringer's Ridge Tunnel	Hamilton	TN	1,001	1950	Tennessee Department of Transportation & Bridge
U.S. 82 Tunnel	Otero	NM	520	1950	New Mexico State Highway & Transportation Department
Washburn Tunnel	Houston	TX	2,909	1950	Harris County Precinct 2
U.S. 6 Tunnel No. 1	Jefferson	CO	883	1951	Colorado Department of Transportation
U.S. 6 Tunnel No. 2	Jefferson	CO	1,068	1951	Colorado Department of Transportation
Broadway Tunnel	San Francisco	CA	1,616	1952	City of San Francisco
Downtown Tunnel No. 1	Norfolk	VA	3,350	1952	Virginia Department of Transportation
Malibu Canyon Road Tunnel	Los Angeles	CA	558	1952	California Department of Transportation
Queen Creek Tunnel	Pinal County	AZ	1,151	1952	Arizona Department of Transportation

HIGHWAY TUNNELS*(continued)

Name	Location	State	Length (ft)†	Year Constructed or Opened	Owner‡
Stevens Canyon Road Tunnel	Mount Rainier National Park	WA	117	1952	Federal Highway Administration for National Parks Service
Baytown Tunnel	Harris	TX	4,110	1953	U.S. Army Corps of Engineers
Boulder Canyon Tunnel	Boulder	CO	350	1953	Colorado Department of Transportation
Gaviota Gorge Tunnel	Santa Barbara	CA	420	1953	California Department of Transportation
Squirrel Hill Tunnel - East	Allegheny	PA	4,225	1953	Pennsylvania Department of Transportation
Squirrel Hill Tunnel - West	Allegheny	PA	4,225	1953	Pennsylvania Department of Transportation
Battery Street Tunnel	Seattle	WA	2,140	1954	Washington State Department of Transportation
Memorial Tunnel	Kanawha	WV	2,802	1954	West Virginia Turnpike
Mount Baldy Road Tunnel	Los Angeles	CA	243	1954	California Department of Transportation
Heart-O-the-Hills Highway Tunnels	Challam	WA	1,213	1955	Federal Highway Administration for National Parks Service
Belle Chasse Tunnel	Belle Chasse	LA	800	1956	Louisiana Department of Transportation & Development
Baltimore Harbor Tunnel - North	Baltimore	MD	6,298	1957	Maryland Transportation Authority
Baltimore Harbor Tunnel - South	Baltimore	MD	6,298	1957	Maryland Transportation Authority
Bluff Mountain Tunnel	Blue Ridge Parkway	VA	630	1957	Federal Highway Administration for National Parks Service
Hampton Roads Tunnel No. 1	Hampton	VA	7,479	1957	Virginia Department of Transportation
Harvey Tunnel	New Orleans	LA	1,080	1957	Louisiana Department of Transportation & Development
Lehigh Tunnel No. 1	Lehigh & Carbon	PA	4,380	1957	Pennsylvania Turnpike Commission
Mount Carmel Tunnel	Zion National Park	UT	410	1957	Federal Highway Administration for National Parks Service
U.S. 6 Tunnel No. 3	Jefferson	CO	769	1957	Colorado Department of Transportation
Knowles Creek Tunnel	Lane County	OR	1,430	1958	Oregon Department of Transportation
Mule Pass Tunnel	Cochis County	AZ	1,433	1958	Arizona Department of Transportation
Ralph Petersen Tunnel	Lane	OR	1,430	1958	Oregon Department of Transportation
Cave Rock Tunnel - West	Douglas County	NV	410	1959	Nevada Department of Transportation
Dewey Square Tunnel - North	Boston	MA	2,378	1959	Massachusetts Highway Department
Dewey Square Tunnel - South	Boston	MA	2,378	1959	Massachusetts Highway Department
Cody to Yellowstone Tunnel No. 1	Cody to Yellowstone	WY	267	1960	Wyoming Department of Transportation
Cody to Yellowstone Tunnel No. 2	Cody to Yellowstone	WY	196	1960	Wyoming Department of Transportation
Cody to Yellowstone Tunnel No. 3	Cody to Yellowstone	WY	3,202	1960	Wyoming Department of Transportation
Fort Pitt Tunnels	Pittsburgh	PA	7,000	1960	Pennsylvania Department of Transportation
Henry E. Kinney Tunnel	Fort Lauderdale	FL	864	1960	Florida Department of Transportation
Callahan Tunnel	Boston	MA	5,071	1961	Massachusetts Turnpike Authority
Fort Snelling Tunnel	Hennepin	MN	300	1961	Minnesota Department of Transportation
Houma Tunnel	Houma	LA	960	1961	Louisiana Department of Transportation & Development
Idaho Springs East	Clear Creek	CO	665	1961	Colorado Department of Transportation
Idaho Springs West	Clear Creek	CO	725	1961	Colorado Department of Transportation
Ravena Ramp South Tunnel	Seattle	WA	152	1961	Washington State Department of Transportation
Midtown Tunnel	Norfolk	VA	4,194	1962	Virginia Department of Transportation
Roosevelt Way Tunnel	Seattle	WA	390	1962	Washington State Department of Transportation
S-E Ramp Tunnel	Seattle	WA	662	1962	Washington State Department of Transportation
Tunnel No. 1 (20/316)	Skagit County	WA	625	1962	Washington State Department of Transportation
Tunnel No. 2 (20/327)	Skagit County	WA	88	1962	Washington State Department of Transportation
Exp Tunnel	Seattle	WA	801	1963	Washington State Department of Transportation
N-W Ramp Tunnel	Seattle	WA	888	1963	Washington State Department of Transportation
Randolph Collier Tunnel	Del Norte County	CA	1,886	1963	California Department of Transportation
Webster Street Tunnel	Alameda	CA	3,350	1963	California Department of Transportation

* This list is by no means exhaustive, but it tries to capture the most significant highway tunnels in the United States with respect to age, length, and so forth.
† Length is usually the as-constructed length. Some tunnel lengths are approximate based on the accuracy of the source information.
‡ Tunnel ownership may have changed.

APPENDIX | U.S. CONSTRUCTED TUNNEL ARCHIVE 519

HIGHWAY TUNNELS*(continued)

Name	Location	State	Length (ft)†	Year Constructed or Opened	Owner‡
5th Expressway Tunnel	Seattle	WA	631	1964	Washington State Department of Transportation
Airport Tunnel	Milwaukee	WI	800	1964	Wisconsin Department of Transportation
Caldecott Tunnel No. 3	Contra Costa County	CA	3,771	1964	California Department of Transportation
Chesapeake Channel Tunnel	Northampton	VA	5,237	1964	Chesapeake Bay Bridge Tunnel
Garfield Tunnel	Garfield County	CO	1,045	1964	Colorado Department of Transportation
No Name Tunnel - East	Garfield County	CO	1,045	1964	Colorado Department of Transportation
No Name Tunnel - West	Garfield County	CO	1,045	1964	Colorado Department of Transportation
Thimble Shoal Tunnel	Northampton	VA	5,551	1964	Chesapeake Bay Bridge Tunnel
Allegheny Tunnel No. 2	Somerset	PA	6,070	1965	Pennsylvania Turnpike Commission
I-40 Tunnel No. 1	Howard	NC	1,212	1965	North Carolina Department of Transportation
I-40 Tunnel No. 3	Howard	NC	1,045	1965	North Carolina Department of Transportation
Prudential Center Complex Tunnel	Boston	MA	3,180	1965	Massachusetts Highway Department
Snow Shed Tunnel	Mineral	CO	379	1965	Colorado Department of Transportation
Green River Tunnel No. 1	Sweetwater	WY	1,138	1966	Wyoming Department of Transportation
Green River Tunnel No. 2	Sweetwater	WY	1,132	1966	Wyoming Department of Transportation
Wheeling Tunnel - East	Ohio	WV	1,518	1966	West Virginia Department of Transportation
Wheeling Tunnel - West	Ohio	WV	1,518	1966	West Virginia Department of Transportation
E-S Ramp 5/Mercer Tunnel	Seattle	WA	600	1967	Washington State Department of Transportation
Highway 55/94 Tunnel	Hennepin	MN	450	1967	Minnesota Department of Transportation
Portland Tunnel	Hennepin	MN	391	1967	Minnesota Department of Transportation
Courthouse Tunnel	Milwaukee	WI	400	1968	Wisconsin Department of Transportation
Kanan Road Tunnel - North	Los Angeles	CA	850	1968	California Department of Transportation
Kanan Road Tunnel - South	Los Angeles	CA	329	1968	California Department of Transportation
Kilbourn Tunnel No. 1	Milwaukee	WI	1,268	1968	Wisconsin Department of Transportation
Kilbourn Tunnel No. 2	Milwaukee	WI	1,309	1968	Wisconsin Department of Transportation
Kittatinny Tunnel No. 2	Franklin	PA	4,727	1968	Pennsylvania Turnpike Commission
N-W I-90 UC Tunnel	Seattle	WA	435	1968	Washington State Department of Transportation
S-E I-90 UC Tunnel	Seattle	WA	305	1968	Washington State Department of Transportation
Tuscarora Tunnel No. 2	Franklin & Huntingdon	PA	5,326	1968	Pennsylvania Turnpike Commission
I-40 Tunnel No. 2	Howard	NC	1,012	1969	North Carolina Department of Transportation
Lowry Hill Tunnel	Hennepin	MN	1,496	1969	Minnesota Department of Transportation
Lytle Tunnel	Cincinnati	OH	1,099	1969	Ohio Department of Transportation
Vista Ridge Tunnel - East	Portland	OR	1,017	1969	Oregon Department of Transportation
Golden Gate Avenue Tunnel	Alameda	CA	632	1970	California Department of Transportation
Lakeview Drive Tunnel	Swain	NC	1,200	1970	North Carolina Department of Transportation
North Matilija Tunnel	Ventura	CA	413	1970	California Department of Transportation
Vista Ridge Tunnel - West	Portland	OR	1,061	1970	Oregon Department of Transportation
Walnut Creek BART Tunnel	Contra Costa County	CA	413	1970	California Department of Transportation
Big Walker Mountain Tunnel No. 1	Bland	VA	4,229	1972	Virginia Department of Transportation
Big Walker Mountain Tunnel No. 2	Bland	VA	4,229	1972	Virginia Department of Transportation
BN UC Tunnel	Seattle	WA	360	1972	Washington State Department of Transportation
Carlin Tunnel - East	Carlin	NV	1,426	1973	Nevada Department of Transportation
Carlin Tunnel - West	Carlin	NV	1,361	1973	Nevada Department of Transportation
CNW RR Tunnel	Milwaukee	WI	709	1973	Wisconsin Department of Transportation
Eisenhower–Johnson Memorial Tunnel - East	Dillon	CO	8,941	1973	Colorado Department of Transportation
George Wallace Tunnel	Mobile	AL	3,000	1973	Alabama Department of Transportation
I-95 & I-495 Tunnels	Washington	DC	3,500	1973	District Department of Transportation

HIGHWAY TUNNELS* (continued)

Name	Location	State	Length (ft)†	Year Constructed or Opened	Owner‡
Mall Tunnel	Washington	DC	3,400	1973	District Department of Transportation
Airport Tunnel - North	Clark	NV	2,729	1974	Nevada Department of Transportation
Airport Tunnel - South	Clark	NV	2,729	1974	Nevada Department of Transportation
East River Mountain Tunnel No. 1	Bland	VA	5,412	1974	Virginia Department of Transportation
East River Mountain Tunnel No. 2	Bland	VA	5,412	1974	Virginia Department of Transportation
Kanan Dume Road Tunnel	Los Angeles	CA	1,044	1974 & 1983	California Department of Transportation
Hampton Roads Tunnel No. 2	Hampton	VA	7,479	1976	Virginia Department of Transportation
N Park Plaza Tunnel	Seattle	WA	289	1976	Washington State Department of Transportation
S-Col Ramp W Plaza UC Tunnel	Seattle	WA	83	1976	Washington State Department of Transportation
S Park Plaza UC Tunnel	Seattle	WA	77	1976	Washington State Department of Transportation
I-564 Tunnel	Norfolk	VA	662	1977	Virginia Department of Transportation
BN UC Tunnel	Seattle	WA	180	1978	Washington State Department of Transportation
Eisenhower–Johnson Memorial Tunnel - West	Dillon	CO	8,941	1979	Colorado Department of Transportation
Minillas Tunnel	San Juan	PR	1,205	1980	Autopistas Metropolitanas de Puerto Rico
Beavertail Tunnel - East	Mesa County	CO	697	1983	Colorado Department of Transportation
Beavertail Tunnel - West	Mesa County	CO	627	1983	Colorado Department of Transportation
E Tunnel (Lief Erikson)	St. Louis	MN	639	1983	Minnesota Department of Transportation
Fort McHenry Tunnel - North	Baltimore	MD	8,800	1985	Maryland Transportation Authority
Fort McHenry Tunnel - South	Baltimore	MD	8,800	1985	Maryland Transportation Authority
Riverside Shed Tunnel	Ouray	CO	180	1985	Colorado Department of Transportation
West Tunnel	St. Louis	MO	513	1986	Minnesota Department of Transportation
Downtown Tunnel No. 2	Norfolk	VA	3,813	1987	Virginia Department of Transportation
Lake Place Tunnel	St. Louis	MO	883	1987	Minnesota Department of Transportation
Convention Center Tunnel	Seattle	WA	547	1988	Washington State Department of Transportation
History Center Tunnel	Ramsey	MN	470	1988	Minnesota Department of Transportation
CANA Tunnel	Boston	MA	1,505	1989	Massachusetts Turnpike Authority
First Hill Lid Tunnel	Skamania	WA	2,873	1989	Washington State Department of Transportation
I-90 Mercer Island Tunnel	King	WA	1,113	1989	Washington State Department of Transportation
Lafayette Bluff Tunnel	Lake	MN	852	1989	Minnesota Department of Transportation
Martin L. King Lid Tunnel	King	WA	2,012	1989	Washington State Department of Transportation
Mount Baker Ridge Tunnel - North	King	WA	1,476	1989	Washington State Department of Transportation
N-W Ramp Tunnel	Seattle	WA	1,113	1989	Washington State Department of Transportation
REV E-S Ramp Tunnel	Seattle	WA	566	1989	Washington State Department of Transportation
Reverse Curve Tunnel	Garfield County	CO	582	1989	Colorado Department of Transportation
Rose Garden Tunnel	St. Louis	MN	1,480	1989	Minnesota Department of Transportation
Deck at Central Avenue	Maricopa County	AZ	3,698	1990	Arizona Department of Transportation
Papago Freeway Tunnel	Phoenix	AZ	2,887	1990	Arizona Department of Transportation
Lehigh Tunnel No. 2	Lehigh & Carbon	PA	4,380	1991	Pennsylvania Turnpike Commission
E-S Ramp 90/80th Ave Tunnel	Seattle	WA	280	1992	Washington State Department of Transportation
Hanging Lake Tunnel - East	Garfield County	CO	3,390	1992	Colorado Department of Transportation
Hanging Lake Tunnel - West	Garfield County	CO	3,834	1992	Colorado Department of Transportation
Monitor-Merrimac Memorial Bridge-Tunnel	City of Newport News	VA	4,800	1992	Virginia Department of Transportation
S-E Ramp Tunnel	Seattle	WA	371	1992	Washington State Department of Transportation
Silver Cliff Tunnel	Lake	MN	1,344	1992	Minnesota Department of Transportation
E105-N405 Tunnel	Los Angeles	CA	1,350	1993	California Department of Transportation

* This list is by no means exhaustive, but it tries to capture the most significant highway tunnels in the United States with respect to age, length, and so forth.
† Length is usually the as-constructed length. Some tunnel lengths are approximate based on the accuracy of the source information.
‡ Tunnel ownership may have changed.

HIGHWAY TUNNELS*(continued)

Name	Location	State	Length (ft)†	Year Constructed or Opened	Owner‡
Hospital Rock Tunnels	O'ahu	HI	707	1993	Hawaii State Department of Transportation
RMA Deck Tunnel	Richmond	VA	400	1994	Richmond Metropolitan Authority
E-N Ramp Tunnel	Seattle	WA	466	1995	Washington State Department of Transportation
Ted Williams Tunnel	Boston	MA	8,115	1995	Massachusetts Turnpike Authority
Cumberland Gap Tunnel - North	Bell/Claiborne	KY	4,600	1996	National Park Service
Cumberland Gap Tunnel - South	Bell/Claiborne	KY	4,600	1996	National Park Service
Sioux Falls Tunnel	Sioux Falls	SD	800	1996	South Dakota Department of Transportation
H-3 Tunnel No. 1	O'ahu	HI	4,890	1997	Hawaii State Department of Transportation
H-3 Tunnel No. 2	O'ahu	HI	5,165	1997	Hawaii State Department of Transportation
Tetsuo Harano Tunnels	O'ahu	HI	10,145	1997	Hawaii State Department of Transportation
133/N224-S133 Tunnel	Orange County	CA	404	1998	California Department of Transportation
241/N261-N241 Tunnel	Orange County	CA	759	1998	California Department of Transportation
Bobby Hopper (Bunyard) Tunnel - East	Washington County	AR	1,595	1998	Arkansas State Highway & Transportation Department
Bobby Hopper (Bunyard) Tunnel - West	Washington County	AR	1,595	1998	Arkansas State Highway & Transportation Department
Portage Lake Tunnel	Prince William Sound	AK	440	1998	State of Alaska Department of Transportation & Alaska Railroad Corporation
Provo Canyon - North Tunnel	Provo Canyon	UT	347	1998	Utah Department of Transportation
Provo Canyon - South Tunnel	Provo Canyon	UT	304	1998	Utah Department of Transportation
Addison Airport Toll Tunnel	Addison	TX	1,600	1999	North Texas Tollway Authority
Hiawatha Tunnel	Hennepin	MN	127	1999	Minnesota Department of Transportation
Route 29 Tunnel	Trenton	NJ	2,400	2000	New Jersey Department of Transportation
St. Francis Tunnel	Milwaukee	WI	403	2000	Wisconsin Department of Transportation
Atlantic City Expressway Connector Tunnel	Atlantic City	NJ	1,957	2001	South Jersey Transportation Authority
Wolf Creek Pass Tunnel	Mineral	CO	916	2002	Colorado Department of Transportation
Thomas P. O'Neill Jr. Tunnel	Boston	MA	7,920	2003	Massachusetts Department of Transportation
Lindbergh Tunnel	St. Louis	MO	1,400	2006	Missouri Department of Transportation
Bremerton Ferry Tunnel	Bremerton	WA	959	2009	Washington State Department of Transportation
Caldecott Tunnel No. 4	Contra Costa County	CA	3,389	2013	California Department of Transportation
Devil's Slide (Tom Lantos) Tunnels	Near Pacifica	CA	4,200	2013	California Department of Transportation
PortMiami Tunnel	Miami	FL	8,400	2014	Florida Department of Transportation
Presido Parkway Tunnels (Battery & Main Post)	San Francisco	CA	1,800	2014	California Department of Transportation
SR-99 Alaskan Way Viaduct Replacement Tunnel	Seattle	WA	10,560	Projected Completion 2019	Washington State Department of Transportation
			649,922	Feet	
			123	Miles	

WATER TUNNELS*

Name	Project/System/Location	State	Length (ft)†	Year Constructed or Opened	Owner‡
Montgomery Bell Tunnel (Patterson Forge)	Cheatham County	TN	290	1819	National Park Service
Old Croton Aqueduct	New York City	NY	216,480	1842	New York City Department of Environmental Protection
Cochituate Aqueduct	Boston	MA	73,920	1848	Massachusetts Water Resources Authority
Chicago Lake Tunnel - Two-Mile Crib	Chicago	IL	10,560	1865	City of Chicago, Department of Water Management
Cross-Town Tunnel	Chicago	IL	31,680	1874	City of Chicago, Department of Water Management
Intake No. 2 Tunnel	Cleveland	OH	6,662	1874	City of Cleveland
Sudbury Aqueduct	Boston	MA	10,560	1878	Massachusetts Water Resources Authority
Loch Raven Reservoir to Lake Montebello	Baltimore	MD	38,500	1881	City of Baltimore, Department of Public Works
Intake No. 3 Tunnel	Cleveland	OH	9,117	1890	City of Cleveland
68th Street Crib	Chicago	IL	10,560	1892	City of Chicago
New Croton Aqueduct	New York City	NY	174,768	1893	New York City Department of Environmental Protection
Northeast Lake Tunnel (formerly Northwest Land & Lake Tunnel)	Chicago	IL	14,200	1900	City of Chicago, Department of Water Management
Intake No. 4 Tunnel	Cleveland	OH	20,000	1904	City of Cleveland
Torresdale Conduit	Philadelphia	PA	13,809	1905	Philadelphia Water Department
Wachusett Aqueduct Tunnel	Boston	MA	10,560	1905	Massachusetts Water Resources Authority
Polk Street Tunnel - Four-Mile Cribb	Chicago	IL	7,500	1907	City of Chicago, Department of Water Management
Blue Island Avenue Tunnel	Chicago	IL	26,550	1909	City of Chicago, Department of Water Management
Gunnison Tunnel	Montrose	CO	30,582	1909	U.S. Bureau of Reclamation
La Poudre Water Diversion Tunnel	Laramie River Basin	CO	11,500	1910	U.S. Bureau of Reclamation
Southwest Land & Lake Tunnel	Chicago	IL	50,950	1911	City of Chicago, Department of Water Management
Elizabeth Tunnel	Los Angeles Aqueduct	CA	26,400	1912	Los Angeles Department of Water & Power
Strawberry Tunnel	Strawberry Valley Project, Utah County	UT	20,064	1912	U.S. Bureau of Reclamation
Los Angeles Aqueduct (142 Tunnels)	Los Angeles Aqueduct	CA	227,040	1913	Los Angeles Department of Water & Power
Third Avenue West Siphon	Seattle	WA	500	1913	City of Seattle
Castkill Aqueduct	New York City	NY	654,720	1915	New York City Department of Environmental Protection
City Tunnel No. 1	New York City	NY	95,040	1915	New York City Department of Environmental Protection
Intake No. 5 Tunnel	Cleveland	OH	14,000	1918	City of Cleveland
Wilson Avenue Tunnel	Chicago	IL	45,510	1918	City of Chicago, Department of Water Management
Priest Tunnel	Hetch Hetchy Water System	CA	1,200	1920	San Francisco Public Utilities Commission
Moccasin Power Tunnel	Hetch Hetchy Water System	CA	5,370	1922	San Francisco Public Utilities Commission
Shandaken Tunnel	New York City	NY	95,040	1924	New York City Department of Environmental Protection
Ward (Florence) Tunnel	Big Creek Hydroelectric Project	CA	71,250	1925	Pacific Gas & Electric
Mountain Tunnel	Hetch Hetchy Water System	CA	100,320	1927	San Francisco Public Utilities Commission
Foothill Tunnel	Hetch Hetchy Water System	CA	83,952	1929	San Francisco Public Utilities Commission
Irvington Tunnel	Hetch Hetchy Water System	CA	17,952	1932	San Francisco Public Utilities Commission
City Tunnel No. 2	New York City	NY	105,600	1933	New York City Department of Environmental Protection
Quabbin Aqueduct	Boston	MA	129,888	1933	Massachusetts Water Resources Authority
Coast Range Tunnel	Hetch Hetchy Water System	CA	132,000	1934	San Francisco Public Utilities Commission
Chicago Avenue Tunnel	Chicago	IL	60,720	1935	City of Chicago, Department of Water Management
Hoover/Boulder Dam Diversion Tunnels	Lake Mead	AZ/NV	16,000	1935	U.S. Bureau of Reclamation
Moffat Tunnel	Englewood	CO	30,000	1936	City of Englewood
Hultman Aqueduct Tunnel	Boston	MA	5,280	1939	Massachusetts Water Resources Authority
San Jacinto Tunnel	Colorado River Aqueduct	CA	68,840	1939	Metropolitan Water District of Southern California

* This list is by no means exhaustive, but it tries to capture the most significant water tunnels in the United States with respect to age, length, and so forth.
 This representative listing includes water distribution or conveyance tunnels, irrigation tunnels, and tunnels associated with dams and hydroelectric projects.
† Length is usually the as-constructed length. Some tunnel lengths are approximate based on the accuracy of the source information.
‡ Tunnel ownership may have changed.

APPENDIX | U.S. CONSTRUCTED TUNNEL ARCHIVE

WATER TUNNELS* *(continued)*

Name	Project/System/Location	State	Length (ft)†	Year Constructed or Opened	Owner‡
Mono Craters Tunnel	Los Angeles Aqueduct	CA	59,812	Early 1940s	Los Angeles Department of Water & Power
Gunpowder Falls - Montebello Tunnel	Baltimore	MD	35,900	1941	City of Baltimore, Department of Public Works
Delaware Aqueduct	New York	NY	448,800	1943	New York City Department of Environmental Protection
Alva B. Adams Tunnel	Colorado Big Thompson Project	CO	69,168	1947	U.S. Bureau of Reclamation
79th Street Tunnel	Chicago	IL	26,400	1950s	City of Chicago, Department of Water Management
Wilson Avenue to Central Water Filtration Plant Tunnel	Chicago	IL	26,400	1950s	City of Chicago
Patapsco - Montebello Tunnel	Baltimore	MD	67,000	1950	City of Baltimore, Department of Public Works
Telocote Tunnel	Cachuma Lake Project	CA	31,680	1950	Goleta Water District
City Tunnel	Boston	MA	25,260	1951	Massachusetts Water Resources Authority
Neversink Tunnel	New York City	NY	31,680	1954	New York City Department of Environmental Protection
East Delaware Tunnel	New York City	NY	132,000	1955	New York City Department of Environmental Protection
Glen Canyon Dam Diversion Tunnels	Coconino County	AZ	5,400	1957	U.S. Bureau of Reclamation
Malden Tunnel	Boston	MA	5,266	1958	Massachusetts Water Resources Authority
City Tunnel Extension	Boston	MA	37,511	1961	Massachusetts Water Resources Authority
Harold D. Roberts Tunnel	Summit County	CO	122,912	1962	Denver Water
Oahe Dam Outlet Tunnels (6)	Pierre	SD	20,987	1962	U.S. Army Corps of Engineers
West Delaware Tunnel	New York City	NY	232,320	1964	New York City Department of Environmental Protection
Canyon Power Tunnel	Hetch Hetchy Water System	CA	54,912	1965	San Francisco Public Utilities Commission
Cosgrove Tunnel	Boston	MA	36,960	1965	Massachusetts Water Resources Authority
Crystal Springs Tunnel	Hetch Hetchy Water System	CA	17,160	1968	San Francisco Public Utilities Commission
San Juan – Chama, Blanco Tunnel	Colorado River Storage Project	NM	45,600	1969	U.S. Bureau of Reclamation
Tehachapi Tunnels	California Aqueduct	CA	16,368	1969	Los Angeles Department of Water & Power
San Bernardino Tunnel	California Aqueduct	CA	21,120	1970s	Los Angeles Department of Water & Power
San Fernando (Sylmar) Tunnel	Sylmar	CA	29,040	1970s	Los Angeles Department of Water & Power
Lake Mead Intake Tunnel No. 1	Lake Mead, near Boulder City	NV	1,400	1970	Southern Nevada Water Authority
Richmond Tunnel	New York City	NY	9,000	1970	New York City Department of Environmental Protection
River Mountain Tunnel No. 1	Las Vegas Valley	NV	19,970	1970	Southern Nevada Water Authority
San Juan – Chama, Azotea Tunnel	Colorado River Storage Project	NM	67,500	1970	U.S. Bureau of Reclamation
San Juan – Chama, Oso Tunnel	Colorado River Storage Project	NM	26,600	1970	U.S. Bureau of Reclamation
Dorchester Tunnel	Boston	MA	33,430	1978	Massachusetts Water Resources Authority
Beacon Hill Waterline Tunnel	Seattle	WA	150	1984	Seattle Public Utilities
Helms Pumped Storage Plant	Fresno County	CA	22,364	1984	Pacific Gas & Electric
Bath County Pumped Storage Project	Bath County	VA	28,470	1985	Dominion Generation & FirstEnergy
Terror Lake Hydroelectric Power Plant Tunnel	Kodiak Island	AK	27,000	1985	Kodiak Electric Association
Eklutna Raw Water Transmission Tunnel	Eklutna Water Project	AK	8,500	1988	Anchorage Water & Wastewater Utility
Moose River Hydroelectric Project	Lewis County	NY	5,000	1990	Fortis U.S. Energy Corporation
Bad Creek Pumped Storage Plant	Lake Jocassee	SC	8,760	1991	Duke Energy
Bradley Lake Power Plant Power Tunnel	Kenai Peninsula, near Homer	AK	18,930	1991	Alaska Energy Authority
Rocky Mountain Pumped Storage Plant	Rome	GA	3,960	1995	Ogelthorpe Power & Georgia Power
Benbrook Tunnel	Crowley (Fort Worth)	TX	19,600	1997	Tarrant Regional Water District
City Tunnel No. 3 - Stage 1	New York City	NY	68,640	1998	New York City Department of Environmental Protection
79th Street Tunnel Extension	Chicago	IL	19,044	1999	City of Chicago, Department of Water Management
Lake Mead Intake Tunnel No. 2	Lake Mead, near Boulder City	NV	1,600	2000	Southern Nevada Water Authority
Borman Park Intake	Borman Park Water Filtration Plant	IN	15,817	2003	Indiana-American Water Company
Metrowest Supply Tunnel	Boston	MA	92,928	2003	Massachusetts Water Resources Authority
Mill to Bull Creek Tunnel	Pollock Pines	CA	10,360	2003	El Dorado Irrigation District
Upper Diamond Fork	Utah County	UT	18,400	2004	Central Utah Water Conservancy District

WATER TUNNELS* (continued)

Name	Project/System/Location	State	Length (ft)†	Year Constructed or Opened	Owner‡
Arrowhead East & West Tunnels	Inland Feeder/San Bernardino	CA	42,338	2009	Metropolitan Water District of Southern California
New Crystal Springs Bypass Tunnel	San Francisco	CA	4,200	2010	San Francisco Public Utilities Commission
San Vicente Pipeline Tunnel	Emergency Storage Project	CA	58,228	2011	San Diego County Water Authority
Lake Hodges Pipeline	Emergency Storage Project	CA	5,800	2012	San Diego County Water Authority
City Tunnel No. 3 - Stage 2	New York City	NY	29,040	2013	New York City Department of Environmental Protection
Lake Mead Intake Tunnel No. 3	Lake Mead, near Boulder City	NV	18,408	2014	Southern Nevada Water Authority
Bay Tunnel	San Francisco	CA	26,400	2015	San Francisco Public Utilities Commission
New Irvington Tunnel	Irvington	CA	18,480	2015	San Francisco Public Utilities Commission
New York Harbor Siphon	New York City	NY	9,504	2015	New York City Department of Environmental Protection
			4,813,661	Feet	
			912	Miles	

* This list is by no means exhaustive, but it tries to capture the most significant water tunnels in the United States with respect to age, length, and so forth. This representative listing includes water distribution or conveyance tunnels, irrigation tunnels, and tunnels associated with dams and hydroelectric projects.
† Length is usually the as-constructed length. Some tunnel lengths are approximate based on the accuracy of the source information.
‡ Tunnel ownership may have changed.

APPENDIX | U.S. CONSTRUCTED TUNNEL ARCHIVE

WASTEWATER TUNNELS*

Name	Location	State	Length (ft)†	Year Constructed or Opened	Owner‡
Chesbrough's Tunnel	Chicago	IL	10,560	1864	Unknown
Lake Union Sewer Tunnel	Seattle	WA	5,700	1894	Unknown
Chicago TARP Tunnels	Chicago	OH	577,632	1975–2006	Metropolitan Water Reclamation District
Heights-Hilltop Tunnels	Cleveland	OH	123,000	1985–2006	Northeastern Ohio Regional Sewer District
North Outfall Replacement Sewer	Los Angeles	CA	42,240	1990s	City of Los Angeles
Milwaukee Deep Tunnel Program - Phases 1, 2, & 3	Milwaukee	WI	153,120	1994	Milwaukee Metropolitan Sewer District
Southwest Interceptor Tunnel	Cleveland	OH	52,800	1996	Northeastern Ohio Regional Sewer District
Boston Harbor Outfall	Boston	MA	52,800	2000	Massachusetts Water Resources Authority
Columbia Slough Consolidated Conduit	Portland	OR	17,424	2000	City of Portland
Deer Island Outfall Tunnel	Boston	MA	50,160	2000	Massachusetts Water Resources Authority
Denny CSO Storage Tunnel	Seattle	WA	6,212	2002	King County
Henderson CSO Storage Tunnel	Seattle	WA	3,105	2002	King County
Chattahoochee Tunnel	Cobb County	GA	50,160	2004	Cobb County Water System
East Central Interceptor Sewer Tunnel	Los Angeles	CA	58,080	2004	City of Los Angeles
Northeast Interceptor Sewer Tunnel	Los Angeles	CA	27,984	2004	City of Los Angeles
Nancy Creek Tunnel	Atlanta	GA	42,300	2005	City of Atlanta
Big Walnut Augmentation Rickenbacker Interceptor Tunnel	Columbus	OH	21,648	2006	City of Columbus
West Side CSO Tunnel	Portland	OR	22,000	2006	City of Portland
Big Walnut Outfall Augmentation Sewer Tunnel	Columbus	OH	13,200	2007	City of Columbus
West Area CSO Tunnels	Atlanta	GA	44,880	2007	City of Atlanta
North Dorchester Bay Tunnel	Boston	MA	11,088	2009	Massachusetts Water Resources Authority
No Business Creek Tunnel	Gwinnette	GA	16,000	2010	City of Atlanta
Brightwater Sewage Tunnel	Seattle	WA	68,640	2011	King County
East Side CSO Tunnel	Portland	OR	29,260	2011	City of Portland
Detroit River Outfall Tunnel No. 2	Detroit	MI	36,900	2012	Detroit Water & Sewage Department
Mill Creek Storage Tunnels	Cleveland	OH	43,000	2012	Northeastern Ohio Regional Sewer District
Western Regional Conveyance Tunnel	Bellevue	KY	35,000	2012	Northern Kentucky Sanitation District No. 1
Euclid Creek Storage Tunnel	Cleveland	OH	18,044	2013	Northeastern Ohio Regional Sewer District
Blue Plains Tunnel	Washington	DC	24,300	2016	DC Water
First Street Tunnel	Washington	DC	2,800	2016	DC Water
Anacostia River Tunnel	Washington	DC	12,500	Projected Completion November 2017	DC Water
Dugway Storage Tunnel	Cleveland	OH	14,763	Projected Completion July 2019	Northeastern Ohio Regional Sewer District
Olentangy Scioto Interceptor Sewer Augmentation & Relief Sewer Tunnels	Columbus	OH	23,317	Projected Completion 2019	City of Columbus
			1,700,057	Feet	
			322	Miles	

* This list is by no means exhaustive, but it tries to capture the most significant highway tunnels in the United States with respect to age, length, and so forth.
† Length is usually the as-constructed length. Some tunnel lengths are approximate based on the accuracy of the source information.
‡ Tunnel ownership may have changed.

ESSENTIAL UNDERGROUND PUBLICATIONS

Many books and publications have been written over the last century regarding the work that has been showcased in this book. They include textbooks, everyday working manuals and best practices of the day, and guidelines written by the industry, for the industry. Others chronicle key achievements on actual projects or true stories from the past. For those who were previously unaware of the tunneling industry and now possess a new interest or passion, are looking to broaden an existing interest, or simply cannot get enough of reading about all things tunneling, below is a list of useful current publications that may be of further interest or reference.

This list is not exhaustive and is merely aimed at providing a sample of the material available for satisfying your newly found or existing interest in appreciating the depth and breadth of the tunneling industry.

TECHNICAL AND CONTRACTUAL

The Art of Tunneling
(Széchy 1973) *Akadémiai Kiadó*

AUA Guidelines for Backfilling and Contact Grouting of Tunnels and Shafts
(Henn 2003) *ASCE Press and Thomas Telford*

Concrete for Underground Structures: Guidelines for Design and Construction
(Goodfellow 2011) *UCA of SME*

Construction Dispute Review Board Manual
(Matyas, Mathews, Smith, and Sperry 1996) *McGraw-Hill*

Design-Build Subsurface Projects, 2nd Edition
(Brierley, Corkum, and Hatem 2010) *UCA of SME*

Earth Tunneling with Steel Supports
(Proctor and White 1977) *Commercial Shearing & Stamping*

Engineering Rock Mass Classifications: A Complete Manual for Engineers and Geologists in Mining, Civil, and Petroleum Engineering
(Bieniawski 1989) *Wiley*

Geotechnical Baseline Reports for Construction: Suggested Guidelines
(Essex 2007) *ASCE*

Guidelines for Improved Risk Management: On Tunnel and Underground Construction Projects in the United States of America
(O'Carroll and Goodfellow 2015) *UCA of SME*

Guidelines for Tunnel Lining Design
(O'Rourke 1984) *ASCE*

Introduction to Tunnel Construction
(Chapman, Metje, and Stärk 2010) *CRC Press*

Mechanised Shield Tunnelling, 2nd Edition
(Maidl, Herrenknecht, Maidl, and Wehrmeyer 2012) *Ernst & Sohn*

Practical Guide to Grouting of Underground Structures
(Henn 1996) *ASCE Press*

Practical Tunnel Construction
(Hemphill 2012) *Wiley*

Practical Tunnel Driving
(Richardson and Mayo 1976) *McGraw-Hill*

Recommended Contract Practices for Underground Construction
(Edgerton 2008) *UCA of SME*

Rock Tunneling with Steel Supports
(White, Terzaghi, and Proctor 1946) *Commercial Shearing & Stamping*

Soil Movements Induced by Tunneling Their Effects on Pipelines and Structures
(Attewell, Yeates, and Selby 1986) *Chapman & Hall*

Sprayed Concrete Lined Tunnels
(Thomas 2008) *CRC Press*

Support of Underground Excavations in Hard Rock
(Hoek, Kaiser, and Bawden 2014) *CRC Press*

TBM Tunnelling in Jointed and Faulted Rock
(Barton 2000) *CRC Press*

Technical Manual for Design and Construction of Road Tunnels
(American Association of State Highway and Transportation Officials [AASHTO], National Highway Institute, and Parsons, Brinckerhoff, Quade & Douglas 2010) *AASHTO*

Tunnel Engineering Handbook, 2nd Edition
(Bickel, Kuesel, and King 2004) *Chapman & Hall*

Underground Engineering for Sustainable Urban Development
(National Research Council and National Academies 2013) *National Academies Press*

Underground Excavations in Rock
(Hoek and Brown 2005) *E&FN Spon*

Underground Structures: Design and Instrumentation
(Sinha 1989) *Elsevier Science*

HISTORICAL

The Big Dig
(McNichol and Ryan 2000) Silver Lining

The Blue Ridge Tunnel: A Remarkable Engineering Feat in Antebellum Virginia
(Lyons 2014) History Press

Builders of the Hoosac Tunnel
(Schexnayder 2015) *Peter E. Randall*

Chesapeake Bay Bridge-Tunnel
(Warren and Holland 2015) *Arcadia Publishing*

Cincinnati's Incomplete Subway: The Complete History
(Mecklenborg 2010) *History Press*

Conquering Gotham: Building Penn Station and Its Tunnels
(Jonnes 2008) *Penguin*

Encyclopedia of Bridges and Tunnels
(Johnson and Leon 2002) *Facts on File*

Great Railroad Tunnels of North America
(Putnam 2011) *McFarland*

The Great Society Subway: A History of the Washington Metro
(Schrag 2006) *Johns Hopkins University Press*

Highway Under the Hudson: A History of the Holland Tunnel
(Jackson 2011) *New York University Press*

The Moffat Line: David Moffat's Railroad Over and Under the Continental Divide
(Sells 2011) *iUniverse*

Nothing Like it in the World: The Men Who Built the Transcontinental Railroad, 1863–1869
(Ambrose 2005) *Simon & Schuster*

The Race Underground: Boston, New York, and the Incredible Rivalry That Built America's First Subway
(Most 2014) *St. Martin's Press*

Rails Under the Mighty Hudson: The Story of the Hudson Tubes, the Pennsy Tunnels, and Manhattan Transfer, 2nd Edition
(Cudahy 2002) *Fordham University Press*

Trapped Under the Sea: One Engineering Marvel, Five Men, and a Disaster Ten Miles into the Darkness
(Swidey 2014) *Crown*

Tunnel Visions: The Rise and Fall of the Superconducting Super Collider
(Riordan, Hoddeson, and Kolb 2015) *University of Chicago Press*

CONTRIBUTORS AND ACKNOWLEDGMENTS

Gratitude and appreciation are extended to the following individuals who expended so much of their valuable time and resources to researching, writing, illustrating, and developing this project. Without their perseverance and dedication, *The History of Tunneling in the United States* would not have been possible. The editors and authors acknowledge and thank the various individuals and organizations for their input, support, and permissions for material in the preparation of their chapters.

CHAPTER 1: THE BUILDING OF A NATION

Author

Colin A. Lawrence, Mott MacDonald

CHAPTER 2: SOCIETAL BENEFITS

Authors

Robert J.F. Goodfellow, Aldea Services LLC
Priscilla P. Nelson, Colorado School of Mines

CHAPTER 3: RAILROAD TUNNELS

Author

Robert A. Robinson, Shannon & Wilson Inc.

Contributing Authors

Gary Brierley, Doctor Mole Inc.
Brian L. Garrod, Hatch Mott MacDonald
Paul M. Godlewski, Shannon & Wilson Inc.
Kyle R. Ott, Parsons Brinckerhoff
Michael F. Roach, Traylor Bros. Inc.
Andrew J. Thompson, Mott MacDonald
Klaus G. Winkler, Shannon & Wilson Inc.

CHAPTER 4: TRANSIT TUNNELS

Author

William H. Hansmire, WSP | Parsons Brinckerhoff

Contributing Authors

Matthew Fowler, WSP | Parsons Brinckerhoff
Frank P. Frandina, Mott MacDonald
James Parkes, WSP | Parsons Brinckerhoff
Michael F. Roach, Traylor Bros. Inc.
Paul A. Roy, AECOM
Henry A. Russell Jr., WSP | Parsons Brinckerhoff
Richard A. Sage, Sound Transit
Vincent Tirolo Jr., Arup
Brian H. Zelenko, WSP | Parsons Brinckerhoff

Acknowledgments

Harald C. Cordes, WSP | Parsons Brinckerhoff
Mitchell Fong, WSP | Parsons Brinckerhoff
Letitia Ivins, Los Angeles County Metropolitan Transit Authority
John S. Prizner, AECOM
Zachary M. Schrag, George Mason University
Timothy P. Smirnoff, HDR Inc.

CHAPTER 5: HIGHWAY TUNNELS

Author
Lee W. Abramson, Mott MacDonald

Acknowledgments
Moe Amini, Retired
Christine Baker, Pennsylvania Turnpike
Darryl Brogan, Mott MacDonald
Richard Buck, Mott MacDonald
Les Dixon, Cumberland Gap National Historic Park
Amanda Elioff, WSP | Parsons Brinckerhoff
David Field, Mott MacDonald
Frank Frandina, Mott MacDonald
William Hall, California Department of Transportation
William H. Hansmire, WSP | Parsons Brinckerhoff
Russell Hubbard, Arkansas State Highway and Transportation Department
Bruce Hull, Ohio Department of Transportation
Jeremy Menzies, San Francisco Municipal Transportation Agency
Christopher Preto, Mott MacDonald
Robert A. Robinson, Shannon & Wilson Inc.
Stephen Taylor, Mott MacDonald
David Young, Mott MacDonald

CHAPTER 6: WATER TUNNELS

Author
Michael P. Bruen, MWH, now part of Stantec

Contributing Authors
Brigid A. Baty, Metropolitan Water District of Southern California
Steven W. Hunt, CH2M Hill
John Shamma, Metropolitan Water District of Southern California
David F. Tsztoo, San Francisco Public Utility Commission
Lawrence A. Williamson, Mott MacDonald

Acknowledgments
Sarah Acheson, New York City Department of Environmental Protection
Barbara Allen, Massachusetts Water Resources Authority
Andrew Gahan, U.S. Bureau of Reclamation
Marcus Jensen, Southern Nevada Water Authority
Alex Margevicius, Cleveland Department of Water
Erika Moonin, Southern Nevada Water Authority
Burt Rezko, Chicago Department of Water
Robin Rockey, Southern Nevada Water Authority
Maureen Russell, MWH, now part of Stantec
Robin Scheswohl, San Francisco Public Utility Commission
Ross Sweeney, Chicago Department of Water
Richard Wiltshire (retired), U.S. Bureau of Reclamation
Carlos Zambrano, MWH, now part of Stantec
Walter Zeisl, Los Angeles Department of Water and Power
Lin Zhao, MWH, now part of Stantec

CHAPTER 7: WASTEWATER TUNNELS

Author

Michael Vitale, Mott MacDonald

Acknowledgments

Olga Beltsar, Mott MacDonald
Justin Brown, Metropolitan Water Reclamation District of Greater Chicago
Tammy Cleys, City of Portland Bureau of Environmental Services
Matt Dalrymple, Mott MacDonald
William Edgerton, McMillen Jacobs Associates
Larry Ellis, Milwaukee Metropolitan Sewerage District
Jennifer Elting, Northeast Ohio Regional Sewer District
Dave Fergusson, Traylor Bros. Inc.
Kevin Fitzpatrick, Metropolitan Water Reclamation District of Greater Chicago
Allison Fore, Metropolitan Water Reclamation District of Greater Chicago
John Gonzalez, Northeast Ohio Regional Sewer District
Bill Graffin, Milwaukee Metropolitan Sewerage District
Julie Irick, Seattle Office of the City Clerk
Richard Lanyon, Metropolitan Water Reclamation District of Greater Chicago (retired)
Karen Martinek, City of Portland Bureau of Environmental Services
Jerome McGovern, Metropolitan Water Reclamation District of Greater Chicago (retired)
Kellie Rotunno, Northeast Ohio Regional Sewer District
Michael Uva, Northeast Ohio Regional Sewer District
Daniel Wendt, Metropolitan Water Reclamation District of Greater Chicago
Moussa Wone, DC Water
Thomas Zimmerman, Milwaukee Metropolitan Sewerage District (retired)

CHAPTER 8: INNOVATIONS IN TUNNELING

Author

David R. Klug, David R. Klug & Associates Inc.

Contributing Authors

W. Brian Fulcher, Kenny Construction
Colin A. Lawrence, Mott MacDonald
Michael Mooney, Colorado School of Mines
Dennis Ofiara, The Robbins Company
John Reilly, John Reilly Associates
Paul Schmall, Moretrench American Corp.
William Warfield, Mining Engineer

Acknowledgments

Christophe Delus, Optimas Solutions
Louis Falco, Concrete Systems Inc.
Jonathan Klug, David R. Klug & Associates Inc.
Paul Madsen, Kiewit Infrastructure
Richard McLane, Traylor Bros. Inc.
Karl Mitterndorfer, Antraquip Corp.
Pamela Moran, Wisko America Inc.
Jack Mulvoy, DSI Tunneling, LLC
Matt Pope, Mining Equipment Ltd.
Stephen Price, The Walsh Group
Gordon Revey, Revey Associates Inc.
Rick Robinson, Sandvik Mining and Rock Technology
Heiner Sander, HNTB
Andrew J. Thompson, Mott MacDonald
George Yoggy, GCS, LLC

CHAPTER 9: THE FUTURE OF TUNNELING

Author

Amanda Elioff, WSP | Parsons Brinckerhoff

Contributing Authors

Werner Burger, Herrenknecht AG
William Edgerton, McMillen Jacobs Associates
Colin A. Lawrence, Mott MacDonald
Priscilla Nelson, Colorado School of Mines
Tesse Roberts, HDR Inc.
Steve Steir, Traylor Bros. Inc.

Acknowledgments

Jay Mezher, WSP | Parsons Brinckerhoff

APPENDIX: U.S. CONSTRUCTED TUNNEL ARCHIVE

Authors

David Field, Mott MacDonald
Robert A. Robinson, Shannon & Wilson Inc.

TUNNELING MILESTONES IN THE U.S.

Authors

David Field, Mott MacDonald
Priscilla P. Nelson, Colorado School of Mines

HISTORICAL BIOGRAPHIES

Author

David Field, Mott MacDonald

ILLUSTRATION CREDITS

UCA of SME, the editors, and the authors thank the following individuals and organizations for providing maps, photographs, and graphics, and for granting permission to reproduce copyrighted material. While every effort has been made to trace and acknowledge copyright holders, we apologize for any errors or omissions.

TUNNELING MILESTONES IN THE U.S. (LEFT TO RIGHT)

MARTA's Peachtree Center Station: David Sailors © 1988 by WSP | Parsons Brinckerhoff; **Midtown Hudson Tunnel, New York, 1936:** © Port Authority of New York and New Jersey; **Blue Plains TBM, Washington, D.C.:** © District of Columbia Water and Sewer Authority; **SR 99 TBM, the Largest in the World:** © WSP | Parsons Brinckerhoff.

FRONT DUST JACKET (CLOCKWISE FROM TOP LEFT)

Blue Plains TBM, Washington, D.C.: © District of Columbia Water and Sewer Authority; **Ventilation System of Wawona Tunnel, Yosemite National Park:** Todd A. Croteau, National Park Service; **Cascade Tunnel Vicinity Map, Washington:** The Eight-Mile Cascade Tunnel, Great Northern Railway, A Symposium, Paper No. 1809, *ASCE Transactions Vol. 96*, p. 918, with permission from ASCE; **Beach Hydraulic Tunneling Shield:** © 1890 by Scientific American; **Typical Cross Section of a Highway Tunnel:** © Mott MacDonald.

BACK DUST JACKET (CLOCKWISE FROM TOP LEFT):

Typical Cross Section of a Highway Tunnel: © Mott MacDonald; **Connection Tunnel near Chicago:** © Metropolitan Water Reclamation District of Greater Chicago; **TBM Cutterhead for Euclid Creek Tunnel, Cleveland, Ohio:** © Northeast Ohio Regional Sewer District; **East Portal of Weehawken Tunnel, New Jersey:** © WSP | Parsons Brinckerhoff; **Cross Section of SR 99 Tunnel, Seattle:** © Washington State Department of Transportation; **Plan View for Chicago Lake and Land Tunnels:** © Chicago Department of Water Management; **Parmley Reinforced Concrete Segments:** SewerHistory.org.

CHAPTER 1: THE BUILDING OF A NATION

Chapter Opener: © San Francisco Public Utilities Commission, Robin Scheswohl; **Figure 1.1:** © Fulcher/Elioff Collections; **Figure 1.2:** Library of Congress; **Figures 1.3, 1.4:** © Kenny Construction Company; **Figure 1.5:** © Mining Equipment Inc. Collection; **Figure 1.6:** Dominick I. Drummond (ca. 1830–1899) and C. Frank King (Printing attributed to Charles H. Crosby & Company), "Rapid Transit. Save Time & Distance. Take the Hoosac Tunnel Route, 1877." Chromolithographic advertisement. 29¾ × 23¾ inches (sheet). Boston Athenæum; **Figure 1.7:** Library of Congress; **Figure 1.8:** © Metropolitan Water Reclamation District of Greater Chicago; **Figure 1.9, 1.10, 1.11:** Library of Congress; **Figure 1.12:** © WSP | Parsons Brinckerhoff; **Figures 1.13, 1.14:** © Port Authority of New York and New Jersey; **Figure 1.15:** © Fulcher/Elioff Collections; **Figure 1.16:** © Underhill, N.Y.C. via Library of Congress; **Figure 1.17:** © 2013 by Metropolitan Transportation Authority of the State of New York, Patrick Cashin, CC BY 2.0; **Figure 1.18:** public domain; **Figure 1.19:** © Fulcher/Elioff Collections; **Figure 1.20:** Kiewit Infrastructure

Co.; **Figures 1.21, 1.22, 1.23:** © Fulcher/Elioff Collections; **Figures 1.24, 1.25:** © The Robbins Company; **Figure 1.26:** © New York City Metropolitan Transit Authority; **Figure 1.27:** HNTB and Washington State Department of Transportation; **Figures 1.28:** © Fulcher/Elioff Collections; **Figure 1.29:** Traylor Bros. Inc.; **Figure 1.30:** © Washington Metropolitan Area Transit Authority; **Figure 1.31:** © Traylor Bros. Inc.

CHAPTER 2: SOCIETAL BENEFITS

Chapter Opener: © Mott MacDonald; **Figure 2.1:** © 2013 by Shannon1, CC BY-SA 2.5; **Figure 2.2:** Owen Bissell/KQED Science; **Lincoln Biography:** public domain; **Figure 2.3:** Azusa Pacific University Special Collections; **Figure 2.4:** Environmental Protection Agency; **Figures 2.5, 2.6:** NYC Department of Environmental Protection; **Figure 2.7:** © Metropolitan Water District of Southern California; **Figure 2.8:** © Margaret Lazzari, CC BY-ND 1.0; **Figure 2.9:** Los Angeles Times Photographic Archive, Department of Special Collections, Charles E. Young Research Library, UCLA, CC BY 4.0; **Figure 2.10:** © 1954 by Denver Water; **Figure 2.11:** © 1960 by Denver Water; **Figure 2.12:** Centers for Disease Control and Prevention; **Figure 2.13:** State Library of Massachusetts, Photo 361, Legislator's Photographs Collection; **Figure 2.14:** public domain; **Figure 2.15:** © Nathan Morton; **Figure 2.16:** © Mike B. Sturmovik; **Figure 2.17:** public domain; **Figure 2.18:** © 1989 by Fred Bauhof, Shannon & Wilson Inc.; **Figure 2.19:** Detroit Publishing Co.; **Figures 2.20, 2.21:** public domain; **Figure 2.22:** © 1913 by Scientific American; **Figure 2.23:** © Joe McKendry; **Figure 2.24:** Boston Public Library. Picture taken 1897; **Figure 2.25:** McGraw Publishing Co., New York City; **Figure 2.26:** public domain; **Figure 2.27:** National Park Service; **Figure 2.28:** © Jonathan Warren, CC BY-SA 3.0; **Figure 2.29:** © 2014 by Visitor7, CC BY-SA 3.0; **Figure 2.30:** © Fulcher/Elioff Collections; **Figure 2.31:** www.gypsynester.com; **Figure 2.32:** © CardCow.com; **Figure 2.33:** public domain; **Figure 2.34:** © 2008 by Patrick Pelster, CC BY-SA 3.0; **Figure 2.35:** public domain; **Figure 2.36:** © Route 82 at English Wikipedia, CC BY-SA 2.5; **Figure 2.37:** © Daniel Azoulay; **Figure 2.38:** © 2015 by MrJARichard, CC BY 4.0; **Figure 2.39:** © 2008 by gconservancy, CC BY 2.0; **Figure 2.40:** Fred Bottomer, Cleveland Press Collection, Michael Schwartz Library, Cleveland State University; **Figure 2.41:** Eric Crawford; **Figure 2.42:** U.S. Environmental Protection Agency; **Figure 2.43:** © Fulcher/Elioff Collections; **Figure 2.44:** © Kenny Construction Company; **Figure 2.45:** © Fulcher/Elioff Collections; **Figure 2.46:** © Traylor Bros. Inc.; **Figure 2.47:** © Fulcher/Elioff Collections; **Figure 2.48:** © The Robbins Company; **Figure 2.49:** © Salini-Impregilo; **Figure 2.50:** © Fulcher/Elioff Collections; **Nixon Biography:** public domain.

CHAPTER 3: RAILROAD TUNNELS

Chapter Opener: © Drew Mitchem; **Figure 3.1:** from H. Drinker, 1893, *Tunneling, Explosive Compounds, and Rock Drills*, 3rd ed., New York, John Wiley & Sons.; **Figure 3.2:** © OpenStreetMap; **Figure 3.3:** Sarver Maps; **Figure 3.4:** Jack Boucher, National Park Service; **Figure 3.5:** public domain; **Figure 3.6:** 1869 photo: A.J. Russell; 2003 photo: Centpacrr, CC BY-SA 3.0; **Figure 3.7:** public domain; **Figure 3.8:** © OpenStreetMap; **Figure 3.9:** David Brossard 2011, CC BY-SA 2.0; **Figure 3.9 inset:** Sean Lamb 2005, CC BY-SA 2.0; **Figure 3.10:** Library of Congress, Geography and Map Division; **Figure 3.11:** Frank Jay Haynes; **Figure 3.12:** *Poor's Manual of the Railroads of the United States*, 1895; **Figure 3.13:** postcard by Lowman & Hanford, Seattle; photo by Asahel Curtis and Miller; **Figure 3.14:** Clyde Osmer DeLand, artist; **Figure 3.15:** Detroit Publishing Co.; **Figure 3.16:** public domain; **Figure 3.17:** from H. Drinker, 1893, *Tunneling, Explosive Compounds, and Rock Drills*, 3rd ed., New York, John Wiley & Sons; **Figure 3.18:** Lee Pickett; **Moffat Biography:** public domain; **Figure 3.19:** © Robert A. Robinson, Shannon & Wilson Inc.; **Figure 3.20:** © P. Godlewski, Shannon & Wilson Inc.; **Figure 3.21:** University of Washington Libraries, Special Collections, A. Curtis 04390; **Figure 3.22:** © P. Godlewski, Shannon & Wilson Inc.; **Figure 3.23:** © 1890 by Scientific American; **Hobson Biography:** public domain; **Figures 3.24, 3.25, 3.26, 3.27 3.28:** © P. Godlewski, Shannon & Wilson Inc; **Figure 3.29:** © W. Hultman, Shannon & Wilson Inc.; **Figure 3.30:** © P. Godlewski, Shannon & Wilson Inc.; **Figure 3.31:** © 2012 by Chris van der Heide; **Figure 3.32:** U.S. Department of Transportation; **Figures 3.33, 3.34:** © Robert A. Robinson, Shannon & Wilson Inc.; **Figure 3.35:** Historic American Engineering Record PA-520-1; **Figures 3.36, 3.37:** © M. Kucker, Shannon & Wilson Inc.; **Figures 3.38, 3.39:** © Robert A. Robinson, Shannon & Wilson Inc.; **Figure 3.40:** © 2006 by jpmueller99, CC BY 2.0; **Figures 3.41, 3.42, 3.43, 3.44, 3.45:** public domain; **Figure 3.46:** photCL 184 (256), The Huntington Library, San Marino, California; **Figure 3.47:** photCL 184 (204), The Huntington Library, San Marino, California; **Figure 3.48:** © Richard Steinheimer; **Figure 3.49:** © Shirley Burman; **Figure 3.50:** © Klaus Winkler, Shannon & Wilson Inc.; **Figures 3.51, 3.52, 3.53:** © Robert Nordlund, Shannon & Wilson Inc.; **Figure 3.54:** © Shannon & Wilson Inc.; **Figure 3.55:** C.E. Watkins, photCL 74(446), The Huntington Library, San Marino, California; **Figure 3.56:** California State Railroad Museum, negative no. 387-1693, used with permission; **Figure 3.57:** Southern Pacific, John Signor Collection; **Figure 3.58:** © Klaus Winkler, Shannon & Wilson Inc.; **Figure 3.59:** © Frontier-Kemper Constructors; **Figures 3.60, 3.61, 3.62, 3.63:** © WSP | Parsons Brinckerhoff; **Figures 3.64, 3.65, 3.66, 3.67:** public domain; **Figures 3.68, 3.69, 3.70:** © Traylor Bros. Inc.; **Figure 3.71:** University of Washington Libraries, Special Collections, UW11210; **Figure 3.72:** University of Washington Libraries, Special Collections, A. Curtis 04387; **Figure 3.73:**

University of Washington Libraries, Special Collections, A. Curtis 04391; **Figure 3.74:** University of Washington Libraries, Special Collections, A. Curtis 04393; **Figures 3.75, 3.76:** © Robert A. Robinson, Shannon & Wilson Inc.; **Figure 3.77:** The Eight-Mile Cascade Tunnel, Great Northern Railway, A Symposium. Paper No. 1809, *ASCE Transactions* Vol. 96, p. 918, with permission from ASCE; **Figure 3.78:** With permission from Transportation Club of Seattle; **Figure 3.79:** University of Washington Libraries, Special Collections, A. Curtis 27082; **Figure 3.80:** © P. Godlewski, Shannon & Wilson Inc.; **Figure 3.81:** The Eight-Mile Cascade Tunnel, Great Northern Railway, A Symposium. Paper No. 1809, *ASCE Transactions* Vol. 96, p. 958, with permission from ASCE; **Figure 3.82:** The Eight-Mile Cascade Tunnel, Great Northern Railway, A Symposium. Paper No. 1809, *ASCE Transactions* Vol. 96, p. 941, with permission from ASCE; **Figure 3.83:** © P. Godlewski, Shannon & Wilson Inc.; **Figure 3.84:** The Eight-Mile Cascade Tunnel, Great Northern Railway, A Symposium. Paper No. 1809, *ASCE Transactions* Vol. 96, p. 945, with permission from ASCE; **Figure 3.85:** University of Washington Libraries, Special Collections, L. Pickett 3821; **Figure 3.86:** University of Washington Libraries, Special Collections, L. Pickett 3279; **Figure 3.87:** The Eight-Mile Cascade Tunnel, Great Northern Railway, A Symposium. Paper No. 1809, *ASCE Transactions* Vol. 96, p. 944, with permission from ASCE; **Figure 3.88:** University of Washington Libraries, Special Collections, L. Pickett 3920; **Figure 3.89:** © Metropolitan Transportation Authority; **Figures 3.90, 3.91, 3.92, 3.93:** © Andy Thompson, Metropolitan Transportation Authority.

CHAPTER 4: TRANSIT TUNNELS

Chapter Opener: © Shu-Hung Liu via Shutterstock; **Figure 4.1:** David Sailors © 1992 by WSP | Parsons Brinckerhoff; **Figure 4.2:** © 2006 by Willem van Bergen, CC BY 2.0; **Figure 4.3:** © Massachusetts Bay Transportation Authority; **Figure 4.4:** Boston Transit Commission; **Figure 4.5:** © Henry A. Russell; **Figure 4.6:** Boston Transit Commission; **Figures 4.7, 4.8:** © Henry A. Russell; **Figure 4.9:** Boston Transit Commission; **Figure 4.10:** public domain; **Figure 4.11:** © Henry A. Russell; **Figure 4.12:** © 2010 by Michael Hicks, CC BY 2.0; **Figure 4.13:** © Henry A. Russell; **Figure 4.14:** © Paul Roy; **Figure 4.15:** © John Livzey; **Figure 4.16:** © 2011 by Dmitry Avdeev, CC BY-SA 3.0; **Figure 4.17:** © WSP | Parsons Brinckerhoff; **Figure 4.18:** © 2017 by Jake Berman, CC BY-SA 3.0, maps.complutense.org; **Brunel Biography:** public domain; **Figure 4.19:** © Joseph Brennan; **Figures 4.20, 4.21, 4.22:** from Robert Ridgeway, "Subway Construction in New York City," 1921; **Figure 4.23:** *Engineering News* 1914; **Parsons Biography:** public domain; **Figures 4.24, 4.25:** © WSP | Parsons Brinckerhoff; **Figures 4.26, 4.27:** © Skanska USA; **Figure 4.28:** David Sailors © 2010 by WSP | Parsons Brinckerhoff; **Figure 4.29:** David Sailors © 2013 by WSP | Parsons Brinckerhoff; **Figures 4.30, 4.31, 4.32:** © Fulcher/Elioff Collections; **Figure 4.33 (top):** © WSP | Parsons Brinckerhoff; **(bottom):** David Sailors © 2009 by WSP | Parsons Brinckerhoff; **Figures 4.34, 4.35:** David Sailors © 2015 by WSP | Parsons Brinckerhoff; **Figure 4.36:** David Sailors © 2014 by WSP | Parsons Brinckerhoff; **Figure 4.37:** © 2006 by Christian Mehlführer, CC BY 2.5; **Figure 4.38:** Matthew Fowler © WSP | Parsons Brinckerhoff; **Figure 4.39:** public domain; **Figure 4.40:** © *Civil Engineering* magazine, Vol. 38, No. 6, pp. 52–55; **Figure 4.41:** Matthew Fowler © WSP | Parsons Brinckerhoff; **Figure 4.42:** © SFMTA Photo | sfmta.com/photo; **Figure 4.43:** © 1972 American Institute of Mining, Metallurgical and Petroleum Engineers; **Figures 4.44, 4.45:** © WSP | Parsons Brinckerhoff; **Kuesel Biography:** © *TunnelTalk*.com; **Figure 4.46:** Mliu92, CC BY-SA 4.0; **Figure 4.47:** Randy Burton and Matthew Fowler © WSP | Parsons Brinckerhoff; **Figure 4.48:** © SFMTA, Robert Pierce; **Figure 4.49:** Matthew Fowler © WSP | Parsons Brinckerhoff; **Figure 4.50:** Randy Burton and Matthew Fowler © WSP | Parsons Brinckerhoff; **Figure 4.51:** © SFMTA, Robert Pierce; **Figure 4.52:** Randy Burton and Matthew Fowler © WSP | Parsons Brinckerhoff; **Figure 4.53:** Matthew Fowler © WSP | Parsons Brinckerhoff; **Figure 4.54:** © 2007 by Chuck Koehler, CC BY 2.0; **Figure 4.55:** public domain; **Figure 4.56:** David Sailors © 1996 by WSP | Parsons Brinckerhoff; **Figure 4.57:** David Sailors © 1988 by WSP | Parsons Brinckerhoff; **Figure 4.58:** © WSP | Parsons Brinckerhoff; **Figure 4.59:** © 2011 by UrbanRail.Net, R. Schwandl; **Figure 4.60:** U.S. Department of Transportation; **Figures 4.61, 4.62, 4.63:** Ken Merrill © WSP | Parsons Brinckerhoff; **Figure 4.64:** John Shillabeer; **Figure 4.65:** © Niagara Frontier Transportation Authority; **Figures 4.66, 4.67, 4.68:** Frank Frandina; **Figures 4.69, 4.70:** © Niagara Frontier Transportation Authority; **Figure 4.71:** © 2015 by Dllu, CC BY-SA 4.0; **Figure 4.72:** adapted from Port Authority of Allegheny County map; **Figures 4.73, 4.74:** David Sailors © 1987 by WSP | Parsons Brinckerhoff; **Figure 4.75:** AECOM; **Figure 4.76:** © Paul Roy; **Figure 4.77:** AECOM; **Figure 4.78:** © Paul Roy; **Figure 4.79:** © John Livzey; **Figure 4.80:** © JeffreyKatzPhotography.com; **Figure 4.81:** © Paul Roy; **Figure 4.82:** 2005, U.S. Navy photo by Chief Photographer's Mate Johnny Bivera; **Figure 4.83:** © Washington Metropolitan Area Transit Authority; **Peck Biography:** public domain; **Figure 4.84:** © 1972 by William H. Hansmire; **Figures 4.85, 4.86, 4.87, 4.88, 4.89:** © William H. Hansmire; **Figures 4.90, 4.91, 4.92, 4.93:** © W. Brian Fulcher; **Figure 4.94:** © Michael F. Roach; **Figure 4.95:** © William H. Hansmire; **Figures 4.96, 4.97, 4.98:** © Michael F. Roach; **Figure 4.99:** © GZ Consultants; **Figure 4.100:** © WSP | Parsons Brinckerhoff; **Figures 4.101, 4.102:** © Washington Metropolitan Area Transit Authority; **Figure 4.103:** © 2009 by Ben Schumin, CC BY-SA 3.0; **Figures 4.104, 4.105:** © Washington Metropolitan Area Transit Authority; **Figure 4.106:** © 2007 by Navid Serrano, CC BY-SA 3.0; **Figure 4.107:** © Traylor Bros. Inc.; **Figure 4.108:** from p. 349 in "Tunnel Digging Is Begun in Los Angeles," *Electric Railway*

Journal, by C.A. Elliott, Sept. 6, 1924; **Figure 4.109:** © Matt Brown; **Figure 4.109 inset:** University of Southern California Libraries and California Historical Society, CC BY 3.0; **Figure 4.110:** Metro Los Angeles, © 2016 by LACMTA; **Figure 4.111:** © William H. Hansmire; **Figure 4.112:** public domain; **Figure 4.113:** © Rachelle Andrews; **Figure 4.114:** David Sailors © 1992 by WSP | Parsons Brinckerhoff; **Figure 4.115:** David Sailors © 2009 by WSP | Parsons Brinckerhoff; **Figures 4.116, 4.117:** © Traylor Bros. Inc; **Figure 4.118:** David Sailors © 1992 by WSP | Parsons Brinckerhoff; **Figures 4.119, 4.120:** © Traylor Bros. Inc.; **Figures 4.121, 4.122:** Metro Los Angeles © 2016 by LACMTA; **Figures 4.123, 4.124:** © William H. Hansmire; **Figure 4.125:** Metro Los Angeles © 2016 by LACMTA; **Figure 4.126:** The Jon B. Lovelace Collection of California Photographs in Carol M. Highsmith's America Project, Library of Congress, Prints and Photographs Division; **Figure 4.126 inset:** © Rob Young; **Figure 4.127 and inset:** The Jon B. Lovelace Collection of California Photographs in Carol M. Highsmith's America Project, Library of Congress, Prints and Photographs Division; **Figure 4.128:** Civic Center Station, *People Portraits in Creativity, Performing, Sports & Fashion*, Faith Ringgold, artist. Courtesy of Metro (Los Angeles County Metropolitan Transportation Authority); **Figure 4.129: Universal City/Studio City Station**, *Universal Delights*, Stephen Johnson, artist. Courtesy of Metro (Los Angeles County Metropolitan Transportation Authority); **Wilshire/Western Station**, *People Coming People Going*, Richard Wyatt, artist. Courtesy of Metro (Los Angeles County Metropolitan Transportation Authority); **Soto Station**, *Landings*, Nobuho Nagasawa, artist. Courtesy of Metro (Los Angeles County Metropolitan Transportation Authority); **Hollywood/Highland Station**, *Underground Girl*, Sheila Klein, artist. Courtesy of Metro (Los Angeles County Metropolitan Transportation Authority); **Figure 4.130:** © 2010 Daniel Schwen, CC BY-SA 4.0; **Figure 4.131:** © Sound Transit; **Figure 4.132:** © The Robbins Company; **Figure 4.133:** © 2009 Oran Viriyincy, CC BY-SA 2.0; **Figures 4.134, 4.135:** © 2016 William H. Hansmire; **Figure 4.136:** © Robert A. Robinson, Shannon and Wilson Inc.; **Figure 4.137:** © 2016 Mott MacDonald; **Figures 4.138, 4.139, 4.140, 4.141, 4.142, 4.143, 4.144:** © Sound Transit; **Figure 4.145:** Michael DiPonio, © Jay Dee Contractors Inc.; **Figure 4.146:** © Sound Transit.

CHAPTER 5: HIGHWAY TUNNELS

Chapter Opener: Alex Proimos, CC BY 2.0; **Figure 5.1:** public domain; **Figure 5.2:** © Mott MacDonald; **Figure 5.3:** public domain; **Figure 5.4:** © SFMTA Photo | sfmta.com/photo; **Figure 5.5:** Todd A. Croteau, National Park Service; **Figure 5.6:** © Christine Baker, Pennsylvania Turnpike; **Figure 5.7:** © Mott MacDonald; **Figure 5.8:** Famartin, CC BY-SA 4.0; **Figure 5.9:** public domain; **Holland Biography:** public domain; **Figures 5.10, 5.11:** © Port Authority of New York and New Jersey; **Figure 5.12:** Photographer unknown / Museum of the City of New York, X2010.11.13642; **Figure 5.13:** from C.C. Gray and H.F. Hagen, 1927, *The Eighth Wonder*, Boston, B.F. Sturtevant Co.; **Singstad Biography:** public domain; **Figure 5.14:** Hoboken Historical Museum Collection; **Figure 5.15:** public domain; **Figures 5.16, 5.17:** © Port Authority of New York and New Jersey; **Figures 5.18, 5.19:** public domain; **Figures 5.20, 5.21:** © Port Authority of New York and New Jersey; **Figures 5.22, 5.23:** © Christine Baker, Pennsylvania Turnpike; **Figure 5.24:** © Mott MacDonald; **Figures 5.25, 5.26:** © Christine Baker, Pennsylvania Turnpike; **Figures 5.27, 5.28, 5.29:** © Mott MacDonald; **Figure 5.30:** © California Department of Transportation. All rights reserved; **Figure 5.31:** © Mott MacDonald; **Figures 5.32, 5.33, 5.34, 5.35, 5.36:** © California Department of Transportation. All rights reserved; **Figure 5.37:** © 2013 by California Department of Transportation. All rights reserved; **Eisenhower Biography:** public domain; **Figures 5.38, 5.39, 5.40, 5.41:** © Chesapeake Bay Bridge-Tunnel; **Figure 5.42:** Ohio Department of Transportation; **Figure 5.43:** © Mott MacDonald; **Figure 5.44:** © Colorado Department of Transportation; **Figure 5.45:** © Frontier-Kemper; **Figure 5.46:** David Sailors © 1992 by WSP | Parsons Brinckerhoff; **Figure 5.47:** Cumberland Gap National Historic Park; **Figure 5.48:** © Russell Hubbard, Arkansas State Highway and Transportation Department; **Figure 5.49:** David Sailors © 1998 by WSP | Parsons Brinckerhoff; **Figure 5.50:** © Frontier-Kemper; **Figures 5.51, 5.52:** David Sailors © 1997 by WSP | Parsons Brinckerhoff; **Figure 5.53:** © William H. Hansmire; **Figure 5.54:** David Sailors © 1986 by WSP | Parsons Brinckerhoff; **Figure 5.55:** © Robert A. Robinson; **Figure 5.56:** © Fulcher/Elioff Collections; **Figure 5.57:** David Sailors © 2003 by WSP | Parsons Brinckerhoff; **Figures 5.58, 5.59:** © Mott MacDonald; **Figures 5.60, 5.61:** © Fulcher/Elioff Collections; **Figures 5.62, 5.63:** © Mott MacDonald; **Figure 5.64:** © MAT Concessionaire, LLC; **Figure 5.65:** © Washington State Department of Transportation; **Figures 5.66, 5.67:** © Traylor Bros. Inc.

CHAPTER 6: WATER TUNNELS

Chapter Opener: © Southern Nevada Water Authority; **Jervis Biography:** Minisink Valley Historical Society; **Figure 6.1:** © San Francisco Public Utilities Commission; **Mulholland Biography:** Los Angeles Department of Water and Power; **Figure 6.2:** City of New York Department of Environmental Protection Archives; **Figure 6.3:** Massachusetts State Archives; **Figure 6.4:** City of New York Department of Environmental Protection Archives; **Figure 6.5:** © Massachusetts Water Resources Authority; **Figure 6.6:** City of New York Department of Environmental Protection Archives; **Figures 6.7, 6.8:** © Massachusetts Water Resources Authority; **Figures 6.9, 6.10, 6.11, 6.12:** City of New York Department of Environmental Protection Archives; **Figures 6.13, 6.14, 6.15:** © Massachusetts Water Resources Authority; **Figure 6.16:** City of New York Department of Environmental Protection Archives; **Figure 6.17:** Frontier-Kemper/Schiavone/Picone, JV; **Figure 6.18:** Chicago History Museum, ICHi-85692, John M. Wing; **Figures 6.19, 6.20, 6.21, 6.22:** © Chicago Department of Water Management; **Figures 6.23, 6.24:** © MWH, now part of Stantec; **Figures 6.25, 6.26:** © Cleveland Department of Water Management; **Figures 6.27 (left and right), 6.28:** SewerHistory.org; **Figure 6.29:** Cleveland State University, Michael Schwartz Library, Special Collections; **Figure 6.30:** © San Francisco Public Utilities Commission; **Morgan Biography:** Cleveland State University, Michael Schwartz Library, Special Collections; **Figures 6.31, 6.32, 6.33, 6.34, 6.35, 6.36, 6.37, 6.38, 6.39, 6.40, 6.41:** © San Francisco Public Utilities Commission; **Figures 6.42, 6.43:** © San Francisco Public Utilities Commission, Photographer Robin Scheswohl; **Figure 6.44:** © San Francisco Public Utilities Commission, Photographer Katherine DuTiel; **Figure 6.45:** © Metropolitan Water District of Southern California; **Figures 6.46, 6.47, 6.48, 6.49, 6.50, 6.51:** Historical Photo Collection of the Department of Water and Power, City of Los Angeles; **Figures 6.52, 6.53, 6.54, 6.55, 6.56, 6.57, 6.58:** © Metropolitan Water District of Southern California; **Figure 6.59:** public domain; **Figure 6.60:** © SME; **Figures 6.61, 6.62, 6.63, 6.64:** © Metropolitan Water District of Southern California; **Figure 6.65:** © Southern Nevada Water Authority; **Figure 6.66:** © Southern Nevada Water Authority and Gene Hertzog; **Figures 6.67, 6.68, 6.69, 6.70, 6.71, 6.72:** © Southern Nevada Water Authority; **Figure 6.73:** Herrenknecht and Vegas Tunnel Constructors; **Figure 6.73 inset:** © Southern Nevada Water Authority; **Figure 6.74:** © Eric Jamison and Southern Nevada Water Authority; **Figures 6.75, 6.76, 6.77, 6.78, 6.79:** U.S. Bureau of Reclamation; **Figures 6.80, 6.81, 6.82:** Rosemary Allen Bond Collection, Charles H. Baker Papers and Puget Sound Energy Historical Archives; **Figures 6.83, 6.84:** Michael J. Semas Collection; **Figure 6.85:** U.S. Department of Interior; **Figures 6.86, 6.87, 6.88:** U.S. Bureau of Reclamation; **Figures 6.89, 6.90:** © The Robbins Company; **Figures 6.91, 6.92:** © Pacific Gas and Electric; **Figures 6.93, 6.94, 6.95:** © MWH, now part of Stantec.

CHAPTER 7: WASTEWATER TUNNELS

Chapter Opener: © Fulcher/Elioff Collections; **Figure 7.1:** public domain; **Figure 7.2 (top):** From "Discussion" by H.G. Payrow, "Historic Review of the Development of Sanitary Engineering in the United States During the Past One Hundred and Fifty Years: A Symposium," in *Transactions of the American Society of Civil Engineers*, Vol. 92 (1928), with permission from ASCE; **(bottom):** Village of Holly, Michigan; **Figure 7.3:** Tennessee State Library and Archives; **Figure 7.4:** public domain; **Waring Biography:** public domain; **Figure 7.5:** Seattle Municipal Archives (Photograph Collection Item No. 6229); **Figure 7.6:** Seattle Municipal Archives (Photograph Collection Item No. 11039); **Chesbrough Biography:** public domain; **Figures 7.7, 7.8, 7.9:** © Metropolitan Water Reclamation District of Greater Chicago; **Figure 7.10:** Cleveland Press Collection, Michael Schwartz Library, Cleveland State University; **Figures 7.11, 7.12, 7.13, 7.14, 7.15, 7.16:** © Metropolitan Water Reclamation District of Greater Chicago; **Figures 7.17, 7.18, 7.19, 7.20, 7.21:** © Milwaukee Metropolitan Sewerage District; **Figures 7.22, 7.23, 7.24, 7.25, 7.26, 7.27:** © Northeast Ohio Regional Sewer District; **Figure 7.28:** © City of Los Angeles Department of Sanitation; **Figure 7.29:** © Michael McKenna and Richard Calvo; **Figures 7.30, 7.31:** © Traylor Bros. Inc.; **Figures 7.32, 7.33, 7.34:** © City of Portland, Bureau of Environmental Services; **Figures 7.35, 7.36, 7.37:** © District of Columbia Water and Sewer Authority.

CHAPTER 8: INNOVATIONS IN TUNNELING

Chapter Opener: © Mott MacDonald; **Figure 8.1:** from H.S. Drinker, 1893, *Tunneling, Explosive Compounds, and Rock Drills*, 3rd ed., New York, John Wiley & Sons; **Figure 8.2:** from L. Von Rosenberg, 1887, *The Vosberg Tunnel*, New York; **Figure 8.3:** © 1905 Brown Bros.; **Figure 8.4:** © 1890 Scientific American; **Figure 8.5:** © Traylor Bros. Inc.; **Figures 8.6, 8.7:** © The Robbins Company; **Figures 8.8, 8.9:** © Fulcher/Elioff Collections; **Figure 8.10:** © Kenny Construction Company; **Figure 8.11:** from C.M. Jacobs, 1894, *Chief Engineer's General Report Upon the Initiation and Construction of the Tunnel Under the East River, New York, to the President and Directors of the East River Gas Company*, New York; **Terzaghi Biography:** public domain; **Figure 8.12:** public domain; **Figure 8.13:** © Mott MacDonald; **Figure 8.14:** from Keuffel & Esser Co., 1913, *Survey and Drafting Catalogue*, New York; **Figure 8.15:** © David R. Klug & Associates; **Figure 8.16:** Benjamin Henry Latrobe, Engineer, Library of Congress, https://www.loc.gov/item/00650597/; **Figure 8.17:** © ER_09, Shutterstock; **Figures 8.18, 8.19, 8.20:** © Mott MacDonald; **Figure 8.21:** JF Shea Construction; **Figure 8.22:** © Fulcher/Elioff Collections; **Figure 8.23:** public domain; **Figure 8.24:** Granger Vintage Images; **Figure 8.25:** from S.D.V. Burr, 1885, *Tunneling Under the Hudson River*, New York, John Wiley & Sons;

Figures 8.26, 8.27, 8.28, 8.29: © Moretrench; **Figures 8.30, 8.31, 8.32:** © Hayward Baker, a Keller Company; **Figure 8.33:** © Atlas Copco; **Figure 8.34:** © Sandvik; **Figure 8.35:** from P. Benjamin, 1892, *Modern Mechanism*, New York, MacMillan and Co.; **Figure 8.36:** from G.H. Gilbert, L.I. Wightman, and W.L. Saunders, 1912, *The Subways and Tunnels of New York: Methods and Costs*, New York, John Wiley & Sons; **Figure 8.37:** © Atlas Copco; **Figure 8.38:** © Ingersoll Rand; **Figure 8.39:** © Atlas Copco; **Figure 8.40:** Sandvik; **Figure 8.40 (inset):** © Atlas Copco; **Figure 8.41:** public domain; **Figure 8.42:** © 2008 Pbroks13, CC BY-SA 3.0; **Figure 8.43:** from G.H. Gilbert, L.I. Wightman, and W.L. Saunders, 1912, *The Subways and Tunnels of New York: Methods and Costs*, New York, John Wiley & Sons; **Figures 8.44, 8.45, 8.46:** © Mott MacDonald; **Figure 8.47:** © Revey Associates; **Figure 8.48:** © The Robbins Company; **Figures 8.49, 8.50:** © 1890 Scientific American; **Figure 8.51:** © The Robbins Company; **Figure 8.52:** public domain; Robbins Biography: © The Robbins Company; **Figures 8.53, 8.54, 8.55, 8.56, 8.57, 8.58:** © The Robbins Company; **Figure 8.59:** © Fulcher/Elioff Collections; **Figures 8.60, 8.61:** *Engineering News*, 1890; **Figure 8.62:** © Mott MacDonald; **Figures 8.63, 8.64:** © Herrenknecht AG; **Figure 8.65:** © Washington State Department of Transportation; **Figure 8.66:** © Sandvik; **Figure 8.67:** © Antraquip Corp.; **Figure 8.68:** © DSI Tunneling LLC; **Figures 8.69, 8.70:** © David R. Klug & Associates; **Figure 8.71:** © Traylor Bros. Inc.; **Figures 8.72, 8.73, 8.74:** © George D. Yoggy; **Figures 8.75, 8.76, 8.77, 8.78, 8.79:** © DSI Tunneling LLC; **Figures 8.80, 8.81, 8.82:** from B. Stillborg, 1994, *Professional Users Handbook for Rock Bolting*, 2nd ed., Clausthal-Zellerfeld, Germany, Trans Tech Publications; **Figure 8.83:** © David R. Klug & Associates; **Figures 8.84, 8.85, 8.86:** © WIKSO America Inc.; **Figure 8.87:** © Moretrench; **Figure 8.88:** © Traylor Bros. Inc.; **Figures 8.89, 8.90, 8.91, 8.92:** © Fulcher/Elioff Collections; **Figure 8.93:** © Kiewit Infrastructure Co.; **Figure 8.94:** © David R. Klug & Associates; **Figure 8.95:** from "The New York Tunnel Extension of the Pennsylvania Railroad: The East River Tunnels" (Paper No. 1159) in *Transactions of the American Society of Civil Engineers*, Vol. LXVIII, September 1910; **Figure 8.96:** © CSI Tunnel Systems; **Figures 8.97, 8.98:** © Fulcher/Elioff Collections; **Figure 8.99:** © Michael F. Roach; **Figure 8.100:** © CSI Tunnel Systems; **Figures 8.101, 8.102, 8.103:** © David R. Klug & Associates; **Figure 8.104:** © CSI Tunnel Systems; **Figures 8.105, 8.106:** © Optimas–Sofrasar Tunnel Products; **Figures 8.107, 8.108:** © Herrenknecht Formwork Technology GmbH; **Figure 8.109:** © Fulcher/Elioff Collections; **Figure 8.110:** © Traylor Bros. Inc.

CHAPTER 9: THE FUTURE OF TUNNELING

Chapter Opener: © Washington State Department of Transportation; **Figure 9.1:** © SME; **Figure 9.2:** © Fulcher/Elioff Collections; **Figures 9.3, 9.4:** © WSP | Parsons Brinckerhoff; **Figure 9.5:** © Mott MacDonald; **Figure 9.6:** © McMillan Jacobs; **Figure 9.7:** © Mike McKenna; **Figure 9.8:** © Fulcher/Elioff Collections; **Figure 9.9:** © Prof. Markus Thewes; **Figure 9.10:** © Washington State Department of Transportation; **Figure 9.10 inset:** © WSP | Parsons Brinckerhoff; **Figures 9.11, 9.12:** © China Railway Equipment Group Co. Ltd.; **Figures 9.13, 9.14:** © Herrenknecht AG; **Figure 9.15:** © CH2M; **Figure 9.16:** © *Tunnelling Journal*; **Figure 9.17:** © Herrenknecht AG; **Figure 9.18:** © New York City Metropolitan Transportation Authority; **Figure 9.19:** © Hatch Mott MacDonald; **Figures 9.20, 9.21, 9.22, 9.23:** © Fulcher/Elioff Collections; **Figures 9.24, 9.25:** © Washington State Department of Transportation; **Figures 9.26, 9.27, 9.28, 9.29, 9.30:** © Fulcher/Elioff Collections; **Figures 9.31, 9.32:** © Metro (Los Angeles Metropolitan Transportation Authority); **Figures 9.33, 9.34:** California High-Speed Rail Authority; **Figure 9.35:** © Metropolitan Water District of Southern California; **Figure 9.36:** © WSP | Parsons Brinckerhoff; **Figure 9.37:** © Heller Manus; **Figure 9.38:** © Henry Russell; **Figure 9.39 (left):** © Fulcher/Elioff Collections; **(right):** © Washington State Department of Transportation; **Figure 9.40:** © Los Angeles Times; **Figure 9.41:** © Mott MacDonald; **Figure 9.42:** © SME; **Figure 9.43:** © Cargocap.com; **Figure 9.44:** © Traylor Bros. Inc.

APPENDIX: U.S. CONSTRUCTED TUNNEL ARCHIVE

Appendix Opener: © Traylor Bros. Inc.

INDEX

Note: *f.* indicates figure; *t.* indicates table

A. Guthrie & Company, 115
Adams, Alva B., 342
adits, 219
Alabama
   Bankhead Tunnel (Mobile River), 248
   George Wallace Tunnel (Mobile), 260
Alaska
   Anton Anderson Memorial Tunnel, 251
   Whittier Tunnel, 251, 251*f.*–252*f.*
Alaska Railroad, 251
Alaskan Way Viaduct (Seattle), 274, 476*f.*, 484, 484*f.*
Allegheny Mountain Tunnel (Pennsylvania), 249*f.*, 250
Allegheny Tunnel (Gallitzin), 68, 82–83, 82*f.*–83*f.*
Alva B. Adams Tunnel (Colorado), 280, 342, 342*f.*–343*f.*
American Association of State Highway and Transportation Officials (AASHTO), 234–235
American Locomotive Company, 66
American Recovery and Reinvestment Act (2009), 77
American Road and Transportation Builders Association (ARTBA), 235
American Society for Testing and Materials (ASTM), 235
American Society of Civil Engineers (ASCE), 267, 268*f.*
Anacostia River tunnel (Washington, D.C.), 190–192
Anton Anderson Memorial Tunnel (Alaska), 251
aqueducts, 7. *see also* water tunnels; specific aqueducts
Arch Cape Tunnel (Oregon), 238, 240*f.*
Arizona
   Margaret T. Hance Park (Phoenix), 266
   Papago Freeway Tunnel (Phoenix), 266
Arizona Department of Transportation, 479
Arkansas
   Bobby Hopper Tunnel, 264, 264*f.*
Armstrong Tunnels (Pittsburgh), 241
Arrowhead Tunnels (Northern California), 331–332, 332*f.*, 394*f.*, 460, 460*f.*
Arroyo Seco Parkway tunnels (Los Angeles), 238
Atlanta transit system. *see* Metropolitan Atlanta Rapid Transit Authority (MARTA) (Georgia)
Auburn Canal Tunnel (Pennsylvania), 53, 53*f.*
automobiles, 234

Baker, Charles, 344
Baker, Wallace Hayward, 413
Ballinger, Richard, 305
Baltimore & Ohio (B&O) Railroad, 63, 66
Baltimore Central Light Rail Line (Maryland), 166
Baltimore Harbor Tunnel (Patapsco River), 248
Baltimore Metro Subway (Maryland), 165–169, 166*f.*, 168*f.*–169*f.*
Baltimore Region Rapid Transit System (Maryland), 165
Bankhead Tunnel (Mobile River), 248
Barlow, Peter W., 71, 140, 404
Basilica Cistern (Istanbul), 461*f.*, 462
Bath County Pumped Storage Project (Virginia), 351, 351*f.*–352*f.*
Battery Tunnel (San Francisco), 273
Bay Area Rapid Transit (BART) (San Francisco), 37, 150–156, 152*f.*, 483, 483*f.*–484*f.*
Bay Bridge. *see* San Francisco-Oakland Bay Bridge (California)
Bay Delta Tunnel (San Francisco), 483, 483*f.*
Bay Tunnel (Hetchy Hetch), 314, 315*f.*
Beach, Alfred Ely, 35, 71–72, 139, 140*f.*, 423
Beacon Hill Project (Seattle), 219, 220*f.*–223*f.*, 222
Belmont, August, Jr., 143
Ben C. Gerwick Company, 154
Berkeley Hills Tunnels (San Francisco), 154–155
Bernick, Michael, 483
Bickford, William, 419
Big Creek Hydroelectric System, 345
Big Dig (Boston), 42, 43*f.*, 268–269, 269*f.*–271*f.*, 410–411, 410*f.*
Big Thompson tunnel (Colorado), 25
Big Walker Mountain Tunnel (Wytheville), 259–260
black powder, 419
Black Rock Tunnel (Phoenixville), 54, 78–80, 79*f.*–81*f.*
Blue Island Avenue Tunnel (Chicago), 28, 298–299, 299*f.*
Blue Plains Tunnel (Washington, D.C.), 489*f.*
Blue River Tunnel. *see* Harold D. Roberts Tunnel (Colorado)
Bobby Hopper Tunnel (Arkansas), 264, 264*f.*
Bolton Hills Tunnel (Baltimore), 413
Bonnema, Janet, 260
Boone, Daniel, 263

Boston Elevated Railway Company (BERY), 128, 131
Boston Main Drainage System (BMDS), 357–358
Boston subway. *see* Massachusetts Bay Transportation Authority (MBTA)
Boston Transit Commission, 128–130
Boulder Dam. *see* Hoover (Boulder) Dam (Lake Mead)
Bouygues Civil Works Florida, 431, 479
Broadway Tunnel (San Francisco), 236, 237*f.*
Brooklyn Bridge (New York City), 405–406, 405*f.*
Brooklyn-Battery Tunnel (New York City), 246
Brown, Douglas R., 412
Brown, Frank O., 423
Brunel, Marc Isambard, 71, 404, 422–423
Buffalo Metro Rail (New York), 170–171, 172*f.*–173*f.*
building information modeling (BIM) software, 398–399, 399*f.*–400*f.*
Bureau of Los Angeles Aqueduct. *see* Los Angeles Aqueduct (Owens Valley)
Burleigh, Charles, 66–67, 85
Burlington Northern Railway, 97, 104
Burlington Northern Santa Fe (BNSF) Railway, 104, 108

caisson disease, 140, 405–406
Caldecott, Thomas E., 255
Caldecott Tunnel (Berkeley Hills), 255, 257*f.*
California
   Arrowhead Tunnels (Northern California), 331–332, 332*f.*, 394*f.*, 460, 460*f.*
   Arroyo Seco Parkway tunnels (Los Angeles), 238
   Battery Tunnel (San Francisco), 273
   Bay Area Rapid Transit (BART), 37, 150–156, 152*f.*, 483, 483*f.*–484*f.*
   Bay Delta Tunnel (San Francisco), 483, 483*f.*
   Bay Tunnel (Hetchy Hetch), 314, 315*f.*
   Berkeley Hills Tunnels (San Francisco), 154–155
   Broadway Tunnel (San Francisco), 236, 237*f.*
   Caldecott Tunnel (Berkeley Hills), 255, 257*f.*
   California Aqueduct (Northern California), 317, 326–332
   California High-Speed Rail Authority, 482, 482*f.*
   Canyon Power Tunnel (Hetchy Hetch), 312
   Central Subway (San Francisco), 157–161, 157*f.*–161*f.*, 431
   Coast Range Tunnel (Hetchy Hetch), 310, 312, 312*f.*
   Colorado River Aqueduct (Colorado River), 24, 25*f.*, 317, 321–326
   Crenshaw/LAX Project (Los Angeles), 210*f.*, 211
   Devil's Slide (Tom Lantos) Tunnel (San Francisco), 272–273, 272*f.*, 448, 449*f.*
   Donner Pass Tunnels (Nevada County), 58, 58*f.*–59*f.*, 88–90, 88*f.*–91*f.*, 92*t.*–93*t.*
   Doyle Drive Tunnel (San Francisco), 273, 273*f.*
   East Central Interceptor Sewer (ECIS) (Los Angeles), 374, 374*f.*
   Elizabeth Tunnel (Owens Valley), 24, 282, 318–319, 318*f.*–319*f.*
   Foothill Tunnel (Hetchy Hetch), 307, 309–310, 309*f.*–311*f.*
   Helms Pumped Storage Plant (Sierra Nevadas), 350–351, 350*f.*
   Hetch Hetchy Aqueduct (San Francisco), 20*f.*, 25

Hetch Hetchy Regional Water System (San Francisco), 281–282, 304–315, 305*f.*
Hollywood and Vine Station (Los Angeles), 211, 211*f.*–213*f.*
"Hollywood Line," 37, 198, 199*f.*
Hollywood Water Quality Improvement Project (Los Angeles), 460, 460*f.*
Irvington Tunnel (Hetchy Hetch), 1*f.*, 310, 312*f.*
Islais Creek tunnels (San Francisco), 413
Kennedy Tunnel (Berkeley Hills), 255, 256*f.*
Lake Eleanor, 304–305
Los Angeles Aqueduct (Owens Valley), 24, 280, 317–321
Los Angeles Metro Rail system, 37, 197–215, 200*f.*, 479, 480*f.*–481*f.*
Main Post Tunnel (San Francisco), 273
Market Street Stations (San Francisco), 152*f.*–153*f.*, 154
Metro Gold Line Eastside Extension (Los Angeles), 204–206, 206*f.*
Metro Gold Line (Los Angeles), 17*f.*
Metro Red Line Tunnels (Los Angeles), 202–203, 203*f.*, 205*f.*
Mono Craters Tunnel (Owens Valley), 319–321, 320*f.*–321*f.*
Mountain Tunnel (Hetchy Hetch), 306–307, 307*f.*–308*f.*
Mulholland water supply system (Los Angeles), 24
New Crystal Springs Bypass Tunnel (Hetchy Hetch), 314, 314*f.*
New Irvington Tunnel (Hetchy Hetch), 21*f.*, 315, 316*f.*
North Outfall Sewer (Los Angeles), 374
Northeast Interceptor Sewer (NEIS) (Los Angeles), 374, 374*f.*–375*f.*
O'Shaughnessy Dam (Hetchy Hetch), 282, 304, 312, 313*f.*
Posey Tube (Oakland-Alameda), 247
Purple Line (Los Angeles), 207, 208*f.*
Regional Connector Transit Corridor Project (Los Angeles), 207, 208*f.*–210*f.*, 211
San Bernardino Tunnel (Northern California), 330–331, 330*f.*
San Fernando (Sylmar) Tunnel (Northern California), 59, 329–330
San Francisco Municipal Railway (Muni), 156*f.*, 157–161
San Francisco-Oakland Bay Bridge, 252, 253*f.*
San Jacinto Tunnel (Colorado River), 323–326, 323*f.*–328*f.*
Second Street Tunnel (Los Angeles), 238
State Route 710 (SR-710) Gap Closure (California), 485, 485*f.*
Stockton Street Tunnel (San Francisco), 235–236, 236*f.*, 282
subway stations, 211–215
Summit Tunnel (Donner Pass), 58, 88
Tehachapi Loop (Northern California), 60*f.*–61*f.*, 61
Tehachapi Tunnels (Northern California), 94, 94*f.*–97*f.*, 97, 326, 329, 329*f.*
Third Street Tunnel (Los Angeles), 236
Transbay Tube (San Francisco), 154, 155*f.*
Tunnel 17 (Los Angeles), 282
Tuolumne River (California), 304
Twin Peaks Tunnels (San Francisco), 282
Waldo Tunnel (San Francisco), 255
Ward (Florence) Tunnel (Big Creek), 345, 345*f.*
water and aqueduct systems, 317
Wawona Tunnel (Yosemite National Park), 238, 238*f.*–239*f.*

Webster Street Tube (Oakland-Alameda), 247
Yerba Buena Island Tunnel, 252, 254f.–255f.
California Aqueduct (Northern California), 317, 326–332
California Department of Water Resources, 326
California High-Speed Rail Authority, 482, 482f.
California State Water Project, 330
California WaterFix, 483
Callahan Tunnel (Boston Harbor), 247
Canada
    Detroit-Windsor Tunnel, 247–248
    Mount Macdonald railroad tunnel (British Columbia), 67–68
    Niagara Tunnel Portal (Ontario), 428, 428f.–429f.
    South Saskatchewan Dam, 425, 425f.
    St. Clair River Tunnels (Michigan/Ontario), 31, 32f.–33f., 72, 76–77, 76f., 100–103, 100f.–103f., 428, 429f.–430f.
Canadian National Railway, 102
canal tunnels, 3, 53, 53f.
Canyon Power Tunnel (Hetchy Hetch), 312
Capitol Hill Station (Seattle), 225, 228f.–229f.
CargoCap, 485–486, 486f.
Carnegie, Andrew, 41f.
cars, 234
Cascade Tunnel (Stevens Pass), 29–31, 30f.–31f., 108–115, 108f.–115f.
Catskill Aqueduct (New York), 24, 285
Cave Rock Tunnel (Lake Tahoe), 241, 242f.
Center Leg Freeway (Washington, D.C.), 266
Central Artery/Tunnel (Boston), 42, 43f., 268–269, 269f.–271f., 410–411, 410f., 484, 484f.
Central Pacific Railroad, 57–58, 88–90
Central Subway (San Francisco), 157–161, 157f.–161f., 431
Chadwick, Edwin, 359
Chapin Mining Company, 409
Chattahoochee CSO Tunnel (Atlanta), 46f., 49f.
Chesapeake Bay Bridge-Tunnel Commission, 257
Chesapeake Bay Bridge-Tunnel (Virginia), 41, 257–259, 258f.–259f.
Chesapeake Bay Ferry Commission, 257
Chesbrough, Ellis Sylvester, 281, 297, 361
Chestnut Hill Reservoir (Boston), 283
Chicago Avenue Tunnel (Chicago), 300
Chicago Lake Tunnel (Lake Michigan), 297, 298f.
Chicago Sanitary and Ship Canal (Illinois), 362, 362f.–363f.
cholera, 27–28, 28f., 356–357, 357f.
City and South London Railway Tunnel, 405
City of Miami, 479
City Tunnel (Boston), 293
City Tunnel Extension (Boston), 293
City Water Tunnel No. 1 (New York City), 285, 286f.
City Water Tunnel No. 2 (New York City), 286, 287f., 288
City Water Tunnel No. 3 (New York City), 292, 410
Civil War, 52, 56, 82
Clean Rivers Project (Washington, D.C.), 378, 379f.
Clean Water Act (1972), 42
Clean Water Act (1977), 365
Clean Water Act (2000), 365
Clemente, Lewis M., 88
Cleveland, Moses, 363
Cleveland Lake Tunnels (Lake Erie), 301–304, 302f.–304f.
Cleveland Regional Sewer District, 372
Coast Range Tunnel (Hetchy Hetch), 310, 312, 312f.
Cobble Mountain Reservoir (Springfield), 286
Cochituate Aqueduct (Boston), 282
Cochrane, Thomas, 139, 404, 423
Colorado
    Alva B. Adams Tunnel, 280, 342, 342f.–343f.
    Big Thompson tunnel, 25
    Eisenhower-Johnson Memorial Tunnel, 13f., 41, 41f., 260–261, 261f., 440
    Glenwood Canyon tunnel project, 260–263, 262f.–263f.
    Gunnison Tunnel, 25, 336–340, 340f.
    Hanging Lake Tunnels (Glenwood Canyon), 261
    Harold D. Roberts Tunnel, 25, 26f.–27f.
    Moffat Tunnel, 68
    No Name Tunnel (Glenwood Springs), 260
    Reverse Curve Tunnel (Glenwood Canyon), 261
    Wolf Creek Pass Tunnel, 269, 272, 272f.
Colorado Department of Transportation, 479
Colorado River Aqueduct (Colorado River), 24, 25f., 317, 321–326
Columbia Slough Consolidated Conduit (CSCC) (Portland), 375, 376f.
compaction grouting, 167–168, 167f., 412
compensation grouting, 412f.–413f., 413
compressed air, use of, 139–140, 141f., 404–406, 425
Connecticut
    Heroes Tunnel (New Haven), 241
    Vernon Tunnel (Vernon), 241
construction manager/general contractor at risk, 401–402
contracting, 401–404, 401f.
Cooper, Peter, 63
Cosgrove Tunnel (Boston), 293
Couch, J.J., 66–67
Crenshaw/LAX Project (Los Angeles), 210f., 211
Crocker, Alvah, 28, 29f.
Cross-Town Tunnel (Lake Michigan), 298
Croton Aqueduct (New) (New York), 285, 296, 297f.
Croton Aqueduct (Old) (New York), 3f., 24, 282–283, 285
CSO (combined sewer overflow) environmental tunnels. *see also* wastewater tunnels; *specific wastewater systems*
    history of, 7, 28, 42–45, 365

map of U.S. communities with, 45f.
Cumberland Gap Tunnel (Appalachian Range), 263–264, 264f.
Cuyahoga River (Cleveland), 42, 43f.–44f., 364–365, 364f.

dams, 283, 344–351
Dart, J.P., 304
Davies, J.V., 407
Davis, Phineas, 63
Deere, Don U., 186, 394
Delaware & Hudson Canal Company, 63
Delaware Aqueduct (Catskills), 24, 280, 288, 288f.–290f.
delivery tunnels, 485–486, 485f.
Department of Environmental Protection (New York), 296
Department of Water Resources, 317
design technology, 392–400, 395f.–400f.
design-bid-build (DBB), 391–392, 400–401
design-build (DB), 401
Detroit-Windsor Tunnel, 247–248
Devil's Slide (Tom Lantos) Tunnel (San Francisco), 272–273, 272f., 448, 449f.
dewatering systems, 407–408, 408f.
dispute review boards (DRBs), 403
District of Columbia Water and Sewer Authority (DC Water), 376, 378
Donner Pass Tunnels (Nevada County), 58, 58f.–59f., 88–90, 88f.–91f., 92t–93t.
Dorchester Tunnel (Boston), 293
Dowd, Charles, 58–59
Downtown Crossing (Ohio River), 274
Downtown Pittsburgh Subway (Pennsylvania), 176, 177f.
Downtown Seattle Transit Project (DSTP) (Washington), 217–219
Downtown Tunnel (Portsmouth-Norfolk), 257
Doyle Drive Tunnel (San Francisco), 273, 273f.
Drano Tunnel (Hood), 51f.
drifts, 219
Drinker, Henry Sturgis, 21
Dulles International Airport (Washington, D.C.), 19f.
Dupont Circle Station (Washington, D.C.), 194, 195f.
dynamite, 385, 385f., 419–420, 419f.–420f.

Eads, James E., 139
earthquakes
    Kern County earthquake (1952), 94, 97
    Loma Prieta earthquake (1989), 156, 312
    San Francisco earthquake (1906), 304, 306f.–307f.
    seismic design for, 155–156, 201, 314
    Sylmar/San Fernando earthquake (1971), 329–330
East Central Interceptor Sewer (ECIS) (Los Angeles), 374, 374f.
East End Crossing (Ohio River), 274–275
East Link Extension (Bellevue), 223

East River Mountain Tunnel (Bluefield), 259–260
East River Tunnel (New York City), 392f.–393f.
East Side Access (ESA) Tunnels (New York), 76–77, 116–119, 116f.–121f., 428, 430, 446f., 447, 470f.
East Side CSO Tunnel (Portland), 376, 376f., 378f., 459, 459f.
East Side Trolley Tunnel (Providence), 241
Eastern Massachusetts Street Railway Company, 128
Eaton, Fred, 282
Edison, Thomas, 141
Eisenhower, Dwight D., 40, 40f., 256–257, 260
Eisenhower-Johnson Memorial Tunnel (Colorado), 13f., 41, 41f., 260–261, 261f., 440
Elizabeth Tunnel (Owens Valley), 24, 282, 318–319, 318f.–319f.
emulsion explosives, 420
engineering, 471–472
engineering technology, 392–400, 395f.–400f.
England
    City and South London Railway Tunnel, 405
    Kilsby Tunnel, 404, 404f.
    Metropolitan District Railway (London), 139
    Thames Tunnel (London), 71, 139, 404, 422–423
    Tower Subway Tunnel (London), 71
Environmental Protection Agency (EPA), 7, 42–43, 49, 365
equipment. *see also* tunnel boring machines (TBMs); tunnel shields
    compressed-air-powered shovels, 67, 67f.
    drill carriages, 85f.
    drill jumbos, 12f., 387, 388f.–389f., 417f.
    drills, 66–67, 67f., 347, 384, 384f., 416–417
    MEMCO cutting wheel machine, 152, 153f.
    pneumatic percussion drills, 66–67, 67f.
    roadheaders, 432–434, 433f.
    rotating cutterheads, 189, 189f.
    shallow diggers, 466, 467f.
    valves, 288, 289f.
escrow bid documents, 403
Euclid Creek Tunnel (Cleveland), 372, 374, 374f.
excavation. *see* underground construction
explosives, 58, 85, 384–385, 385f., 417–422, 419f.–421f.

face loss, 188
Federal Aid Road Act (1916), 234
Federal Highway Act (1921), 234
Federal Highway Administration, 234, 256, 264
Federal Water Pollution Control Act Amendments (1972), 365
Federal-Aid Highway Act (1938), 234
Federal-Aid Highway Act (1944), 234
Federal-Aid Highway Act (1956), 182, 241, 256, 484
Five-Mile Crib (Lake Erie), 303

Flat Rock Tunnel (West Manayunk), 54, 78–80, 80f.
Flathead Railroad Tunnel (Montana), 68, 76–77
Florida
    PortMiami Tunnel, 41, 42f., 273, 274f., 411, 431, 459, 459f., 478–479
Foothill Tunnel (Hetchy Hetch), 307, 309–310, 309f.–311f.
Fort McHenry Tunnel (Baltimore), 266, 267f.
Fort Pitt Tunnel (Pittsburgh), 256
fracture grouting, 413–414
Francois, Albert, 411
Freeman, John Ripley, 305
Freeman, Milton, 244
friction bolts, 442, 442f.
Fulton Street Transit Center (New York City), 147, 149f.

Gale, Sarah Fister, 479, 482
Gallery Place/Chinatown Station (Washington, D.C.), 196f.
Gallitzin Tunnel (Gallitzin), 74, 74f., 82–83
Garfield, James, 305
Garlock Fault, 329
gas intrusion, 201–202
Gasden Purchase (1853), 59
gasket sealing systems, 451–452, 452f.–453f.
Gateway Center Station (Pittsburgh), 178
General Electric, 66
George Wallace Tunnel (Mobile), 260
Georgia
    Chattahoochee CSO Tunnel (Atlanta), 46f., 49f.
    Metropolitan Atlanta Rapid Transit Authority (MARTA), 37, 162–164, 163f.–164f.
    Peachtree Center Station (Atlanta), 163, 164f.
geotechnical baseline reports (GBRs), 403
geotechnical instrumentation and monitoring, 186, 393–394, 467, 467f.
Glenmont Station (Washington, D.C.), 194–195, 194f.
Glenwood Canyon tunnel project (Colorado), 260–263, 262f.–263f.
Gloria Alitto Majewski Reservoir (Chicago), 370
Gotthard Tunnels (Switzerland), 440
Graf, Ed, 412
Grand Trunk Railway, 72, 100
Great Lakes, 296–304
Great Northern Railway, 62–63, 63f.
Great Northern Railway Tunnel (Seattle), 68, 70, 70f., 104–107, 105f.–107f.
Greathead, James Henry, 71, 140, 404–405
green bonds, 479, 482
green infrastructure, 45, 379
ground freezing, 133–134, 134f., 409–411, 409f.–410f.
ground settlements, 167–168
grouted bolts, 442–443, 442f.
grouting, 167–168, 167f., 185–186, 188–189, 331, 411–414

guaranteed maximum allowable construction cost (GMACC), 402
gunite, 73
Gunnison Tunnel (Colorado), 25, 339–340, 340f.

Hallandsås Tunnel (Sweden), 412
Hampton Roads Bridge-Tunnel (Virginia), 257
Hanging Lake Tunnels (Glenwood Canyon), 261
Harold D. Roberts Tunnel (Colorado), 25, 26f.–27f.
Harrison Crib (Lake Michigan), 301, 301f.
Harry Weese & Associates, 194
Haskin, Dewitt Clinton, 70, 139, 406–407
Hawaii
    Hawaii Route 61, 257
    Interstate H-3 (John A. Burns Freeway), 264, 265f., 266
    John H. Wilson Tunnel (Oahu), 256–257
    Likelike Highway (Oahu), 256–257
    Pali Highway (Oahu), 257
    Tetsuo Harano Tunnels (Oahu), 264, 266, 266f.–267f.
Hayward Fault, 154–155
Hearst, William Randolph, 306
heavy-rail transit, 127
Helms Pumped Storage Plant (Sierra Nevadas), 350–351, 350f.
Henderson Street Sewer Tunnel (Seattle), 361f.
Heroes Tunnel (New Haven), 241
Herrenknecht, 336
Herrick, Rensselaer, 363
Hetch Hetchy Aqueduct (San Francisco), 20f., 25
Hetch Hetchy Regional Water System (San Francisco), 281–282, 304–315, 305f.
high-density polyethylene, 202, 202f.
High-Speed Rail system, 77, 77f.
highway tunnels. *see also specific tunnels*
    anatomy of, 234–235, 235f.
    future of, 275, 460, 484
    history of, 7, 40–42
    list of, 515–521
    safety features, 235
Hill, James Jerome, 62, 108
Hobson, Joseph, 72, 100–101, 140
Holland, Clifford Milburn, 243–244
Holland Tunnel (Hudson River), 40, 40f., 242–244, 243f.–245f.
Hollywood and Vine Station (Los Angeles), 211, 211f.–213f.
Hollywood Fault, 201
"Hollywood Line," 37, 198, 199f.
Hollywood Water Quality Improvement Project (Los Angeles), 460, 460f.
Homer M. Hadley Memorial Bridge (Lake Washington), 256
Hoosac Tunnel (Hoosac Mountain), 6f., 13, 28–29, 66–67, 84–87, 84f.–87f.
Hoover (Boulder) Dam (Lake Mead), 22f., 346–347, 346f.–347f., 349

Howard, Edward, 419
Howard Street Tunnel (Baltimore), 34, 35*f.*
Howden, James, 58
Hudson River Railroad Company, 406
Hudson River Railroad Tunnel, 70, 406–407, 406*f.*
Hudson River Vehicular Tunnel. *see* Holland Tunnel (Hudson River)
Hugh L. Carey Tunnel (New York City), 246
Hultman Aqueduct (Boston), 294–296
hydropower, 344–351

Illinois
    Blue Island Avenue Tunnel (Chicago), 28, 298–299, 299*f.*
    Chicago Avenue Tunnel (Chicago), 300
    Chicago Lake Tunnel (Lake Michigan), 297, 298*f.*
    Chicago land and lake tunnels, 299*f.*
    Chicago Sanitary and Ship Canal, 362, 362*f.*–363*f.*
    Chicago wastewater system, 3*f.*, 6*f.*, 361–362
    Cross-Town Tunnel (Lake Michigan), 298
    Gloria Alitto Majewski Reservoir (Chicago), 370
    Harrison Crib (Lake Michigan), 301, 301*f.*
    Intramural Railway (Chicago), 141
    LaSalle Street Tunnel (Chicago), 234, 234*f.*
    McCook Reservoir (Chicago), 370
    North Shore Extension (Lake Michigan), 298
    Northeast Lake Tunnel (Lake Michigan), 301
    Polk Street Tunnel (Lake Michigan), 298
    79th Street Tunnel (Chicago), 300, 301*f.*
    Southside Elevated Railroad (Chicago), 142
    Southwest Land and Lake Tunnel (Chicago), 299–300, 300*f.*
    Thornton Reservoir (Chicago), 368*f.*–370*f.*, 370
    Tunnel and Reservoir Plan (TARP) system (Chicago), 17*f.*, 43, 46*f.*, 365–366, 365*f.*–370*f.*, 370, 427
    Two-Mile Crib (Lake Michigan), 297–298
    Washington Street Tunnel (Chicago), 234
    Wilson Avenue to Central Water Filtration Plant Tunnel (Chicago), 300
immersed-tube tunnels, 193, 193*f.*, 247, 257, 266
Indiana
    Downtown Crossing (Ohio River), 274
    East End Crossing (Ohio River), 274–275
infrastructure, and tunneling, 10–13. *see also* underground infrastructure
Ingersoll, Samuel, 384
Ingersoll-Rand, 66
initial supports, 439
intake cribs, 301
Interborough Rapid Transit (IRT), 143. *see also* New York City Subway
International Society of City and Regional Planners (ISOCARP), 471
International Tunnelling and Underground Space Association (ITA), 471
Interstate H-3 (John A. Burns Freeway) (Oahu), 264, 265*f.*, 266

Interstate Highway System, 40–41
Intramural Railway (Chicago), 141
irrigation tunnels, 339–343
Irvington Tunnel (Hetchy Hetch), 1*f.*, 310, 312*f.*
Ischy, 411
Islais Creek tunnels (San Francisco), 413
ITA's Committee on Underground Space (ITACUS), 471

J.D. and G. Brunton, 423
Jacobs, Charles M., 140, 407
James Appleton Construction, 79
Jervis, John B., 280–281
jet grouting, 412–413
John F. Fitzgerald Expressway (Boston), 268
John H. Wilson Tunnel (Oahu), 256–257
Johnson, Edwin C., 260
Judah, Theodore D., 57, 88

Kennedy Tunnel (Berkeley Hills), 255, 256*f.*
Kentucky
    Cumberland Gap Tunnel (Appalachian Range), 263–264
    Downtown Crossing (Ohio River), 274
    East End Crossing (Ohio River), 274–275
    Ohio River Bridges Project, 274–275
Kern County earthquake (1952), 94, 97
Keystone Tunnel (South Dakota), 40, 40*f.*
Kilsby Tunnel (England), 404, 404*f.*
Kuesel, Thomas R., 155

Lacey V. Murrow Memorial Bridge (Lake Washington), 255–256
Lake Eleanor (California), 304–305
Lake Erie, 301–304
Lake Granby, 342
Lake Mead Intake No. 1 (Nevada), 333–334, 333*f.*–334*f.*
Lake Mead Intake No. 2 (Nevada), 334–335, 334*f.*–335*f.*
Lake Mead Intake Tunnel No. 3 (Nevada), 280*f.*, 335–336, 335*f.*–339*f.*, 432, 460
Lake Michigan, 296–301
Lake Union Sewer Tunnel (Seattle), 359
Lake Washington Floating Bridge. *see* Lacey V. Murrow Memorial Bridge (Lake Washington)
LaSalle Street Tunnel (Chicago), 234, 234*f.*
Leheigh Tunnels (Pennsylvania), 250, 250*f.*
Liberty Tunnels (Pittsburgh), 238, 241, 241*f.*
light-rail transit, 127
Likelike Highway (Oahu), 256–257
Lincoln, Abraham, 9, 9*f.*, 21, 29, 40, 56–57, 241
Lincoln Highway, 241
Lincoln Tunnel (Hudson River), 4*f.*–5*f.*, 10*f.*, 244, 245*f.*–247*f.*

Line 9 (Barcelona), 477, 477f.
locomotives, 63–66, 63f.–66f. see also railroads
Loma Prieta earthquake (1989), 156, 252, 312
Long Island Sound Link (New York), 485, 485f.
Los Angeles Aqueduct (Owens Valley), 24, 280, 317–321
Los Angeles County Metropolitan Transportation Authority (LACMTA). see Los Angeles Metro Rail system (California)
Los Angeles County Transportation Commission (LACTC), 199
Los Angeles Department of Water and Power, 317
Los Angeles Metro Rail system (California), 37, 197–215, 200f., 479, 480f.–481f.
Los Angeles Metropolitan Transit Authority (LAMTA), 198
Los Angeles Water Department, 318
lost ground, 188
Lovat Tunnel Equipment, 102
Lowry Hill Tunnel (Minneapolis), 260
Lytle Tunnel (Cincinnati), 260, 260f.

MacLennan, Duncan, 102
Main Post Tunnel (San Francisco), 273
"Main Street Across America," 241
Main Street Railroad Company, 198
Malaysia
    Stormwater Management and Road Tunnel (SMART), 460, 460f.
Malden Tunnel (Boston), 293
Mall Tunnel (Washington, D.C.), 266
Mangla Dam Project (Pakistan), 426f., 427
Manson, Marsden, 305
Margaret T. Hance Park (Phoenix), 266
Market Street Stations (San Francisco), 152f.–153f., 154
Maryland
    Baltimore Central Light Rail Line, 166
    Baltimore Harbor Tunnel (Patapsco River), 248
    Baltimore Metro Subway, 165–169, 166f., 168f.–169f.
    Bolton Hills Tunnel (Baltimore), 413
    Fort McHenry Tunnel (Baltimore), 266, 267f.
    Howard Street Tunnel (Baltimore), 34, 35f.
    Seagirt Marine Terminal (Patapsco River), 266
Maryland Transportation Authority, 266
mass transit, 127
Massachusetts
    Big Dig (Boston), 42, 43f., 268–269, 269f.–271f., 410–411, 410f.
    Boston Main Drainage System (BMDS), 357–358
    Boston water supply system, 294f.–295f.
    Callahan Tunnel (Boston Harbor), 247
    Central Artery/Tunnel (Boston), 42, 43f., 268–269, 269f.–271f., 410–411, 410f., 484, 484f.
    Chestnut Hill Reservoir (Boston), 283
    City Tunnel (Boston), 293
    City Tunnel Extension (Boston), 293
    Cobble Mountain Reservoir (Springfield), 286
    Cochituate Aqueduct (Boston), 282
    Cosgrove Tunnel (Boston), 293
    Dorchester Tunnel (Boston), 293
    Hoosac Tunnel (Hoosac Mountain), 6f., 13, 28–29, 66–67, 84–87, 84f.–87f.
    Hultman Aqueduct (Boston), 294–296
    John F. Fitzgerald Expressway (Boston), 268
    Malden Tunnel (Boston), 293
    Massachusetts Bay Transportation Authority (MBTA), 35–36, 36f., 128–134, 128f.–134f.
    Metropolitan Sewerage System (Boston), 358
    MetroWest Water Supply Tunnel (Boston), 295, 296f.
    Mystic River Tunnel (Boston), 284f., 285
    Quabbin Tunnel (Boston), 286, 286f.
    Sudbury Aqueduct (Boston), 283, 295
    Sumner Tunnel (Boston Harbor), 246–247
    Ted Williams Tunnel (Boston), 268–269, 269f.
    Thomas P. O'Neill Jr. Tunnel (Boston), 269
    Wachusett Aqueduct (Boston), 283, 285
Massachusetts Bay Transportation Authority (MBTA), 35–36, 36f., 128–134, 128f.–134f.
Massachusetts Water Resources Authority, 294
Mathews, A.A., 186
Maximum Design Earthquake, 201
MBTA (Boston). see Massachusetts Bay Transportation Authority (MBTA)
McAdoo, William G., 140
McClellan, George B., 24
McCook Reservoir (Chicago), 370
mechanical bolts, 442, 442f.
megaprojects, 459, 482–483
MEMCO cutting wheel machine, 152, 153f.
mercury fulminate, 419, 419f.
methane, 201
Metro Gold Line Eastside Extension (Los Angeles), 204–206, 206f.
Metro Gold Line (Los Angeles), 17f.
Metro Red Line Tunnels (Los Angeles), 202–203, 203f., 205f.
Metro System Station (Washington, D.C.), 2f., 18f., 39f.
Metropolitan Atlanta Rapid Transit Authority (MARTA) (Georgia), 37, 162–164, 163f.–164f.
Metropolitan District Railway (London), 139
Metropolitan Sewerage System (Boston), 358
Metropolitan Transit Authority (MTA), 128, 147. see also New York City Subway
Metropolitan Water District Act (1928), 321
Metropolitan Water District of Southern California, 317, 321, 325, 332
Metropolitan Water Reclamation District of Greater Chicago (MWRDGC), 365–366, 482
MetroWest Water Supply Tunnel (Boston), 295, 296f.

Miami Dade County, 479
Michigan
　Detroit-Windsor Tunnel, 247–248
　St. Clair River Tunnels (Port Huron/Ontario), 31, 32f.–33f., 72, 76–77, 76f., 100–103, 100f.–103f., 428, 429f.–430f.
Midtown Hudson Tunnel. *see* Lincoln Tunnel (Hudson River)
Midtown Tunnel (Portsmouth-Norfolk), 257
Milwaukee Deep Tunnel System, 43, 43f., 370–371, 370f.–371f.
Milwaukee Metropolitan Sewerage District (MMSD), 370–371
Minnesota
　Lowry Hill Tunnel (Minneapolis), 260
Mittry Construction Company, 349, 425
mobile drilling gantries, 387
Modesto Irrigation District, 305–306
Moffat, David Halliday, 68
Moffat Tunnel (Colorado), 68
Moir, Ernest, 140, 407
Monitor-Merrimac Memorial Bridge Tunnel (Virginia), 259
Mono Craters Tunnel (Owens Valley), 319–321, 320f.–321f.
Montana
　Flathead Railroad Tunnel, 68, 76–77
Moore, Thomas, 407–408
Morgan, Garrett Augustus, 304, 304f., 305
Morton, R., 404
Mount Baker Ridge Highway Tunnel (Seattle), 16f., 255–256, 267, 268f., 479, 479f.
Mount Carmel Tunnel (Zion National Park), 233f., 241
Mount Lebanon Tunnel (Pittsburgh), 176–177, 177f.
Mount Macdonald railroad tunnel (British Columbia), 67–68
Mountain Tunnel (Hetchy Hetch), 306–307, 307f.–308f.
Mowbray, George, 85
Muir, John, 305–306
Mulholland, William, 24, 25f., 282, 318, 321
Mulholland water supply system (Los Angeles), 24
Municipality of Metropolitan Seattle, 216
Musconetcong Tunnel (Pennsylvania/New York), 21
Muskie, Edward, 42
Mystic River Tunnel (Boston), 284f., 285

National Capital Planning Act (1952), 182
National Capital Transportation Agency, 182
National Environmental Policy Act (1970), 365
Nevada
　Cave Rock Tunnel (Lake Tahoe), 241, 242f.
　Hoover (Boulder) Dam (Lake Mead), 22f., 346–347, 346f.–347f., 349
　Lake Mead Intake No. 1 (Nevada), 333–334, 333f.–334f.
　Lake Mead Intake No. 2 (Nevada), 334–335, 334f.–335f.
　Lake Mead Intake Tunnel No. 3 (Nevada), 280f., 335–336, 335f.–339f., 432, 460
　water supply system, 332–339

new Austrian tunneling method (NATM). *see* sequential excavation method (SEM)
New Crystal Springs Bypass Tunnel (Hetchy Hetch), 314, 314f.
New Irvington Tunnel (Hetchy Hetch), 21f., 315, 316f.
New Jersey
　Holland Tunnel (Hudson River), 40, 40f., 242–244, 243f.–245f.
　Hudson River Railroad Tunnel, 70, 406–407, 406f.
　Lincoln Tunnel (Hudson River), 4f.–5f., 10f., 244, 245f.–247f.
　Route 29 Tunnel, 272
　Weehawken Tunnel (Weehawken), 98–99, 98f.–99f.
New Jersey Interstate Bridge and Tunnel Commission, 242
New Portage Tunnel (Gallitzin), 82–83
New York
　Brooklyn Bridge (New York City), 405–406, 405f.
　Brooklyn-Battery Tunnel (New York City), 246
　Buffalo Metro Rail, 170–171, 172f.–173f.
　Catskill Aqueduct, 24, 285
　City Water Tunnel No. 1 (New York City), 285, 286f.
　City Water Tunnel No. 2 (New York City), 286, 287f., 288
　City Water Tunnel No. 3 (New York City), 292, 410
　Croton Aqueduct (New), 285, 296, 297f.
　Croton Aqueduct (Old), 3f., 24, 282–283, 285
　Delaware Aqueduct (Catskills), 24, 280, 288, 288f.–290f.
　East River Tunnel (New York City), 392f.–393f.
　East Side Access (ESA) Tunnels, 76–77, 116–119, 116f.–121f., 446f., 447, 470f.
　Fulton Street Transit Center (New York City), 147, 149f.
　Holland Tunnel (Hudson River), 40, 40f., 242–244, 243f.–245f.
　Hudson River Railroad Tunnel, 70, 406–407, 406f.
　Hugh L. Carey Tunnel (New York City), 246
　Lincoln Tunnel (Hudson River), 4f.–5f., 10f., 244, 245f.–247f.
　Long Island Sound Link, 485, 485f.
　Musconetcong Tunnel, 21
　New York City Subway, 36, 37f.–38f., 136–149, 138f.
　New York City wastewater system, 358
　No. 7 Line Extension (New York City), 147, 147f.–149f.
　Port Authority Trans-Hudson (PATH) tunnels (Manhattan), 408
　Queens-Midtown Tunnel (East River), 244–245, 247f., 396f.–397f.
　Rapid Transit Subway (New York City), 142–143
　Rondout Tunnel (Catskills), 285, 291f.–292f.
　Second Avenue Subway (New York City), 16f., 144–146, 402f., 471f.
　subaqueous tunnels, 143, 242–247
　tunnels excavated circa 1900, 419f.
　water supply system, 23–24, 23f.–24f., 282–283, 283f., 292f.
New York City Subway, 36, 37f.–38f., 136–149
New York State Bridge and Tunnel Commission, 242
Niagara Tunnel Portal (Ontario), 428, 428f.–429f.
Nicolls, W.J., 79
Nippon Yusen Kaisha, 108

nitroglycerin, 58, 85, 384, 419
Nixon, Richard M., 49
No. 7 Line Extension (New York City), 147, 147f.–149f.
No Name Tunnel (Glenwood Springs), 260
Nobel, Alfred, 385, 419
North Outfall Sewer (Los Angeles), 374
North Shore Connector (NSC) (Pittsburgh), 177–179, 177f.–181f.
North Shore Extension (Lake Michigan), 298
North Trunk Sewer Tunnel (Seattle), 359, 360f.
Northeast Interceptor Sewer (NEIS) (Los Angeles), 374, 374f.–375f.
Northeast Lake Tunnel (Lake Michigan), 301
Northeast Ohio Regional Sewer District (NEORSD), 372
Northern Pacific Railroad, 61–62, 62f.
Northgate Link Extension (Seattle), 223
Northridge earthquake (1994), 201
Northwest Land and Lake Tunnel. *see* Northeast Lake Tunnel (Lake Michigan)

Oahe Dam (Missouri River), 348f.–349f., 349
Obama, Barack, 77
Occupational Safety and Health Act (1970), 49
Occupational Safety and Health Administration (OSHA), 49, 474–475
Ohio
    Cleveland Lake Tunnels (Lake Erie), 301–304, 302f.–304f.
    Cleveland wastewater system, 363
    Cuyahoga River (Cleveland), 42, 43f.–44f., 364–365, 364f.
    Euclid Creek Tunnel (Cleveland), 372, 374, 374f.
    Five-Mile Crib (Lake Erie), 303
    Lytle Tunnel (Cincinnati), 260, 260f.
    Ohio River Bridges Project, 274–275
    Project Clean Lake (Cleveland), 372, 372f.–374f., 374
    West Side Tunnel (Cleveland), 303, 303f.
Ohio River Bridges Project, 274–275
O'Moriarty, Mr., 79
Operating Design Earthquake, 201
Oregon
    Arch Cape Tunnel, 238, 240f.
    Columbia Slough Consolidated Conduit (CSCC) (Portland), 375, 376f.
    East Side CSO Tunnel (Portland), 376, 376f., 378f., 459, 459f.
    West Side CSO Tunnel (Portland), 376, 376f.–377f.
O'Shaughnessy, Michael, 281–282, 281f., 305–306, 312f.
O'Shaughnessy Dam (Hetchy Hetch), 282, 304, 312, 313f.
Otis Elevator Company, 86

Pacific Electric Railroad Company, 198, 198f.
Pacific Gas and Electric Company, 350
Pacific Railway Act (1862), 9, 21, 29, 57, 88
Pakistan
    Mangla Dam Project, 426f., 427

Pali Highway (Oahu), 257
Papago Freeway Tunnel (Phoenix), 266
Parallel Crossing Project (Chesapeake Bay), 259
Parallel Thimble Shoal Tunnel (Chesapeake Bay), 259
Parmley, Walter, 303
Parsons Brinkerhoff, 142, 155
Parsons, William Barclay, 142–143, 386
partnering, 403
Peachtree Center Station (Atlanta), 163, 164f.
Peck, Ralph Brazelton, 186
Pennsylvania
    Allegheny Mountain Tunnel, 249f., 250
    Allegheny Tunnel (Gallitzin), 68, 82–83, 82f.–83f.
    Armstrong Tunnels (Pittsburgh), 241
    Auburn Canal Tunnel, 53, 53f.
    Black Rock Tunnel (Phoenixville), 54, 78–80, 79f.–81f.
    Downtown Pittsburgh Subway, 176, 177f.
    Flat Rock Tunnel (West Manayunk), 54, 78–80, 80f.
    Fort Pitt Tunnel (Pittsburgh), 256
    Gallitzin Tunnel (Gallitzin), 74, 74f., 82–83
    Gateway Center Station (Pittsburgh), 178
    Leheigh Tunnels, 250, 250f.
    Liberty Tunnels (Pittsburgh), 238, 241, 241f.
    Mount Lebanon Tunnel (Pittsburgh), 176–177, 177f.
    Musconetcong Tunnel, 21
    New Portage Tunnel (Gallitzin), 82–83
    North Shore Connector (NSC) (Pittsburgh), 177–179, 177f.–181f.
    Pennsylvania Turnpike, 41, 41f.–42f., 248–251, 248f.
    Pittsburgh light rail ("the T"), 175–179, 176f.
    Squirrel Hill Tunnel, 251
    Stanwix Street tunnel (Pittsburgh), 178–179, 179f.–181f.
    Staple Bend Tunnel, 53–54, 78
Pennsylvania Railroad Company (PRR), 66, 407
Pennsylvania Turnpike, 41, 41f.–42f., 248–251, 248f.
permeation grouting, 411–412, 412f.
Persson, Per-Anders, 420
Phelan, James D., 304
Philadelphia & Reading (P&R) Railroad, 78
Pioneer Tunnel (Stevens Pass), 108, 110, 110f.–111f., 113–115
Pittsburgh light rail ("the T"), 175–179, 176f.
pneumatic tube subway, 139, 140f., 423
Poetsch, F.H., 409
Polk Street Tunnel (Lake Michigan), 298
polyvinyl chloride (PVC) membranes, 202, 444, 444f.–446f., 447
Port Authority of Allegheny County, 175
Port Authority of New York and New Jersey, 242
Port Authority Trans-Hudson (PATH) tunnels (Manhattan), 408
PortMiami Tunnel (Florida), 41, 42f., 273, 274f., 411, 431, 459, 459f., 478–479

Posey, George, 247
Posey Tube (Oakland-Alameda), 247
Potash Company of America, 409
Poulter, John, 412
Project Clean Lake (Cleveland), 372, 372f.–374f., 374
project delivery practices, 391–392, 400–404, 478–479, 482
project sponsors, 9
Promontory Summit (Utah), 21, 29, 29f., 57–58, 58f., 89
public acceptance of tunneling construction, 477
public health, 27–28, 28f., 356–357, 357f.
Public Works Standards Inc. (PWSI), 235
public-private partnerships, 478
public-private partnerships (PPP or P3), 9, 400, 402
Puerto Rico
    Tren Urbano, Rio Piedras Station, 14f.
Purple Line (Los Angeles), 207, 208f.

Quabbin Tunnel (Boston), 286, 286f.
Queens-Midtown Tunnel (East River), 244–245, 247f., 396f.–397f.

Rabcewicz, Ladislaus von, 434
railroad tunnels. *see also specific tunnels*
    clearance requirements, 70f., 74–75
    constructed by 1850 (selected), 55t.
    history of, 3, 6, 9, 28–34, 52–55
    list of, 490–512
    map of selected, 54f.
railroads. *see also* locomotives
    current usage and growth, 75–77
    electrification of, 141–142
    expansion of, 56t.
    future expansion, 77, 77f.
    history of, 53–57, 56t.
    transcontinental, 57–63, 59f.
Rails-to-Trails program, 57, 77
Raker, John Edward, 306, 306f.
Raker Act (1913), 25, 306
rapid transit, 127
Rapid Transit Subway (New York City), 142–143
recycling water, 462
Regional Connector Transit Corridor Project (Los Angeles), 207, 208f.–210f., 211
Regional Water Quality Control Board (Los Angeles), 374
Reverse Curve Tunnel (Glenwood Canyon), 261
Rhode Island
    East Side Trolley Tunnel (Providence), 241
Ritter, Wilhelm, 393
river tunnels. *see* subaqueous tunnels

Robbins, James S., 349, 387, 390f., 425–428
Robbins Company, 67
Roberts Tunnel. *see* Harold D. Roberts Tunnel (Colorado)
rock bolting, 73, 387, 390, 440–443, 442f.–443f.
Rock Quality Designation (RQD) Index, 394
Rock Structure Rating (RSR), 77
Rockefeller, John D., 363
Roebling, John, 405
Rolph, James, Jr., 281
Rondout Tunnel (Catskills), 285, 291f.–292f.
roof bolting, 441
Roosevelt, Franklin D., 234
Rosslyn Station (Arlington), 192–193, 193f.
Route 29 Tunnel (New Jersey), 272
Russia
    Silberwald tunnel (Moscow), 460

S. Pearson & Son, 407
safety, 15, 330, 391, 392f., 405–406, 474–476, 474f.
San Andreas Fault, 304, 319, 332
San Bernardino Mountains, 330
San Bernardino Tunnel (Northern California), 330–331, 330f.
San Fernando Fault, 329
San Fernando (Sylmar) Tunnel (Northern California), 59, 329–330
San Francisco Bay Area Rapid Transit Commission, 151
San Francisco earthquake (1906), 304, 306f.–307f.
San Francisco Municipal Railway (Muni), 156f., 157–161
San Francisco subway. *see* Bay Area Rapid Transit (BART) (San Francisco)
San Francisco-Oakland Bay Bridge (California), 252, 253f.
San Jacinto Tunnel (Colorado River), 323–326, 323f.–328f.
sandhogs, 243, 285, 405
Sandstrom, Gosta E., 84, 87
Sanitary District of Chicago, 361
sanitation tunnels. *see* wastewater tunnels
Santa Fe Railway, 97
Santa Susana thrust fault, 329–330
Seagirt Marine Terminal (Patapsco River), 266
Second Avenue Subway (New York City), 16f., 144–146, 402f., 471f.
Second Street Tunnel (Los Angeles), 238
seismic design, 155–156, 201, 314
sequential excavation method (SEM), 177, 186, 192, 434, 434f.–435f.
settlement control, 167–168
79th Street Tunnel (Chicago), 300, 301f.
sewers. *see* CSO (combined sewer overflow) environmental tunnels;
    wastewater tunnels
shaft sinking, 468, 469f.
Shanley Brothers, 85
shield loss, 188

shotcrete, 73, 73f., 436f.–438f., 437, 439
Sierra Ditch and Water Company, 305
Silberwald tunnel (Moscow), 460
Singstad, Ole, 244–245, 247
Smith, J. Waldo, 288
Snoqualmie Falls Underground Powerhouse (Cascades), 344–345, 344f.–345f.
Sobrero, Ascanio, 419
soil fracture grouting, 413–414
soil-structure interaction, 156
soldier pile-tremie concrete (SPTC), 157–158
Sound Move (Seattle), 218–219
Sound Transit (Seattle), 216–228, 217f.
South Dakota
    Keystone Tunnel, 40, 40f.
    Oahe Dam (Missouri River), 348f.–349f., 349
South Saskatchewan Dam (Canada), 425, 425f.
Southern California Rapid Transit District (SCRTD), 198
Southern Nevada Water Authority (SNWA), 333–334
Southern Pacific Railroad, 59, 89–90, 94, 97
Southside Elevated Railroad (Chicago), 142
Southwest Land and Lake Tunnel (Chicago), 299–300, 300f.
Spain
    Line 9 (Barcelona), 477f.
Sprague, Frank, 140–142
Sprague Electric Railway and Motor Company, 141
Spring Valley Water Company, 305–306
Squirrel Hill Tunnel (Pennsylvania), 251
St. Clair River Tunnels (Michigan/Ontario), 31, 32f.–33f., 72, 76–77, 76f., 100–103, 100f.–103f., 428, 429f.–430f.
St. Paul & Pacific Railroad, 62
Staggers Rail Act (1980), 76
Stampede Pass Tunnel (Cascade Mountains), 62, 62f.
Stanwix Street tunnel (Pittsburgh), 178–179, 179f.–181f.
Staple Bend Tunnel (Pennsylvania), 53–54, 78
State of Florida Department of Transportation (FDOT), 478–479
State Route 99 (SR 99) Tunnel (Seattle), 274, 275f.–276f., 431, 432f., 457f., 459
State Route 710 (SR-710) Gap Closure (California), 485, 485f.
station caverns, 145f.–146f., 146, 194–195, 194f.–196f.
steam engines, 140–142
steel sets, 69f., 73–74, 73f., 439–440, 440f.–441f.
steel-fiber-reinforced shotcrete (SFRS), 437
Stephenson, George, 404
Stephenson, Robert, 404
Stevens, John Frank, 31, 31f., 62–63
Stevens Pass (Cascades), 31, 63
Stevens Pass Tunnel (Cascades), 63
Stockton Street Tunnel (San Francisco), 235–236, 236f., 282
Stokes, Carl, 364

Stokes, Louis, 365
Stormwater Management and Road Tunnel (SMART) (Malaysia), 460, 460f.
Straight Creek Tunnel. *see* Eisenhower-Johnson Memorial Tunnel (Colorado)
Strawberry Tunnel (Utah), 340–342, 340f.–341f.
subaqueous tunnels, 31, 139–140, 143, 242–248, 404–408
subways. *see* transit tunnels
Sudbury Aqueduct (Boston), 283, 295
Summit Tunnel (Donner Pass), 58, 88
Summit Tunnel (Pennsylvania). *see* Allegheny Tunnel (Gallitzin)
Sumner, William, 246
Sumner Tunnel (Boston Harbor), 246–247
sunken-tube tunnels, 193
superhighways, 234
Sweden
    Hallandsås Tunnel, 412
Switzerland
    Gotthard Tunnels, 440
Sylmar/San Fernando earthquake (1971), 329–330

tail loss, 188
team alignment, 403
Ted Williams Tunnel (Boston), 268–269, 269f.
Tehachapi Loop (Northern California), 60f.–61f., 61
Tehachapi Tunnels (Northern California), 94, 94f.–97f., 97, 326, 329, 329f.
Tennessee
    Cumberland Gap Tunnel (Appalachian Range), 263–264, 264f.
    Memphis wastewater system, 359
Terzaghi, Karl von, 393
Tetsuo Harano Tunnels (Oahu), 264, 266, 266f.–267f.
Texas
    Washburn Tunnel (Houston), 248
Thames Tunnel (London), 71, 139, 404, 422–423
Third Street Tunnel (Los Angeles), 236
Thomas P. O'Neill Jr. Tunnel (Boston), 269
Thompson, Reginald H., 359
Thoresen, Soren, 247
Thornton Reservoir (Chicago), 368f.–370f., 370
timber support, 68, 69f., 73–74
time zones, 58–59
toll roads, 41
Tom Carrell Memorial Tunnel and Mine Safety Act (1972), 330
Tower Subway Tunnel (London), 71
train tunnels. *see* railroad tunnels
trains. *see* locomotives
Transbay Tube (San Francisco), 154, 155f.
Transcontinental Railroad, 9, 21, 29, 29f., 40, 57–58
transcontinental railroads, 57–63, 59f.

Transfield Services Australia, 479
transit tunnels. *see also specific transit systems*
    art in, 127, 214, 214*f*.–215*f*.
    defined, 127
    future of, 229, 459
    history of, 6, 34–39
    list of, 513–514
    necessity of, 126–127
transverse ventilation, 243, 251
Tren Urbano, Rio Piedras Station (Puerto Rico), 14*f*.
Triborough Bridge and Tunnel Authority (TBTA), 244, 246
Triger, Jacques, 139
Tunnel 17 (Los Angeles), 282
Tunnel and Reservoir Plan (TARP) system (Chicago), 17*f*., 43, 46*f*., 365–366, 365*f*.–370*f*., 370, 427
tunnel boring machines (TBMs). *see also* tunnel shields
    Arrowhead Tunnels (Northern California), 331, 331*f*.
    "Bertha" (Seattle), 274, 275*f*., 431, 465*f*.–466*f*.
    "Big Alma" (North Beach), 161*f*.
    City Water Tunnel No. 3 (New York City), 292–293, 293*f*.
    earth pressure balance (EPB), 72, 158*f*., 190–192, 206*f*., 224–225, 226*f*.–229*f*., 428, 431, 431*f*.
    "Excalibore" (St. Clair River), 428, 429*f*.
    history of, 13–15, 67–68, 118–119, 387
    "Lady Bird" (Washington, D.C.), 378, 379*f*.–381*f*.
    "Lady Bird" (Washington, D.C.), 47*f*.
    Lake Mead (Nevada), 48*f*.
    "Mackenzie" (Cleveland), 372, 373*f*.
    "The Mole" (Boston), 293, 294*f*.
    Niagara Tunnel Portal (Ontario), 428, 428*f*.–429*f*.
    Oahe Dam (Missouri River), 349, 349*f*., 425
    pressurized-face, 190–191, 190*f*.–192*f*., 204–205, 206*f*., 408
    Robbins TBMs, 425–428, 425*f*.–427*f*.
    slurry pressure balance (SPB), 119*f*., 190, 428, 430, 430*f*.–431*f*.
    soft-ground, 428, 429*f*.
    technological growth, 15*f*., 463–467, 464*f*.–466*f*., 473*f*.
    Tunnel and Reservoir Plan (TARP) system (Chicago), 48*f*.
tunnel linings
    cast-in-place, 167, 446*f*.–455*f*., 447–454
    designing, 399
    history of, 68, 70, 71*f*.
    lattice girder support systems, 439, 439*f*.
    one-pass, 190, 191*f*., 204, 450–454
    polyvinyl chloride (PVC) membranes, 202, 444, 444*f*.–446*f*., 447
    ribs and lagging system, 390, 391*f*.
    segmental precast concrete, 207*f*., 303, 390–391, 391*f*., 448–454, 449*f*.–455*f*., 468, 468*f*.

    sprayed shotcrete, 73, 73*f*., 436*f*.–438*f*., 437, 439
    steel, 153, 153*f*.
    technological growth, 467–468, 468*f*.
    two-pass, 187–188, 188*f*., 204
    waterproofing innovations, 443–447
tunnel shields. *see also* tunnel boring machines (TBMs)
    Barlow-Greathead shield, 404
    Beach's shield, 71*f*., 423
    Brown's EPB shield, 423–424, 423*f*.
    Brunel's shield, 422–423, 422*f*.
    and compressed air, 139–140, 141*f*., 404–406, 425
    digger shields, 187, 187*f*., 203*f*., 218*f*., 329, 387, 390*f*., 408, 424, 424*f*.
    history of, 71–72, 71*f*.
    Zokor "Big John" digger shield, 424, 424*f*.
tunneling. *see also* underground construction
    compressed air, 139–140, 141*f*., 404–406, 425
    rock, 66–68, 192–193, 387
    sequential excavation method (SEM), 177, 186, 192, 434, 434*f*.–435*f*.
    shaft sinking, 468, 469*f*.
    shield, 71–72, 139, 152–153, 153*f*., 187–189, 189*f*., 385, 386*f*., 404–406
    soft-ground, 186–187, 386–387, 390
    soil, 68, 70–72
    technological growth, 469, 471
tunnels. *see also* highway tunnels; railroad tunnels; transit tunnels; underground construction; underground infrastructure; wastewater tunnels; water tunnels
    canal, 3, 53, 53*f*.
    cargo and delivery, 485–486, 485*f*.–486*f*.
    designing, 15, 385–387, 392–400
    enlarging, 74–75, 75*f*.
    immersed-tube, 193, 193*f*., 247, 257, 266, 386*f*.
    inspecting and rehabilitating, 12, 48, 469
    irrigation, 339–343
    multi-use, 458*f*., 460, 460*f*.
    rehabilitation of, 72–75, 73*f*.–75*f*., 485
    seismic design, 155–156, 201, 314
    and societal development, 10–11, 20–23
    subaqueous, 31, 139–140, 143, 242–248, 404–408
    support and reinforcement, 68, 69*f*., 73–74, 73*f*., 439–440, 440*f*.–441*f*.
    and urban planning, 3–9, 23, 34–39, 45–46
    ventilation in, 201–202, 235, 238*f*.–239*f*., 243–245, 251
Tuolumne County Water Company, 304
Tuolumne River (California), 304
Turkey
    Basilica Cistern (Istanbul), 461*f*., 462
Turlock Irrigation District, 305–306
Tweed, William "Boss," 139

Twin Peaks Tunnels (San Francisco), 282
Two-Mile Crib (Lake Michigan), 297–298
typhoid, 27–28, 28f.
underground construction. *see also* equipment; tunneling; tunnels
    and active faults, 155–156, 201
    cut-and-cover, 142f., 144f., 145–146
    drill-and-blast, 118, 387, 388f.–389f., 414–417, 414f.–418f., 420f.
    explosives technology, 417–422, 419f.–421f.
    financing for, 9–10, 16–17, 45–46, 391–392, 400–402, 478–479, 482
    future of, 16–18, 471–473, 486–487, 487f.
    and gassy environments, 201
    ground freezing, 133–134, 134f., 409–411, 409f.–410f.
    ground supports, 69f., 73–74, 73f., 387, 390, 439–443, 440f.–443f.
    historic methods, 384–392
    industrialized tunneling, 463
    innovation in, 14–16, 48, 455
    labor force, 471–473
    mined cavern stations, 145f.–146f., 146, 194–195, 194f.–196f.
    public acceptance of, 476f.–477f., 477
    technological growth, 13–14, 48, 185–186, 462
    waterproofing, 443–454
    worker safety, 15, 330, 391, 392f., 405–406, 474–476, 474f.
underground infrastructure. *see also* tunnels; underground construction
    CSOs, 42–45
    development of, 23
    future of, 458–462, 486–487
    green, 45, 379
    Interstate Highway System, 40–42
    public transit, 34–39
    railroads, 28–34
    water and wastewater management, 23–28
underwater tunnels. *see* subaqueous tunnels
Union Pacific Railroad, 57–58, 89–90, 97
United Nations, 471
University Link Tunnels (Seattle), 224–225, 224f.–229f.
University of Illinois, Urbana-Champaign, 186
urban planning, and tunneling, 3–9, 23, 34–39, 45–46. *see also* underground infrastructure
U.S. Bureau of Mines, 441
U.S. Bureau of Reclamation, 334, 339
U.S. Department of the Interior, National Park Service, 264, 305
U.S. Reclamation Service, 339
Utah
    Mount Carmel Tunnel (Zion National Park), 233f., 241
    Promontory Summit, 21, 29, 29f., 57–58, 58f., 89
    Strawberry Tunnel, 340–342, 340f.–341f.

Vernon Tunnel (Vernon), 241
Virginia
    Bath County Pumped Storage Project, 351, 351f.–352f.
    Big Walker Mountain Tunnel (Wytheville), 259–260
    Chesapeake Bay Bridge-Tunnel, 41, 257–259, 258f.–259f.
    Downtown Tunnel (Portsmouth-Norfolk), 257
    East River Mountain Tunnel (Bluefield), 259–260
    Hampton Roads Bridge-Tunnel, 257
    Midtown Tunnel (Portsmouth-Norfolk), 257
    Monitor-Merrimac Memorial Bridge Tunnel, 259
    Parallel Crossing Project (Chesapeake Bay), 259
    Parallel Thimble Shoal Tunnel (Chesapeake Bay), 259
    Rosslyn Station (Arlington), 192–193, 193f.
    Virginia Avenue Tunnel (Washington, D.C.), 34

Wachusett Aqueduct (Boston), 283, 285
Waldo Tunnel (San Francisco), 255
Walker, Thomas, 263
Ward (Florence) Tunnel (Big Creek), 345, 345f.
Waring, George E., Jr., 359
Warner, James, 412
Washburn Tunnel (Houston), 248
Washington (state)
    Alaskan Way Viaduct (Seattle), 274, 476f., 484, 484f.
    Beacon Hill Project (Seattle), 219, 220f.–223f., 222
    Capitol Hill Station (Seattle), 225, 228f.–229f.
    Cascade Tunnel (Stevens Pass), 29–31, 30f.–31f., 108–115, 108f.–115f.
    Downtown Seattle Transit Project (DSTP), 217–219, 218f.–219f.
    Drano Tunnel (Hood), 51f.
    East Link Extension (Bellevue), 223
    Great Northern Railway Tunnel (Seattle), 68, 70, 70f., 104–107, 105f.–107f.
    Henderson Street Sewer Tunnel (Seattle), 361f.
    Homer M. Hadley Memorial Bridge (Lake Washington), 256
    Lacey V. Murrow Memorial Bridge (Lake Washington), 255–256
    Lake Union Sewer Tunnel (Seattle), 359
    Mount Baker Ridge Highway Tunnel (Seattle), 16f., 255–256, 267, 268f., 479, 479f.
    North Trunk Sewer Tunnel (Seattle), 359, 360f.
    Northgate Link Extension (Seattle), 223
    Pioneer Tunnel (Stevens Pass), 108, 110, 110f.–111f., 113–115
    Seattle wastewater system, 359–361
    Snoqualmie Falls Underground Powerhouse (Cascades), 344–345, 344f.–345f.
    Sound Move (Seattle), 218–219
    Sound Transit (Seattle), 216–228, 217f.
    Stampede Pass Tunnel (Cascade Mountains), 62, 62f.
    State Route 99 (SR 99) Tunnel (Seattle), 274, 275f.–276f., 431, 432f., 457f., 459
    Stevens Pass (Cascades), 31, 63

Stevens Pass Tunnel (Cascades), 63
University Link Tunnels (Seattle), 224–225, 224f.–229f.
Washington, D.C.
    Anacostia River tunnel, 190–192
    Blue Plains Tunnel, 489f.
    Center Leg Freeway, 266
    Clean Rivers Project, 378, 379f.
    Dulles International Airport, 19f.
    Dupont Circle Station, 194, 195f.
    Gallery Place/Chinatown Station, 196f.
    Glenmont Station, 194–195, 194f.
    Mall Tunnel, 266
    Metro System Station, 2f., 18f., 39f.
    Virginia Avenue Tunnel, 34
    Washington Metro, 182–196, 184f.
    Washington Metropolitan Area Transit Authority (WMATA), 37, 39, 182, 185, 444
    Wheaton Station, 444
Washington Metro (Washington, D.C.), 182–196, 184f.
Washington Metropolitan Area Transit Authority (WMATA), 37, 39, 182, 185, 444
Washington Street Tunnel (Chicago), 234
wastewater tunnels. *see also* CSO (combined sewer overflow) environmental tunnels; *specific wastewater systems*
    design elements, 374
    future of, 378–379, 459
    history of, 6, 27–28, 356–357, 356f., 358f.
    list of, 525
water distribution systems, 280–281
Water Supply Act (1905) (New York), 24

water tunnels. *see also* aqueducts; *specific water systems*
    future of, 460, 462
    history of, 6–7, 23–28, 352
    list of, 522–524
    and the water distribution system, 280–281
Wawona Tunnel (Yosemite National Park), 238, 238f.–239f.
Webster Street Tube (Oakland-Alameda), 247
Weehawken Tunnel (Weehawken), 98–99, 98f.–99f.
wellpoints, 407–408, 408f.
West Rock Tunnel (New Haven). *see* Heroes Tunnel (New Haven)
West Side CSO Tunnel (Portland), 376, 376f.–377f.
West Side Tunnel (Cleveland), 303, 303f.
Wheaton Station (Washington, D.C.), 444
Whitney, Asa, 57
Whitney, Henry, 141
Whittier Tunnel (Alaska), 251, 251f.–252f.
Wilderness Trail, 263
Williams, Benezette, 359
Wilson, John H., 256
Wilson, Woodrow, 25, 306
Wilson Avenue to Central Water Filtration Plant Tunnel (Chicago), 300
Wisconsin
    Milwaukee Deep Tunnel System, 43, 45f., 370–371, 370f.–371f.
Wolf Creek Pass Tunnel (Colorado), 269, 272, 272f.
work trains, 73, 74f.

Yerba Buena Island Tunnel (California), 252, 254f.–255f.